FROM THE MICROSCOPIC TO THE MACROSCOPIC

Selected Papers of

Joel Lebowitz

FROM THE MICROSCOPIC TO THE MACROSCOPIC

Selected Papers of Joel Lebowitz

Joel Lebowitz
Rutgers University, USA

World Scientific

NEW JERSEY • LONDON • SINGAPORE • BEIJING • SHANGHAI • HONG KONG • TAIPEI • CHENNAI • TOKYO

Published by

World Scientific Publishing Co. Pte. Ltd.
5 Toh Tuck Link, Singapore 596224
USA office: 27 Warren Street, Suite 401-402, Hackensack, NJ 07601
UK office: 57 Shelton Street, Covent Garden, London WC2H 9HE

British Library Cataloguing-in-Publication Data
A catalogue record for this book is available from the British Library.

FROM THE MICROSCOPIC TO THE MACROSCOPIC
Selected Papers of Joel Lebowitz

ISBN 978-981-98-1955-3 (hardcover)
ISBN 978-981-98-1956-0 (ebook for institutions)
ISBN 978-981-98-1957-7 (ebook for individuals)

For any available supplementary material, please visit
https://www.worldscientific.com/worldscibooks/10.1142/14483#t=suppl

Preface

I take this opportunity to express my gratitude to my teachers, colleagues and students who deserve much of the credit for what is valuable in this book. They have made doing science a pleasurable social activity. As is evident from the number of co-authors on the papers in this volume, I love doing research as a cooperative activity. I have had a total of more than 200 collaborators and have learned much from most of them. Many of them have also become dear friends. Fortunately, there are too many to list them all here.

I would like to dedicate this book to my parents and sister, who died in the Holocaust.

Joel L. Lebowitz
Rutgers University, USA
June 2025

Contents

Introduction

Joel Louis Lebowitz is Distinguished George William Hill Professor of Mathematics and Physics at Rutgers University. He has published over six hundred research articles and a few dozen other writings. Joel's persistent fascination concerns the link of the small, atoms and molecules, to the large, macroscopic phenomena. This branch of physics is known as Statistical Mechanics, covering both static and time-dependent behavior.

Joel worked on many aspects of Statistical Physics. He kept emphasizing the need for mathematical clarity and whenever possible tried to provide proofs. A considerable fraction of his research uses non-rigorous methods to advance our physical understanding. Fine examples are his early work on the equilibrium theory of classical fluids and the pioneering use of numerical Monte-Carlo methods. Joel studied equilibrium phase transitions. As two examples, he discovered powerful inequalities for a general class of ferromagnetic Ising models, now known as Lebowitz inequalities. Jointly with A. Mazel and E. Presutti he established rigorously the existence of a liquid-gas phase transition for systems of classical particles with short-ranged interaction potentials. While considerably less studied, he initiated the investigation of phase transitions for steady states far from thermal equilibrium. Jointly with Elliott Lieb, Joel proved the stability of matter, which refers to the existence of an extensive free energy for charged particles interacting through the Coulomb potential.

In the mid 1980s, many-particle systems with stochastic time evolution came into focus, like exclusion processes and interacting Brownian motions. Perhaps less fundamental from a physics perspective, such models are widely used as effective descriptions on mesoscopic scales. While more accessible than mechanical systems, the challenge to understand the transition from the small to the large still persists. Joel outstandingly joined the intense efforts. This naturally leads to another field which Joel had always been interested in. On the kinetic scale the Boltzmann equation is the appropriate model. But often it suffices to use a coarser scale on which Euler and Navier-Stokes equations become valid, which again leads to a challenging scale separation problem. Of course, Statistical Physics is not a static research area. Over recent decades new topics have been added, as soft matter, active matter, and living matter. Joel is very open to and interested in such new fields of application.

Joel has a lifelong interest in the conceptual foundations of Statistical Mechanics. Well remembered is his 1998 encounter with Ilya Prigogine, taking place at the large auditorium of the Université Sorbonne as part of the StatPhys20 conference at Paris.

At that occasion Joel vividly defended the Boltzmann point of view. More lately he stressed and illuminated the central role of typicality.

In chronological order, Joel received the Boltzmann Medal of IUPAP, the Delmer S. Fahrney Medal by the Franklin Institute, the AAAS Award for Scientific Freedom and Responsibility, the Henri Poincaré Prize of IAMP, the APS Dwight Nicholson Medal for Outreach, the Max Planck Medal of DPG, the Volterra Award of the Academia Lincei Roma, the Grande Médaille d'Or of the French Academy of Sciences, the Dannie Heineman Prize for Mathematical Physics of APS, and the Dirac Medal of ICTP.

To capture Joel's stellar influence in Statistical Physics it does not suffice to count the number of publications and to mention the truly impressive list of prizes and awards. The most obvious further evidence is the biannual Statistical Mechanics Conference, first at Stevens Institute of Technology, then at Yeshiva University, and finally at Rutgers University. Just now we count the Conference number 128 plus 5 which are missing because of Covid. Joel himself organized and chaired every single meeting. These Conferences glued together the Statistical Mechanics community in the US and worldwide.

Joel commands a huge network of collaborators and he is curious about every single topic. Even after a long intermediate time span, he would return to the commonly studied problem exactly at the point where one had been stuck in the first place.

On a more global view, Joel is co-chair of the Committee of Concerned Scientists, which is a human rights organization promoting academic and personal freedom for their colleagues worldwide. Through him, in Statistical Mechanics Conferences it has become common practice to organize a human rights session. As Joel would say, after days of technical presentations and discussions, we are morally obliged to invest at least one hour to learn about colleagues in less favorable circumstances.

From his many research publications Joel himself singled out what he regards as his twenty-five most important contributions. These articles are reprinted here as SELECTA. Before starting to browse the publications, I strongly recommend to also consult the sources listed below. A glimpse on Joel's life trajectory is accounted for in [1; 2], see also the AIP life interview [3]. The Statistical Mechanics Conferences are enthusiastically described in [4; 5; 6]. Not to be missed is the moving Laudatio by David Ruelle [7] at the occasion of Joel being awarded with the Boltzmann Medal.

Herbert Spohn
Technical University of Munich, Germany
June 2025

References

1. J. L. Lebowitz and L. Bonolis, A life in statistical mechanics: Part 1: From Chedar in Taceva to Yeshiva University in New York. The European Physical Journal H **42**, 1–21 (2017).
2. J. L. Lebowitz, Human Rights and Science: Biographical Notes. Annual Review of Condensed Matter Physics **16**, 25–39 (2025).
3. Oral History Interview of Joel Lebowitz on 2021 June 8, American Institute of Physics.
4. T. Feder, One man, one hundred meetings, and a physics subfield. Physics Today, **61**(10), 30–31 (2008).
5. R. P. Crease, Joel's conference. Physics World **30**(2), p. 18 (2017).
6. M. Kiessling, One, two, three, and still counting: Joel Lebowitz's unparalleled series of biannual statistical mechanics meetings. IAMP Bulletin April 2023.
7. Laudatio for Joel L. Lebowitz by David Ruelle (IHES, Paris). `https://www.iamp.org/poincare/jl00-laud.html`.

PHYSICAL REVIEW VOLUME 99, NUMBER 2 JULY 15, 1955

New Approach to Nonequilibrium Processes*

PETER G. BERGMANN AND JOEL L. LEBOWITZ
Department of Physics, Syracuse University, Syracuse, New York
(Received March 23, 1955)

In this paper we propose a new model for the description of irreversible processes, which permits the construction of a Gibbs-type ensemble and the employment of the general techniques of statistical mechanics. The internal dynamics of the system that is engaged in the process is assumed to be described fully by its Hamiltonian. Its interaction with the driving reservoirs is described in terms of impulsive interactions (collisions). The reservoirs themselves possess definite temperatures, are inexhaustible, and free of internal gradients. The ensemble obeys an integro-differential equation in Γ space, containing both the terms of the Liouville equation and a stochastic integral term that describes the collisions with the reservoirs. It is shown in this paper (1) that the ensemble will approach canonical distribution in the course of time in the presence of a single driving reservoir, (2) that it will approach a stationary nonequilibrium distribution in the presence of several reservoirs at different temperatures, and (3) that in the latter case, and for small temperature differences, Onsager's reciprocal relations are satisfied by the stationary distribution.

1. INTRODUCTION

THE purpose of this paper is to propose a statistical-mechanical model for nonequilibrium processes that leads naturally to the existence of a stationary state, approached by an ensemble in the course of time, which is independent of the initial ensemble distribution. In the stationary state, the ensemble is not in thermodynamic equilibrium. It remains in its state only because its surroundings maintain gradients inside the thermodynamic system; in the simplest case, these are temperature gradients.

In statistical mechanics, the concept of temperature is based on the canonical distribution with respect to the energy of the whole system. If the ensemble is not canonical, temperatures may still be definable on canonical or quasi-canonical distributions of partial energies; we speak of local temperature, kinetic temperature, temperature of the nuclear spins, etc. The first pioneering work in the theory of irreversible processes was based on the assumption that such partial temperatures could be defined significantly.[1,2] Other workers succeeded in formulating general rules by adopting modified versions of Boltzmann's assumptions: That is that certain fluctuating variables are always statistically uncorrelated to each other.[2] Later refinements of Boltzmann's "Stosszahlansatz," in which the single-particle occupation numbers are assumed to be uncorrelated, make similar assumptions about higher (n-particle) distribution functions.[3] None of these theories clarifies completely the nature of irreversibility of a process which in its internal dynamics obeys deterministic laws of motion. Some of the assumptions made, i.e., Boltzmann's original assumptions, are demonstrably inconsistent with the classical laws of motion. It appeared impossible, with a closed system, to overcome Zermelo's classical argument, either in classical or in quantum mechanics, according to which the solutions of Liouville's equation are either periodic or almost periodic and, hence, cannot exhibit true irreversibility.[4] The argument that Poincaré cycles have such enormous periods that in practice they cannot be observed is not wholly unsatisfactory. Nevertheless, the fact that a Poincaré cycle may be longer than the total period since the development of the present physical universe (10^9 to 10^{10} years) does not entitle us to the positive conclusion that the specific assumptions made by various workers[2,3] will lead to valid approximations for realistically observable periods of time.

The most satisfactory approach, so it appears to us, is to construct models that permit treatment in terms of Gibbs-type ensembles, in which the openness of the system, and with it the possibility of true irreversible behavior, is explicitly introduced. Klein and Prigogine have considered such a model,[5] which represents a perfect one-dimensional crystal. In this model the mean free path of phonons is infinite; hence there can be no internal gradient. The spectral distribution of phonons is everywhere the same; hence the Klein-Prigogine model does not lend itself to a study of internal gradients and internal dissipation. Our paper will present a new and somewhat more general attempt to construct a system that interacts with its surroundings, in such a fashion that Gibbsian ensembles can be constructed; their development in the course of time, i.e., their dynamics, becomes the subject of mathematical investigation.

For a truly stationary process to be possible, the reservoirs that maintain the gradients within the system must be inexhaustible. They must possess an infinite

* This work was supported by the Office of Scientific Research, Research and Development Command, United States Air Force.

[1] L. Onsager, Phys. Rev. 37, 405 (1931); 38, 2265 (1931).

[2] For other authors who have used similar approaches leading essentially to macroscopic (phenomenological) laws, see E. W. Montroll and M. S. Green, Ann. Rev. Phys. Chem. 5, 449 (1954). The same review article also contains an extensive bibliography of work based on different approaches, *viz.*, kinetic theory, Markov chains, etc.

[3] M. Born and H. S. Green, *A General Kinetic Theory of Liquids* (Cambridge University Press, Cambridge, 1949); J. G. Kirkwood, J. Chem. Phys. 14, 180 (1946); 15, 72 (1947).

[4] P. and T. Ehrenfest, Encykl. d. Math. Wiss. 4, 2ii, September, 6 (1911).

[5] G. Klein and I. Prigogine, Physica 19, 74, 89 (1953).

number of degrees of freedom, in such a manner that they present to the system to be considered always the same appearance. Hence their canonical coordinates cannot be fully included in the description of the process in phase space (or in Hilbert space, in a quantum description). What should be included is the (fluctuating) interaction between system and reservoirs. Again, if in this interaction we disregard the effects of fluctuations, i.e., if we simply add to the internal Hamiltonian of the system the interaction energy with the reservoirs, averaged at each point of system-phase space over the possible values of the reservoir coordinates, then the resulting Liouville equation in system-phase space will not differ from one with fixed external forces, and no disspation will result.

How, then, is the action of the reservoir(s) to be introduced? There appear to be two possibilities. They represent, respectively, continuous and impulsive interaction. In both it is assumed that any part of the surroundings that will interact with the system will be statistically independent of the state of the system at the beginning of such interaction. But once the interaction has begun, the statistical independence will be lost; in fact one can show that with a continuous interaction Hamiltonian the correlation function between system and reservoir(s) satisfies a number of differential equations that preclude continued statistical independence. At any instance the interaction term in the Hamiltonian will depend on the previous history. We have been unable as yet to discover a simple mathematical scheme that would represent such finite interaction.

The second possibility is to treat the interaction impulsively. If the reservoir interacts for brief spans of time only, but then strongly, it may be assumed that the net effect of such impulsive interaction will be to move the representative point in system-phase space a finite distance, that depends both on its original location and on the state of the reservoir just prior to interaction. If we further simplify the reservoir by assuming that it consists of a sensibly infinite number of disconnected and similar parts, that each such part interacts with the system but once, and that prior to interaction there is statistical independence between that reservoir component and the system, then we can average over the possible states of the reservoir (i.e., of that portion of the reservoir about to interact impulsively with the system). The result of this averaging operation will be the determination of a contingent probability $K(x',x)dx'dt$, which tells us the probability that the representative point of the system, known to be at the location x (in system-phase space) will be thrown into the volume element dx' (a finite distance from x) within the time interval dt. Once we know the function K, then the Liouville equation for the probability density μ in system-phase space will take the form

$$\frac{\partial\mu(x,t)}{\partial t}+(\mu,H)_x = \int_{x'} [K(x,x')\mu(x')-K(x',x)\mu(x)]dx'. \quad (1.1)$$

The remainder of this paper will be concerned with an investigation of the structure of the function K (Sec. 2), with the approach toward thermal equilibrium in the presence of a single reservoir at a definite temperature (Sec. 3), with a proof that in the presence of several reservoirs ensembles obeying Eq. (1.1) approach a stationary state in the course of time (Sec. 4), and with the Onsager reciprocal relations (Sec. 5).

Before we enter these somewhat more detailed discussions, it may be in order to justify further the form of Eq. (1.1). We are well aware of the fact that impulsive interaction represents an asymptotic limit, which is not always realistic. But it is possible to construct model interactions that approach this limit arbitrarily closely. By assuming this particular form of interaction, and by suppressing the influence that the several components of the reservoir exert on each other, we are able to introduce into the mathematical formulation the notion of the inexhaustible reservoir. We need not neglect recoil (i.e., the reaction of the system back on the reservoir); and we are able to maintain the reservoirs at definite temperatures (or other thermodynamic potentials) without allowing for thermal or other gradients within the reservoirs.

2. STOCHASTIC KERNEL

The expression $K(x',x)$ in Eq. (1.1) represents a transition probability density in the phase space of the system. It is a so-called contingent probability, dealing with the rate of transitions from the state x to the state x' (we shall always write the arguments of such transitions going from right to left) on the assumption that the system is known to be initially in the state x. Hence the value of K will be independent of the probability distribution of the ensemble of systems in the system phase space.

The likelihood of such a transition will depend on the distribution of the reservoir (or reservoir components) as well as on the intrinsic probability of a collision (something like the collision cross section) if both system and reservoir are known to be in suitable states for a collision of the type considered. We shall have to consider both these factors.

As for the distribution of the reservoir, we shall consider it as part of the basic concept of reservoir that its parts are in a canonical distribution, characterized by a definite temperature. Considering the components of such a reservoir in their own phase space (i.e., the phase space of any one of the infinitely many identical reservoir components, usually called the μ space of such a composite), we may write this distribution in the standard form

$$\mu_r = \frac{1}{Z_r} e^{-E_r/kT} = \frac{1}{Z_r} e^{-\beta E_r}. \tag{2.1}$$

The subscript r indicates that these quantities belong to a reservoir component (and not to our system).

Next, we come to the much more delicate problem of the intrinsic reaction constant (or collision cross section). All microscopic laws, both classical and quantum theoretical, are reversible with respect to the direction of time. That is to say, there exists a transformation involving (in a suitable coordinate system) a change in the signs of the time coordinate and of all the momentum coordinates with respect to which these laws are invariant. In classical mechanics, the Hamiltonian is an even function of the canonical momentum components; in quantum mechanics (in a wave mechanical representation), the same transformation is accomplished if we change the sign of the time coordinate and at the same time replace the wave function by its complex conjugate. Hence, under this transformation the Hamiltonian remains the same function of its arguments. Each point of phase space (or of Hilbert space) is mapped on another point. We shall denote this mapping or transformation by a bar: $x \to \bar{x}$. Given a point in classical phase space with the coordinates x (standing for both q_k and p_k), there will be a point $\bar{x}$. If a trajectory in phase space is given by the function $x(t)$, then $\bar{x}(-t)$ is a trajectory, also consistent with the equations of motion.

Microreversibility must also apply to the impulsive interactions between system and reservoir. If a certain initial state of the system x_s and initial state of the reservoir x_r go over into a final state x_s', x_r' through discontinuous impulsive interaction, then the initial state $\bar{x}_s'$, $\bar{x}_r'$ must go over into the final state $\bar{x}_s$, $\bar{x}_r$ by means of the exactly reversed impulsive interaction, and the cross sections for these two corresponding processes must be equal. We may assume that the energy of the reservoir component associated with the point x_r equals the energy associated with the point $\bar{x}_r$, and likewise the energies at the two points x_r' and $\bar{x}_r'$ equal each other. We may, therefore, conclude that the ratio between the transition rates from x to x' and from $\bar{x}'$ to $\bar{x}$, (contingent on the initial state of the system but not on that of the reservoir) equals $\exp[\beta(E_r'-E_r)]$, because of Eq. (2.1). But impulsive interactions conserve the energy; hence the difference in reservoir energies equals the negative difference in system energies. Accordingly, we shall find that the ratio of the contingent proabilities of transition of the system from x_s to x_s' and transition from $\bar{x}_s'$ to $\bar{x}_s$ equals $\exp[\beta(E_s-E_s')]$,

$$K(x',x) = e^{\beta(E_s-E_s')} K(\bar{x},\bar{x}'), \quad E_s = \bar{E}_s, \quad E_s' = \bar{E}_s'. \tag{2.2}$$

This result appears unexceptionable, but it is insufficient to achieve our purpose. We have, therefore, gone one step further and assumed that microreversibility (i.e., equal cross sections) holds not only for the transition from x to x' and the "properly reversed" transition $\bar{x}'$ to $\bar{x}$, but also for the "directly reversed" transition x' to x. This additional assumption is sufficient to assure the monotonic decrease of the Helmholtz potential of a system in thermal contact with a single temperature reservoir until the ensemble of systems becomes canonical. However, our assumption may not be necessary for this result to hold. In fact though one can conceive of many forms of interaction in which the strengthened form of symmetry between a transition and its "direct reverse" holds, because of some special *geometric* symmetry of the model, it appears likely that the assumption made in this paper will have to be generalized, i.e., weakened, so that the approach proposed here is applicable to a wider class of models and phenomena. This possibility is now being investigated.

In what follows we shall accept the strongest form of symmetry of interaction, that is to say, equality between the cross sections of any transition and its direct reverse. If there are several reservoirs present, each with its own peculiar coupling to the system, then the kernel K of Eq. (1.1) will be the sum of such individual kernels K_a,

$$K = \sum_a K_a, \tag{2.3}$$

each representing the effects of one (the ath) reservoir. The kernel K_a will correspond to both the requirements of strong microreversibility (equality between the cross sections of direct reverses) and of canonical distribution of each reservoir. It will satisfy the equality:

$$\begin{gathered} K_a(x,x') = e^{\beta_a[H(x')-H(x)]} K_a(x',x), \\ K_a(x,x') = e^{\beta_a H(x')} L_a(x,x'), \quad L_a(x,x) = L_a(x'x) \geq 0. \end{gathered} \tag{2.4}$$

This condition on the stochastic kernel introduced into the Liouville equation (1.1) represents the point of departure of this paper. In the next section we shall show that with only one reservoir present the assumption (2.3) brings about the monotonic approach of the system governed by Eq. (1.1) toward the canonical distribution.

3. APPROACH TO THERMAL EQUILIBRIUM

We shall first consider an ensemble of systems each of which is in thermal contact with a reservoir at a definite temperature T. In that case the Liouville equation (1.1), with (2.4), assumes the form

$$\frac{\partial \mu}{\partial t} + (\mu, H) = \int_{x'} [e^{\beta H(x')} \mu(x') - e^{\beta H(x)} \mu(x)] L(x,x') dx'. \tag{3.1}$$

We shall now examine the solutions of this integro-differential equation with arbitrary initial conditions.

Of course, we cannot require that the entropy of the system will always increase as a result of its interaction with the reservoir. After all, energy is being exchanged; depending on initial conditions, it is quite possible that the system on the average transfers energy to the reservoir. We shall show, instead, that as the result of the interaction the total entropy, i.e., the sum of the entropies of the system and the reservoir, will increase. True, we have no detailed description of the reservoir. But as our reservoir is inexhaustible and, at the same time, possesses a definite temperature, it is to be assumed that the change in its entropy equals the amount of energy transferred to it, divided through by its absolute temperature. In other words, the rate of change of total entropy equals

$$dS_{\text{total}}/dt=\sigma=\dot{S}-k\beta\dot{U}. \tag{3.2}$$

Of course, the Hamiltonian of the system is assumed to be explicitly time-independent, i.e., no work is being performed on the system.

Our conjecture is, then, that the quantity $(S-k\beta U)$ will increase monotonically with time. If we multiply this expression by T and reverse the sign, we obtain the Helmholtz potential F,

$$F=U-ST=\int\mu(H+\beta^{-1}\ln\mu)dx, \tag{3.3}$$

which is to decrease monotonically in the course of time. We shall now proceed to prove this assertion.

The form of Eq. (3.1) suggests that instead of the variable μ we introduce ν,

$$\nu=\mu e^{\beta H}. \tag{3.4}$$

In terms of ν, Eqs. (3.1) and (3.3) read

$$\frac{\partial\nu}{\partial t}+(\nu,H)=e^{\beta H}\int_{x'}[\nu(x')-\nu(x)]L(x,x')dx',$$
$$F=\beta^{-1}\int_x e^{-\beta H}\nu\ln\nu dx, \tag{3.5}$$

respectively. The time derivative of the Helmholtz potential then becomes

$$\begin{aligned}\dot{F}&=\beta^{-1}\int_x e^{-\beta H}(\ln\nu+1)\dot{\nu}dx\\ &=\beta^{-1}\int_x\int_{x'}[\ln\nu(x)+1][\nu(x')-\nu(x)]L(x,x')dx'dx\\ &\quad+\beta^{-1}\int_x e^{-\beta H}(\ln\nu+1)(H,\nu)dx. \end{aligned}\tag{3.6}$$

It can be shown immediately that the second integral on the right-hand side vanishes. We make use of the fact that the integral over a Poisson bracket can be converted into a surface integral through integration by parts and hence vanishes when extended over the whole phase space, provided only the variables under the integrand satisfy reasonable boundary conditions (the analogous statement in quantum mechanics is that the trace of a commutator vanishes). All we need to do then is to carry out the obvious transformation of the integrand

$$\begin{aligned}e^{-\beta H}(\ln\nu+1)(H,\nu)&\equiv((\ln\nu+1)H,e^{-\beta H}\nu)\\&=((\ln\nu+1)H,\mu),\end{aligned}\tag{3.7}$$

and the desired result follows as a matter of course. The rate of change in the Helmholtz potential is therefore determined wholly by the first integral on the right-hand side of Eq. (3.6). By interchanging the role of the two sets of variables of integration, x and x', and by adding the resulting expression to the original form of the integral, we obtain this expression for

$$\begin{aligned}\dot{F}=(2\beta)^{-1}\int_x\int_{x'}&[\ln\nu(x)-\ln\nu(x')]\\&\times[\nu(x')-\nu(x)]L(x,x')dxdx'\\ \leqslant 0.\end{aligned}\tag{3.8}$$

The integrand on the right is manifestly negative-definite. Thus we have shown that the Helmholtz potential is bound to decrease with time. This decrease will be nonzero as long as there is any stochastic interchange between any two regions in phase space at which the ratio of the densities has not yet assumed the canonical value corresponding to the temperature $T=(\beta k)^{-1}$.

It remains to be shown that the Helmholtz potential assumes its absolute minimum for this canonical distribution. To examine the expression (3.3) for minima, we form its variation,

$$\delta F=\int\left[H+\frac{1}{\beta}(\ln\mu+1)\right]\delta\mu dx, \tag{3.9}$$

but subject to the restriction that the variations are not to violate the normalization requirement,

$$\int\delta u dx=0. \tag{3.10}$$

It follows that the only distributions stationary with respect to the variation (3.9), (3.10) are proportional, and therefore equal, to the canonical distribution. We can form the second variation as well, in order to determine whether the stationary solution represents a maximum, minimum, or saddle point. We have

$$\begin{aligned}\delta^2F&=\int[(H+\beta^{-1}\ln\mu)\delta^2\mu+\beta^{-1}(\delta\mu)^2]dx\\ &=\beta^{-1}\int[(\delta\mu)^2-\ln Z\delta^2\mu]dx\\ &=\beta^{-1}\int(\delta\mu)^2dx>0.\end{aligned}\tag{3.11}$$

The absence of any additional stationary solutions guarantees that the canonical distribution is not only a relative but the absolute minimum. The value of this minimum is, of course $F=-\beta^{-1}\ln Z$.

Our conclusion is, then, that the Helmholtz potential decreases monotonically in the course of time and that its absolute minimum corresponds to the canonical distribution at the temperature of the (single) reservoir. This result is not quite equivalent to the assertion that the canonical distribution is reached asymptotically. It is possible, first of all, that certain regions of phase space do not interact with each other through the stochastic kernel. Such a contingency is equivalent to the existence of quantities that are absolute constants of the motion, in the sense that they change their value neither under the influence of the internal forces nor under the impact of collisions with the surroundings. Such absolute constants of the motion are known, e.g., the symmetry character of the wave function with respect to indistinguishable particles in quantum mechanics. The existence of such absolute constants of the motion will make it possible to subdivide the phase space into invariant subspaces, each of which is ergodic with respect to the motion of Eq. (3.1).

Another possibility is that the stochastic kernel L converges to zero near certain points or regions and that there is insufficient non-stochastic flux into them, (see the discussion at the end of Sec. 4). This possibility is well known in a special case, namely the interaction of a Knudsen gas with the walls. Molecules of very low velocity take a long time to reach the wall. If the number of collisions per second drops too rapidly with velocity, then the approach to canonical distribution will not be uniform or even certain. Apart from these two possibilities, our system will approach the canonical distribution.

4. APPROACH TO STATIONARY NONEQUILIBRIUM STATES

In this section we shall prove that for a wide class of kernels $K(x,x')$ the distribution $\mu(x,t)$ will approach asymptotically some stationary distribution $\mu_s(x)$, which is independent of the initial distribution $\mu(x,0)$,

$$\lim_{t\to\infty}\mu(x,t)=\mu_s(x). \tag{4.1}$$

Our proof will be based on a general theorem by Doeblin,[6,7] which applies to all Markoffian processes.

If the initial distribution of the ensemble in phase space corresponds to a definite location, at the point x',

$$\mu(x,0)=\delta(x-x'), \tag{4.2}$$

we shall denote the distribution at any subsequent time t by the symbol $P(x, x'/t)$. For an arbitrary initial distribution $\mu(x,0)$, we then have

$$\mu(x,t)=\int_\Gamma P(x, x'/t)\mu(x',0)dx', \tag{4.3}$$

$$P(x, x'/t)\geqslant 0,\quad \int_\Gamma P(x, x'/t)dx=1,\quad t\geqslant 0. \tag{4.4}$$

The probability that at the time t the system will be in some particular region in phase space A (i.e., A might be the region between two energy surfaces E_1 and E_2) will be denoted by $\mu(A,t)$; it is given by

$$\mu(A,t)=\int_A\mu(x,t)dx=\int_\Gamma P(A, x'/t)\mu(x',0)dx'. \tag{4.5}$$

The integral

$$P(A, x'/t)=\int_A P(x, x'/t)dx \tag{4.6}$$

represents the (contingent) probability that a system, initially at the point x', will be found in the domain A at the time t. It is usually called a transition probability. Obviously $P(\Gamma, x'/t)=1$ for all t. It follows from the time independence of the Hamiltonian, and of the kernel $K(x,x')$ in Eq. (1.1) obeyed by $\mu(x,t)$, that $\mu(x, t+\tau)$ is determined by $\mu(x,\tau)$, again through Eq. (4.3); i.e.,

$$\begin{aligned}\mu(x, t+\tau)&=\int_\Gamma P(x, x'/t)\mu(x',\tau)dx'\\&=\int\int P(x, x'/t)P(x', x''/\tau)\mu(x'',0)dx''dx'\\&=\int P(x, x''/t+\tau)\mu(x'',0)dx''.\end{aligned} \tag{4.7}$$

Because $\mu(x'',0)$ is arbitrary, it follows that

$$P(x, x''/t+\tau)=\int P(x, x'/t)P(x', x''/\tau)dx'. \tag{4.8}$$

Equation (4.8) is a special case of the Chapman-Kolmogorov equation, which holds for all stationary Markov transition functions:

$$P(A, x''/t+\tau)=\int_\Gamma P(A, x'/t)P(dx', x''/\tau). \tag{4.9}$$

In this equation, $P(dx', x''/\tau)$ is the probability that the system will pass from the initial point x'' to the "infinitesimal" set $(x', x'+dx')$ in time τ, and the integral is a Lebesgue integral. In the case where a transition density exists, $P(dx', x''/t)=P(x', x''/t)dx'$.

If the Hamiltonian of Eq. (1.1) were explicitly time-dependent, then we should have a nonstationary

[6] W. Doeblin, Ann. Sci. Ecole Norm. Sup. (3) 57, 61 (1940).
[7] J. L. Doob, *Stochastic Processes* (John Wiley and Sons, Inc., New York, 1953).

Markov process, for which Eqs. (4.7), (4.8), and (4.9) do not hold. In that case, there could be no approach to stationary equilibrium in the strict sense, though it is not excluded that in the presence of a periodic Hamiltonian (e.g., spin resonance in the presence of dissipative processes) the solution of Eq. (1.1) may approach periodic behavior. In this paper we are not concerned with that possibility.

We shall now state Doeblin's theorem for the function $P(A, x'/t)$ that satisfies Eq. (4.9): If there exists an $\epsilon>0$, an $s>0$, and a set measure $\Psi(A)$ in Γ space, normalized so that $\Psi(\Gamma)=1$, and if these quantities can be chosen so that

$$P(A, x'/s)<1-\epsilon, \tag{4.10}$$

whenever $\Psi(A)<\epsilon$, for all x', then the limit

$$\lim_{t\to\infty}\{P(A, x'/t)\}$$

exists. The convergence is uniformly exponentially fast.

When the above conditions for the existence of the limit are satisfied for some time s, then they will also be satisfied for all times $t>s$, because

$$P(A, x'/t)=\int_\Gamma P(A, x''/s)P(dx'', x'/t-s) \leqslant \max\{P(A, x''/s)\}<1-\epsilon. \tag{4.11}$$

Intuitively, the conditions on $P(A, x/t)$ in Doeblin's theorem require that the stochastic process spread distributions that are concentrated initially in regions of measure zero into domains possessing a nonvanishing measure. This requirement excludes, for instance, the case of an isolated system, one that does not interact with a random reservoir. The ensemble distribution of such a system in its phase space obeys the unmodified Liouville equation, which maps distributions initially concentrated in a point on another point without spreading. Such a distribution, as is well known, will never become stationary if it is not stationary to begin with.

The requirement in Doeblin's theorem that $\psi(A)$ be normalized means that the measure of infinitely far regions in phase space is zero with respect to ψ. Hence with any transition function $P(A, x'/t)$ which satisfies the Doeblin conditions we must be able to find a finite region B in Γ space such that the probability of finding the system in B at any time $t>s$ is uniformly bounded away from zero (i.e., the bound is independent of the initial state of the system).

The assumptions made in Eq. (4.10) are sufficient but not necessary conditions for the asymptotic approach to a stationary distribution. As an illustration of this point we may use an example cited by Doob[7] for this purpose. Let

$$P(A, x'/t)=[2\pi(1-\rho^{2t})]^{-\frac{1}{2}}\int_A \exp\left[-\frac{(x-\rho^t x')^2}{2(1-\rho^{2t})}\right]dx, \quad 0<\rho<1. \tag{4.12}$$

Then

$$\lim_{t\to\infty} P(A, x'/t)=\frac{1}{(2\pi)^{\frac{1}{2}}}\int_A \exp(-\tfrac{1}{2}x^2)dx. \tag{4.13}$$

This convergence is not uniform in x': if A is any finite set, $P(A, x'/t)$ can be made arbitrarily small for any t, by choosing x' sufficiently large. Hence the conditions under which Doeblin's theorem was proved are not satisfied in this case.

We shall now turn to the conditions to be satisfied by the transition rate $K(x,x')$, Eq. (1.1), in order that Doeblin's theorem be applicable to the distribution $P(A, x'/t)$ whose time dependence obeys the generalized Liouville equation (1.1), We shall expand $P(x, x'/t)$ into a series with respect to the number of collisions between the reservoir and the system during the time t.

The system can pass from x' to x in time t without any collisions only if $x=x_t'$, where x_t' is the position at time t of a system initially at x', according to its Hamiltonian equations of motion. The probability of zero collisions, $\eta(x',t)$, is

$$\eta(x',t)=\exp\left(-\int_0^t p(x_s')ds\right), \tag{4.14}$$

where

$$p(x')=\int_\Gamma K(y,x')dy \tag{4.15}$$

is the rate at which collisions with the reservoir occur at the point x'. Hence

$$P^0(x, x'/t)=\left[\exp\left(-\int_0^t p(x_s')dx\right)\right]\delta(x-x_t'). \tag{4.16}$$

The probability that the system will experience exactly one collision in the interval t, given that at the beginning of the interval the system was at x' and at the end of the interval at x, will be a product of the following three probabilities: (1) there was no collision up to some time λ, $0\leq\lambda\leq t$, which is given by $\eta(x',\lambda)$; (2) there was a collision during the time interval $(\lambda, \lambda+d\lambda)$, which caused the system to make a discontinuous jump from x_λ' to $x_{(\lambda-t)}$ [the probability for this event is $K(x_{(\lambda-t)}, x_\lambda')d\lambda$]; (3) there was no collision between $(t-\lambda)$ and t, $\eta(x, \lambda-t)=\eta(x_{(\lambda-t)}, t-\lambda)$. Since λ is arbitrary, we have to add the different possibilities. Hence

$$P^{(1)}(x, x'/t)=\int_0^t d\lambda \exp\left[-\int_{\lambda-t}^0 p(x_s)ds\right] \times K(x_{\lambda-t}, x_\lambda')\exp\left[-\int_0^\lambda p(x_s')ds\right]. \tag{4.17}$$

By analogous reasoning we find the probability for the system to have two collisions with the reservoir to be

$$P^{(2)}(x, x'/t)$$
$$=\int_\Gamma dy\int_0^t d\beta\int_0^{t-\beta} d\lambda\left\{\exp\left[-\int_{\lambda+\beta-t}^0 p(x_s)ds\right]\right.$$
$$\times K(x_{\lambda+\beta-t}, y_\beta)\exp\left[-\int_0^\beta p(y_s)ds\right]$$
$$\left.\times K(y,x_\lambda')\exp\left[-\int_0^\lambda p(x_s')ds\right]\right\}. \quad (4.18)$$

Following up this analysis, we can get all the terms $P^{(n)}(x, x'/t)$. The total transition density $P(x, x'/t)$ is the sum of all these different possibilities

$$P(x, x'/t)=\sum_{n=0}^{\infty} P^{(n)}(x, x'/t). \quad (4.19)$$

This expansion can also be obtained by means of an entirely different, nonprobabilistic, scheme. It is the result of a perturbation expansion of $\mu(x,t)$ in powers of $K(x,x')$. The same type of perturbation theory is commonly used in quantum mechanics for the wave function $\psi(q,t)$ which, like $\mu(x,t)$, obeys a first-order differential equation in time. Let us rewrite Eq. (1.1) in an operator form thus

$$\partial\mu/\partial t=(H,\mu)-p(x)\mu+\int K(x,x')\mu(x')dx'$$
$$=(\mathcal{H}+V)\mu, \quad (4.20)$$

where

$$\mathcal{H}\mu=(H,\mu)-p(x)\mu, \quad V\mu=\int K(x,x')\mu(x')dx'.$$

The operator $\mathcal{H}$ is singular, but V is regular, bounded, and positive definite, hence we shall expand $\mu(x,t)$ as a power series in V. For this purpose we shall first convert the integro-differential equation for $\mu(x,t)$ into an integral equation,

$$\mu(x,t)=\exp(t\mathcal{H})\mu(x,0)+\int_0^t \exp[(t-\lambda)\mathcal{H}]V\mu(x,\lambda)d\lambda. \quad (4.21)$$

Expanding now in V we get

$$\mu(x,t)=\exp(t\mathcal{H})\mu(x,0)+\int_0^t d\lambda\{\exp[(t-\lambda)\mathcal{H}]V\exp(\lambda\mathcal{H})\}\mu(x,0)$$
$$+\int_0^t d\gamma\left\{\exp[(t-\gamma)\mathcal{H}]V\int_0^\gamma d\lambda[\exp((\gamma-\lambda)\mathcal{H})V\exp(\lambda\mathcal{H})]\right\}\mu(x,0)+\cdots. \quad (4.22)$$

Now $\exp(t\mathcal{H})f(x,0)=f(x,t)$ is the solution of the equation

$$\partial f(x,t)/\partial t=(H,f)-p(x)f. \quad (4.23)$$

More explicitly, this solution has the form

$$f(x,t)=\exp\left[-\int_0^t p(x_{s-t})ds\right]f(x_{-t},0). \quad (4.24)$$

Substituting this expression back into Eq. (4.21) for the case when $\mu(x,0)=\delta(x-x')$ and hence $\mu(x,t)=P(x, x'/t)$, we see that the expansions of Eqs. (4.22) and (4.19) are identical.

We can now prove that if $K(x,x')$ is bounded by an integrable function $v(x)$[8]

$$K(x,x')<Mv(x), \quad \int_\Gamma v(x)dx=1, \quad (4.25)$$

then our $P(x, x'/t)$ satisfies the Doeblin conditions. There are no additional restrictions imposed on $K(x,x')$ if $v(x)$ is made a function of $H(x)$ alone, $v(x)=v(H(x))$. This last assumption implies that $v(x_t)=v(x)$ for all t.

[8] This condition is somewhat more restrictive than absolutely necessary but it simplifies the proof considerably.

It now follows from (4.17), (4.18) that

$$P^{(1)}(x, x'/t)<M\int_0^t v(x_{\lambda-t})d\lambda=Mtv(x),$$
$$P^{(2)}(x,x'/t)<\frac{(Mt)^2}{2!}v(x), \quad (4.26)$$

and in general

$$P^{(n)}(x, x'/t)<\frac{(Mt)^n}{n!}v(x), \quad n\geq 1. \quad (4.27)$$

Hence, from (4.19),

$$P(x, x'/t)<\exp\left[-\int_0^t p(x_s')ds\right]\delta(x-x_t')$$
$$+(e^{Mt}-1)v(x). \quad (4.28)$$

In order that our system should approach a stationary state independent of its initial position, x', in phase space, we must, of course, assume that there are no trajectories in Γ space on which the system can move without ever getting into a region where it can collide with the reservoir. Otherwise if the system is initially in a state x' which lies on such a trajectory, it would behave as if there were no reservoir, and the density $\mu(x,t)$ would always be $\delta(x-x_t')$ without ever becoming stationary. More precisely, we assume that there exists a time t_1, such that for any time $t>t_1$, the system will spend at least an interval τ, $\tau=\alpha(t-t_1)$, $\alpha>0$ in such a region in phase space where the probability per unit time, $p(x)$, of its undergoing a collision is greater than some positive number m. Hence

$$P(x, x'/t)<\exp[-m\alpha(t-t_1)]\delta(x-x_t') +(e^{Mt}-1)v(x), \quad t>t_1, \tag{4.29}$$

and

$$P(A, x'/t)<\exp[-m\alpha(t-t_1)]+(e^{Mt}-1)\Psi(A), \tag{4.30}$$

where

$$\Psi(A)=\int_A v(x)dx. \tag{4.31}$$

When $\Psi(A)<\epsilon$, then

$$P(A, x'/t)<\exp[-m\alpha(t-t_1)]+(e^{Mt}-1)\epsilon, \tag{4.32}$$

which is less than $(1-\epsilon)$ whenever

$$\epsilon<e^{-Mt}(1-\exp[-m\alpha(t-t_1)]), \quad t>t_1. \tag{4.33}$$

Thus whenever $K(x,x')$ satisfies (4.25), $P(A, x'/t)$ satisfies (4.10); Doeblin's theorem applies, and

$$\lim_{t\to\infty} P(A, x'/t)=P(A,x') \tag{4.34}$$

exists, The limiting distribution $P(A,x')$ will be independent of x' if there are no noninteracting regions in phase space. Otherwise, the stationary state will depend on the initial distribution between these invariant subspaces. This requirement of ergodicity is the same as in the case of a single reservoir discussed at the end of Sec. 3.

5. STATIONARY STATE WITH SEVERAL RESERVOIRS: ONSAGER RELATIONS

We shall now consider an ensemble of systems in contact with n reservoirs at temperatures $T_i=(k\beta_i)^{-1}$, $i=1, \cdots, n$. Equation (1.1) then assumes the form

$$\frac{\partial\mu}{\partial t}+(\mu,H)=\sum_{i=1}^{n}\int_{x'}[e^{\beta_iH(x')}\mu(x')-e^{\beta_iH(x)}\mu(x)] \times L_i(x,x')dx'. \tag{5.1}$$

The rate of entropy production σ will be given, in analogy with Eq. (3.2), by

$$\sigma=\dot{S}-k\sum_{i=1}^{n}\beta_iJ_i, \tag{5.2}$$

where J_i is the average rate of energy flow from the ith reservoir to the system. The rate of change of the mean energy U is equal to the sum of the flows

$$\dot{U}=\int_\Gamma H\frac{\partial\mu}{\partial t}dx=\sum_{i=1}^{n}J_i=\sum_{i=1}^{n}\int_x dx \times\int_{x'}dx'[e^{\beta_iH(x')}\mu(x')-e^{\beta_iH(x)}\mu(x)]H(x)L_i(x,x')$$

$$=\sum_{i=1}^{n}\int_x dx\mu(x)\int_{x'}dx'\, e^{\beta_iH(x)}L_i(x,x')[H(x')-H(x)], \tag{5.3}$$

$$J_i=\int_x dx\mu(x)\int_{x'}dx'e^{\beta_iH(x)}L_i(x,x')[H(x')-H(x)] =\int_x dx\mu(x)f_i(x). \tag{5.4}$$

It can be shown by methods analogous to those used in the derivation of Eq. (3.8) that $\sigma\geqslant 0$. The equality can hold only if all the β_i are equal to each other.

When the ensemble becomes stationary, $\partial\mu/\partial t=0$, the internal entropy remains constant, $\dot{S}=0$, and Eq. (5.4) reduces to

$$\sigma=-k\sum_{i=1}^{n}\beta_iJ_i\geqslant 0. \tag{5.5}$$

Since in the stationary state the mean energy U is constant, it follows that

$$\dot{U}=\sum_{i=1}^{n}J_i=0. \tag{5.6}$$

Here we may solve for some particular flow J_m in terms of the other flows

$$J_m=-\sum_{i\neq m}{}' J_i \tag{5.7}$$

in a stationary ensemble. Substituting this expression into Eq. (5.5), we get

$$\sigma=k\sum_{i\neq m}{}'(\beta_m-\beta_i)J_i=\sum_{i\neq m}{}'\gamma_iJ_i(\gamma_l), \tag{5.8}$$

where the J_i are now independent flows.

It follows from this expression for the total entropy production that in our model, the J_i and the γ_i, are the currents and the forces respectively, which are used in the Onsager theory.[1] Onsager's reciprocal relations state that in the linear approximation the dependence of J_l on γ_k is the same as that of J_k on γ_l. In other words, when the reservoir temperatures are close to each other (i.e., where the γ_i's are small), when we can set

$$J_l=\sum_{k\neq m}{}' A_{lk}\gamma_k, \tag{5.9}$$

then

$$A_{lk}=\left(\frac{\partial J_l}{\partial \gamma_k}\right)_{\gamma_i=0}=\left(\frac{\partial J_k}{\partial \gamma_l}\right)_{\gamma_i=0}=A_{kl}. \quad (5.10)$$

We shall now prove the validity of Eq. (5.10) for our model. Since the J_m chosen in Eq. (5.7) was arbitrary, the symmetry relations should be independent of that choice. They will be if and only if

$$J_{lk}=\left(\frac{\partial J_l(\beta_i)}{\partial \beta_k}\right)_{\beta_i=\beta}=\left(\frac{\partial J_k(\beta_i)}{\partial \beta_l}\right)_{\beta_i=\beta}=J_{kl}, \quad (5.11)$$

for l, $k=1, \cdots, m, \cdots n$. We shall, therefore, prove Eq. (5.11) rather than Eq. (5.10). Generally, $J_{lk}=-A_{lk}$, l, $k\neq m$.

From Eq. (5.3), we get

$$\frac{\partial J_l}{\partial \beta_k}=\int_x dx\left[\frac{\partial \mu(x)}{\partial \beta_k}f_l(x)\right], \quad l\neq k, \quad (5.12)$$

since $f_l(x)$ depends only on β_l. Solving for $\partial\mu/\partial\beta_k$, in the stationary state, from the time independent form of Eq. (5.1), we get

$$\left(\frac{\partial \mu}{\partial \beta_k}, H\right)=\sum_{i=1}^{n}\int_{x'}\left\{e^{\beta_i H(x')}\frac{\partial \mu(x')}{\partial \beta_k}-e^{\beta_i H(x)}\frac{\partial \mu(x)}{\partial \beta_k}\right\}L_i(x,x')dx'$$
$$+\int_{x'}\left\{\mu(x')e^{\beta_k H(x')}\left[H(x')L_k(x,x')+\frac{\partial L_k}{\partial \beta_k}\right]-\mu(x)e^{\beta_k H(x)}\left[H(x)L_k+\frac{\partial L_k}{\partial \beta_k}\right]\right\}dx'. \quad (5.13)$$

We shall now evaluate this equation when all the β_i's equal to β. Under these conditions $\mu(x)=(1/Z)e^{-\beta H(x)}$, as derived in Sec. 3, and Eq. (5.13) reduces to

$$(\mu_k,H)=\int_x[\mathcal{L}(x,x')e^{\beta H(x')}\mu_k(x')]dx'-q(x)\mu_k(x)+\frac{1}{Z}e^{-\beta H(x)}f_k^0(x), \quad (5.14)$$

where

$$\mu_k=\left(\frac{\partial \mu}{\partial \beta_k}\right)_{\beta_i=\beta}, \quad \mathcal{L}(x,x')=\sum_{i=1}^{n}[L_i(x,x')]_{\beta_j=\beta}=\mathcal{L}(x',x),$$

$$q(x)=\int_{x'}\mathcal{L}(x,x')e^{\beta H(x)}dx'=p(x)|_{\beta_i=\beta}.$$

Multiplying Eq. (5.14) by $Ze^{\beta H(x)}\mu_l(x)$ and integrating over Γ, we get

$$J_{kl}=\int_x f_k^0(x)\mu_l(x)dx$$
$$=Z\int_x\left\{e^{\beta H(x)}\mu_l(x)(\mu_k(x),H(x))-\int_{x'}\mathcal{L}(x,x')e^{\beta[H(x)+H(x')]}\mu_k(x')\mu_l(x)dx'+\mu_k(x)\mu_l(x)q(x)e^{\beta H(x)}\right\}dx. \quad (5.15)$$

Hence

$$J_{kl}-J_{lk}=Z\int_x e^{\beta H(x)}[\mu_l(x)(\mu_k(x),H(x))-\mu_k(x)(\mu_l(x),H(x))]dx$$
$$=2Z\int_x e^{\beta H}\mu_l(\mu_k,H)dx-\beta^{-1}Z\int_x(\mu_k\mu_l,e^{\beta H})dx. \quad (5.16)$$

The last term vanishes. In order to evaluate the remainder in (5.16), it is convenient to separate $\mu_i(x)$ into a symmetric and an antisymmetric part with respect to the momentum variables of the system:

$$\mu_i(x)=E_i(x)+O_i(x), \quad E_i(x)=E_i(\bar{x}), \quad O_i(x)=-O_i(\bar{x}). \quad (5.17)$$

A Poisson bracket between even functions is odd, and between an even and an odd function even, etc. Because H is an even function, its Poisson bracket with $E_i(x)$ is odd, and with $O_i(x)$ even. Since the integral of an odd function over the whole phase space vanishes, it follows that

$$J_{kl}-J_{lk}=2Z\int_\Gamma[E_l(O_k,H)+O_l(E_k,H)]e^{\beta H}dx$$
$$=2Z\int_\Gamma[O_l(e^{\beta H}E_k,H)-O_k(e^{\beta H}E_l,H)]dx. \quad (5.18)$$

We can now solve for $(e^{\beta H}E_k,H)$ from Eq. (5.14), remembering that it follows from the way we arrive at Eq. (2.3) that $L(x,x')=L(\bar{x},\bar{x}')$. Thus

$$(e^{\beta H}E_k,H)=\int_{x'}\mathcal{L}(x,x')e^{\beta[H(x)+H(x')]}O_k(x')dx'-e^{\beta H(x)}q(x)O_k(x). \quad (5.19)$$

Substituting (5.19) into (5.18) we arrive at our desired result,

$$J_{kl}-J_{lk}=0. \quad (5.20)$$

This derivation of the Onsager relations requires no reference to fluctuation theory, nor does it involve an assumption that the laws connecting the macroscopic variables are Markoffian.

6. CONCLUSION

In this paper we have developed in classical (non-quantum) mechanics a model for a stationary irreversible process that is capable of being treated with the methods of statistical mechanics due to Gibbs. Our model is based on the assumption that no essential features of the real process are lost if the interaction of the system with the driving reservoirs is pictured in terms of instantaneous impulsive interactions. The reservoirs themselves are described as infinitely large composites consisting of identical noninteracting components in canonical distribution. Thus each reservoir has a definite temperature, infinite heat capacity, and vanishing internal heat conductivity. No special assumptions are made concerning the structure of the system; its internal dynamics are governed by some nonsingular Hamiltonian.

We have succeeded in showing that our model, with arbitrary initial ensemble distribution, will approach the canonical distribution if driven by a single reservoir, will approach a stationary (noncanonical) distribution if driven by several reservoirs at different temperatures, and in the stationary state will obey the Onsager relations if the driving temperature gradients are small.

Further work will be devoted to a more detailed investigation of the stochastic kernels that represent the action of the reservoirs, the introduction of more general thermodynamic forces than temperature gradients, and the transition to quantum mechanics.

THE JOURNAL OF CHEMICAL PHYSICS VOLUME 31, NUMBER 2 AUGUST, 1959

Statistical Mechanics of Rigid Spheres

H. REISS AND H. L. FRISCH, *Bell Telephone Laboratories, Inc., Murray Hill, New Jersey*

AND

J. L. LEBOWITZ,* *Stevens Institute of Technology, Hoboken, New Jersey*

(Received January 15, 1959)

An equilibrium theory of rigid sphere fluids is developed based on the properties of a new distribution function $G(r)$ which measures the density of rigid sphere molecules in contact with a rigid sphere solute of arbitrary size. A number of *exact* relations which describe rather fully the functional form of $G(r)$ are derived. These are based on both geometrical considerations and the virial theorem. A knowledge of $G(a)$ where a is the diameter of a rigid sphere enables one to arrive at the equation of state. The resulting analytical expression which is exact up to the third virial coefficient gives the fourth virial coefficient within 3% and the fifth, insofar as it is known, within 5%. Furthermore over the entire range of fluid density, the equation of state derived from theory agrees with that computed using machine methods. Theory also gives an expression for the surface tension of a hard sphere fluid in contact with a perfectly repelling wall. The dependence of surface tension on curvature is also given. The expressions obtained correlate nicely with those adduced by other thermodynamic and statistical mechanical theories. They also suggest that macroscopic consideration on surface tension can sometimes be successfully extrapolated to molecular dimensions.

1. INTRODUCTION

FLUIDS of rigid spheres have been the subject of many investigations. Through the application of cluster theory[1] and a certain amount of numerical analysis[2,3] one is now in possession of the first five virial coefficients for any such fluid. Recently the equation of state has been investigated by machine calculations, both by application of the Monte Carlo method[4,5] and by direct integration of the equations of motion of several hundreds of interacting rigid spheres.[6]

Aside from the application of cluster theory the problem has been attacked analytically by the distribution function method.[7] This has the disadvantage of invoking the superposition approximation[8] whose accuracy has never been clearly defined. Furthermore, the calculation of the *entire* radial distribution function is not required for the determination of the equation of state. For rigid spheres this equation can be shown through use of the virial theorem[9] to have the form[10]

$$(p/kT)=\rho+\tfrac{2}{3}\pi a^3\rho^2 g(a). \qquad (1.1)$$

Here p is the pressure, T the temperature, ρ the number density of particles, k the Boltzmann constant, a the diameter of a rigid sphere, and $g(r)$ the radial distribution function at separation r. It is evident from (1.1) that $g(r)$ is not required for all values of r and that all we need is $g(a)$ the value of g corresponding to two spheres in contact, i.e., when their centers are separated by the distance a.

If one is only interested in the equation of state, it would seem worthwhile to look for new methods which give information about $g(a)$. The present paper will offer an analysis of this sort. It will appear that a certain number of exact relations can be derived which go a long way toward determining the equation of state. Furthermore, the degree by which they fail in the full determination can be understood.

2. AN INTEGRAL EQUATION

We shall begin with a fluid of N rigid spheres of diameter a contained in the volume V. To this system another sphere of the same type can be added by means of a coupling parameter.[11] The more usual type[12] of coupling parameter has the following properties. Assume that the potential energy Φ_{N+1} of the system is composed of terms u_{ij} signifying the pairwise interaction of particles i and j,

$$\Phi_{N+1}=\sum_{1\leq i<j\leq N+1} u_{ij}. \qquad (2.1)$$

By assigning to say particle 1 the parameter ξ, in the sense that each pairwise term containing the index 1 in (2.1) is to be multiplied by ξ, we obtain

$$\Phi_{N+1}(\xi)=\sum_{j=1}^{j=N} u_{1j}(\xi)+\sum_{2\leq i<j\leq N+1} u_{ij},$$
$$u_{1j}(\xi)=\xi u_{1j}(1). \qquad (2.2)$$

By permitting ξ to go to unity we recover (2.1) while by setting $\xi=0$ we are left with only the second term on

* Supported by the Air Force Office of Scientific Research.

[1] T. L. Hill, *Statistical Mechanics* (McGraw-Hill Book Company, Inc., New York, 1956), Chapter 5.

[2] R. Majumdar, Bull. Calcutta Math. Soc. **21**, 107 (1929).

[3] Hirschfelder, Curtiss, and Bird, *Molecular Theory of Gases and Liquids* (John Wiley & Sons, Inc., New York, 1954), p. 157.

[4] M. N. Rosenbluth and A. W. Rosenbluth, J. Chem. Phys. **22**, 881 (1954).

[5] W. W. Wood and J. D. Jacobsen, J. Chem. Phys. **27**, 1207 (1957).

[6] T. E. Wainwright and B. J. Alder, AEC Report Contract No. W-7405-eng-48, Radiation Laboratory at Livermore, University of California.

[7] See reference 1, Chapter 6.

[8] See reference 1, p. 195.

[9] See reference 3, pp. 41, 134.

[10] See reference 1, p. 216.

[11] L. Onsager, Chem. Revs. **13**, 73 (1933); J. G. Kirkwood, J. Chem. Phys. **3**, 300 (1935).

[12] See reference 1, pp. 191, 192.

the right of (2.2) which clearly represents the potential energy ϕ_N of the system of N particles, which remains if particle 1 is removed. Thus by allowing ξ to go from zero to unity we can formally add an additional particle to a system of N particles. Evidently ξ is the coupling parameter and it should be noticed that with this type of coupling

$$\partial u_{1j}(\xi)/\partial\xi = u_{1j}(1). \qquad (2.3)$$

Now the chemical potential μ for a system of particles interacting in a pairwise fashion can be written in terms of the coupling parameter in the following form[13]

$$\mu = kT\ln\rho\Lambda^3 + \rho\int_0^1 d\xi\int_0^\infty \frac{\partial u}{\partial\xi}4\pi r^2 g(r,\xi)dr, \qquad (2.4)$$

where u stands for any u_{1j}, the same for all j, and where we have replaced the $u(1)$ in the formula of reference 13 by $\partial u/\partial\xi$ in accordance with (2.3). Λ^3 is the reciprocal of the momentum partition function and is given by

$$\Lambda^3 = h^3/(2\pi mkT)^{\frac{3}{2}}, \qquad (2.5)$$

where h is Planck's constant and m is the mass of a particle.

Equation (2.4) is true generally in the case of systems involving pairwise interaction for any coupling parameter which is able to remove the contribution of a given particle from Φ_{N+1}. The upper limit on the integration over the coupling parameter may not be 1 as in (2.4), however, but some appropriate number which signifies the complete coupling of the added particle to the rest of the system. Suppose for example that we wish to add an additional rigid sphere of diameter b to a fluid of N spheres of diameter a. The pairwise interaction of this sphere with any of the original is specified by

$$u=\infty, \qquad r\leq(a+b)/2=\lambda$$
$$u=0, \qquad r>(a+b)/2=\lambda. \qquad (2.6)$$

Notice that u, could be made zero everywhere by setting $\lambda=0$. Therefore, λ can be made to play the role of a coupling parameter; complete coupling being achieved when $\lambda=a$.

It should be observed that $\lambda=0$ implies $b=-a$. In other words it is not sufficient to set $b=0$ in order to decouple the added particle but we must assign it a negative diameter. This is so because with $b=0$ we still have a *point* molecule which is capable of excluding the centers of other molecules from a sphere of radius $a/2$ surrounding it. Making b negative need not be regarded as adding a molecule with negative dimensions but merely as adding a point molecule capable of penetrating the rigid cores of the original molecules.

It will be convenient to think of the added molecule as a *solute* when $b\neq a$, while the original molecules constitute the *solvent*.

[13] See reference 1, Eq. (30.36).

Inserting (2.6) into the integral on the right of (2.4) gives

$$\rho\int_0^a d\lambda\int_0^\infty \frac{\partial u}{\partial\lambda}4\pi r^2 g(r,\lambda)dr. \qquad (2.7)$$

Now

$$(\partial/\partial\lambda)[\exp(-u/kT)]$$
$$=-(1/kT)(\partial u/\partial\lambda)\exp(-u/kT)=-\delta(\lambda-r), \qquad (2.8)$$

where $\delta(\lambda-r)$ is the Dirac delta function.[14] From Eq. (2.6) and the left member of (2.7) it is seen that the delta function is properly normalized so that its integral over the infinite range is unity. Solving for $\partial u/\partial\lambda$ from the right-hand equation of (2.8) and substitution in (2.7) yields, after the latter is placed in (2.4),

$$\mu = kT\ln\rho\Lambda^3 + kT\rho\int_0^a 4\pi\lambda^2 g(\lambda,\lambda)d\lambda. \qquad (2.9)$$

Now $g(\lambda,\lambda)$ is the value of the radial distribution function for solvent molecules around a solute molecule on contact between solvent and solute, i.e., when their centers are separated by the distance $\lambda=(a+b)/2$. It is convenient to denote $g(\lambda,\lambda)$ simply by $G(\lambda)$. Thus $\rho G(\lambda)$ measures the average density of solvent molecules in contact with the solute, and in particular

$$G(a)=g(a). \qquad (2.10)$$

For the case of rigid spheres it is instructive to derive (2.9) by another method which provides more physical insight. Suppose the added solute molecule of diameter b is placed with its center at some *fixed* point **R** in the fluid. Then about **R** will exist a spherical region of radius $\lambda=(a+b)/2$ from which the centers of all N solvent molecules will be excluded. The presence of the solute at **R** is evidently equivalent to the existence of a spherical cavity at **R** in the solvent of at least radius λ. For this reason the reversible work which would be expended in the production of such a cavity must be identical with that required for the addition of the fixed solute molecule. If this work is increased by the free energy of mixing which would attend the liberation of the solute so that it could wander from the point **R**, the result would equal the reversible work involved in the addition of a real solute molecule, and so would equal the chemical potential of the solute. In the special case $b=a$ we would derive μ, the chemical potential of the solvent, given presumably by (2.9). Our program then will be to compute μ through consideration of the reversible work required for the production of a cavity of radius a.

It is a general result[15] of statistical mechanics that if $W(\lambda)$ is the reversible work required for the creation of a cavity of radius at least λ, then the probability

[14] V. Rojansky, *Introductory Quantum Mechanics* (Prentice-Hall, Inc., Englewood Cliffs, New Jersey, 1942), p. 68.
[15] R. C. Tolman, *The Principles of Statistical Mechanics* (Oxford University Press, London, 1938), Sec. 141.

$p_0(\lambda)$ of observing a fluctuation in which such a cavity forms is given by

$$p_0(\lambda)=\exp(-W(\lambda)/kT). \quad (2.11)$$

Thus knowledge of $p_0(\lambda)$ is equivalent to knowledge of $W(\lambda)$.

Now $p_0(\lambda)$ can be expressed in terms of $G(\lambda)$ by the following device. Introduce the auxiliary function $h(\lambda)$ which has the following meaning; $h(\lambda)d\lambda$ is the probability that at an arbitrary point in the fluid the center of the *nearest* molecule is a distance λ away, in the range $d\lambda$. Then $h(\lambda)d\lambda$ must be given by

$$h(\lambda)d\lambda=p_0(\lambda)4\pi\lambda^2\rho G'(\lambda)d\lambda. \quad (2.12)$$

The factor $p_0(\lambda)$ on the right is the probability that the *nearest* molecule is *at least* a distance λ away (i.e., the probability of having a spherical cavity of radius at least λ) while $4\pi\lambda^2\rho G'(\lambda)d\lambda$ is the conditional probability that the spherical shell of volume $4\pi\lambda^2 d\lambda$ enveloping the cavity of radius λ contains the center of at least one molecule. From the definition of $h(\lambda)$, $\rho G'(\lambda)$ is the average density of molecular centers just outside the cavity of radius λ (or the average density of molecules in contact with the solute of radius b) or $G'(\lambda)=G(\lambda)$.

It should also be clear that

$$p_0(\lambda)=1-\int_0^\lambda h(\lambda')d\lambda' \quad (2.13)$$

since the integral on the right is the probability that the center of at least one molecule *is* in the spherical region of radius λ. Differentiating (2.13) with respect to λ we have

$$h(\lambda)=-\partial p_0/\partial\lambda. \quad (2.14)$$

Substituting this result into (2.12) then yields

$$G(\lambda)=-(1/4\pi\lambda^2\rho)(\partial \ln p_0/\partial\lambda) \quad (2.15)$$

or, since $p_0(0)=1$,

$$-\ln p_0(\lambda)=\rho\int_0^\lambda 4\pi\lambda'^2 G(\lambda')d\lambda' \quad (2.16)$$

which in combination with (2.11) gives

$$W(\lambda)=kT\rho\int_0^\lambda 4\pi\lambda'^2 G(\lambda')d\lambda'. \quad (2.17)$$

The free energy of mixing of the solute molecule with solvent is the well-known expression $kT\ln(\Lambda_b^3/V)$. The factor $(1/V)$ is essentially the density ρ_b of solute molecules. In case there is more than one solute molecule present the free energy of mixing can still be expressed as $kT\ln(\rho_b\Lambda_b^3)$ provided the solution is dilute enough so that solutes do not ineract. Adding this contribution to (2.17) we obtain for μ_b the chemical potential of the solute

$$\mu_b=kT\ln(\rho_b\Lambda_b^3)+kT\rho\int_0^{(a+b)/2} 4\pi\lambda^2 G(\lambda)d\lambda. \quad (2.18)$$

Equation (2.18) is the expression for the chemical potential of a molecule of radius $b/2$ in a mixture of b and a molecules with $\rho_b b^3 \ll 1$.

If our single additional "solute" molecule has a diameter a and is indistinguishable from the "solvent" molecules then clearly its entropy of mixing is $kT\ln(\Lambda_a^3\rho)$. Hence its chemical potential, which is now identically that of a "solvent molecule," is given by (2.18), (without reference to any solutions), when we set $b=a$, $\rho_b=\rho$. When these replacements are made (2.18) becomes identical with (2.9) which completes the alternative mode of derivation.

Combining (2.16) with (2.12) and (2.13) yields

$$h(\lambda)=4\pi\lambda^2\rho G(\lambda)\exp\left\{-\int_0^\lambda 4\pi\lambda'^2\rho G(\lambda')d\lambda'\right\}. \quad (2.19)$$

The function $h(\lambda)$ is related to the Hertz or *nearest neighbor* distribution.[16] This function and its close relatives have found application in other problems in physics.[17–19]

The next step involves computing the pressure p from μ given by (2.9). In this connection we employ the straightforward thermodynamic relation

$$p=\int_0^\rho \rho\left(\frac{\partial\mu}{\partial\rho}\right)_T d\rho. \quad (2.20)$$

Substitution of (2.9) into (2.20) yields, after an integration by parts,

$$\frac{p}{kT}=\rho+\rho^2\int_0^a 4\pi r^2 G(r,\rho)dr-\int_0^\rho \rho' d\rho'\int_0^a 4\pi r^2 G(r,\rho')dr, \quad (2.21)$$

in which we now use r in place of λ. Introduction of (2.10) into (1.1) and equating the right-hand sides of (1.1) and (2.21) then gives

$$\tfrac{2}{3}\pi a^3\rho^2 G(a,\rho)=\rho^2\int_0^a 4\pi r^2 G(r,\rho)dr - \int_0^\rho \rho' d\rho'\int_0^a 4\pi r^2 G(r,\rho')dr, \quad (2.22)$$

which is an integral equation on $G(r,\rho)$.

It is worthwhile to pause at this stage to consider the implications of (2.22). First we notice that because it contains definite integrals it cannot determine $G(a,\rho)$ uniquely. Nevertheless, it may provide a considerable amount of information concerning $G(a,\rho)$. It is *exact*, and together with other exact conditions may lead to a powerful method for getting at $G(a,\rho)$ and

[16] S. Chandrasekhar, Revs. Modern Phys. **15**, 1 (1943).

[17] Holtsmark, Ann. Physik **58**, 577 (1919); Physik. Z. **20**, 162 (1919); **25**, 73 (1924); S. Chandrasekhar, Revs. Modern Phys. **15**, 1 (1943).

[18] R. M. Fuoss, Trans. Faraday Soc. **30**, 967 (1934).

[19] H. Reiss, J. Chem. Phys. **25**, 400, 408 (1956).

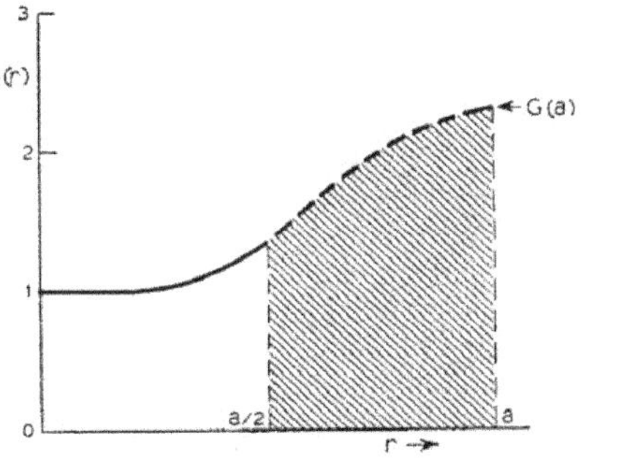

FIG. 1. The behavior of the distribution function $G(r)$ in $0 \leq r \leq a$.

thus, through (1.1), at the equation of state. We shall see that this is the case.

On the other hand there is a deeper significance to which attention should be directed. This is that Eq. (2.22) has been derived with the aid of the virial theorem. Conventionally, the virial theorem has been used to obtain information concerning the *macroscopic* properties of a system. In (2.22) it provides information on the *microstructure* of the system. In fact by using other types of coupling parameters together with the virial theorem it is actually possible to derive other integral equations which give information on various aspects of the radial distribution function. The elucidation of these is not however within the domain of this paper. It suffices to mention that it may be profitable to investigate them.

3. FURTHER CONDITIONS ON $G(r, \rho)$

Additional exact conditions on $G(r, \rho)$ can be obtained. Consider for example $p_0(r)$ for $r \leq a/2$. A spherical region whose radius lies in this range can contain at the most the center of *one* rigid sphere at a time. The probability that it is so occupied is therefore $(\frac{4}{3})\pi r^3 \rho$, and the probability that it is empty is just one minus this quantity, so

$$p_0(r) = 1 - \tfrac{4}{3}\pi r^3 \rho, \quad r \leq a/2. \tag{3.1}$$

Substitution of this result into (2.14) then yields

$$G(r) = (1 - \tfrac{4}{3}\pi r^3 \rho)^{-1}, \quad r \leq a/2. \tag{3.2}$$

Therefore, $G(r, \rho)$ is actually known up to $r = a/2$.

It has been pointed out that the condition $r = a/2$ corresponds to the presence of a point molecule. It may be of interest to compute the work $W(a/2)$ expended in the introduction of a fixed point molecule into the fluid. This is accomplished by inserting (3.2) in (2.17), setting the limit λ equal to $a/2$.

$$W(a/2) = -kT \ln[1 - (\pi a^3 \rho/6)]. \tag{3.3}$$

According to (3.2) $G(r)$ is a monotone increasing function of r in the range $r \leq a/2$ [except when $(\frac{4}{3})\pi r^3 \rho > 1$, a condition which cannot be realized at densities less than close packing]. There are strong grounds for believing that $G(r)$ is monotone increasing over the whole range of r as long as the system remains in the fluid state, but we have not been able to prove this. For large values of r, however, a strong plausibility argument can be given. This is based on the fact that for large r the cavity is essentially a bubble and its work of formation can be represented as the sum of surface and volume terms. Thus

$$W(r) = \omega v - \gamma v^{2/3}, \quad r \text{ large} \tag{3.4}$$

where $v = (\frac{4}{3})\pi r^3$ is the volume of the cavity and ω and γ are positive and independent of v. The term in v is the volume term and that in $v^{2/3}$ is the surface work. The surface work is negative because for hard spheres the surface tension is negative.

Differentiation of (2.17), with $v = (\frac{4}{3})\pi r^3$, and also (3.4),

$$\partial W/\partial v = kT\rho G = P$$

$$\partial^2 W/\partial v^2 = kT\rho(\partial G/\partial v) = \tfrac{2}{9}\gamma v^{-4/3} > 0. \tag{3.5}$$

In the first of these equations P is the stress normal to the surface of the cavity which since it is wholly kinetic is equal to kT times the density ρG at the surface of the cavity. We could have derived the relation between p_0 and G directly from this fact. The second of Eqs. (3.5) shows that $\partial G/\partial v$ is positive and therefore G increases monotonically with v or r.†

If the cavity is allowed to grow to infinite size, its bounding surface will behave like a plane wall, and $\rho G(\infty, \rho)$ is therefore the average density ρ_{wall} of molecules next to a plane wall. Now according to (3.5)

$$p/kT = \rho_{\text{wall}} = \rho G(\infty, \rho). \tag{3.6}$$

Comparing this with (1.1) after taking notice of (2.8) yields

$$G(\infty, \rho) = 1 + \tfrac{2}{3}\pi a^3 \rho G(a, \rho) \tag{3.7}$$

which is another exact condition on the behavior of $G(r, \rho)$.

Some perspective on the entire problem can now be had. For the equation of state we require $G(a, \rho)$, and this is available from Eq. (2.22) provided the integrals on the right are known. In essence we need to know the area under the curve in Fig. 1. Now we are exactly cognizant of this area up to the point $r = a/2$ in view of (3.2). What is lacking is the shaded area in Fig. 1. But if $G(r)$ is a fairly smooth and simple curve we might connect the points $a/2$ and a by some elementary function having undetermined parameters which do not depend on r. If the parameters are then determined by whatever exact information on $G(r)$ is available we arrive at a fairly accurate specification of $G(a)$.

† The authors are indebted to G. H. Wannier for suggesting that a proof of monotone behavior might be obtained through consideration of the work.

We have not exhausted the store of possible exact conditions on $G(r, \rho)$. Suppose $p_1(r)$, $p_2(r)$, etc., represent the probabilities that the centers of *exactly* one particle, *exactly* two particles, etc., lie in the spherical region of radius r. Then

$$p_0+p_1+p_2+p_3+\cdots=1. \tag{3.8}$$

This can also be written as

$$\begin{aligned} p_0+\sum_{m=1}^{\infty}p_m &= p_0-\sum_{m=1}^{\infty}\{[1+(-1)]^m-1\}p_m \\ &= p_0-\sum_{m=1}^{\infty}p_m\sum_{n=1}^{\infty}(-1)^nC_n^m \\ &= p_0-\sum_{n=1}^{\infty}(-1)^n\sum_{m=n}^{\infty}C_n^m p_m=1 \end{aligned} \tag{3.9}$$

where C_n^m are the binomial coefficients.

Solving for p_0 we have

$$p_0=1-(p_1+2p_2+3p_3+\cdots)+(p_2+3p_3+6p_4+\cdots)-(p_3+4p_4+\cdots)+\cdots. \tag{3.10}$$

Now in (3.10) the first sum in brackets represents the average number of individual particles to be found in the region of radius r. The second sum represents the average number of pairs, the third the average number of triplets, etc. From this it is obvious that

$$\begin{aligned} p_0(r) &= 1+\sum_{m=1}^{\infty}\frac{(-1)^m\rho^m}{m!}\int_{\text{(region of radius } r)} g^{(m)}d\tau_1 d\tau_2\cdots d\tau_m \\ &= 1+\sum_{m=1}^{\infty}(-1)^m F_m \end{aligned} \tag{3.11}$$

where $g^{(m)}$ is the m-particle correlation function[20] and $d\tau_1$ etc., are the volume elements for the coordinates of the different particles. The F_m are quantities defined by the last equality. It will be noticed that $g^{(2)}$ is just g, the radial distribution function.

Equation (3.11) is derivable by purely statistical mechanical means, beginning with the partition function, but we have chosen to use probabilistic arguments because they are simpler to develop,‡ and also furnish more insight into the problem. The equation in more general form was derived essentially, by using statistical mechanics, by Mayer and Montroll.[21] It is in fact Eq. (60) of their paper if for the activity which they call Z we substitute $(p_0\Lambda^3)^{-1}$ which can be shown to represent the activity.

Equation (3.11) is exceedingly useful for the development of exact conditions on $G(r, \rho)$. Thus suppose r is just large enough so that the region can accommodate two but not three particles. This condition arises when

$$a/2<r\leq a/\sqrt{3}. \tag{3.12}$$

Then $g^{(m)}$ for $m\geq 3$ must be zero everywhere in the spherical region, so that according to (3.11)

$$\begin{aligned} p_0 &= 1-\tfrac{4}{3}\pi r^3\rho+(\rho^2/2)\int_{\text{(region of radius } r)} g\,d\tau_1 d\tau_2 \\ &= 1-F_1+F_2 \end{aligned} \tag{3.13}$$

where we have used the fact that $g^{(1)}=1$.

Although the functions F_m defined in (3.11) relate to integrals over the limited region of radius r they can be given an alternative form in terms of integrals taken over the whole of the volume V in which the system is contained. This form is in fact used in Eq. (60) of reference 21. Consider m particles of diameter a with their centers respectively at the points $\mathbf{R}_1, \mathbf{R}_2, \cdots, \mathbf{R}_m$. Around each of these centers describe a sphere of radius r. Then $\Omega_m(\mathbf{R}_1, \mathbf{R}_2, \cdots, \mathbf{R}_m, r)$ is defined as the total volume of overlap of the m spheres of radius r when their centers are at the stated positions. We will now show that

$$F_m=(\rho^m/m!)\int\cdots\int_V g^{(m)}(\mathbf{R}_1, \mathbf{R}_2, \cdots, \mathbf{R}_m)\times\frac{\Omega_m(\mathbf{R}_1, \mathbf{R}_2, \cdots, \mathbf{R}_m, r)}{V}d\mathbf{R}_1 d\mathbf{R}_2\cdots d\mathbf{R}_m, \tag{3.14}$$

where the integration now proceeds over the entire volume V.

Equation (3.14) can be justified by a simple interpretation in terms of $p_0(r)$. If a sphere of radius r is drawn about the center of each molecule in the system, then the probability $p_0(r)$ that there is no molecule within a distance r of an arbitrary point in the fluid is just equal to the fraction of V outside all the spheres. If the spheres did not overlap this fraction would be

$$p_0(r)=\frac{V-N\frac{4}{3}\pi r^3}{V}=1-\rho\int_V \Omega_1(\mathbf{R}_1, r)d\mathbf{R}_1. \tag{3.15}$$

This is exactly the case for $r\leq a/2$ since then the spheres of radius r are all inside the rigid sphere molecules and cannot possibly overlap. Since $N/V=\rho$, (3.15) is identical, as expected, with (3.1).

For $r>a/2$ there will be some overlap to be accounted for. Thus to $V-N\frac{4}{3}\pi r^3$ in (3.15) would have to be added the volume of overlap of all pairs of spheres (some of these pairs may be part of a triple overlap configuration etc.) which has been subtracted twice through the term $N\frac{4}{3}\pi r^3$. But now the volume of triple overlaps will have been added once too often and we have to subtract it. The process of adding and subtracting may be continued through consideration of all higher

[20] See reference 1, p. 183.
‡ Equation (3.11) is not limited to a gas of rigid spheres but holds for cavities in all fluids.
[21] J. E. Mayer and E. Montroll, J. Chem. Phys. 9, 2 (1941).

order overlappings with the result [in extension of (3.15)],

$$p_0(r)=\frac{V-\rho\int_V\Omega_1(\mathbf{R}_1,r)d\mathbf{R}_1+(\rho^2/2)\iint_V g^{(2)}(\mathbf{R}_1,\mathbf{R}_2)\Omega_2(\mathbf{R}_1,\mathbf{R}_2,r)d\mathbf{R}_1d\mathbf{R}_2+\cdots}{V},$$

$$=1+\sum_{m=1}^{\infty}(-1)^mF_m \qquad (3.16)$$

which identifies F_m in the form (3.14).

In particular since both $g^{(2)}$ $(\mathbf{R}_1, \mathbf{R}_2)$ and $\Omega_2(\mathbf{R}_1, \mathbf{R}_2, r)$ depend on $\mathbf{R}_1$ and $\mathbf{R}_2$ only through $|\mathbf{R}_1-\mathbf{R}_2|$ it follows that

$$F_2(r)=\rho^2/2\iint_V g(\mathbf{R}_1,\mathbf{R}_2)\frac{\Omega_2(\mathbf{R}_1,\mathbf{R}_2)}{V}d\mathbf{R}_1d\mathbf{R}_2$$

$$=(\rho^2/2)\int_0^{2r}g(s)\Omega_2(s,r)4\pi s^2ds. \qquad (3.17)$$

Here $s=|\mathbf{R}_1-\mathbf{R}_2|$ and the upper limit $2r$ is derived from the fact that $\Omega(s, r)=0$ for $s\geq 2r$ since two spheres of radius r cannot overlap when their centers are separated by a distance greater than $2r$. Differentiation of (3.17) with respect to r yields

$$\partial F_2/\partial r=(\rho^2/2)\int_0^{2r}g(s)(\partial\Omega_2/\partial r)4\pi s^2ds$$
$$+\rho^2g(2r)\Omega(2r,r)16\pi r^2. \qquad (3.18)$$

The second term on the right vanishes because, as mentioned, $\Omega(2r, r)=0$. Furthermore, $\partial\Omega_2/\partial r$ in the first term is nothing more than $S_2(s, r)$, the surface area of the overlapping volume of two spheres. This last follows from the fact that by $(\partial\Omega_2/\partial r)$ dr we mean a volume increment achieved by moving the surface of overlap a distance r normal to itself. Thus we have

$$\partial F_2/\partial r=(\rho^2/2)\int_0^{2r}g(s)S_2(s,r)4\pi s^2ds, \qquad (3.19)$$

and it is shown easily by simple geometry that

$$S_2(s,r)=4\pi r[r-(s/2)]. \qquad (3.20)$$

From this we see that S_2 $(2r, r)=0$, so that by reasoning analogous to that used in deriving (3.19) we obtain

$$\partial^2F_2/\partial r^2=(\rho^2/2)\int_0^{2r}g(s)(\partial S_2/\partial r)4\pi s^2ds. \qquad (3.21)$$

Since

$$g(s)=0,\ s\leq a \qquad (3.22)$$

it follows from (3.19) and (3.21) that

$$(\partial F_2/\partial r)_{r=a/2}=(\partial^2F_2/\partial r^2)_{r=a/2}=0. \qquad (3.23)$$

This means that in spite of the sudden appearance of of F_2 in p_0 [see (3.13)] at $r=a/2$ both $\partial p_0/\partial r$ and $\partial^2p_0/\partial r^2$ are continuous at $r=a/2$. Differentiating (3.21) making use of (3.20) and (3.22) yields

$$(\partial^3F_2/\partial r^3)_{r=a/2}=8\pi^2\rho^2a^3g(a) \qquad (3.24)$$

so that $\partial^3p_0/\partial r^3$ is not continuous at $r=a/2$, but contains a jump proportional to $g(a)$.

Since $G(r)$ is determined according to (2.15) by the derivative of p_0, (3.23) implies that both G and its first derivative are continuous at $r=a/2$, while (3.24) requires that the second derivative of G be discontinuous. The exact relations for G and its derivatives are

$$G=\frac{1}{1-(\pi a^3\rho/6)},\quad r=a/2 \qquad (3.25)$$

$$\left(\frac{\partial G}{\partial r}\right)_{r=a/2}=\frac{\pi a^2\rho}{[1-(\pi a^3\rho/6)]^2} \qquad (3.26)$$

$$\left(\frac{\partial^2G}{\partial r^2}\right)_{r=a/2}=\frac{4\pi a\rho}{[1-(\pi a^3\rho/6)]^2}+\frac{2\pi^2a^4\rho^2}{[1-(\pi a^3\rho/6)]^3}$$
$$-\frac{8\pi a\rho}{[1-(\pi a^3\rho/6)]}G(a). \qquad (3.27)$$

The derivative in (3.27) is taken to the right of $r=a/2$, the last term representing the discontinuity proportional to $G(a)$. In (3.27) $G(a)$ has been used for $g(a)$. Both (3.26) and (3.27) are further *exact* conditions on the behavior of $G(r, \rho)$.

The function G is obviously not analytic since discontinuities in its higher derivatives occur at points where r becomes large enough so that the spherical region can accommodate more particles. These are the points $a/2$, $a/\sqrt{3}$, etc., at which two, three, etc., particles can be accommodated. The discontinuities are associated with the appearance of the new processes associated with the additional particles. Between $a/2$ and a, for example, $G(r)$ is not analytic at twelve points, at least, because up to twelve particles can be accommodated.

Nevertheless G is a smooth function because the discontinuity occurs in high order derivatives. For example, at $a/2$ it first occurs in the second derivative. By an analysis similar to that given previously it can be shown that at $a/\sqrt{3}$

$$F_3(r)=A[r-(a/\sqrt{3})]^{11/2}g^{(3)}(a,a,a) \qquad (3.28)$$

plus higher order terms in $[r-(a/\sqrt{3})]$, and where A is independent of r. From this it follows that at $r=a/\sqrt{3}$ the first five derivatives of $p_0(r)$ are continuous, the discontinuity occurring in the sixth. Therefore, the first four derivatives of G are continuous, the discontinuity occurring in the fifth. In general the order of the derivative at which the discontinuity occurs increases by at least two at each successive value of r at which another particle can be accommodated.

The results contained in (3.26) and (3.27) could have been deduced by performing integrations by parts in the integral of (3.13). These integrations have been performed by Hill[22] and yield§

$$p_0(r)=1-\tfrac{4}{3}\pi r^3\rho+\frac{\rho^2}{2}\int_0^{2r}4\pi s^2 g(s)\left[\frac{\pi r^3}{3}\left\{4-\frac{3s}{r}+\frac{s^3}{4r^3}\right\}\right]ds \tag{3.29}$$

$[\Omega_2(s,r)$ is equal to the terms in the bracket under the integral sign in (3.29)$]$ which can be differentiated to yield (3.26) and (3.27). Although (3.29) will prove useful later in a discussion of the relation between $g(r)$ and $G(r)$ for a range of values of r it is not the most efficient means of arriving at (3.26) and (3.27), and in fact a similar technique could probably not be employed for the derivation of (3.28).

In closing this section we shall find $G(r,\rho)$ to the first order in ρ for all values of r. This will also give an illustration of the validity of Eq. (2.22). This is accomplished by finding an exact expression for $G(r,\rho)$ to the appropriate order in ρ. It is convenient to work in terms of the reduced variables||

$$\xi=r/a.$$
$$x=a^3\rho, \tag{3.30}$$

instead of r and ρ. When these new variables are substituted into (2.22) and the integrations up to $a/2$ (or up to $\frac{1}{2}$ in terms of ξ) are performed with the aid of (3.2) we obtain

$$\frac{x^2}{6}G(1,x)+\frac{3}{2\pi^2}\left\{\ln\left(1-\frac{\pi x}{6}\right)+\frac{\pi x}{6}\right\}$$
$$=x^2\int_{\frac{1}{2}}^{1}\xi^2G(\xi,x)d\xi-\int_0^x zdz\int_{\frac{1}{2}}^{1}\xi^2G(\xi,z)d\xi. \tag{3.31}$$

According to (3.11) $p_0(r)$ is determined up to terms in ρ^2 *for all values of* r by the part of (3.11) which appears in (3.29). This follows from the fact that the integrals in $g^{(m)}$ for $m\geq 3$ in (3.11) are preceded by ρ raised to a higher power than 2. Furthermore, g can be expanded[23] in a power series in ρ the first term of which is always unity. We will only be interested in $p_0(r)$ to terms in ρ^2 so that in (3.29) we can use

$$g=0,\quad s\leq a$$
$$g=1,\quad s>a \tag{3.32}$$

and obtain a value of $p_0(r)$ up to powers in ρ^2 and valid for $r>a/2$. Using this in (2.15) provides a value for $G(r,\rho)$ up to the first power in density, and valid for $r>a/2$. Written in terms of ξ, x this is

$$G(\xi,x)=1+(\pi x/24)[16-(6/\xi)]+0(x^2),\ \xi>\tfrac{1}{2}. \tag{3.33}$$

Substitution of (3.33) into the integrals of (3.31) with expansion of the logarithmic term in powers of x and subsequent collection of powers yields up to terms in x

$$G(1,x)=1+\tfrac{5}{12}\pi x+0(x^2)=g(a), \tag{3.34}$$

which is the same as (3.33) for $\xi=1$. This also agrees with the well-known virial expansion for $g(a)$.[24]

4. EVALUATION OF $G(1,x)$ BY POLYNOMIAL METHOD

In this section we shall attempt to evaluate $G(1,x)$ and hence the equation of state through manipulation of the exact relations derived in the preceding sections. In particular we shall make use of (3.31), (3.25), (3.26), and (3.27). As we have mentioned earlier in connection with Fig. 1, we require information concerning the area below the curve $G(r,\rho)$ between $r=a/2$ and $r=a$. This region corresponds to the integrals in (3.31). It has been shown that $G(r,\rho)$ is a smooth curve with discontinuities only in the higher derivatives, and also is probably monotone increasing. We shall therefore try to approximate $G(r,\rho)$ in the short range between $a/2$ and a by some simple function of r with four parameters which may be functions of ρ.

Almost any simple function will do and for convenience we choose the cubic polynomial

$$G(\xi,x)=\alpha_0(x)+\alpha_1(x)(\xi-\tfrac{1}{2})+[\alpha_2(x)/2!](\xi-\tfrac{1}{2})^2$$
$$+[\alpha_3(x)/3!](\xi-\tfrac{1}{2})^3, \tag{4.1}$$

where the reduced variables ξ and x are used in place of r and ρ. By the straightforward substitution of (4.1) into (3.31), (3.25), (3.26), and (3.27) a set of simultaneous equations for the coefficients α_0, α_1, α_2, and α_3 are obtained which can be solved, and in fact $G(1,x)$ can be represented by a single quadrature.

$$G(1,x)=t^{-\delta}\exp[H_1(t)]\int_0^t u^\delta$$
$$\times\exp[-H_1(u)][(\delta/u)+H_2(u)]du, \tag{4.2}$$

where

$$t=\pi x/6, \tag{4.3}$$

$$\delta=111/31=3.5806452, \tag{4.4}$$

$$H_1(t)=\ln(1-t)+[4783/31(47)]\ln[1+(47t/31)], \tag{4.5}$$

$$H_2(t)=\frac{6718+11776t-12830t^2+5217t^3}{31(1-t^3)(31+47t)}. \tag{4.6}$$

---

[22] T. L. Hill, J. Chem. Phys. **28**, 1179 (1958).

§ The present authors wish to signify their indebtedness to Hill's paper for having planted some of the germinal ideas of the present work. There is apparently however an error in Hill's equation which appears to express $p_0(r)$ exclusively in terms of $g(r)$ for all values of r.

|| The ξ introduced here should not be confused with the coupling parameter mentioned in Sec. 2. The latter does not appear in this paper beyond Sec. 2.

[23] J. DeBoer, Repts. Progr. in Phys. **12**, 305 (1949).

[24] See reference 1, p. 211.

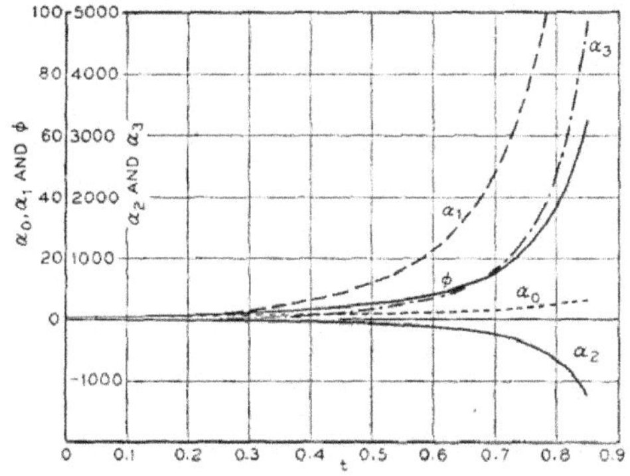

FIG. 2. Plots of the coefficients of the polynomial method [Eq. (4.1)] *versus* the reduced density.

The quadrature in (4.2) was performed numerically using the IBM 704 computer at Bell Laboratories. With $G(1, x)$ known the set of simultaneous equations permits the immediate evaluation of α_0, α_1, α_2, and α_3 by algebraic manipulation. Figure 2 contains plots of $G(1, t)$, $\alpha_0(t)$, $\alpha_1(t)$, $\alpha_2(t)$, and $\alpha_3(t)$ determined in this manner up to $t=0.85$ or $x=1.62$. Knowledge of the α's is not only equivalent to $G(1, x)$ but also to $G(\xi, x)$ according to (4.1), at least in the interval $a/2$ to a. This means that we can treat rigid sphere *solutions* with solutes having diameters $b<a$, through application of (2.17). As a matter of interest we give in Fig. 3 a plot of $G(\xi, x)$ for $x=0.574$ between $\xi=0$ and $\xi=1$. Between $\xi=\frac{1}{2}$ and $\xi=1$ the curve is concave downward and does not depart appreciably from a straight line.

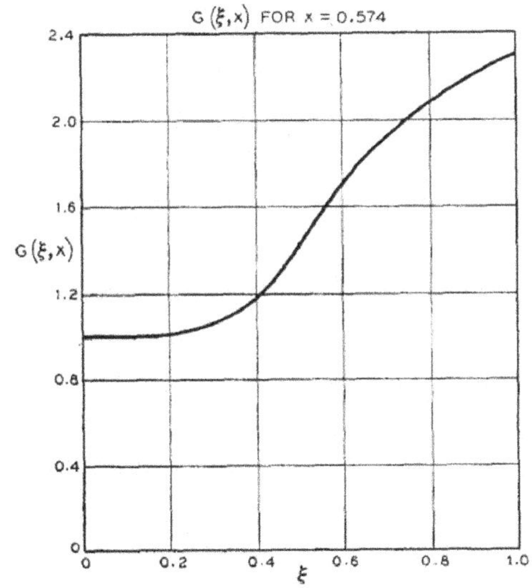

FIG. 3. Plot of $G(\xi, x)$ from the polynomial method.

Our value of $G(1, x)=g(a)$ can now be compared with the most reliable previous estimates. These are the Monte Carlo calculations of Rosenbluth and Rosenbluth[4] and Wood and Jacobsen[5] and the dynamical equations calculations of Alder and Wainwright.[6] Wood and Jacobsen and Alder and Wainwright are in agreement with respect to their computed equations of state, but Rosenbluth and Rosenbluth arrive at values of $p/\rho kT$ which in the high-density range fall from 5 to 10% below those of the first authors. Since Wood and Jacobsen eliminated some possible sources of error in the procedure of Rosenbluth and Rosenbluth, and in addition agree with Alder and Wainwright it is reasonable to assume that both they and Alder and Wainwright are in possession of the more reliable equation of state.

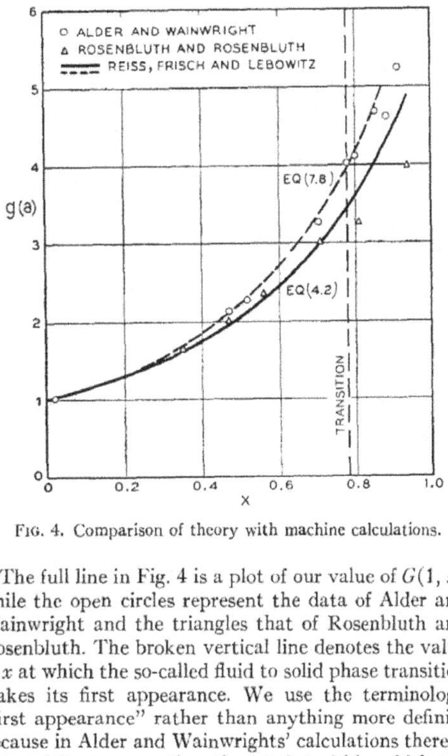

FIG. 4. Comparison of theory with machine calculations.

The full line in Fig. 4 is a plot of our value of $G(1, x)$ while the open circles represent the data of Alder and Wainwright and the triangles that of Rosenbluth and Rosenbluth. The broken vertical line denotes the value of x at which the so-called fluid to solid phase transition makes its first appearance. We use the terminology "first appearance" rather than anything more definite because in Alder and Wainwrights' calculations there is a transition *range* rather than *point* within which the system fluctuated with time between fluid and solid states. These fluctuations are presumably a result of the small number of particles used for the calculation. In Fig. 4 those results of Alder and Wainwright which lie to the right of the dashed line refer to the system in the "metastable" *fluid* condition.

Rosenbluth and Rosenbluth also saw signs of a transition to the right of the dashed line. However, the change of state appeared to be gradual, the calculated radial distribution function changing smoothly with increasing density from that characteristic of a fluid to that for a solid.

As can be seen from Fig. 4 our results agree with those of Rosenbluth and Rosenbluth to within about 1%, below the transition line. At densities beyond this line, however, Rosenbluth and Rosenbluth's results fall below ours. Presumably this happens because the latter authors have observed the transition while due to the smoothing effects inherent in our use of the analytic polynomial (4.1) we have removed the possibility of transition. Thus our pressure [and therefore $g(a)$] refers to a metastable fluid while Rosenbluth and Rosenbluth's is that of a solid and should be lower. For the purpose of orientation it should be mentioned that close packing corresponds to the density $x=\sqrt{2}$.

It may be of interest to expand $G(1, x)$ in powers of x to see how our first five coefficients compare with the known exact ones. Equation (4.7) is the exact expression, while (4.8) is derived from our $G(1, x)$. For comparison we have also shown in (4.9) the expansion which results when in place of (4.1) the linear expression

$$G(\xi, x)=\alpha_0(x)+\alpha_1(x)\xi$$

is used, together with Eqs. (3.31) and (3.25). This corresponds to connecting the points at $a/2$ and a by a straight line.

$$G(1, x)=g(a, \rho)=1+0.417\pi x+0.1275(\pi x)^2 + (0.0341\pm0.0015)(\pi x)^3 \quad (4.7)$$

$$G(1, x)=g(a, \rho)=1+0.399\pi x+0.1174(\pi x)^2 +0.0266(\pi x)^3 \quad (4.8)$$

$$G(1, x)=g(a, \rho)=1+0.31\pi x+0.079(\pi x)^2 +0.0185(\pi x)^3. \quad (4.9)$$

Although only terms up to x^3 are included in the above expressions they represent expansions including the fifth virial coefficient since in (1.1) $g(a)$ is multiplied by the density squared. Comparison of (4.8) and (4.9) shows the expected improvement as we go from a linear to a cubic polynomial, but the final result is still low by a few percent. Since our $G(1, x)$ agrees with that of Rosenbluth and Rosenbluth this adds support to the contention that theirs is low.

5. A MORE GENERAL INTEGRAL EQUATION

We have indicated that Eq. (2.22) suffers from the defect that the integrals over r are definite, and therefore $G(r, \rho)$ is only determined to a limited degree. In this section we shall derive an equation with indefinite integrals. This can be employed together with other information to derive a more precise value of $G(1, x)$. We shall in fact obtain analytic forms for both $G(1, x)$ and $G(\xi, x)$. Furthermore we also obtain expressions for the surface tension σ_0 of a rigid sphere fluid as well as for the distance δ separating the Gibbs surface of tension[25] from the dividing surface for which the superficial density of matter[25] vanishes.

Consider a fluid containing solutes of diameter b and solvent molecules of diameter a. In particular assume that the solution is dilute so that the concentration of solute ρ_b is small. The generalization of Eq. (1.1) for the pressure of this solution is easily shown to be[26]

$$\frac{p}{kT}=\rho_a+\rho_b+\tfrac{2}{3}\pi a^3\rho_a^2 g_{aa}(a)+\tfrac{4}{3}\pi\left(\frac{a+b}{2}\right)^3\rho_a\rho_b g_{ab}\left(\frac{a+b}{2}\right) +\tfrac{2}{3}\pi b^3\rho_b^2 g_{bb}(b), \quad (5.1)$$

where $g_{aa}(a)$ is the radial distribution function on contact between two solvent molecules, $g_{bb}(b)$ the radial distribution function on contact between two solute molecules, and $g_{ab}[(a+b)/2]$ the similar quantity between a solute and solvent. ρ_a is the concentration of the solvent, formerly denoted by ρ.

The chemical potential μ_b of the solute is given by (2.18) with ρ replaced by ρ_a. Introducing (2.16) into (2.18) yields the alternative form

$$\mu_b=kT\ln\{\rho_b\Lambda^3/p_0[(a+b)/2]\}. \quad (5.2)$$

By straightforward thermodynamic methods it can be shown that

$$(\partial p/\partial\rho_b)_{\rho_a}=\rho_b(\partial\mu_b/\partial\rho_b)_{\rho_a}+\rho_a(\partial\mu_b/\partial\rho_a)_{\rho_b}. \quad (5.3)$$

Then by differentiation of (5.2) and substitution of (5.3) we have

$$-\frac{1}{kT}\left(\frac{\partial\mu_b}{\partial\rho_a}\right)_{\rho_b}=\left(\frac{\partial\ln p_0}{\partial\rho_a}\right)_{\rho_b}=-\frac{1}{kT\rho_a}\left\{\left(\frac{\partial p}{\partial\rho_b}\right)_{\rho_a}-\rho_b\left(\frac{\partial\mu_b}{\partial\rho_b}\right)_{\rho_a}\right\}. \quad (5.4)$$

Substitution of (5.1) into the right-hand member of (5.4) yields

$$\left(\frac{\partial\ln p_0}{\partial\rho_a}\right)=-\frac{1}{\rho_a}\left\{1+\tfrac{2}{3}\pi a^3\rho_a^2\left(\frac{\partial g_{aa}(a)}{\partial\rho_b}\right)_{\rho_a} +\tfrac{4}{3}\pi\left(\frac{a+b}{2}\right)^3\rho_a g_{ab}\left(\frac{a+b}{2}\right) +\tfrac{4}{3}\pi\left(\frac{a+b}{2}\right)^3\rho_a\rho_b\left(\frac{\partial g_{ab}[(a+b)/2]}{\partial\rho_b}\right)_{\rho_a}-\frac{\rho_b}{kT}\left(\frac{\partial\mu_b}{\partial\rho_b}\right)_{\rho_a}\right\}, \quad (5.5)$$

where the term in g_{bb} has been omitted because we shall be interested in the case $\rho_b\to0$ and it is of higher order in ρ_b. In the limit $\rho_b\to0$

$$g_{ab}\left(\frac{a+b}{2}\right)\to G\left(\frac{a+b}{2}, \rho\right)$$

$$g_{aa}(a)\to G(a, \rho) \quad (5.6)$$

[25] R. C. Tolman, J. Chem. Phys. 16, 758 (1948); 17, 118, 333 (1949).

[26] H. S. Green, *The Molecular Theory of Fluids* (North-Holland Publishing Company, Amsterdam, 1952), p. 171.

and

$$(\partial\mu_b/\partial\rho_b)_{\rho_a}\to(kT/\rho_b), \tag{5.7}$$

since μ_b is given by (5.2) and in the limit $p_0[(a+b)/2]$ is independent of ρ_b. Substituting (5.6) and (5.7) into (5.5) and writing ρ for ρ_a we then have in the limit $\rho_b\to 0$.

$$\frac{\partial \ln p_0}{\partial\rho}=-\tfrac{2}{3}\pi a^3\rho\left(\frac{\partial G(a,\rho)}{\partial\rho_b}\right)_{\rho_b=0}-\tfrac{4}{3}\pi\left(\frac{a+b}{2}\right)^3G\left(\frac{a+b}{2},\rho\right)$$
$$=\frac{Q(b,a,\rho)}{\rho}-\tfrac{4}{3}\pi\left(\frac{a+b}{2}\right)^3G\left(\frac{a+b}{2},\rho\right). \tag{5.8}$$

It is evident from (5.8) that

$$Q(b,a,\rho)=-\tfrac{2}{3}\pi a^3\rho^2\left(\frac{\partial G(a,\rho)}{\partial\rho b}\right)_{\rho_b=0}. \tag{5.9}$$

At this stage it is again convenient to introduce the transformation

$$\xi=(a+b)/2a,$$
$$x=a^3\rho. \tag{5.10}$$

From (5.8) and (2.15) one can then derive

$$Q(\xi,x)=x\frac{\partial \ln p_0}{\partial x}-\frac{\xi}{3}\frac{\partial \ln p_0}{\partial\xi}$$
$$=-x\frac{\partial}{\partial x}\left\{x\int_0^\xi 4\pi\xi^2G(\xi,x)d\xi\right\}+\tfrac{4}{3}\pi x\xi^3G(\xi,x). \tag{5.11}$$

This is an integral equation on $G(\xi,x)$ containing indefinite integrals.¶ However the function $Q(\xi,x)$ defined by (5.4) is not immediately available in useful form, and the utility of (5.11) will depend on how explicitly $Q(\xi,x)$ can be determined.

Introducing (3.1) into (5.11) together with the additional knowledge that $\partial p_0/\partial\xi$ and $\partial^2p_0/\partial\xi^2$ are continuous at $\xi=\frac{1}{2}$ we obtain

$$Q=\partial Q/\partial\xi=0,\quad \xi\leq\tfrac{1}{2}. \tag{5.12}$$

This implies that an expansion of Q in powers of $(\xi-\frac{1}{2})$ begins with a quadratic term whose coefficient is $\frac{1}{2}(\partial^2Q/\partial\xi^2)_{\xi=\frac{1}{2}}$. In fact this coefficient can be determined by differentiating (5.11) three times with respect to ξ and making use of (3.27). The result for the quadratic term is then

$$\lim_{\xi=\frac{1}{2}}Q(\xi,x)=-\frac{2\pi^2x^2}{3[1-(\pi x/6)]}G(1,x)(\xi-\tfrac{1}{2})^2. \tag{5.13}$$

It can also be shown that the dominant term in Q as $\xi\to\infty$ is

$$Q_{\rm dom}=-\tfrac{4}{3}\pi\xi^3x^2[\partial G(\infty,x)]/\partial x. \tag{5.14}$$

This varies with the third power of ξ.

¶ For $\xi=1$ (5.11) is equivalent to (2.22) since according to (5.9) $Q(1,x)=-\frac{2}{3}\pi x^2[\partial G(1,x)]/\partial x$.

Although Q and G are not analytic functions of ξ it may be possible to approximate them closely by analytic functions. An analytic approximation to Q which interpolates the proper limiting behavior indicated by (5.13) and (5.14) is

$$Q(\xi,x)=\omega_2(x)(\xi-\tfrac{1}{2})^2+\omega_3(x)(\xi-\tfrac{1}{2})^3,\quad \xi>\tfrac{1}{2} \tag{5.15}$$

where $\omega_2(x)$ and $\omega_3(x)$ are the coefficients of the powers of $\xi-\frac{1}{2}$ (and ξ) in (5.13) and (5.14).

If (5.15) is substituted into (5.11), and the result differentiated once with respect to ξ, the following equation is obtained

$$\frac{1}{2\pi}\omega_2(x)(\xi-\tfrac{1}{2})+\frac{3}{4\pi}\omega_3(x)(\xi-\tfrac{1}{2})^2$$
$$=-x^2\xi^2\frac{\partial G}{\partial x}+\tfrac{1}{3}x\xi^3\frac{\partial G}{\partial\xi}. \tag{5.16}$$

Now it is easily shown that the solution of (5.16) must be of the form (if we assume in approximation that G is analytic in $\xi>\frac{1}{2}$).

$$G(\xi,x)=\phi(\xi^3x)+\frac{\alpha_{-2}(x)}{\xi^2}+\frac{\alpha_{-1}(x)}{\xi}+\alpha_0(x) \tag{5.17}$$

where ϕ is some function of the single variable ξ^3x. This can be proved by direct substitution of an arbitrary form of $G(\xi,x)$ containing all possible powers of ξ as well as the part shown in (5.17).

The function of $\phi(\xi^3x)$ must vanish as $\xi\to\infty$ since $G(\infty,x)$ is a function of x alone, that is $\alpha_0(x)=G(\infty,x)$. But the limit of (5.17) as $\xi\to\infty$, $x\to 0$ in such a way that $\xi^3x\to y$, a finite but arbitrary number, shows that $\phi(y)$ must vanish. This leaves the suggested form

$$G(\xi,x)=\alpha_0(x)+\frac{\alpha_{-1}(x)}{\xi}+\frac{\alpha_{-2}(x)}{\xi^2}. \tag{5.18}$$

We shall see that thermodynamic considerations suggest the same form.

6. THERMODYNAMIC CONSIDERATIONS, A MORE PRECISE $G(1, x)$ AND SURFACE TENSION

By differentiation of (2.17), expressed in terms of ξ and x, one obtains

$$G(\xi,x)=(1/4\pi\xi^2kTx)(\partial W/\partial\xi). \tag{6.1}$$

When the spherical cavity is sufficiently large (ξ large) W can be represented by[25,27,28]

$$W=a^3p\tfrac{4}{3}\pi\xi^3+a^24\pi\xi^2\sigma_0(x)\left(1-\frac{2\delta(x)}{\xi}\right). \tag{6.2}$$

Here the first term on the right is the pressure-volume work expended in creating the cavity and the second term is the surface work. The quantity

$$\sigma_0[1-(2\delta/\xi)] \tag{6.3}$$

[27] J. S. Kirkwood and F. P. Buff, J. Chem. Phys. 17, 338 (1949).
[28] F. P. Buff, J. Chem. Phys. 23, 419 (1955).

is the surface tension. $\delta(x)$ is the distance, (divided by a), between the dividing surface for which the superficial density of matter vanishes and the Gibbs surface of tension,[25,27,28] while $\sigma_0(x)$ is the surface tension at infinite ξ, i.e. infinite radius of curvature.

The pressure in (6.2) can be replaced by a function of $G(1, x)$ through use of (1.1). When this is done and (6.2) is substituted into (6.1) we obtain

$$G(\xi, x) = \{1+\tfrac{2}{3}\pi x G(1, x)\} + \left\{\frac{2a^2\sigma_0(x)}{kTx}\right\}\frac{1}{\xi} - \left\{\frac{2a^2\sigma_0(x)\delta(x)}{kTx}\right\}\frac{1}{\xi^2}. \quad (6.4)$$

This is identical in form with (5.18), but is exact in the region in which ξ is *very large*, while (5.18) refers to ξ *small*. This lends support to the idea that (5.18) is a good approximation over the entire range between $\xi=\frac{1}{2}$ and $\xi=\infty$. Equation (3.7) shows that the first bracketed expression on the right of (6.4) is nothing more than $G(\infty, x)$ so that G given by (6.4) approaches the correct limit for $\xi\rightarrow\infty$.

If we do assume that (6.4) is valid right down to $\xi=\frac{1}{2}$ we can substitute it into the exact relation (3.25) and (3.26). The result is two equations on the three unknowns $G(1, x)$, $\sigma_0(x)$, and $\delta(x)$. A third equation is obtained by simply setting ξ in (6.4) equal to unity.** The three equations are

$$\frac{1}{1-(\pi x/6)} = 1+\tfrac{2}{3}\pi x G(1, x) + \frac{4a^2\sigma_0(x)}{kTx} - \frac{8a^2\sigma_0(x)\delta(x)}{kTx}. \quad (6.5)$$

$$\frac{\pi x}{[1-(\pi x/6)]^2} = -\frac{8a^2\sigma_0(x)}{kTx} + \frac{32a^2\sigma_0(x)\delta(x)}{kTx}. \quad (6.6)$$

$$(1-\tfrac{2}{3}\pi x)G(1, x) = 1 + \frac{2a^2\sigma_0(x)}{kTx} - \frac{2a^2\sigma_0(x)\delta(x)}{kTx}. \quad (6.7)$$

These can be solved simultaneously to yield

$$G(1, x) = \frac{1}{4[1-(\pi x/6)]} + \frac{3}{4[1-(\pi x/6)]^2} + \frac{\pi x}{8[1-(\pi x/6)]^3}. \quad (6.8)$$

$$\sigma_0(x) = -\frac{kT\pi x^2[1+(\pi x/6)]}{8a^2\,[1-(\pi x/6)]^3}. \quad (6.9)$$

$$\delta(x) = \tfrac{1}{4} - \tfrac{1}{4}\left\{\frac{1-(\pi x/6)}{1+(\pi x/6)}\right\}. \quad (6.10)$$

** *Note added in proof.*—A third equation can also be obtained by merely substituting (6.4) into (3.31). The result is a value for $G(1x)$ different from that given in (6.8), and which in fact gives more accurate vivial coefficients. Of course, before (6.4) is substituted in (3.31), the first term is represented as $G(\alpha, x)$, i.e., (3.7) is not satisfied by the resulting $G(1, x)$.

We have thus not only determined the equation of state, but the surface tension and its dependence on curvature as well.

The expression (6.8) is remarkable in a number of respects: Expanded in powers of x it yields

$$G(1, x) = 1+0.417\pi x+0.1319(\pi x)^2+0.0359(\pi x)^3+\cdots. \quad (6.11)$$

This is to be compared with (4.7). It is seen that up to and including the third virial coefficient (6.11) is exact. (The 0.417 is actually the fraction (5/12) which is known to be exact.) The error in the fourth virial coefficient is only about 3% and in the fifth only 5%. In fact the exact fifth virial coefficient is not known to within better than 5%.[3] Inspection of (6.8) shows that all the coefficients of its expansion in powers of x are positive. This supports the idea that for hard spheres all the virial coefficients are positive.

However, it must be understood that Eq. (6.4) is an interpolation formula whose functional form is suggested by previous considerations. Since this "best" functional form contains only three parameters, we cannot use all the information we possess concerning G (e.g., G'' ($\frac{1}{2}$), Eq. (2.22). Thus we do not expect Eq. (6.4) to give reliably higher derivatives of G. Equation (6.4) does, however, give $G(\xi, x)$ exactly to terms linear in x.

The dashed line in Fig. 4 is a plot of $G(1, x)$ as given by (6.8). The results are almost indistinguishable from the data of Alder and Wainwright over the entire range of density. In fact agreement is maintained even beyond the transition line showing that Alder and Wainwright's points do indeed correspond to a metastable fluid.

Equation (6.9) shows that the surface tension σ_0 for a rigid sphere fluid is indeed negative. Furthermore, for $x\rightarrow 0$,

$$\sigma_0 = -\frac{kT\pi x^2}{8a^2} = -\frac{kT\pi a^4\rho^2}{8}, \quad (6.12)$$

which is identical with the result derived from an equation due to Kirkwood and Buff[27] [Eq. (29) of their paper] when for the gradient of the potential is substituted $-kT\delta(r-a)$ [see Eq. (2.6)]. Since Kirkwood and Buff's result must be exact in the limit of zero density this is further confirmation of the validity of (6.9).

Equation (6.10) shows that δ lies between 0 and $\frac{1}{2}$. This shows according to (6.3) that the surface tension does not vary strongly with curvature until a cavity of molecular dimensions is involved. In fact at zero density $\delta=0$ and within the framework of the approximation there is no dependence on curvature.

7. RELATION BETWEEN $G(r, \rho)$ AND $g(r)$

The foregoing analysis has been aimed primarily at determining, to a sufficient approximation, $g(a)$ from which the thermodynamic properties of the rigid sphere fluid follow. One expects though that a knowledge of $G(r, \rho)$ will not only specify $g(a)$ but to some extent at least $g(r)$ for $r>a$. We will show that in the range $(a/2)\leq r<a/\sqrt{3}$ a knowledge of $G(r, \rho)$ determines $g(r)$ exactly in $a\leq r<2a/\sqrt{3}$.

Consider Eq. (3.29) which applies in this range. $p_0(r)$ on the left-hand side can be written by virtue of (2.16) as a negative exponential of a quadrature of $G(r, \rho)$. Thus (3.29) can be rewritten as

$$F(r)=4\int_0^{2r}s^2g(s)ds-\frac{3}{r}\int_0^{2r}s^3g(s)ds+\frac{1}{4r^3}\int_0^{2r}s^5g(s)ds, \tag{7.1}$$

where

$$F(r)=\frac{p_0(r)-1+\frac{4}{3}\pi r^3\rho}{\frac{2}{3}\pi^2r^3\rho^2}$$
$$=\frac{\exp\left\{-\rho\int_0^r 4\pi s^2G(s,\rho)ds\right\}-1+\frac{4}{3}\pi r^3\rho}{\frac{2}{3}\pi^2r^3\rho^2}. \tag{7.2}$$

After differentiating (7.2) three times with respect to r, one notes that the three derivatives $F'(r)$, $F''(r)$, $F'''(r)$ can be written in terms of $g(2r)$ and the last two integrals over $g(s)$ on the right-hand side of (7.1) (i.e., no other terms containing $g(s)$ are present). On eliminating these integrals we find that

$$g(2r)=[F'''(r)/96]+(1/96r)\{(8/r)F'(r)+7F''(r)\}, \tag{7.3}$$

for

$$a/2\leq r<a/\sqrt{3},$$

as was to be shown.

Although Eq. (7.3) only allows us to determine $g(r)$ exactly up to $r=2a/\sqrt{3}$ in terms of $G(r, \rho)$ it may approximate $g(r)$, for certain purposes, sufficiently well beyond this point. In any case the investigation of the analytic form of $G(r, \rho)$ beyond $r=a/\sqrt{3}$ will be postponed for a later discussion, as will the validity of (7.3) beyond $r=a/\sqrt{3}$ and possible extensions of (7.3).

8. CONCLUDING REMARKS

In this paper we have presented two schemes for arriving at the equation of state for a system of rigid spheres. These are the polynomial method of Sec. 4, and the method based on the second integral equation (5.11). Whereas the polynomial method leads to a less precise result it has the advantage of being entirely free of thermodynamic considerations, while the second method though largely statistical requires some thermodynamic argument to support the form of $G(\xi, x)$ suggested by statistical investigation.

Nevertheless the second method yields far more information than the polynomial method and should, in fact, be capable of giving a completely nonthermodynamic theory. The next step ought to be aimed along these lines. An attempt must be made to express Q defined as (5.9) as a function of $G[(a+b)/2, \rho]$. This would render Eq. (5.11) more determinate.

A considerable amount of interesting discussion concerning solutions can be given. We have omitted this here but intend to include it in a later paper.

Finally we remark that it is possible to treat molecules with attractive potentials by the techniques described here. This too will be reserved for a later paper.

ACKNOWLEDGMENTS

We wish to express our gratitude to Dr. B. J. Alder of the University of California Radiation Laboratory at Livermore and to Drs. G. H. Wannier, C. Herring, M. Lax, and E. Helfand of Bell Laboratories for stimulating discussion in connection with this work.

THE

PHYSICAL REVIEW

A journal of experimental and theoretical physics established by E. L. Nichols in 1893

SECOND SERIES, VOL. 133, NO. 4A 17 FEBRUARY 1964

Exact Solution of Generalized Percus-Yevick Equation for a Mixture of Hard Spheres*

J. L. LEBOWITZ

Belfer Graduate School of Science, Yeshiva University, New York, New York

(Received 15 August 1963)

The Percus-Yevick approximate equation for the radial distribution function of a fluid is generalized to an m-component mixture. This approximation which can be formulated by the method of functional Taylor expansion, consists in setting $\exp[-\beta\varphi_{ij}(r)]C_{ij}(r)$ equal to $g_{ij}(r)[e^{-\beta\varphi_{ij}(r)}-1]$, where C_{ij}, g_{ij}, and φ_{ij} are the direct correlation function, the radial distribution function and the binary potential between a molecule of species i and a molecule of species j. The resulting equation for C_{ij} and g_{ij} is solved exactly for a mixture of hard spheres of diameters R_i. The equation of state obtained from $C_{ij}(r)$ via a generalized Ornstein-Zernike compressibility relation has the form $p/kT=\{[\sum\rho_i][1+\xi+\xi^2]-18/\pi\ \sum_{i<j}\eta_i\eta_j(R_i-R_j)^2 \times[R_i+R_j+R_iR_j(\sum\eta_lR_l^2)]\}(1-\xi)^{-3}$, where $\eta_i=\pi/6$ times the density of the ith component and $\xi=\sum\eta_lR_l^3$. This equation yields correctly the virial expansion of the pressure up to and including the third power in the densities and is in very good agreement with the available machine computations for a binary mixture. For a one-component system our solution for $C(r)$ and $g(r)$ reduces to that found previously by Wertheim and Thiele and the equation of state becomes identical with that found on the basis of different approximations by Reiss, Frisch, and Lebowitz.

RECENTLY, Wertheim[1] and Thiele[2] succeeded in obtaining an exact solution of the Percus-Yevick (P.Y.) integral equation[3,4] for the radial distribution function $g(r)$ of a fluid of hard spheres. The equation of state obtained from their $g(r)$ through the use of the Ornstein-Zernike compressibility relation[5] coincided with that obtained previously by Reiss, Frisch, and Lebowitz[6] (R.F.L.) and is in very good agreement with the machine computations[7] for the whole range of "fluid" densities. The R.F.L. theory which does not involve $g(r)$ directly is based on approximating the work necessary to add a hard sphere of diameter R_2 to a fluid of hard spheres of diameter R_1 and density ρ_1. (The temperature $T=(k\beta)^{-1}$ will not be indicated explicitly in most cases.) This work is essentially the chemical potential $\mu_2(\rho_1,R_1,\rho_2,R_2)$ in the limit $\rho_2\to 0$, which is equal to the chemical potential of the single-component fluid when $R_2=R_1$ and thus yields the equation of state of a single-component fluid also found by Wertheim and Thiele. The P.Y. approximation for a hard-sphere fluid, on the other hand, is completely characterized[8] by assuming that the direct correlation function[4,9] $C(r)$, introduced by Ornstein and Zernike[5] vanishes for r greater than the diameter of the spheres. The identity of the equations of state for a single-component hard-sphere fluid obtained from these apparently unrelated theories led us to a generalization of the P.Y. equation to fluid mixtures. We shall present here the explicit solution of this generalized P.Y. equation for a binary mixture of hard spheres of diameters R_1, R_2 and densities ρ_1 and ρ_2, and briefly the results for a general mixture of m components. The chemical potential μ_2 obtained from this solution agrees, when $\rho_2\to 0$, with that obtained by R.F.L.[6] for all values of R_2. The reason underlying this

* Supported by the U. S. Air Force Office of Scientific Research under Grant No. 62-64 and the U. S. Atomic Energy Commission under Contract AT(30-1)-1480.

[1] M. Wertheim, Phys. Rev. Letters **8**, 321 (1963).
[2] E. Thiele, J. Chem. Phys. **38**, 1959 (1963).
[3] J. K. Percus and G. J. Yevick, Phys. Rev. **110**, 1 (1957).
[4] J. K. Percus, Phys. Rev. Letters **8**, 462 (1962).
[5] L. Ornstein and F. Zernike, Proc. Acad. Sci. Amsterdam **17**, 793 (1914).
[5] L. Ornstein and F. Zernike, Proc. Acad. Sci. Amsterdam **17**, 793 (1914).
[6] H. Reiss, H. L. Frisch, and J. L. Lebowitz, J. Chem. Phys. **31**, 369 (1959).
[7] B. Alder and T. Wainwright, J. Chem. Phys. **33**, 1439 (1960).
[8] G. Stell, Physica **29**, 517 (1963).
[9] J. L. Lebowitz and J. K. Percus, J. Math. Phys. **4**, 116 (1963); **4**, 248 (1963).

agreement and a more complete discussion of this equation will be presented elsewhere.

The P.Y. equation for a mixture may be derived in complete analogy with that of a single-component fluid[4] by using the methods of functional Taylor expansion. We consider a binary system confined to a volume V which is represented by a grand canonical ensemble with chemical potentials μ_1 and μ_2. The particles interact via a pair potential $\varphi_{ij}(r)$ and are subject to an external potential $U_i(r)$ where i, $j=1$, 2. Let Ξ be the grand partition function of this system. It is then easy to show[9] that,

$$\frac{\delta \ln\Xi}{\delta(-\beta U_i(\mathbf{r}))}=n_i(\mathbf{r}), \tag{1}$$

$$\frac{\delta n_i(\mathbf{r})}{\delta(-\beta U_j(\mathbf{r}'))}=\frac{\delta^2 \ln\Xi}{\delta(-\beta U_j(\mathbf{r}'))\delta(-\beta U_i(\mathbf{r}))}=n_i(\mathbf{r})n_j(\mathbf{r}')\times[g_{ij}(\mathbf{r},\mathbf{r}')-1]+n_i(\mathbf{r})\delta(\mathbf{r}-\mathbf{r}')\delta_{ij}, \tag{2}$$

where $n_i(\mathbf{r})$ is the density of the ith species at $\mathbf{r}$ and $g_{ij}(\mathbf{r},\mathbf{r}')$ is the radial distribution function between an i particle at $\mathbf{r}$ and a j particle at $\mathbf{r}'$, $g_{ij}(\mathbf{r},\mathbf{r}')=g_{ji}(\mathbf{r}',\mathbf{r})$. Defining now the direct correlation functions $C_{ij}(\mathbf{r},\mathbf{r}')$ by the equations,

$$\frac{\delta(-\beta U_i(\mathbf{r}))}{\delta n_j(r')}=\delta_{ij}\frac{\delta(\mathbf{r}-\mathbf{r}')}{n_i(r)}-C_{ij}(\mathbf{r},\mathbf{r}'), \tag{3}$$

we obtain immediately the relations[10]

$$[g_{ij}(\mathbf{r},\mathbf{r}')-1]=C_{ij}(\mathbf{r},\mathbf{r}')+\sum_{l=1,2}\int[g_{il}(\mathbf{r},\mathbf{y})-1]n_l(\mathbf{y})C_{lj}(\mathbf{y},\mathbf{r}')d\mathbf{y}, \tag{4}$$

$$C_{ij}(\mathbf{r},\mathbf{r}')=C_{ji}(\mathbf{r}',\mathbf{r}). \tag{5}$$

When there is no external potential, $U_i(\mathbf{r})=0$, and the system is sufficiently large, $V\to\infty$, then $n_i(\mathbf{r})=\rho_i$, the average density, and $g_{ij}(\mathbf{r},\mathbf{r}')$, $C_{ij}(\mathbf{r},\mathbf{r}')$ depend only on $|\mathbf{r}-\mathbf{r}'|$.

We consider now the behavior of the function $n_i(r)e^{\beta[U_i(r)-\varphi_{ij}(r)]}$ as the external potential $U_i(r)$ is "turned on" from zero to its final value $\varphi_{ij}(r)$, i.e., at the end there is a particle of the jth species held fixed at the origin. Then,

$$\begin{aligned} n_i(r)e^{\beta[U_i(r)-\varphi(r)]}&=\rho_i e^{-\beta\varphi_{ij}(r)}, &&\text{initially},\\ &=\rho_i g_{ij}(r), &&\text{finally}. \end{aligned} \tag{6}$$

Expanding the final value of this function about its initial value in a functional Taylor series in the density[4,9] we obtain, using (2) and (3),

$$\rho_i g_{ij}(r)=\rho_i e^{-\beta\varphi_{ij}(r)}\Big[1+\sum_{l=1}^{2}\int C_{il}(\mathbf{r}-\mathbf{r}')\times\rho_l[g_{lj}(r')-1]d\mathbf{r}'+\dots\Big]. \tag{7}$$

Truncating this expansion after the second term and using (4) yields the generalized P.Y. equation for mixtures,

$$g_{ij}(r)(e^{-\beta\varphi_{ij}(r)}-1)=e^{-\beta\varphi_{ij}(r)}C_{ij}(r). \tag{8}$$

Thus, as in the case of a single component fluid, the P.Y. approximation for a mixture states that the range of $C_{ij}(r)$ is equal to the range of $\varphi_{ij}(r)$. In particular for a fluid of hard spheres,

$$\begin{aligned} e^{-\beta\varphi_{ij}(r)}&=0, \quad r<R_{ij}\equiv(R_i+R_j)/2,\\ &=1, \quad r>R_{ij}. \end{aligned} \tag{9}$$

Equations (8) and (9) state that $C_{ij}(r)$ vanishes for $r>(R_i+R_j)/2$ and $g_{ij}(r)$ vanishes for $r<(R_i+R_j)/2$. The approximation is of course in the first statement the latter being rigorously true.[11] We may define now,

$$\begin{aligned} \sigma_{ij}(r)&=-12(\eta_i\eta_j)^{1/2}rC_{ij}(r), \quad r\le R_{ij},\\ &=12(\eta_i\eta_j)^{1/2}rg_{ij}(r), \quad r\ge R_{ij}, \end{aligned} \qquad \eta_i=\pi\rho_i/6, \tag{10}$$

and write Eq. (4) in terms of σ alone. Doing this and going over in the integrations into bipolar coordinates yields,

$$\sigma_{ij}(r)=A_{ij}r-\sum_{l=1,2}\int_{R_{il}}^{\infty}dy\sigma_{il}(y)\int_{|r-y|}^{[r+y,R_{lj}]}\sigma_{lj}(x)dx, \tag{11}$$

where $[r+y, R_{lj}]$ indicates that the smaller of these two numbers is to be taken as the upper limit of the integration. Here

$$A_{ij}=12(\eta_i\eta_j)^{1/2}a_j \tag{12}$$

and

$$a_j=1-\sum_{l=1,2}\rho_l\int C_{lj}(\mathbf{r})d\mathbf{r}=\sum_{l=1,2}\Big[\rho_l\frac{\partial\beta\mu_l(\rho_1,\rho_2)}{\partial\rho_j}\Big]=\beta\frac{\partial p(\rho_1,\rho_2)}{\partial\rho_j}, \tag{13}$$

where p is the pressure. The second equality in (13) follows from (3) upon integration with respect to r' in a uniform system when it is realized that the chemical and external potential enter the grand partition function in the combination μ_i-U_i. The last equality follows from thermodynamics. These equalities may also be derived from consideration of fluctuations in a binary system[10] and are a generalization of the Ornstein-Zernike compressibility relation for a one component system.

[10] F. J. Pearson and G. S. Rushbrooke, Proc. Roy. Soc. Edinburgh **A64**, 305 (1957).

[11] For a one-dimensional mixture of hard rods the exact $C_{ij}(r)$ vanishes for $r>R_{ij}$ and Eq. (8) is exact, [c.f., J. L. Lebowitz, J. K. Percus, and I. J. Zucker, Bull. Am. Phys. Soc. **7**, 415 (1962), where this was shown for a one-component system].

Let us assume, for definiteness that $R_2 \geq R_1$. The upper bound on the second integral in (11) will then be R_{ij} except for the case $(i,j)=(1,2)$. Taking the derivative with respect to r of (11) yields

$$\sigma_{ij}^{(1)}(r)=A_{ij}-\sum_l\Big[\int_{|r-x|\geq R_{il},\ -R_{lj}\leq x\leq 0}\sigma_{il}(r-x)\sigma_{lj}(x)dx$$
$$-\int_{|r-x|\geq R_{il},\ 0\leq x\leq R_{lj}}\sigma_{il}(r-x)\sigma_{lj}(x)dx\Big]-P_{ij}(r),\quad(14)$$

where

$$P_{ij}(r)=\delta_{i1}\delta_{j2}P(r),$$

$$P(r)=\sum_{l=1,2}\int_{R_{1l}+r}^{R_{12}}\sigma_{1l}(r-z)\sigma_{l2}(z)dz,\qquad r\leq(R_2-R_1)/2\equiv\lambda,\quad(15)$$
$$=0,\qquad r\geq\lambda.$$

Taking the Laplace transform of (14) yields

$$s[G_{ij}(s)+F_{ij}(s)]=A_{ij}/s-\sum G_{il}(s)\times[F_{lj}(-s)-F_{lj}(s)]+\Gamma_{ij}(s),\quad(16)$$

where

$$G_{ij}(s)=\int_{R_{ij}}^{\infty}e^{-sr}\sigma_{ij}(r)dr\equiv12(\eta_i\eta_j)^{1/2}\times\int_{R_{ij}}^{\infty}e^{-sr}rg_{ij}(r)dr=G_{ji}(s),\quad(17)$$

$$F_{ij}(s)=\int_0^{R_{ij}}e^{-sr}\sigma_{ij}(r)dr\equiv-12(\eta_i\eta_j)^{1/2}\times\int_0^{R_{ij}}e^{-sr}rC_{ij}(r)dr=F_{ji}(s),\quad(18)$$

$$\Gamma_{ij}(s)=\delta_{i1}\delta_{j2}\Gamma_{12}(s)=\delta_{i1}\delta_{j2}[P(-s)-P(s)],\quad(19)$$

$$P(s)=\int_0^{\lambda}e^{-sr}P(r)dr.\quad(20)$$

Solving for $G_{ij}(s)$ we obtain, in matrix notation,

$$\mathbf{G}(s)=[\mathbf{A}+s\boldsymbol{\Gamma}(s)-s^2\mathbf{F}(s)]\cdot[s^2\mathbf{I}-s\mathbf{F}^{\pm}(s)]^{-1}=\mathbf{H}(s)\cdot\mathbf{K}(s),\quad(21)$$
$$\mathbf{F}^{\pm}(s)=\mathbf{F}(s)-\mathbf{F}(-s),$$

where $\mathbf{I}$ is the unit matrix. More explicitly

$$G_{ij}(s)=N_{ij}(s)/D(s),\quad(22)$$

where

$$D(s)=\{[s^2-sF_{22}^{\pm}(s)][s^2-sF_{11}^{\pm}(s)]-s^2F_{12}^{\pm}(s)F_{21}^{\pm}(s)\},\quad(23)$$

$$N_{ij}(s)=\sum_{l=1,2}\{[A_{il}+s\Gamma_{il}(s)-s^2F_{il}(s)]\times[s^2\delta_{l'j'}-(-1)^{l+j}sF_{l'j'}^{\pm}(s)]\}\quad(24)$$

and $1'=2$, $2'=1$. The requirement that $\mathbf{G}$ and $\mathbf{F}$ be symmetric determines the value of Γ_{12}. This value of Γ_{12} also ensures that we can obtain σ_{11} from σ_{22} by simply interchanging η_1, R_1 with η_2, R_2.

The solutions of Eq. (11) which are physically meaningful are those for which $g_{ij}(r)\rightarrow1$ as $r\rightarrow\infty$ in such a way that $\int r|g_{ij}(r)-1|dr$ exists.[12] This requires that $G_{ij}(s)-12(\eta_i\eta_j)^{1/2}s^{-2}$ have no singularities in the closed right-hand plane of the complex s plane. Now the symmetry of $\mathbf{G}$ and $\mathbf{F}$ imply, using (20), that

$$\mathbf{G}(s)=\mathbf{H}(s)\cdot\mathbf{K}^T(s)(s)=\mathbf{G}^T(s)=\mathbf{K}^T(s)\cdot\mathbf{H}^T(s)=\mathbf{K}(s)\cdot\mathbf{H}^T(s),\quad(25)$$

where the superscript T indicates the transpose of a matrix and the last equality follows from the symmetry of K. Multiplying (25) by $\mathbf{H}^T(-s)$ and noting that $\mathbf{K}(s)=\mathbf{K}(-s)$, gives

$$\mathbf{L}(s)\equiv\mathbf{G}(s)\cdot\mathbf{H}^T(-s)=[\mathbf{H}(s)\cdot\mathbf{K}(s)]\cdot\mathbf{H}^T(-s)=\mathbf{H}(s)\cdot[\mathbf{H}(-s)\cdot\mathbf{K}(-s)]=\mathbf{H}(s)\cdot\mathbf{G}(-s)=\mathbf{L}^T(-s).\quad(26)$$

It is seen from the definition of $\mathbf{H}(s)$ that it (i.e., each of its elements) is an entire function of s. Hence, since $\mathbf{H}(0)=\mathbf{A}$, we have, by our assumption on the nature of $\mathbf{G}(s)$, that

$$\mathbf{L}(s)-s^{-2}\mathbf{A}'=\mathbf{G}(s)\cdot\mathbf{H}^T(-s)-s^{-2}\mathbf{A}'=\mathbf{H}(s)\cdot\mathbf{G}(-s)-s^{-2}\mathbf{A}',\quad(27)$$

where $A_{ij}'=12\sum_l(\eta_i\eta_l)^{1/2}A_{jl}=(12)^2(\eta_i\eta_j)^{1/2}\sum\eta_la_l=A_{ij}'$, has no singularities anywhere in the s plane and is therefore an entire function. It follows from (14) that

$$\sigma_{ij}^{(1)}(0)=A_{ij},\quad i=1,2$$
$$\sigma_{21}(r)=\sigma_{12}(r)=A_{21}r,\quad r\leq\lambda,\quad(28)$$

and $\sigma_{21}^{(1)}(r)$ is continuous at $r=\lambda$. An examination of the behavior of the entire function

$$L_{22}(s)-A_{22}'s^{-2}=\frac{N_{21}(s)H_{21}(-s)+N_{22}(s)H_{22}(-s)}{D(s)}-A_{22}'s^{-2}\quad(29)$$

now shows that it is bounded along every ray in the s plane. Hence it must be a constant[13] which we shall call $2B_{22}$. Thus, using (27),

$$G_{21}(s)F_{21}(-s)+G_{22}(s)F_{22}(-s)=\{A_{21}G_{21}(s)+A_{22}G_{22}(s)\}s^{-2}-12\left\{\frac{A_{21}(\eta_2\eta_1)^{1/2}+A_{22}\eta_2}{s^4}\right\}-\frac{2B_{22}}{s^2}.\quad(30)$$

[12] This method is a generalization of that developed by M. Wertheim for the one-component system [J. Math. Phys. (to be published)].

[13] See, e.g., E. C. Titchmarsh, *The Theory of Functions* (Oxford University Press, London, 1939).

If we take the inverse Laplace transform of (30) then for $r \leq R_{21} < R_2$ the first two terms on the right side will not contribute since $G_{ij}(s)$ goes as $e^{-sR_{ij}}$ for real part of s large positive and has no singularities in the right plane. Thus we can close the contour on that side. The Laplace transform of the left side of (30) is precisely the first term in the bracket on the right side of (14) for $\sigma_{22}^{(1)}(r)$ [cf., first term in bracket on right side of (16)]. On the other hand, the second term in the bracket on the right side of (14) makes no contribution to $\sigma_{22}^{(1)}(r)$ for $r < R_{12}$. Therefore, adding the inversely Laplace transformed Eq. (30) to Eq. (14), we obtain

$$\sigma_{22}^{(1)}(r) = A_{22} + 2B_{22}r + 2[A_{21}(\eta_2\eta_1)^{1/2} + A_{22}\eta_2]r^3, \quad r < R_{21}, \tag{31}$$

while for $R_{21} \leq r \leq R_2$, we find after taking two more derivatives and using (28),

$$\sigma_{22}^{(3)}(r) = 12[A_{21}(\eta_2\eta_1)^{1/2} + A_{22}\eta_2]r = 24D_{22}r. \tag{32}$$

Hence,

$$\sigma_{22}(r) = A_{22}r + B_{22}r^2 + D_{22}r^4, \quad r \leq R_2, \tag{33}$$

which is of the same functional form, i.e., a quartic polynomial in r, as that found by Wertheim and Thiele for a one-component system.

We can carry through a similar analysis for the function

$$L_{21}(s)e^{\lambda s}A_{21}'[s^{-2} + \lambda s^{-1}] = \left[\frac{N_{21}(s)H_{11}(-s) + N_{22}(s)H_{12}(-s)}{D(s)}\right] \times e^{\lambda s} - A_{21}'(s^{-2} + \lambda s^{-1}). \tag{34}$$

This function is also an entire function of s which is bounded on every ray. [From (19) $\Gamma_{12}(s)$ is an even function of s which goes as $s^{-2}e^{\lambda s}$ for real part of s large positive.] Hence, it is equal to a constant; $2B_{21}$. In a manner similar to (30), we now find

$$G_{21}(s)F_{11}(-s) + G_{22}(s)F_{12}(-s) = [A_{11}G_{21}(s) + A_{12}G_{22}(s)]s^{-2} + \Gamma_{12}(s)G_{22}(s)s^{-1} - \left\{24\left(\frac{\eta_1}{\eta_2}\right)^{1/2} D_{22}[s^{-4} + \lambda s^{-3}] + 2B_{21}/s^2\right\}e^{-s\lambda}.$$

Taking the inverse Laplace transform of (35) and adding it to the equation for $\sigma_{21}^{(1)}(r)$ in (14) we obtain, (using $F_{21} = F_{12}$),

$$\begin{aligned}\sigma_{21}(r) = \sigma_{12}(r) &= A_{21}r, \quad r \leq \lambda \\ &= A_{21}r + B_{21}x + 2(R_2 - R_1)\left(\frac{\eta_1}{\eta_2}\right)^{1/2} D_{22}x^3 \\ &\quad + \left(\frac{\eta_1}{\eta_2}\right)^{1/2} D_{22}x^4, \quad \lambda \leq r \leq R_{21},\end{aligned} \tag{36}$$

where $x = r - \lambda$. The coefficients A_{ij} and B_{ij} may be found from (13) and from the continuity of $\sigma_{ij}(r)$ and its first two derivatives at R_{ij}. We find

$$\begin{aligned}-C_{ii}(r) &= a_i + b_i r + dr^3, \quad r < R_i, \quad i = 1, 2 \\ -C_{12}(r) = -C_{21}(r) &= a_1, \quad r \leq \lambda \\ &= a_1 + [bx^2 + 4\lambda dx^3 + dx^4]/r, \quad \lambda \leq r \leq R_{21}\end{aligned} \tag{37}$$

and vanish in this approximation[14] for $r > R_{ij}$. Here,

$$a_i = \frac{\partial[\beta p(\rho_1, \rho_2)]}{\partial \rho_i},$$

$$b_1 = -6[\eta_1 R_1^2 g_{11}^2(R_1) + \eta_2 R_{12}^2 g_{12}^2(R_{12})], \tag{38}$$

$$b = -6[\eta_1 R_1 g_{11}(R_1) + \eta_2 R_2 g_{22}(R_2)]R_{12}g_{12}(R_{12}),$$

$$d = \tfrac{1}{2}[\eta_1 a_1 + \eta_2 a_2],$$

$$\beta p = \left\{(\rho_1 + \rho_2)[1 + \xi + \xi^2] - \frac{18}{\pi}\eta_1\eta_2(R_2 - R_1)^2[(R_1 + R_2) + R_1R_2(\eta_1R_1^2 + \eta_2R_2^2)]\right\}(1 - \xi)^{-3}, \tag{39}$$

$$g_{11}(R_1) = \{[1 + \tfrac{1}{2}\xi] + \tfrac{3}{2}\eta_2R_2^2(R_1 - R_2)\}(1 - \xi)^{-2} = -C_{11}(R_1),$$

$$g_{12}(R_{12}) = [R_2 g_{11}(R_1) + R_1 g_{22}(R_2)]/2R_{12} = -C_{12}(R_{12}), \tag{40}$$

$$\xi = \eta_1 R_1^3 + \eta_2 R_2^3,$$

and b_2, $g_{22}(R_2)$ are obtained from b_1, $g_{11}(R_1)$ by interchanging η_1, R_1 with η_2, R_2. The Laplace transforms of $rg_{ij}(r)$, $[12(\eta_i\eta_j)^{1/2}]^{-1}G_{ij}(s)$ are found from (20) or (26):

$$\begin{aligned}G_{11}(s) &= s[h - L_2(s)e^{sR_2}]/\mathfrak{D}(s), \\ G_{21}(s) = G_{12}(s) &= (\eta_1\eta_2)^{1/2}s^2e^{sR_{12}}\{[\tfrac{3}{4}(\eta_2R_2^3 - \eta_1R_1^3) \\ &\quad \times (R_2 - R_1) - R_{12}(1 + \tfrac{1}{2}\xi)]s - (1 + 2\xi)\}/\mathfrak{D}(s),\end{aligned} \tag{41}$$

where

$$\mathfrak{D}(s) = h - L_1(s)e^{sR_1} - L_2(s)e^{sR_2} + S(s)e^{s(R_1+R_2)}, \tag{42}$$

---

[14] The exact one-dimensional direct correlation functions for a one-dimensional mixture of hard rods are (c.f., Ref. 11),

$$\begin{aligned}-C_{11}(r) &= [1 - \rho_2(R_2 - R_1) - (\rho_1 + \rho_2)r](1 - \rho_1R_1 - \rho_2R_2)^{-2}, \quad r < R_1 \\ -C_{12}(r) &= [1 - \rho_2(R_2 - R_1)](1 - \rho_1R_1 - \rho_2R_2)^{-2}, \quad r \leq \lambda \\ &= [1 - \rho_2(R_2 - R_1)] - (\rho_1 + \rho_2)[r - (R_2 - R_1)/2] \\ &\quad \times (1 - \rho_1R_1 - \rho_2R_2)^{-2}, \quad \lambda \leq r \leq R_{12}.\end{aligned}$$

The resemblance to (37) is very striking.

and

$$
\begin{aligned}
h&=36\eta_1\eta_2(R_2-R_1)^2,\\
L_1(s)&=12\eta_2[(1+\tfrac{1}{2}\xi)+\tfrac{3}{2}\eta_1R_1^2(R_2-R_1)]R_2s^2\\
&\quad+[12\eta_2(1+2\xi)-hR_1]s+h,\\
S(s)&=h+[12(\eta_1+\eta_2)(1+2\xi)-h(R_1+R_2)]s\\
&\quad-18(\eta_1R_1^2+\eta_2R_2^2)^2s^2-6(\eta_1R_1^2+\eta_2R_2^2)\\
&\quad\times(1-\xi)s^3-(1-\xi)^2s^4,
\end{aligned}
\tag{43}
$$

and $G_{22}(s)$, $L_2(s)$ can be found from $G_{11}(s)$ and $L_1(s)$ by interchanging η_1, R_1 with η_2, R_2. We may now verify explicitly our previous assumptions on $G_{ij}(s)$ as well as its correct behavior when η_1 or R_1 vanishes or $R_1=R_2$.

The pressure given in (39), which comes from the compressibility relation (13), yields correctly the first three virial coefficients,[15] i.e., coefficients of $\rho_1^l\rho_2^k$ for $l+k\leq 3$. It is also in very good agreement with the Monte-Carlo computations[16] of the pressure done for $R_1=\frac{3}{5}R_2$, $\rho_1=\rho_2$, $\xi<0.2$. The reduced volume of mixture is always negative which implies that there is no phase separation of the components.[17] The pressure may be obtained from $g_{ij}(r)$, in addition to the compressibility relation (13), also by use of the virial theorem. For a mixture of hard spheres this has the form,[18]

$$\beta p^v=\rho_1+\rho_2+\tfrac{2}{3}\pi\sum_{i,j}\rho_i\rho_jR_{ij}^3g_{ij}(R_{ij}). \tag{44}$$

For the correct g_{ij} the two relations, (13) and (44), will yield the same result. For our approximate g_{ij} we find from (40),

$$\beta p^v=\beta p-\frac{18}{\pi}\frac{\xi}{(1-\xi)^3}(\eta_1R_1^2+\eta_2R_2^2)^3, \tag{45}$$

where we continue to label the compressibility pressure (39) by p. The generalization of the above results to an m-component mixture of hard spheres is immediate. The generalization of Eqs. (39), (45), and (40) are

$$\beta p=\left\{[\sum_{i=1}^{m}\rho_i][1+\xi+\xi^2]-\frac{18}{\pi}\sum_{i<j}\eta_i\eta_j(R_i-R_j)^2\times[2R_{ij}+R_iR_j(\sum\eta_lR_l^2)]\right\}(1-\xi)^{-3}, \tag{46}$$

$$\beta p^v=\beta p-\frac{18\xi}{\pi}[\sum_{l=1}^{m}\eta_lR_l^2]^3(1-\xi)^{-3},\quad \xi=\sum_{l=1}^{m}\eta_lR_l^3, \tag{47}$$

$$g_{ij}(R_{ij})=[R_jg_{ii}(R_i)+R_ig_{jj}(R_j)]/2R_{ij}, \tag{48}$$

$$g_{ii}(R_i)=\{(1-\xi)+\tfrac{3}{2}(\sum_{l=1}^{m}\eta_lR_l^2)R_i\}(1-\xi)^{-2}. \tag{49}$$

ACKNOWLEDGMENTS

I would like to thank M. Wertheim and J. K. Percus for many very valuable discussions, and Jim Steadman for help with some of the computations.

[15] A. G. McLellan and B. J. Alder, J. Chem. Phys. **24**, 115 (1956).

[16] E. B. Smith and K. R. Lea, Nature **186**, 714 (1960).

[17] I am indebted for the above results to Professor J. S. Rowlinson. Professor Rowlinson also obtained independently the pressure (39) for the case $R_1=0$.

[18] *Note added in proof.* B. J. Alder has kindly informed me that both p and p^v are in very good agreement (with p slightly above and p^v slightly below) with Monte-Carlo computations carried out by him and his co-workers for several values of R_2/R_1, ρ_2/ρ_1, and a large range of ξ.

PHYSICAL REVIEW VOLUME 134, NUMBER 6B 22 JUNE 1964

Time Symmetry in the Quantum Process of Measurement*

YAKIR AHARONOV, PETER G. BERGMANN, AND JOEL L. LEBOWITZ
Belfer Graduate School of Science, Yeshiva University, New York, New York

(Received 5 February 1964)

We examine the assertion that the "reduction of the wave packet," implicit in the quantum theory of measurement introduces into the foundations of quantum physics a time-asymmetric element, which in turn leads to irreversibility. We argue that this time asymmetry is actually related to the manner in which statistical ensembles are constructed. If we construct an ensemble time symmetrically by using both initial and final states of the system to delimit the sample, then the resulting probability distribution turns out to be time symmetric as well. The conventional expressions for prediction as well as those for "retrodiction" may be recovered from the time-symmetric expressions formally by separating the final (or the initial) selection procedure from the measurements under consideration by sequences of "coherence destroying" manipulations. We can proceed from this situation, which resembles prediction, to true prediction (which does not involve any postselection) by adding to the time-symmetric theory a postulate which asserts that ensembles with unambiguous probability distributions may be constructed on the basis of preselection only. If, as we believe, the validity of this postulate and the falsity of its time reverse result from the macroscopic irreversibility of our universe as a whole, then the basic laws of quantum physics, including those referring to measurements, are as completely time symmetric as the laws of classical physics. As a by-product of our analysis, we also find that during the time interval between two noncommuting observations, we may assign to a system the quantum state corresponding to the observation that follows with as much justification as we assign, ordinarily, the state corresponding to the preceding measurement.

I. INTRODUCTION

ONE of the perennially challenging problems of theoretical physics is that of the "arrow of time." Everyday experience teaches us that the future is qualitatively different from the past, that our practical powers of prediction differ vastly from those of memory, and that complex physical systems tend to develop in the course of time in patterns distinct from those of their antecedents. On the other hand, all the "microscopic" laws of physics ever seriously propounded and widely accepted are entirely symmetric with respect to the direction of time; they are form-invariant with respect to time reversal.[1,2]

The *de facto* absence of time symmetry in nature enters the formal statement of the laws of nature principally in two areas. One of these is thermodynamics, particularly the second law of thermodynamics; the latter proclaims that the entropy of a thermally isolated system can only increase toward the future. The other area is that of cosmogony; our universe is expanding toward the future. Gold[1] has suggested that these two asymmetric phenomena may well be causally related to each other. A third time-asymmetric effect, the preponderance of outgoing radiation in nature over incoming radiation, may be considered to be a special aspect of the second law.

In quantum theory the dynamical laws of motion, either the Schrödinger or the Heisenberg equations, are time symmetric as are their classical counterparts, Hamilton's equations of motion. It has been suggested, though, that asymmetry in the direction of time, and even thermodynamic irreversibility, enters into quantum theory through the theory of measurement.[3,4] Any measurement performed on a quantum system changes its state discontinuously and in a manner not to be described by the Schrödinger or Heisenberg equations of the isolated system. The performance of a measurement leads to the "reduction of the wave packet." That is to say, if the result of the measurement is known, then the quantum state of the system preceding the measurement has been replaced by the eigenvector of the observable that belongs to the eigenvalue recorded. If the outcome of the measurement is not known, the original state vector must now be replaced by a density matrix diagonal with respect to the eigenvectors of the observables measured, each diagonal element equaling the absolute square of the corresponding component of the original state vector. This density matrix is inequivalent to the original state vector in that all phase relations between the components have been destroyed by the act of measurement, though their norms survive in the density matrix.

Quite aside from entropy considerations, the conventional quantum theory of measurements is concerned exclusively with the prediction of probabilities of specific outcomes of future measurements on the basis of the results of earlier observations. Indeed the reduction of the wave packet has as its operational

* This research was supported by the U. S. Air Force Office of Scientific Research, Aerospace Research Laboratories, the National Science Foundation, and the National Aeronautics and Space Administration.

[1] T. Gold, in Onzième Conseil de l'Institut International de Physique Solvay, *La Structure et l'Evolution de l'Universe* (Edition Stoops, Brussels, 1958), p. 81; Am. J. Phys. **30**, 403 (1962); Proceedings of the Conference on the Arrow of Time, Cornell University, 1963 (to be published).

[2] See, however, O. Penrose and I. C. Percival, Proc. Phys. Soc. (London) **79**, 605 (1962).

[3] J. von Neumann, *Mathematical Foundations of Quantum Mechanics*, transl. by R. T. Beyer (Princeton University Press, Princeton, 1955).

[4] D. Bohm, *Quantum Theory* (Prentice-Hall Inc., Englewood Cliffs, New Jersey, 1951), cf. in particular, p. 608.

contents nothing but this probabilistic connection between successive observations.[5]

In this paper we propose to examine the nature of the time symmetry in the quantum theory of measurement. Rather than delve into the measurement process itself, which involves a specialized interaction between the atomic system and a macroscopic device,[3-5] we shall simply accept the standard expressions for probabilities of values furnished by the conventional theory. Whereas the conventional theory deals with ensembles of quantum systems that have been "preselected" on the basis of some initial observation, we shall deduce from it probability expressions that refer to ensembles that have been selected from combinations of data favoring neither past nor future. A theory that concerns itself exclusively with such symmetrically selected ensembles (the "time-symmetric theory") will contain only time-symmetric expressions for the probabilities of observations. Logically this time-symmetric theory is contained in the conventional theory but lacks one of the latter's postulates. It will be developed in Sec. II.

In Sec. III we shall consider the case that prior to the final selection some observations are performed that completely destroy coherence of any state previously existing; we shall find that any earlier observations obey probability laws that formally resemble the conventional prediction formula. Likewise, if the initial selection ("preselection") is followed by coherence destroying measurements to be succeeded in turn by some other observations, then these latter observations obey the precise time-reflected expression of the conventional prediction formula. This reflected relationship might be called a "retrodiction" formula. Finally, in Sec. IV we shall return to the true prediction and "retrodiction" situations, i.e., to the consideration of ensembles that have been either strictly preselected or postselected. By adding to the time-symmetric theory one postulate that appears to portray accurately the conditions of our universe (and whose time-reflected proposition does not hold), we are able to recover the conventional asymmetric theory. We present an argument that this asymmetry represents the intrusion of the irreversibility of macroscopic processes into the microscopic domain, so that the totality of the basic (microscopic) laws of nature emerges completely time symmetric.

II. SEQUENCES OF OBSERVATIONS

We shall begin by considering systems which are subjected to sequences of measurements, each of which is individually "complete"; that is to say, that each observation determines a quantum state of the system. We make the conventional assumption about the selection of ensembles of such systems (and of their histories), which is to the effect that initially all systems of the ensemble have yielded a specified nondegenerate eigenvalue of an observable J; no other conditions are imposed. Under these circumstances the conventional quantum theory of measurements states that, given two successive measurements, the probability of a particular outcome of the later observation depends on the outcome of the earlier observation by being the absolute square of the scalar product of the two state-vectors belonging to the two respective eigenvalues. We shall denote the observables to be measured by symbols $A_1, A_2, \cdots, A_k, \cdots$, all of whose eigenvalues are nondegenerate; let the eigenvalues of A_k be denoted by d_k. Only when necessary will distinct eigenvalues of A_k be denoted by Greek superscripts $d_k^{(\alpha)}, d_k^{(\beta)}, \cdots$. For the sake of simplicity we shall work in a Heisenberg representation and assume further that all the A_k are constants of the motion, not necessarily explicitly time-independent. At any rate, between measurements both the quantum states of our systems and the matrix elements of our observables will be constant. If the observables A_k are to be measured in any particular sequence, which, in general, will not correspond to the order of the subscripts $\cdots, k, \cdots$, we shall indicate the sequence of measurements by Latin superscripts, thus: A_k^m.

Suppose now that we perform a sequence of observations, $A_m^{-M}, \cdots, A_i^{j-1}$, yielding the measurements $d_m, \cdots, d_i$; then the probability that the next measurement A_k^j will yield the eigenvalue d_k is

$$p(d_k/d_m,\cdots,d_i)=|\langle d_i|d_k\rangle|^2=\mathrm{Tr}(D_iD_k), \quad (2.1)$$

where the symbol D_k denotes the idempotent operator

$$D_k\equiv|d_k\rangle\langle d_k|, \quad (2.2)$$

etc. If the measurement A_k^j is to be followed by $A_l^{j+1}, A_n^{j+2}, \cdots, A_r^{N-1}, A_s^N$, the probability that the respective outcomes will be $d_k, d_l, \cdots, d_r, d_s$ is

$$p(d_k,\cdots,d_r,d_s/d_m,\cdots,d_i)$$
$$=\mathrm{Tr}(D_sD_r\cdots D_kD_iD_k\cdots D_r). \quad (2.3)$$

Equations (2.1) and (2.3) hold irrespective of the outcome of the measurements $A_m^{-M}, \cdots, A_h^{j-2}$, and irrespective of the outcome of the members of the ensemble subsequent to the performance of the specified observation(s). These expressions summarize the quantitative content of the conventional theory of measurement in quantum physics.

In passing let us briefly comment on the need in quantum theory for constructing ensembles with well-defined probability characteristics. If, in classical mechanics, we had to deal with a system possessing a phase space with a finite volume Ω, then we could define an *a priori* probability density on that phase space that would be invariant with respect to canonical transformations: the constant probability density Ω^{-1}. One could then modify this density in conformity with any restrictions imposed on the physical system, so as to obtain contingency probabilities by purely deductive methods. In other words, in a finite phase space one might construct statistical mechanics employing a

[5] E. P. Wigner, Am. J. Phys. 31, 6 (1963).

standard ensemble as the point of departure. Because in every realistic physical system the phase space has an infinite volume, a transformation-invariant standard probability density does not exist, and one is led into constructing or conjecturing probability distributions to fit various conditions imposed on the ensemble.

The situation in quantum theory is analogous. If Hilbert space were finite-dimensional, then there would be one density matrix distinguished as being representation-invariant, the normalized multiple of the unit matrix, from which all other density matrices could be derived in response to various contingencies. But again, for all realistic physical systems Hilbert space is infinite-dimensional; hence, there is no "standard ensemble" existing *a priori* and independently of any information about our physical system. Thus, formally, we are forced to construct ensembles of systems having certain restrictive properties. Whether particular classes of restrictions lead to ensembles with unambiguous probability characteristics cannot be decided affirmatively by formal analysis alone, though internal inconsistencies might rule out some conjectures. It is clear that the assumptions underlying the conventional theory of quantum measurements are logically admissible.

Next we shall consider a sequence of measurements J, A_k^{-M}, $\cdots$, A_j^{-1}, A_i^0, A_l^1, A_m^2, $\cdots$, A_n^N, F, in that order. J and F are to be nondegenerate observables like the others, and their eigenvalues are denoted respectively by a and b. We shall now consider an ensemble of systems whose initial and final states are fixed to correspond to the particular eigenvalues a and b, respectively; we ask for the probability that the outcome of the intervening measurements are d_j, $\cdots$, d_n, respectively. This probability, on the strength of Eq. (2.3), is found to be

$$p(d_j,\cdots,d_n/a,b)=\frac{p(d_j,\cdots,d_n,b/a)}{p(b/a)}=\frac{1}{H(a,b)}\operatorname{Tr}(AD_j\cdots D_nBD_n\cdots D_j)\,, \tag{2.4}$$

where

$$H(a,b)=\textstyle\sum_{j'}\cdots\sum_{n'}\operatorname{Tr}(AD_{j'}\cdots D_{n'}BD_{n'}\cdots D_{j'})\,,\quad A=|a\rangle\langle a|\,,\quad B=|b\rangle\langle b|\,. \tag{2.5}$$

This expression is manifestly time symmetric. If we change the sequence of measurements to F, A_n^{-N}, $\cdots$, A_k^M, J, Eqs. (2.4), (2.5) remain unchanged. In the exceptional case $H(a,b)=0$ the probability $p(d_j\cdots d_n/a,b)$ is not defined.

The probabilities (2.4), (2.5) refer to a sample that has been selected on the basis of required outcomes of specified initial and final observations. This procedure may appear artificial compared to the usual prescription: "Prepare a system so that the value of J (at the beginning) be a." But from a formal point of view we may legitimately specify any selection that could be performed with physical equipment, however complex.

As a matter of fact, in experimental physics selections are frequently based on combinations of initial and final characteristics. Consider a beam of particles that enters a cloud chamber or similar device controlled by a master pulse. For the device to select an event as belonging to a sample to be evaluated statistically, the particle must enter the chamber and, prior to the onset of any manipulation by magnetic fields, etc., satisfy certain requirements. But in order to be counted the particle must also activate the circuits of counters placed below the chamber; thus, we make the selection on the basis of both the initial and the final state. In some experiments even intermediate specifications may be imposed in addition to initial and final conditions. Thus, our formal treatment of initial and final states on an equivalent footing is not inconsistent with experimental procedures used in some investigations.

Equations (2.4), (2.5) may be thought of as providing the foundation for a time-symmetric theory of measurement. If we assumed the existence of ensembles with well-defined probabilities only if selected on the basis of both initial and final states, we should have a logically closed theory, though one that would never permit extrapolations to time intervals lying outside the interstice between initial and final determination. Given ensembles of any kind with well-defined probability dispersions, we can always form subensembles obeying additional restrictions and hence the time-symmetric ensembles can be obtained from those of the conventional theory by means of a deductive process. The reverse does not hold, i.e., we cannot infer the characteristics of broadly defined ensembles from those of more narrowly defined ensembles.

On the basis of Eqs. (2.4), (2.5) we may calculate probabilities involving only some of the measurements between J and F, or we may calculate contingent probabilities referring to partial samples in which the outcomes of some of these measurements are fixed. In particular we can calculate the contingent probability of the outcome d_l^1, given the outcome d_i^0. To obtain this probability we must, of course, sum over all the possible outcomes of the measurements preceding A_i^0 and over all the possible outcomes following A_l^1, keeping, as before, the outcomes of J and F fixed. The result is

$$p(d_l/d_i;a,b)=\frac{\sum_{m'}\cdots\sum_{n'}\operatorname{Tr}(D_iD_lD_{m'}\cdots D_{n'}BD_{n'}\cdots D_{m'}D_l)}{\sum_{l'}\sum_{m'}\cdots\sum_{n'}\operatorname{Tr}(D_iD_{l'}D_{m'}\cdots D_{n'}BD_{n'}\cdots D_{m'}D_{l'})}=|\langle d_i|d_l\rangle|^2\frac{\sum_{m'}\cdots\sum_{n'}\operatorname{Tr}(D_lD_{m'}\cdots D_{n'}BD_{n'}\cdots D_{m'})}{\sum_{l'}\cdots\sum_{n'}\operatorname{Tr}(D_iD_{l'}D_{m'}\cdots B\cdots D_{m'}D_{l'})}\,. \tag{2.6}$$

As expected, the history preceding the measurement A_i^0 drops out of our expression, but the coefficient of the squared matrix element of the conventional prediction (2.1) is sensitive to d_i and to d_l as well as to the subsequent history. In other words, postselection will affect the transition probability from d_i to d_l. That this is unavoidable can be understood easily by the consideration of the extreme case in which all observables $A_m, \cdots, A_n, F$ commute with each other as well as with A_l. Depending on the selection of the eigenvalue b, the transition probability in that case will be either 0 or 1.

It is obvious that the time-reflected relationship to (2.6) also holds. That is to say, if we calculate the contingent probability of d_i knowing the outcome d_l of the observation immediately following, we shall obtain an expression that is independent of the whole history subsequent to the measurement A_l^1, but which will depend on the initial selection a as well as observations scheduled prior to A_i^0.

Let us now consider incomplete measurements. The result (2.4), (2.5) can be generalized immediately if we drop the requirement that each intermediate measurement be complete. According to von Neumann,[3] an incomplete observation projects the initial state not on a particular direction but on a particular (multidimensional) linear subspace of the Hilbert space, and may be represented by an idempotent operator D_k. The form of Eqs. (2.4), (2.5) will remain unchanged under this reinterpretation of the symbols D_k. It should be noted, however, that Eq. (2.6) holds only if A_i is nondegenerate.

The replacement of the initial and final states by mixtures is a bit more involved. If we form an ensemble in which histories beginning with state $|a\rangle$ and ending with state $|b\rangle$ form a fraction c_{ab} of the whole

$$c_{ab}\geq 0, \quad \sum_{a'}\sum_{b'} c_{a'b'}=1, \tag{2.7}$$

then the probability $p(d_j,\cdots,d_n/\{c\})$ will be

$$p(d_j,\cdots,d_n/\{c\})=\sum_{a'}\sum_{b'} c_{a'b'}\,p(d_j,\cdots,d_n/a',b'). \tag{2.8}$$

There exists no simple expression that would depend on the initial and final density matrices. The probabilities (2.8) depend on the fractions of systems within the ensemble passing from specified initial to specified final states, not merely on the initial distribution $\sum_{b'} c_{ab'}$ and the final distribution $\sum_{a'} c_{a'b}$.

III. ASYMPTOTIC PROCEDURES

Whereas we have been able to obtain time-symmetric ensembles from those depending only on initial selection, the reverse procedure is impossible without an additional postulate; that is to say, given a theory of ensembles based on time-symmetric double-selection procedures, we cannot obtain probabilities for ensembles in which the selection is based only on initial (or only on final) observations by deduction alone. In this sense, the time-symmetric theory of Sec. II is more restricted than the conventional theory of measurements.

There is, however, a way to blunt the effects of either pre- or postselection. The method to be described in this section rests on the fact that in quantum theory the type of interference that we call an observation destroys the "coherence" of the state of a system, producing a new situation that is connected with the original situation only by stochastic laws. This stochastic connection, or the lack of a tighter relationship, may be expressed either in terms of the state vector, or its replacement by a density matrix, or purely in terms of probabilistic assertions. Whatever the mode of description, it is possible to sever different portions of the history of a system from each other by the interposition of certain types of measurements. By preceding the final selection in the time-symmetric theory by such "coherence destroying" manipulations, we may formally recover the prediction formula (2.1); by scheduling such procedures following the initial selection of a time-symmetric ensemble, we may obtain the time-reverse of Eq. (2.1), a "retrodiction" formula.

These possibilities are of considerable interest because they present us with a relatively large class of possible procedures all of which lead, asymptotically in most cases, to substantially similar results. Though the interpolation of coherence destroying manipulations, say before the act of final selection within the framework of the time-symmetric theory, does not relieve us of the logical necessity of performing the act of final selection, the particular choice of observable and of its numerical value used for that final selection has no effect on the probabilities of events preceding the coherence destroying acts.

We shall first indicate particular sets of measurements which destroy coherence more or less completely. Such sets of two consecutive measurements may be constructed in closed form if the Hilbert space of a system is finite dimensional, e.g., if the particles in a monochromatic and well-collimated beam can differ only in their states of polarization. Consider, in this case, two observables A_i and A_j whose eigenvectors are related to each other by a unitary matrix U and whose matrix elements all have the same absolute square $1/n$, n being the number of dimensions of the Hilbert space. One possible unitary matrix with this property is, for instance, the following:

$$U_{kl}=[1/(n)^{1/2}]e^{i\theta_{kl}}, \quad \theta_{kl}=(2\pi/n)kl. \tag{3.1}$$

Let us denote the idempotent operators to be constructed from the respective eigenvectors of the two observables by D_1 and D_2, respectively, each of these symbols representing n such different operators. Then the following expression constructed with any density matrix M whatsoever is always a multiple of the unit matrix I:

$$\sum_{d_1'}\sum_{d_2'} D_1'D_2'MD_2'D_1'=(1/n)I. \tag{3.2}$$

As for the infinite-dimensional case, the situation is insofar more involved as there exists no density matrix which is precisely a multiple of the unit matrix. We shall assume that the Hilbert space admits a complete set of commuting operators, each having a continuous range of eigenvalues from $-\infty$ to ∞. We shall call these operators x_s and construct by the usual methods a set of operators p_s which satisfy standard canonical commutation relations with each other and with the x_s. In a somewhat symbolic sense the unitary operators leading from the improper joint eigenfunctions of the x_s to the improper joint eigenfunctions of the p_s, i.e., the Fourier integral operators, possess matrix elements all of the same magnitude as in the previous case. In view of the fact that idempotent operators of the type $x(x_0)$, etc., are not really defined, we introduce idempotent operators $X(x,\Delta)$, defined as integral operators whose kernel equals 1 if $x\epsilon\Delta$, and vanishes otherwise. We cover the space of numerical values of the x_s with a denumerable set of domains Δ without overlap. Similarly, we introduce idempotent operators $P(p,E)$, where the domains E cover the momentum space without overlap. The expression constructed in complete analogy to (3.1) will then not equal a multiple of the unit matrix because of the coarseness of the cell structures established in x space and in p space. However, we may establish a sensible limit if we improve the fineness of both cell structures and if we multiply the expression (3.1) on the left by a factor corresponding to the effective n, eventually becoming infinite, so that the right side can actually tend to the identity operator I (whose trace diverges).

We now return to the expression (2.4) and substitute for a certain number of factors centered on F a multiple of I, both in the numerator and the denominator. The constant of proportionality used is immaterial, as it drops out in any case, and we might use I directly. We then see, almost by inspection, that (2.4) reduces to (2.3), the pure prediction formula, and, likewise, that (2.6) reduces to (2.1). We conclude, then, that because of the asymptotic properties of expressions of type (3.2) the prediction formulas may be recovered from the time-symmetric formulas.

We may derive the corresponding "retrodiction" expression by time reversing the procedure that we have just presented. If we follow the initial selection of an ensemble in the time-symmetric theory by a set of coherence-destroying measurements, then the outcome of subsequent observations is related to the final selection as follows:

$$p(d_k,d_l,\cdots,d_s/a,b)=\mathrm{Tr}(D_kD_l\cdots D_sBD_s\cdots D_l)\,. \quad (3.3)$$

If, in particular, we are concerned with the one observation preceding the final selection, then the probability of the outcome d is

$$p(d/b)=|\langle b|d\rangle|^2\,. \quad (3.4)$$

The coherence destroying properties of the procedure summarized in Eqs. (3.1), (3.2), and of the corresponding asymptotic procedure outlined for the infinite-dimensional Hilbert space may be demonstrated by straightforward computation. It would be of considerable interest if there were a broad range of procedures having the same effect. Generally, sequences of measurements will destroy coherence to a greater or lesser extent provided that they involve all directions of Hilbert space in noncommuting measurements. There are, of course, degrees of noncommutativity: The noncommutavity may involve varying numbers of directions in Hilbert space, and the eigendirections of consecutive operators may differ from each other by various angles. Formally, the extent to which coherence is destroyed by a given sequence may be evaluated in terms of the degree to which matrices of the general form (3.2) approximate a multiple of the unit matrix. That there is some approach to the unit matrix in a sequence of noncommuting measurements is assured by the results to be found in von Neumann.[3] If $D^{(\alpha)}$ is a set of idempotent operators belonging to the same measurement and with properties

$$D^{(\alpha)}D^{(\beta)}=\delta^{\alpha\beta}D^{(\alpha)}\,,\quad \textstyle\sum_\alpha D^{(\alpha)}=I\,, \quad (3.5)$$

and if M is an arbitrary density matrix, then

$$M'=\textstyle\sum_\alpha D^{(\alpha)}MD^{(\alpha)} \quad (3.6)$$

is also a density matrix and approximates a multiple of the unit matrix I at least as well as M in the following respects: (a) If we define the entropy of M as usual by the expression

$$S=-k\,\mathrm{Tr}(M\ln M)\,, \quad (3.7)$$

then

$$S'\geqslant S\,. \quad (3.8)$$

The equality holds only if the idempotent operators commute with M. (b) The range of eigenvalues of M' is not greater than the range of eigenvalues of M; that is to say, the upper limit of its eigenvalues is not larger and the lower limit not smaller. Both entropy and range of eigenvalue spectrum are yardsticks for the approach to λI.

Thus, it appears that we can destroy coherence more or less completely by a wide variety of sequences of measurements and thereby obtain the asymptotic prediction and retrodiction situations within the framework of the time-symmetric theory of measurements.

The existence of the retrodiction formula (3.3), (3.4) suggests that the customary assignment of a state vector to a system on the basis of the most recent preceding observation may be somewhat arbitrary. This assignment is based on the intuitive notion that the measurement is the "cause" and the quantum state the "effect," and that cause must precede effect in time. Also, perhaps, there is the notion that the quantum state of a system embodies the maximum of information available to us about the system at any time; ordinarily, we can know the outcome of all observations in the past but not of those yet in the future.

But, as we have seen, under suitable circumstances the usual prediction formula (2.1) may be replaced by the retrodiction formula (3.4), which bases a probabilistic statement about the outcome of one measurement on the outcome of the measurement next following in time. If the measurement of A (whose eigenvalues are being denoted by d) is preceded by coherence destroying operations as we have assumed in deriving Eqs. (3.3) and (3.4), then we know essentially nothing about the outcome of observations preceding A; that is to say, all possible outcomes of such preceding observations are approximately equally likely. Hence, our probabilistic statement about the outcome of the measurement of A is based primarily on the event immediately following, and the information on which our statement is based ought to be incorporated in an appropriate assignment of quantum state. Thus, we are led into assigning the state $|b\rangle$ to the period of time *preceding* the observation of F yielding the eigenvalue b.

From a purely operational point of view, one might eschew the assignment of quantum states to physical systems altogether and instead rely entirely on probabilistic statements referring to carefully defined ensembles. However, as long as one does assign quantum states to physical systems, it appears defensible to do so either in reliance on the (complete) observation immediately preceding (as is customary) or on the one next following, depending on circumstances. This ambiguity indicates that the quantum state of a system, though undoubtedly containing some elements of "reality" independent of any observer, also has subjective aspects.

We shall conclude this section by pointing out that, in general, the dispersion of probabilities of the outcome of one particular observation will be minimized (i.e., the "negative entropy" associated with this dispersion will be maximized) if we use all information about the system's past and future. This statement is a direct consequence of the properties of the entropy function to be found, e.g., in Khinchin.[6] Hence, if both the initial and final state of a system are known, use of the prediction formulas (2.1) or (2.3) instead of (2.4) will lead to a loss in precision of the probabilistic statement concerning the intermediate observation.

IV. DIRECT PREDICTION

By now we have established that the conventional prediction formulas can be recovered from the time-symmetric expressions (2.4) by means of a model that consists of shielding events close at hand from the terminal selection on which (2.4) is based by the interposition of a series of "coherence destroying" experiments. Each measurement constitutes an interference with the physical system which destroys in a limited and mathematically well-described manner its dynamic behavior as an isolated system.

Normally, the prediction formula (2.1) and its corollary (2.3) are not conceived of as depending on carefully managed follow-up maneuvers, but are assumed to be independent of the subsequent history of the system. That this prediction theory is indeed logically independent of the time-symmetric formula (2.4) may be deduced immediately from the circumstance that in our universe the prediction formula is considered to be universally valid, whereas the time-reflected formula, the retrodiction formula, is not.

Consider an ensemble of similar physical systems of arbitrary provenance and select a sample on the strength of a single complete measurement. The conventional theory of measurement then furnishes us, with respect to this selected subensemble, with relative frequencies of outcomes of a subsequent measurement or of a subsequent series of measurements, regardless of the events that may have preceded the initial selection procedure, as well as of those events that follow on the heels of the specified series of measurements, as long as no further selection is involved. The reverse theory would have to concern itself with the probability of the outcome of certain measurements on an ensemble of similar physical systems, the ensemble to be determined solely on the basis of a pure-state selection immediately following the specified series of measurements; the expression for the probabilities should contain no reference to any events following the terminal selection, nor to the manner in which physically similar systems were collected prior to the onset of the series of measurements. Clearly, in our universe no such "retrodiction theory" would be valid: Suppose we constructed a monochromatic and well-collimated beam of particles possessing nonzero spin, performed some observation referring to the spin distribution of the beam, and then followed up with a Stern-Gerlach experiment singling out those particles in the beam being in a very definite spin state. Suppose we ask for the percentage of particles, from among those passing the postselection test, that had specified outcomes in the antecedent experiment (which should refer to an observable not possessing the specified final state as an eigenstate); surely these probabilities would not be independent of the state of polarization of our beam prior to the performance of the first experiment.

We conclude, therefore, that in order to recover the conventional prediction statement from the time-symmetric formulas of Sec. II, we must adopt a postulate that is logically independent of the time-symmetric theory; the postulate that in our universe ensembles chosen on the basis of an initial complete measurement alone possess unambiguous and reproducible probability characteristics. Once we adopt this

[6] A. I. Khinchin, *Mathematical Foundations of Information Theory*, transl. by R. A. Silverman and M. D. Friedman (Dover Publications, Inc., New York, 1957).

postulate the conventional prediction formulas (2.1), (2.3) follow from the time-symmetric formula (2.4) and from the considerations of Sec. III. We found in that section that there are "coherence destroying" procedures that make the "prediction" expressions (2.6) independent of the particular postselection we choose to perform. But if there are methods by which we can make our probabilities independent of the manner of postselection and if, by our new postulate, there exist unambiguous probabilities even in the absence of any postselection, then these two sets of probabilities should be equal.

Logically, it is conceivable that the time reverse of our new postulate should also hold; this would mean that postselection alone results in an ensemble with well-defined and reproducible probability characteristics. Actually we know that in our universe this proposition is untrue. We are thus confronted with an indubitable asymmetry in time direction. It remains to discuss whether this asymmetry is a property of microphysics proper or whether it represents the intrusion of the macroscopic universe on the microscopic scene. Granting that this question does not lend itself to straightforward logical analysis, it appears to us that the construction of ensembles in the real physical universe is a macroscopic operation and that it depends on the realities of the universe as a whole. Let us return once more to our beam of particles endowed with spin.

If we attempt to analyze the different manner in which past and future histories affect its present characteristics, we find that no matter how we gather our beam, its constituent particles have come from one or several "sources" (e.g., a laboratory device, a distant galaxy, etc.), which determine its properties; there simply is no way of avoiding preselection completely. On the other hand, beams are not collimated toward a "sink," unless we arrange it so in our laboratory. This asymmetry is directly associated with the fact that the origins of all kinds of radiations in the universe are spatially and temporally concentrated, and their destinations are not. The nature of ensembles or beams actually occurring in nature is, in fact, macroscopic, not microscopic; it is determined by the same cause as all macroscopic irreversibility, conceivably by the expansion of the universe.[1]

As for the microscopically determined aspects of quantum measurements, we believe that they can be fairly summarized by the statement that in time-symmetrically constructed ensembles the laws of probability are also time symmetric; further, that to the extent that retrodiction situations may be said to exist, they obey the same laws as the corresponding prediction situations.

ACKNOWLEDGMENTS

The authors wish to acknowledge frequent valuable discussions with O. Penrose and with A. Komar. One of us (P. G. B.) was helped considerably in clarifying his own thinking by the privilege of attending a conference on the arrow of time organized by T. Gold and H. Bondi, which took place in May 1963 at Cornell University. Our postulate as to the sufficiency of preselection in Sec. IV was stimulated by the paper by Penrose and Percival.[2] Their time-asymmetric proposal is, however, concerned with ensembles of classical systems, and not directly related to their construction on the basis of observations. Finally, we wish to acknowledge with thanks receipt of a preprint by W. J. Cocke entitled "Statistical Time-Inversion Invariance." Cocke's paper is, in contrast to ours, primarily concerned with histories of macroscopic systems, which lend themselves to a description in terms of Markov chains. H. Margenau has pointed out that the conventional interpretation of quantum measurements which is associated with the name of von Neumann[3] is currently no longer acceptable to a fair number of physicists.[7] We wish to express our appreciation for his critical reading of our work.

[7] H. Margenau, Phil. Sci. **30**, 1 (1963); Ann. Phys. (N. Y.) **23**, 469 (1963). Further references are to be found in these two papers.

JOURNAL OF MATHEMATICAL PHYSICS VOLUME 7, NUMBER 1 JANUARY 1966

Rigorous Treatment of the Van Der Waals–Maxwell Theory of the Liquid–Vapor Transition*

J. L. Lebowitz and O. Penrose†
Belfer Graduate School of Science, Yeshiva University, New York, New York
(Received 4 March 1965)

Rigorous upper and lower bounds are obtained for the thermodynamic free-energy density $a(\rho, \gamma)$ of a classical system of particles with two-body interaction potential $q(\mathbf{r}) + \gamma^\nu\varphi(\gamma\mathbf{r})$ where ν is the number of space dimensions and ρ the density, in terms of the free-energy density $a^0(\rho)$ for the corresponding system with $\varphi(\mathbf{x}) \equiv 0$. When $\varphi(\mathbf{x})$ belongs to a class of functions, which includes those which are nonpositive and those whose ν-dimensional Fourier transforms are nonnegative, the upper and lower bounds coincide in the limit $\gamma \to 0$ and $\lim_{\gamma\to 0} a(\rho, \gamma)$ is the maximal convex function of ρ not exceeding $a^0(\rho) + \frac{1}{2}\alpha\rho^2$, where $\alpha \equiv \int \varphi(\mathbf{x})\, d\mathbf{x}$. The corresponding equation of state is given by Maxwell's equal-area rule applied to the function $p^0(\rho) + \frac{1}{2}\alpha\rho^2$ where $p^0(\rho)$ is the pressure for $\varphi(\mathbf{x}) \equiv 0$. If $a^0(\rho) + \frac{1}{2}\alpha\rho^2$ is not convex the behavior of the limiting free energy indicates a first-order phase transition. These results are easily generalized to lattice gases and thus apply also to Ising spin systems.

The two-body distribution function is found, in the limit $\gamma \to 0$, to be normally identical with that for $\varphi(\mathbf{x}) \equiv 0$, but if the system has a phase transition it has the form appropriate to a two-phase system.

Some of the upper and lower bounds on $a(\rho, \gamma)$ are simple enough to be useful for finite γ. Also, some of our results remain valid for quantum systems.

I. INTRODUCTION

Phase transitions such as melting and boiling are familiar experiences, but their explanation from the first principles of statistical mechanics still presents a major challenge to the theoretical physicist. One of the earliest steps towards a theory of the gas–liquid phase transition was taken by van der Waals.[1] Seeing the interaction between the molecules of a classical fluid as a competition between two distinct parts of the intermolecular force, a short-range repulsive part and a long-range attractive part, he arrived at the equation of state

$$p = kT\rho/(1 - \rho b) + \tfrac{1}{2}\alpha\rho^2 \equiv p_{vdw}(\rho, T) \qquad (1.1)$$

where p is the pressure, k Boltzmann's constant, T the temperature, ρ the number density, and $-\alpha$ and b are positive constants characterizing long- and short-range parts of the potential, respectively.

When T exceeds the critical temperature $T_c \equiv -4\alpha/27bk$, the van der Waals equation of state (1) gives a good qualitative representation of the isotherms of a real fluid; for $T < T_c$, however, each isotherm includes a section where the compressibility is negative, in violation of the thermodynamic stability principle. The reason for this failure is that the argument leading to (1) assumes a single-phase system; it does not allow for the possibility of coexisting liquid and vapor phases.

Maxwell[2] showed that the coexistence region could be included in the theory by using van der Waals' equation of state for both liquid and vapor phases and using the thermodynamic equilibrium condition that the two phases must have equal pressures and chemical potentials. This leads to the following modification of (1) for $T < T_c$:

$$p = \begin{cases} p_{vdw}(\rho, T) \text{ if } \rho < \rho_v(T) \text{ or } \rho > \rho_l(T) \\ p_{sat}(T) \text{ if } \rho_v(T) < \rho < \rho_l(T) \end{cases} \qquad (1.2)$$

where $\rho_v(T)$, $\rho_l(T)$ and $p_{sat}(T)$ may be determined by the graphical construction shown in Fig. 1.

A very interesting derivation of van der Waals' equation of state with Maxwell's rule was given recently by van Kampen.[3] In this derivation the volume Ω occupied by the system is divided into a large number of cells, each small compared with the range of the long-range attractive force, but large enough to contain many particles. Avoiding the pitfall of assuming a uniform distribution of particles over cells, which leads[4] to a generalized form of the van der Waals equation of state, van Kampen obtained the distribution over cells by minimizing the free energy. His method leads to the modified equation of state (2), which implies a first-order phase transition. When $\rho < \rho_v$ or $\rho_l < \rho$, van Kampen's method indeed gives a uniform distribution over cells, but when $\rho_v < \rho < \rho_l$ it leads

* Work supported by the U. S. Air Force Office of Scientific Research at Yeshiva University, Grant No. 508-64.
† Present address: Imperial College, London, S.W. 7, England.
[1] J. van der Waals, thesis (Leiden, 1873) (cited by Kac, Uhlenbeck, and Hemmer, Ref. 6 of this paper).
[2] J. C. Maxwell, Scientific Papers (Dover Reprint, New York), p. 425.
[3] N. G. van Kampen, Phys. Rev. 135, A362 (1964).
[4] O. Ornstein, thesis (Leiden, 1908) (cited by N. G. van Kampen, Ref. 3 of this paper).

to the conclusion that this distribution is nonuniform, as it should be when two phases coexist.

Van Kampen's treatment is, however, not mathematically rigorous. In particular the conditions to be satisfied by the interaction are not specified, and various limiting processes are hinted at but not carried out explicitly. It is the purpose of the present paper to provide a rigorous treatment similar to van Kampen's and to extend it to a more general class of long-range potentials than the purely attractive potentials considered by van Kampen.

The intermolecular potential we consider has the form proposed by Kac[5] and investigated thoroughly for a one-dimensional system by Kac, Uhlenbeck, and Hemmer,[6]

$$v(\mathbf{r}) = q(\mathbf{r}) + w(\mathbf{r}, \gamma), \tag{1.3}$$

where r represents the separation of a pair of particles, $q(\mathbf{r})$ is a short-range potential, and $w(\mathbf{r}, \gamma)$ is a potential (we call it the *Kac potential*) whose range is proportional to the reciprocal of the parameter γ. Kac, Uhlenbeck, and Hemmer studied a one-dimensional system with

$$q(\mathbf{r}) \equiv \begin{cases} +\infty & \text{if } r < r_0, \\ 0 & \text{if } r > r_0, \end{cases} \tag{1.4}$$

$$w(\mathbf{r}, \gamma) \equiv \tfrac{1}{2}\alpha\gamma \exp(-\gamma r), \tag{1.5}$$

where r_0, γ, and $-\alpha$ are positive parameters. For finite γ they found no phase transition, but in the *van der Waals limit* $\gamma \to 0$ the equation of state approaches as a limit Maxwell's modification (2) of the van der Waals equation of state and thus does show a first-order phase transition. Unfortunately their method is very difficult to generalize to other potentials or to more than one dimension.

An important feature of the work of Kac, Uhlenbeck, and Hemmer was the use of the limit process $\gamma \to 0$ to provide a clear distinction between the short (finite) range of the contribution $q(\mathbf{r})$ to $v(\mathbf{r})$ and the long (infinite as $\gamma \to 0$) range of the contribution $w(\mathbf{r}, \gamma)$. It is this feature which we shall exploit here; but unlike Kac, Uhlenbeck, and Hemmer, we do not restrict the system to one dimension nor the potential to the form defined by (4) and (5). Instead of the special form (5) for $w(\mathbf{r}, \gamma)$, we use

$$w(\mathbf{r}, \gamma) = \gamma^{\nu}\varphi(\gamma\mathbf{r}), \tag{1.6}$$

where ν is the number of dimensions of the space

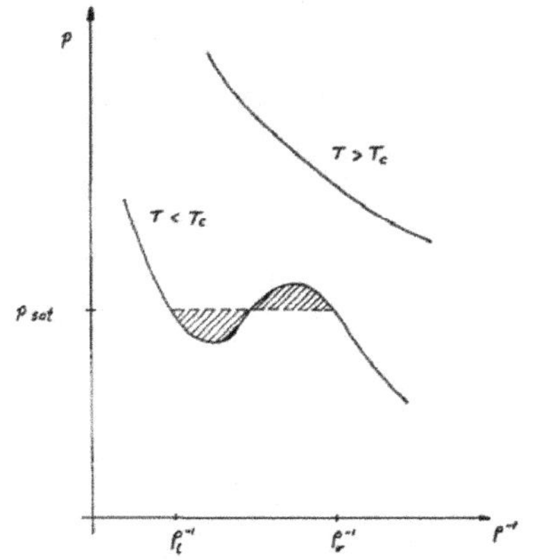

FIG. 1. Typical isotherms for the van der Waals equation of state (solid lines) and Maxwell's modification (dotted line). The shaded areas are equal.

considered; this reduces to (5) if $\nu = 1$ and $\varphi(\mathbf{x}) = \frac{1}{2}\alpha e^{-x}$. If the function $\varphi(\mathbf{x})$ is bounded in a neighborhood of the origin, say in $x < \delta$, then the Kac potential (6) has the property

$$|w(\mathbf{r}, \gamma)| < \gamma^{\nu} \operatorname*{Max}_{x<\delta} |\varphi(\mathbf{x})| \quad \text{if} \quad \gamma < \delta/r \tag{1.7}$$

so that

$$\lim_{\gamma\to 0} w(\mathbf{r}, \gamma) = 0 \quad \text{for all} \quad \mathbf{r}. \tag{1.8}$$

At the same time the integral of the Kac potential over all ν-dimensional space,

$$\int w(\mathbf{r}, \gamma)\, d\mathbf{r} = \int \varphi(\mathbf{x})\, d\mathbf{x} \equiv \alpha, \tag{1.9}$$

is independent of γ. This fact is compatible with (1.8) because the limit operation $\gamma \to 0$ does not commute with the one associated with the infinite region of integration.

The basis of our method is to obtain upper and lower bounds on the free energy

$$A(N, \Omega, \gamma) \equiv -kT \log Z(N, \Omega, \gamma), \tag{1.10}$$

where $Z(N, \Omega, \gamma)$ is the classical partition function for N particles at temperature $T \equiv 1/k\beta$ in a ν-dimensional cube Ω, defined by

$$Z(N, \Omega, \gamma) \equiv (1/N!)(mkT/2\pi\hbar^2)^{\nu N/2} \times \int_{\Omega} \cdots \int_{\Omega} e^{-\beta V}\, d\mathbf{x}_1 \cdots d\mathbf{x}_N \tag{1.11}$$

and

$$V \equiv \sum_{i \leq j \leq N} v(\mathbf{x}_i - \mathbf{x}_j) \tag{1.12}$$

[5] M. Kac, Phys. Fluids **2**, 8 (1959).
[6] M. Kac, G. E. Uhlenbeck, and P. Hemmer, J. Math. Phys. **4**, 216 (1963).

with $v(\mathbf{r})$ defined in (3). The upper and lower bounds are obtained by dividing the cube Ω into M congruent smaller cubes $\omega_1 \cdots \omega_M$ and using estimates of the interactions across cell boundaries to relate the free energy of the cube Ω to the sum of the free energies of the cubes $\omega_1 \cdots \omega_M$.

From these upper and lower bounds, the equation of state in the van der Waals limit is calculated by means of a succession of limit operations. First the thermodynamic free energy is calculated from $A(N, \Omega, \gamma)$ by taking the *thermodynamic limit*. The simplest way of taking this limit is to double the side of the cube Ω repeatedly, adjusting N at each step to the value $\rho\Omega$ where ρ, the density, is a constant, and the symbol Ω is used to represent the volume of the cube as well as the cube itself. The thermodynamic free-energy density, a function of ρ and γ, is then defined as

$$a(\rho, \gamma) \equiv \lim_{\Omega\to\infty} A(\rho\Omega, \Omega, \gamma)/\Omega \qquad (1.13)$$

where $A(N, \Omega, \gamma)$ may be defined for nonintegral N by linear interpolation.[7]

The next operation is to take the van der Waals limit $\gamma \to 0$. This gives the *van der Waals free-energy density*

$$a(\rho, 0+) \equiv \lim_{\gamma\to 0} a(\rho, \gamma)$$
$$= \lim_{\gamma\to 0} \lim_{\Omega\to\infty} A(\rho\Omega, \Omega, \gamma)/\Omega. \qquad (1.14)$$

It is important to take these two limits in the right order. Taking the limit $\Omega \to \infty$ first, as in (14), means that the range of the Kac potential, although very large, is much less than the size of the container. If they are taken in the opposite order, then the Kac potential has a range much larger than the size of the container and in consequence of (8) its effect disappears. In fact, by applying (8) to (10) and (11) we obtain

$$\lim_{\gamma\to 0} A(N, \Omega, \gamma) = A^0(N, \Omega), \qquad (1.15)$$

where A^0 is the free energy for N particles of the *reference system* in a cube Ω. By reference system we mean one for which the interaction potential is $q(\mathbf{r})$ instead of $q(\mathbf{r}) + w(\mathbf{r}, \gamma)$. Taking the thermodynamic limit of (15) we obtain, in constrast to (14), the formula

$$\lim_{\Omega\to\infty} \lim_{\gamma\to 0} A(\rho\Omega, \Omega, \gamma)/\Omega = a^0(\rho) \qquad (1.16)$$

where

$$a^0(\rho) \equiv \lim_{\Omega\to\infty} A^0(\rho\Omega, \Omega)/\Omega \qquad (1.17)$$

is the thermodynamic free-energy density of the reference system at particle density ρ, and is not in general equal to $a(\rho, 0+)$.

The upper and lower bounds on $A(N, \Omega, \gamma)$ which we shall derive lead to upper and lower bounds on $a(\rho, 0+)$. These bounds depend on the volume ω of the cells ω_i used. The bounds can be simplified by a third limit process $\omega \to \infty$, after which, in favorable cases, the upper and lower bounds are equal so that $a(\rho, 0+)$ can be calculated. This triple limit process corresponds to the following relationships:

$$r_0 \ll \omega^{1/\nu} \ll \gamma^{-1} \ll \Omega^{1/\nu} \qquad (1.18)$$

among the four characteristic lengths of our calculation: the range of the short-range potential, the size of the cells, the range of the Kac potential, and the size of the container.

Once $a(\rho, 0+)$ has been found, the equation of state can be calculated by differentiation. The main result of our work is to show rigorously that under suitable conditions the equation of state is indeed given by Maxwell's rule (2) applied to the generalized van der Waals isotherm

$$\tilde{p}_{vdw}(\rho, T) = p^0(\rho, T) + \tfrac{1}{2}\alpha\rho^2 \qquad (1.19)$$

where $p^0(\rho, T)$ is the pressure of the reference system, calculated by differentiation from $a^0(\rho)$. This result can be used to investigate the conditions under which the system will have a first-order phase transition in the van der Waals limit. A further result is to show, by studying the two-particle distribution function, that if there is such a first-order phase transition then two phases of different densities are present during the transition.

For rigorous arguments it is necessary to impose conditions on the short- and long-range potentials. We shall assume that the short-range potential satisfies

$$q(\mathbf{r}) = +\infty \quad \text{for} \quad r < r_0, \qquad (1.20a)$$

$$|q(\mathbf{r})| < D_2 r^{-\nu-\epsilon} \quad \text{for} \quad r_0 < r, \qquad (1.20b)$$

where r_0 (the hard-core diameter), D_2, and ϵ are positive constants; and we shall also assume that the shape function $\varphi(\mathbf{x})$ of the Kac potential satisfies

$$|\varphi(\mathbf{r})| < D_3 r^{-\nu-\epsilon} \quad \text{for all} \quad r, \qquad (1.21a)$$

$$\varphi(\mathbf{r}) \quad \text{is continuous at} \quad r = 0, \qquad (1.21b)$$

$$\int \varphi(\mathbf{r})\, d\mathbf{r} \text{ exists as a Riemann integral.} \qquad (1.21c)$$

Further, when in Sec. V the function $\varphi(\mathbf{r})$ is ex-

[7] M. Fisher, Arch. Rat. Mech. Anals. 17, 377 (1964).

pressed as the sum of two parts satisfying the conditions (5.2), both parts are required to satisfy (1.21). The extent to which the conditions (1.20) and (1.21) can be weakened will be discussed in Sec. IX, along with the possibilities for generalizing this work to quantum systems and to lattice gases.

II. UPPER BOUND ON THE FREE ENERGY

Finding an upper bound on the free energy is equivalent to finding a lower bound on the partition function. To obtain a lower bound, we divide the cube Ω into M smaller cubical regions $\omega_1 \cdots \omega_M$, each of side $(s + t)$ where s and t are positive lengths such that $(s + t)$ is a submultiple of the side of Ω (see Fig. 2). Since the small cubes completely fill Ω its volume, which we also call Ω, is given by

$$\Omega = M(s + t)^\nu. \tag{2.1}$$

For each ω_i, let ω_i' be the cube of side s consisting of all points within ω_i whose distance from the boundary of ω_i is at least $\frac{1}{2}t$. A lower bound on $Z(N, \Omega, \gamma)$ can be obtained by selecting any set of integers $N_1, N_2, \cdots N_M$ which add up to N, and considering only the contribution to the integral in (1.11) from configurations where there are N_1 particles in the cell ω_1', N_2 in ω_2', and so on. The quantity so calculated we denote by $\tilde{Z}(N_1, N_2 \cdots N_M)$. There are $N!/(N_1!N_2! \cdots N_M!)$ ways of choosing the N_1 particles to go in the cell ω_1', N_2 to go in ω_2', etc., and since the particles are identical all these different ways give the same contribution to the integral. Multiplying a typical contribution by $N!/\prod N_i!$ we thus obtain

$$\begin{aligned} Z &\geq \tilde{Z}(N_1, N_2 \cdots N_M) \\ &= \prod_i [(1/N_i!)(m/2\pi\hbar^2\beta)^{\nu N_i/2}] \\ &\quad \times \int_{\omega'_1} \cdots \int_{\omega'_M} e^{-\beta V}\, d\mathbf{x}_1 \cdots d\mathbf{x}_N, \end{aligned} \tag{2.2}$$

where the first N_1 of the N ν-fold integrations are taken over the region ω_1', the next N_2 over ω_2', and so on.

To obtain a lower bound on the integral in (2.2) we write

$$V = V' + V'' \tag{2.3}$$

where V' is the contribution to the total potential energy from pairs of particles that are both in the same cell, and V'' is the contribution from pairs that are in different cells. If $V''_{\max}$ is some upper bound on V'', then (2) implies

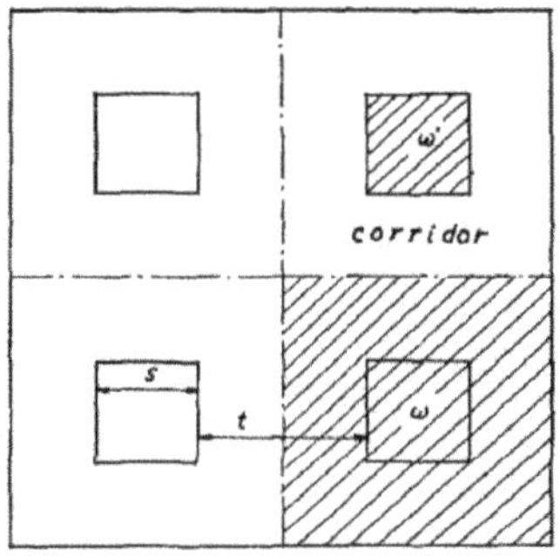

FIG. 2. Division of Ω into cells.

$$\begin{aligned} \tilde{Z} &\geq \prod_i [(1/N_i!)(m/2\pi\hbar^2\beta)^{\nu N_i/2}] \\ &\quad \times \int_{\omega'_1} \cdots \int_{\omega'_M} e^{-\beta V' - \beta V''_{\max}}\, d\mathbf{x}_1 \cdots d\mathbf{x}_N \\ &= [\prod_i Z(N_i, \omega', \gamma)] e^{-\beta V''_{\max}}, \end{aligned} \tag{2.4}$$

where $Z(N_i, \omega', \gamma)$ is the partition function for N_i particles in a cube ω', of side s.

To obtain a lower bound on the exponential factor in (4) we consider the short- and long-range contributions separately, writing

$$V'' = Q'' + W''. \tag{2.5}$$

A convenient upper bound for W'', the long-range contribution to V'', is

$$W'' \leq \sum_{i<j} N_i N_j w_{\max}(\mathbf{k}_{ij}) \tag{2.6}$$

where

$$\begin{aligned} w_{\max}(\mathbf{k}_{ij}) &\equiv \operatorname*{Max}_{\substack{\mathbf{x} \in \omega_i \\ \mathbf{y} \in \omega_j}} w(\mathbf{x} - \mathbf{y}, \gamma) \\ &= \operatorname*{Max}_{\mathbf{r} \in \omega_0} w(\mathbf{k}_{ij} + 2\mathbf{r}, \gamma) \end{aligned} \tag{2.7}$$

where ω_0 is a cube of side $(s + t)$ centered at the origin, and $\mathbf{k}_{ij}$ is the vector from the center of ω_i to that of ω_j. For an upper bound on Q'', the short-range contribution to V'', we may use the same method, in conjunction with the condition (1.20), and obtain, if $t > r_0$, the upper bound

$$\begin{aligned} Q'' &\leq D_2 \sum_{i<j} N_i N_j (r_{ij,\min})^{-\nu-\epsilon} \\ &= \tfrac{1}{2} D_2 \sum_i N_i \sum_j{}' N_j (r_{ij,\min})^{-\nu-\epsilon}, \end{aligned} \tag{2.8}$$

where $r_{ij,\min}$ is the least distance from ω_i' to ω_j' and $\sum_j'$ means a sum with the $j = i$ term omitted. To estimate this last sum we may group the cubes ω_j' into shells centered on ω_i'. The first shell contains

$3^\nu - 1$ cubes, all points of which are a distance at least t from all points of ω_i', the next shell contains $5^\nu - 3^\nu$ cubes, each one at least $2t + s$ from ω_i'; and so on. Writing also N_{max} for the maximum value of N_i, we obtain

$$Q'' \leq \tfrac{1}{2}D_2 \sum_i N_i \sum_{n=1}^{\infty} N_{max}[(2n+1)^\nu - (2n-1)^\nu]/(nt + (n-1)s)^{\nu+\epsilon}$$
$$\leq \tfrac{1}{2}D_2 N N_{max} J_\nu t^{-\nu-\epsilon} \tag{2.9}$$

where

$$J_\nu \equiv \sum_{n=1}^{\infty} n^{-\nu-\epsilon}[(2n+1)^\nu - (2n-1)^\nu] < \infty. \tag{2.10}$$

Substituting (9) and (6) into (5), (4), and (2), we obtain

$$A(N, \Omega, \gamma) \leq \sum_i A(N_i, \omega', \gamma)$$
$$+ \sum_{i<j} N_i N_j w_{max}(\mathbf{k}_{ij}) + \tfrac{1}{2}D_2 J_\nu N N_{max} t^{-\nu-\epsilon}, \tag{2.11}$$

where $A(N, \Omega, \gamma)$ is defined in (1.10). The upper bound (11) holds for any choice of $N_1, N_2, \cdots N_M$ whose sum is N, and in particular for the choice which minimizes the right side.

To obtain an upper bound on the thermodynamic free energy in the van der Waals limit, we apply the triple limit process described in Sec. 1 to the inequality (11). This is simplest for the case where all of $N_1, N_2 \cdots N_M$ are equal:

$$N_1 = N_2 = \cdots = N_M = \rho(s+t)^\nu. \tag{2.12}$$

Replacing $\sum_{i<j}$ in (2.11) by $\frac{1}{2}\sum_i \sum_j'$ and dividing both sides by Ω we obtain

$$A(\rho\Omega, \Omega, \gamma)/\Omega \leq [M/\Omega] A(\rho(s+t)^\nu, \omega', \gamma)$$
$$+ \tfrac{1}{2}[\rho^2(s+t)^{2\nu}/\Omega] \sum_i \sum_j' w_{max}(\mathbf{k}_{ij})$$
$$+ \tfrac{1}{2}D_2 J_\nu \rho^2 (s+t)^\nu t^{-\nu-\epsilon}. \tag{2.13}$$

To take the limit $\Omega \to \infty$ we require the following lemma:

$$\lim_{\Omega\to\infty} 1/M \sum_i \sum_j' w_{max}(\mathbf{k}_{ij}) = \sum_{\mathbf{k}}' w_{max}(\mathbf{k}) \tag{2.14}$$

where $\sum_{\mathbf{k}}'$ is an infinite sum over the complete infinite lattice of possible vectors $\mathbf{k}_{ij}$ except $\mathbf{k} = 0$. To prove (14) it is sufficient to show that the difference between its two sides vanishes. This difference can be written as the limiting value, for large M, of the expression

$$1/M \sum_i \sum_j'' w_{max}(\mathbf{k}_{ij}) \tag{2.15}$$

with $\sum_j''$ meaning the sum over an infinite network of cells outside Ω, continuing the pattern established by the cells inside Ω, but not including the cells inside Ω. Let δ be any small positive number. By (1.6) and (1.21a), the infinite sum on the right of (14) converges absolutely, and therefore we can find a number σ such that

$$\sum_{\substack{\mathbf{k} \\ |\mathbf{k}|>\sigma}}' |w_{max}(\mathbf{k})| < \tfrac{1}{2}\delta. \tag{2.16}$$

The sum over i in (15) may be divided into two parts, the first part including all those cells whose distance from the boundary of Ω is greater than σ, and the second, those for which it is not. For each value of i in the first part of the i-summation, the sum over j covers only a subset of the values of k covered in the sum (15), and hence this part of the i-summation contributes at most $\frac{1}{2}\delta$ to the expression (16). In the second part of the i-summation, the number of terms is at most $[\Omega - (\Omega^{1/\nu} - 2\sigma)^\nu]/(s+t)^\nu$, and this number does not exceed $2\nu\sigma\,\Omega^{1-1/\nu}(s+t)^{-\nu}$ because

$$x^\nu - y^\nu \leq \nu x^{\nu-1}(x-y) \quad \text{if} \quad x \geq y \geq 0. \tag{2.17}$$

Consequently, by (1), the second part of the i-summation contributes to the expression (15) an amount not exceeding $2\nu\sigma\,\Omega^{-1/\nu} \sum' |w_{max}(\mathbf{k})|$, which can be made $< \frac{1}{2}\delta$ by making Ω large enough. Thus the complete expression (15) is less than δ for all large Ω, and since δ is arbitrarily small, the result (14) follows from the definition of a limit, Q.E.D.

Taking the thermodynamic limit of (13) with the help of (1) and (14) we obtain

$$a(\rho, \gamma) \leq A(\rho(s+t)^\nu, \omega', \gamma)/(s+t)^\nu$$
$$+ \tfrac{1}{2}\rho^2(s+t)^\nu \sum_{\mathbf{k}}' w_{max}(\mathbf{k})$$
$$+ \tfrac{1}{2}D_2 J_\nu \rho^2 (s+t)^\nu / t^{\nu+\epsilon}. \tag{2.18}$$

The second limiting process is the van der Waals limit $\gamma \to 0$. To evaluate the van der Waals limit of the middle term on the right in (18) we substitute from (7) and (1.6), obtaining

$$(s+t)^\nu \sum_{\mathbf{k}}' w_{max}(\mathbf{k}) = 2^{-\nu} \sum_{\mathbf{n}}' \Delta \operatorname*{Max}_{\mathbf{x}\in\Delta_\mathbf{n}} \varphi(\mathbf{x}) \tag{2.19}$$

where the sum goes over all nonzero vectors $\mathbf{n}$ with integer components. $\Delta_\mathbf{n}$ stands for the cube of side $2\gamma(s+t)$ centered at the point $\gamma(s+t)\mathbf{n}$ with its sides parallel to those of Ω, and $\Delta \equiv 2^\nu\gamma^\nu(s+t)^\nu$ is the volume of one of the cubes $\Delta_\mathbf{n}$. The network of overlapping cubes $\Delta_\mathbf{n}$ can be disentangled into 2^ν separate nonoverlapping networks, each of them just filling ν-dimensional space (except for one network where the cube centered on the origin is missing). In the limit $\gamma \to 0$ these networks become infinitesimally fine and, by Riemann's definition of an integral,[8] the contribution of each of them to the second sum in (19) tends to $\int \varphi(\mathbf{x})\, d\mathbf{x}$; the van der Waals limit of (2.19) is therefore

$$(s+t)^{\nu} \lim_{\gamma\to 0} \sum_{\mathbf{k}}{}' w_{\max}(\mathbf{k}) = \int \varphi(\mathbf{x})\, d\mathbf{x} \equiv \alpha. \tag{2.20}$$

Using this result, and also (1.15), in (18) we obtain

$$a(\rho, 0+) \le A^0(\rho(s+t)^{\nu}, \omega')/(s+t)^{\nu} + \tfrac{1}{2}\rho^2\alpha + \tfrac{1}{2}D_2J_{\nu}\rho^2(s+t)^{\nu}/t^{\nu+\epsilon}. \tag{2.21}$$

The final limiting process is to make the cell size infinite by making $s \to \infty$. As Fisher has shown[7] the last term in (21), which represents the short-range interactions between cells, can be eliminated if we make t depend on s in such a way that

$$t/s \to 0 \quad \text{and} \quad s^{\nu}/t^{\nu+\epsilon} \to 0 \quad \text{as} \quad s \to \infty. \tag{2.22}$$

This can be done, for example, by making $t \propto s^{\eta}$ where η is a constant satisfying

$$\nu/(\nu+\epsilon) < \eta < 1. \tag{2.23}$$

Applying this limit on both sides of (21) and using the continuity of $a^0(\rho)$, we obtain, since the volume of the cell ω' is s^{ν},

$$a(\rho, 0+) \le a^0(\rho) + \tfrac{1}{2}\rho^2\alpha. \tag{2.24}$$

That is, in the van der Waals limit the increase in free-energy density brought about by introducing a Kac potential into a system with short-range forces cannot exceed the increase that one would calculate by treating the particles as a smoothed-out uniform medium.

The result (2.24) can be strengthened in some cases by using the fact[7,9] that $a(\rho, \gamma)$ is a convex function[10] of ρ, so that $a(\rho, 0+)$, being the limit of a sequence of convex functions, is also convex. Because of this (2.24) implies

$$a(\rho, 0+) \le \mathrm{CE}\,\{a^0(\rho) + \tfrac{1}{2}\alpha\rho^2\} \tag{2.25}$$

where CE $\{f(\rho)\}$ means, for any function $f(\rho)$, the *convex envelope* of that function, defined as the maximal convex function not exceeding f:

$$\mathrm{CE}\,\{f(\rho)\} \equiv \mathrm{Max}\,\phi(\rho) \quad \text{for each value of } \rho \quad \left\{\begin{array}{l} \phi(\cdot) \\ \phi(\cdot) \quad \text{is convex} \\ \phi(\xi) \le f(\xi) \quad \text{for all} \quad \xi \end{array}\right\}. \tag{2.26}$$

Since the maximum of any family of convex functions is[11] itself convex, the function CE $\{f(\rho)\}$ is convex. If $f(\rho)$ is convex, then CE $\{f(\rho)\}$ and $f(\rho)$ coincide; otherwise the graph of CE $\{f(\rho)\}$ consists partly of convex segments of the graph of $f(\rho)$ and partly of segments of double tangents of this graph (Fig. 3). The construction of CE $\{f(\rho)\}$ from $f(\rho)$ is sometimes called the *double tangent construction.*

It will be shown in Sec. VI that the replacement of $a^0(\rho) + \frac{1}{2}\alpha\rho^2$ by its convex envelope is equivalent to the replacement of van der Waals' equation of state by Maxwell's modification (1.2).

III. LOWER BOUNDS ON THE FREE ENERGY: NONNEGATIVE-DEFINITE KAC POTENTIALS

A lower bound on the free energy is most easily found when the shape function $\varphi(\mathbf{x})$ defining the Kac potential (1.6) is *nonnegative definite*: that is to say, when its ν-dimensional Fourier transform

$$\Phi(\mathbf{p}) \equiv \int \varphi(\mathbf{x}) \exp(2\pi i\mathbf{p}\cdot\mathbf{x})\, d\mathbf{x} \tag{3.1}$$

is nonnegative. In this case it is possible to find a lower bound $W_{\min}$ on the long-range contribution

$$W \equiv \sum_{i<j} \gamma^{\nu}\varphi[\gamma(\mathbf{x}_i - \mathbf{x}_j)] = \tfrac{1}{2}\sum_i \sum_j \gamma^{\nu}\varphi[\gamma(\mathbf{x}_i - \mathbf{x}_j)] - \tfrac{1}{2}N\gamma^{\nu}\varphi(0) \tag{3.2}$$

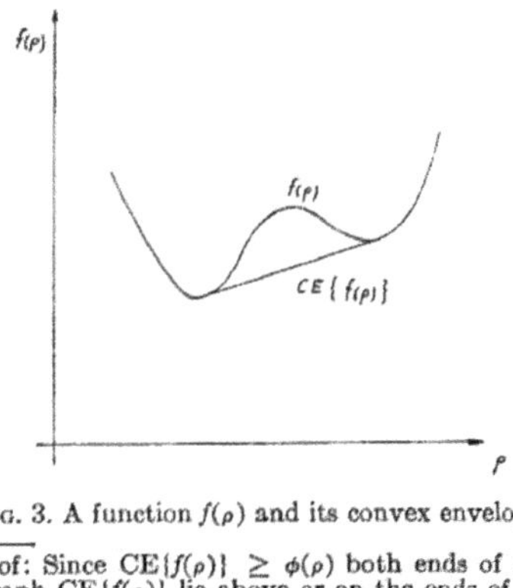

FIG. 3. A function $f(\rho)$ and its convex envelope.

[8] W. Rogosinski, *Volume and Integral* (Oliver and Boyd, London, 1952), Theorem 58.

[9] D. Ruelle, Helv. Phys. Acta **36**, 183 (1963).

[10] G. Hardy, J. E. Littlewood, and G. Polya, *Inequalities* (Cambridge University Press, London and New York, 1959), Chap. 3.

[11] Proof: Since CE$\{f(\rho)\} \ge \phi(\rho)$ both ends of any chord of the graph CE$\{f(\rho)\}$ lie above or on the ends of the corresponding (same end ordinates) chord of any $\phi(\rho)$ and hence above or on the graph of $\phi(\rho)$ itself; consequently the chord of CE$\{f(\rho)\}$ also lies above or on the curve Max $\phi(\rho) \equiv$ CE$\{f(\rho)\}$.

to the total potential energy V, and by substituting such a lower bound into the basic definitions (1.10) and (1.11) we obtain

$$A(N, \Omega, \gamma) \geq A^0(N, \Omega) + W_{\min}. \quad (3.3)$$

To find a suitable $W_{\min}$ we substitute the Fourier inversion formula of (1) into (2), obtaining

$$W = -\tfrac{1}{2}N\gamma^\nu\varphi(0) + \tfrac{1}{2}\gamma^\nu \int d\mathbf{p}\, \Phi(\mathbf{p}) \left| \sum_{j=1}^{N} \exp(2\pi i\gamma\mathbf{p}\cdot\mathbf{x}_j)\right|^2 \quad (3.4)$$

for almost all configurations $\mathbf{x}_1 \cdots \mathbf{x}_N$. The "exceptional" configurations for which formulas such as (3.4) may fail will be ignored since they form a set of zero measure and therefore do not affect the configurational integral in (1.11). Let δ be any positive number. Since $\Phi(\mathbf{p})$ is[12] a continuous function of $\mathbf{p}$, and $\Phi(0) \geq 0$, we can find a positive number p_0 (depending on δ) such that

$$\Phi(\mathbf{p}) \geq (1 - \delta)\Phi(0) \quad (3.5)$$

whenever p is inside a cube of side $2p_0$ centered at the origin. Since $\Phi(\mathbf{p}) \geq 0$ when $\mathbf{p}$ is outside this cube, it follows from (4) that

$$W + \tfrac{1}{2}N\gamma^\nu\varphi(0) \geq \tfrac{1}{2}(1-\delta)\Phi(0)\gamma^\nu \times \int d\mathbf{p} \left|\Psi(\mathbf{p}) \sum_{j=1}^{N} \exp(2\pi i\gamma\mathbf{p}\cdot\mathbf{x})\right|^2 \quad (3.6)$$

where

$$\Psi(\mathbf{p}) \equiv \begin{cases} \prod_{n=1}^{\nu} (1 - |p_n|/p_0) & \text{if } |p_n| < p_0 \text{ for all } n, \\ 0 & \text{otherwise,} \end{cases} \quad (3.7)$$

and $p_1 \cdots p_\nu$ are the components of the vector $\mathbf{p}$. Applying Parseval's theorem to (6) we obtain

$$W + \tfrac{1}{2}N\gamma^\nu\varphi(0) \geq (1-\delta)\Phi(0)\tfrac{1}{2}\gamma^\nu \times \int d(\gamma\mathbf{x})\left\{\sum_{j=1}^{N} \psi[\gamma(\mathbf{x}-\mathbf{x}_j)]\right\}^2 \quad (3.8)$$

where

$$\psi(\mathbf{y}) \equiv \int \Psi(\mathbf{p}) \exp(-2\pi i\mathbf{p}\cdot\mathbf{y})\, d\mathbf{p} = \prod_{n=1}^{\nu} \frac{\sin^2(\pi y_n p_0)}{\pi^2 y_n^2 p_0} \quad (3.9)$$

and $y_1 \cdots y_\nu$ are the components of the vector $\mathbf{y}$.

A lower bound on the integral in (8) can be obtained by confining the integration to a cube Ω^* concentric with Ω and similarly oriented but having a side of length $\Omega^{1/\nu} + l$ where l is an arbitrary positive quantity. Applying the Schwartz inequality to the resulting integral and writing Ω^* for the volume of the cube Ω^* we obtain

$$\begin{aligned} W + \tfrac{1}{2}N\gamma^\nu\varphi(0) &\geq \tfrac{1}{2}(1-\delta)\Phi(0)\left[\sum_{j=1}^{N}\int_{\Omega^*} \psi[\gamma(\mathbf{x}-\mathbf{x}_j)]\, d(\gamma\mathbf{x})\right]^2/\Omega^* \\ &\geq \tfrac{1}{2}(1-\delta)\Phi(0)\left[N\int_{y<\gamma l} \psi(\mathbf{y})\, dy\right]^2/\Omega^* \end{aligned} \quad (3.10)$$

since the integrand is nonnegative and every point $\mathbf{x}$ inside the sphere $|\mathbf{x} - \mathbf{x}_j| < l$ is also within Ω^*.

Let us choose l so that $l \to \infty$ and $l/\Omega^{1/\nu} \to 0$ in the thermodynamic limit; for example, we may take $l \propto \Omega^{1/2\nu}$. Then in this limit we have $\Omega^*/\Omega \to 1$ and $\int_{y<\gamma l} \psi(\mathbf{y})\, dy \to 1$, and the lower bound on W/Ω implied by (10) has the thermodynamic limit $-\tfrac{1}{2}\rho\gamma^\nu\varphi(0) + \tfrac{1}{2}(1-\delta)\Phi(0)\rho^2$.

Since this result holds for arbitrarily small δ, it is also true in the limit $\delta \to 0$; that is to say, there exists a sequence of lower bounds on W, call them $W_{\min}$, with the property[13]

$$\begin{aligned} \lim_{\Omega\to\infty} W_{\min}/\Omega &= -\tfrac{1}{2}\rho\gamma^\nu\varphi(0) + \tfrac{1}{2}\Phi(0)\rho^2 \\ &= -\tfrac{1}{2}\rho\gamma^\nu\varphi(0) + \tfrac{1}{2}\alpha\rho^2 \end{aligned} \quad (3.11)$$

in the notation of (1.8).

This result enables us to take the thermodynamic limit of (3), obtaining

$$a(\rho, \gamma) \geq a^0(\rho) - \tfrac{1}{2}\rho\gamma^\nu\varphi(0) + \tfrac{1}{2}\alpha\rho^2 \quad (3.12)$$

for nonnegative-definite potentials. In the van der Waals limit this formula reduces to

$$a(\rho, 0+) \geq a^0(\rho) + \tfrac{1}{2}\alpha\rho^2 \quad (3.13)$$

which when combined with (2.24) gives

$$a(\rho, 0+) = a^0(\rho) + \tfrac{1}{2}\alpha\rho^2 \quad (3.14)$$

for nonnegative-definite Kac potentials.

IV. LOWER BOUND ON THE FREE ENERGY: NONPOSITIVE KAC POTENTIALS

Another case where a lower bound on the free energy can be found fairly easily arises when the Kac potential is nonpositive; that is when

$$\varphi(\mathbf{x}) \leq 0 \quad \text{for all} \quad \mathbf{x}. \quad (4.1)$$

In this case it is again necessary to divide the cube Ω into cubical cells $\omega_1 \cdots \omega_M$, each of side $s + t$

[12] This follows from the existence of $\int |\varphi(\mathbf{x})|\, d\mathbf{x}$.

[13] A proof of (3.11) for systems with periodic boundary conditions was given by E. Lieb, Phys. Rev. **130**, 2518 (1963). We are indebted to E. Newman and M. Austin for advice on the proof presented here.

(see Fig. 2). The definition (1.11) of the partition function may be written

$$Z(N, \Omega, \gamma) = \sum_{N_1 \cdots N_M} Z(N_1, \cdots, N_M) \qquad (4.2)$$

where the sum is over all sets of M nonnegative integers adding up to N and $Z(N_1, \cdots, N_M)$ means the contribution to (1.11) from configurations with exactly N_i particles in cell $\omega_i (i = 1, 2 \cdots M)$. Since there are $(N + M - 1)!/N!(M - 1)!$ terms in the sum it has the upper bound

$$Z(N, \Omega, \gamma) \leq [(N + M - 1)!/N!\,(M - 1)!] \times \operatorname*{Max}_{N_1 \cdots N_M} Z(N_1, \cdots, N_M), \qquad (4.3)$$

the maximum being taken over all sets of nonnegative integers $N_1 \cdots N_M$ which add up to N. The combinatorial argument which led to (2.2) gives, when applied to $Z(N_1, \cdots, N_M)$, the formula

$$Z(N_1, \cdots, N_M) = \prod_i [(1/N_i!)(m/2\pi\hbar^2\beta)^{\nu N_i/2}] \times \int_{\omega_1} \cdots \int_{\omega_M} e^{-\beta V}\, d\mathbf{x}_1 \cdots d\mathbf{x}_N \qquad (4.4)$$

where the first N_1 of the N integrations are over the cell ω_1, [not ω_1' as in (2.2)] the next N_2 over ω_2, and so on.

To obtain an upper bound on the integral in (4) we separate the potential energy V into three parts:

$$V = Q' + \tilde{Q} + W \qquad (4.5)$$

where Q' is the contribution to V from short-range interactions between particles that are in the same cell, $\tilde{Q}$ the contribution from short-range interactions between particles that are in different cells, and W is the total contribution from long-range interactions. If $\tilde{Q}_{\min}$ and $W_{\min}$ are lower bounds on $\tilde{Q}$ and W, then (4) and (5) lead to the inequality, analogous to (2.4),

$$Z(N_1, \cdots, N_M) \leq \{\prod_i Z^0(N_i, \omega)\} \times \exp[-(\tilde{Q}_{\min} + W_{\min})/kT] \qquad (4.6)$$

where $Z^0(N_i, \omega)$ is the partition function for N_i particles of the reference system in a cube of side $s + t$.

To find a suitable lower bound on $\tilde{Q}$, we split it into two parts:

$$\tilde{Q} = Q'' + Q''' \qquad (4.7)$$

where Q'' is, as in (2.5), the contribution to $\tilde{Q}$ from pairs of particles whose centers are both within the inner cells of side s denoted by $\omega_i' (i = 1, 2 \cdots M)$ in Sec. 2, and Q''' is the contribution of pairs of particles, at least one of which is in the "corridor" consisting of points that are not in any of the ω_i' (see Fig. 2). The argument used to prove (2.9) proves at the same time that

$$Q'' \geq -\tfrac{1}{2} D_2 N N_{\max} J_\nu t^{-\nu-\epsilon}. \qquad (4.8)$$

The contribution Q''' has the lower bound

$$Q''' \geq -N_{\text{corr}} \Phi' \qquad (4.9)$$

where $-\Phi'$ is a lower bound on the interaction of a given particle with all its neighbors, whose existence is a consequence[14] of (1.20), and N_{corr} is the number of particles in the corridor. It can be shown[15] that the number of particles in a given region, multiplied by the reciprocal close-packing density ρ_0^{-1}, does not exceed the volume occupied by all points whose distance from the region is at most r_0. Hence, $N_{\max}$ and N_{corr} have (provided $s \geq 2r_0$,) the upper bounds

$$N_{\max}\rho_0^{-1} < (s + 2r_0)^\nu \qquad (4.10)$$

$$\begin{aligned} N_{\text{corr}}\rho_0^{-1} &\leq (\Omega^{1/\nu} + 2r_0)^\nu - M(s - 2r_0)^\nu \\ &= M\{(s + t + 2r_0M^{-1/\nu})^\nu - (s - 2r_0)^\nu\} \\ &\leq \nu M(s + t + 2r_0M^{-1/\nu})^{\nu-1} \\ &\qquad \times [t + 2r_0(1 + M^{-1/\nu})] \end{aligned} \qquad (4.11)$$

by (2.1) and (2.17).

Substituting (10) and (11) into (7), (8), and (9), we obtain

$$\tilde{Q} \geq \tilde{Q}_{\min}, \qquad (4.12)$$

where

$$\begin{aligned} \tilde{Q}_{\min} = &-\tfrac{1}{2} D_2 N \rho_0 (s + 2r_0)^\nu J_\nu t^{-\nu-\epsilon} \\ &- \rho_0 \nu M(s + t + 2r_0M^{-1/\nu})^{\nu-1} \\ &\times [t + 2r_0(1 + M^{-1/\nu})]\Phi'. \end{aligned} \qquad (4.13)$$

The details of this formula are unimportant; all that is important is its behavior under the triple limiting process described in the introduction. This behavior is

$$\begin{aligned} &\lim_{s\to\infty} \lim_{\gamma\to 0} \lim_{\Omega\to\infty} (\tilde{Q}_{\min}/\Omega) \\ &= \lim_{s\to\infty} \lim_{\gamma\to 0} \{-\tfrac{1}{2} D_2 \rho\rho_0 (s + 2r_0)^\nu J_\nu t^{-\nu-\epsilon} \\ &\qquad - \nu(s + t)^{-1}(t + 2r_0)\Phi'\} \\ &= 0 \end{aligned} \qquad (4.14)$$

by (2.1) and (2.22).

[14] O. Penrose, J. Math. Phys. 4, 1312 (1963).
[15] O. Penrose, Phys. Letters 11, 224 (1964).

A simple lower bound on W, the total long-range interaction, is given by

$$W \geq \tfrac{1}{2} \sum_i \sum_j N_i N_j w_{\min}(\mathbf{k}_{ij}) \qquad (4.15)$$

where

$$w_{\min}(\mathbf{k}_{ij}) \equiv \underset{\substack{\mathbf{x}\in\omega_i \\ \mathbf{y}\in\omega_j}}{\mathrm{Min}}\, w(\mathbf{x}-\mathbf{y},\gamma) = \underset{\mathbf{r}\in\omega_0}{\mathrm{Min}}\, w(\mathbf{k}_{ij}+2\mathbf{r},\gamma). \qquad (4.16)$$

the γ-dependence of $w_{\min}(\mathbf{k})$ not being explicitly shown. These formulas are analogous to (2.6) and (2.7) but the $i = j$ terms are now included. Since $N_i N_j \leq \frac{1}{2}N_i^2 + \frac{1}{2}N_j^2$, and $w_{\min}(\mathbf{k}_{ij}) \leq 0$ by (1), we may deduce from (4.15) the inequality

$$W \geq -\tfrac{1}{2} \sum_i \sum_j (\tfrac{1}{2}N_i^2 + \tfrac{1}{2}N_j^2)\, |w_{\min}(\mathbf{k}_{ij})|$$

$$= -\tfrac{1}{2} \sum_i N_i^2 \sum_j |w_{\min}(\mathbf{k}_{ij})|. \qquad (4.17)$$

The sum over j may be extended, without destroying the validity of (17), to include the infinite network of cells continuing outside Ω the pattern established inside it by the cells $\omega_1 \cdots \omega_M$. In the notation of (2.14) the resulting inequality is $W \geq W_{\min}$ with

$$W_{\min} = \tfrac{1}{2} \sum_i N_i^2 \sum_{\mathbf{k}} w_{\min}(\mathbf{k}). \qquad (4.18)$$

Substituting this into (6) and using (3) and (1.10), we obtain

$$A(N,\Omega,\gamma) \geq kT \log [N!\,(M-1)!/(N+M-1)!] + \underset{N_1\cdots N_M}{\mathrm{Min}} \sum_{i=1}^{M} \{A^0(N_i,\omega) + \tfrac{1}{2}N_i^2 \sum_{\mathbf{k}} w_{\min}(\mathbf{k})\} + \tilde{Q}_{\min}. \qquad (4.19)$$

The second term on the right can be simplified by means of a simple property of the convex envelope of a function, defined in (2.27). This property is

$$M^{-1} \sum_{i=1}^{M} f(N_i) \geq M^{-1} \sum_i \mathrm{CE}\,\{f(N_i)\} \geq \mathrm{CE}\,\{f(M^{-1} \sum_i N_i)\} \qquad (4.20)$$

where the first inequality follows from the fact that CE $\{f\}$ is a lower bound on f, and the second from the fact that CE $\{f\}$ is convex.

Using (4.20) in (4.19) we obtain

$$A(N,\Omega,\gamma) \geq kT \log [N!\,(M-1)!/(N+M-1)!] + M\,\mathrm{CE}\,\{A^0(N/M,\omega) + \tfrac{1}{2}(N/M)^2 \sum_{\mathbf{k}} w_{\min}(\mathbf{k})\} + \tilde{Q}_{\min} \qquad (4.21)$$

where CE $\{\ \}$ is the convex envelope of the quantity in braces regarded as a function of N/M.

To apply the triple limiting process described in Sec I we first divide (4.21) by Ω and take the thermodynamic limit $\Omega \to \infty$, using Stirling's approximation for the factorials. In the notation of (1.13) the result is

$$a(\rho,\gamma) \geq -kT\{(s+t)^{-\nu} \log [1 + \rho(s+t)^{\nu}] + \rho \log [1 + \rho^{-1}(s+t)^{-\nu}]\} + \mathrm{CE}\,\{(s+t)^{-\nu} A^0(\rho(s+t)^{\nu},\omega) + \tfrac{1}{2}\rho^2(s+t)^{\nu} \sum w_{\min}(\mathbf{k})\} + \lim_{\Omega\to\infty} \tilde{Q}_{\min}/\Omega \qquad (4.22)$$

since

$$N/M = \rho\Omega/M = \rho(s+t)^{\nu}.$$

In order to carry through the other two limit processes we require the following lemma:

Let $f_n(\xi)$ be a sequence of functions converging uniformly on an interval to $f(\xi)$ as $n \to \infty$; then we have

$$\lim_{n\to\infty} \mathrm{CE}\,\{f_n(\xi)\} = \mathrm{CE}\,\{f(\xi)\}. \qquad (4.23)$$

Proof: Given any $\delta > 0$ there exists an $\mathfrak{N}$ such that for $n \geq \mathfrak{N}$, $|f_n(\xi) - f(\xi)| < \delta$ for all ξ in the interval. We then have from the definition (2.26) of the convex envelope

$$\mathrm{CE}\,\{f_n(\xi)\} \geq \mathrm{CE}\,\{f(\xi)\} - \delta \quad \text{for all } \xi \text{ in the interval and } n \geq \mathfrak{N}$$

since CE $\{f(\xi)\} - \delta$ is a convex function which is $\leq f_n(\xi)$. Similarly

$$\mathrm{CE}\,\{f(\xi)\} \geq \mathrm{CE}\,\{f_n(\xi)\} - \delta \quad \text{for all } \xi \text{ in the interval and } n \geq \mathfrak{N}.$$

Hence

$$|\mathrm{CE}\,\{f_n(\xi)\} - \mathrm{CE}\,\{f(\xi)\}| \leq \delta \quad \text{for all } \xi \text{ in the interval and } n \geq \mathfrak{N}$$

which, since δ is arbitrary, proves the lemma. Q.E.D.

In our case we consider an interval, $0 \leq \rho \leq \rho_1 < \rho_0$ where ρ_1 is any density less than the close packing density ρ_0. Taking the limit $\gamma \to 0$ of (22) we obtain

$$a(\rho, 0+) \geq -kT\{(s+t)^{-\nu} \log [1 + \rho(s+t)^{\nu}] + \rho \log [1 + \rho^{-1}(s+t)^{-\nu}]\} + \mathrm{CE}\,\{(s+t)^{-\nu} A^0[\rho(s+t)^{\nu},\omega] + \tfrac{1}{2}\alpha\rho^2\} + \lim_{\Omega\to\infty} \tilde{Q}_{\min}/\Omega \qquad (4.24)$$

since, by an argument like that which led to (2.20),

$$(s + t)^\nu \lim_{\gamma\to 0} \sum_{\mathbf{k}} w_{\min}(\mathbf{k}) = \int \varphi(\mathbf{x})\, d\mathbf{x} \equiv \alpha$$

and the convergence is uniform because the difference between the sum and the integral is independent of ρ. Finally, taking the limit $s \to \infty$, where the convergence is again uniform (Sec. 7k of Ref. 7), and using (14), we obtain

$$a(\rho, 0+) \geq \text{CE } \{a^0(\rho) + \tfrac{1}{2}\alpha\rho^2\}. \qquad (4.25)$$

Combined with (2.25) this gives

$$a(\rho, 0+) = \text{CE } \{a^0(\rho) + \tfrac{1}{2}\alpha\rho^2\} \qquad (4.26)$$

for nonpositive Kac potentials. The result obtained by Kac, Uhlenbeck, and Hemmer[6] is an example of (4.26).

V. LOWER BOUND TO THE FREE ENERGY: MORE GENERAL KAC POTENTIALS

When the Kac potential is neither nonnegative-definite nor nonpositive, a lower bound can still be obtained, though the method is more complicated than before. We express $\Phi(\mathbf{p})$, the ν-dimensional Fourier transform of $\varphi(\mathbf{x})$ defined in (3.1) as the sum of two parts

$$\Phi(\mathbf{p}) = \Phi_+(\mathbf{p}) + \Phi_-(\mathbf{p}) \qquad (5.1)$$

so chosen that

$$\left.\begin{array}{l} \Phi_+(\mathbf{p}) \geq 0 \quad \text{for all} \quad p, \\ \Phi_-(0) = \operatorname{Min}_{\mathbf{p}} \Phi_-(\mathbf{p}) = \min_{\mathbf{p}} \Phi(\mathbf{p}) \equiv \Phi_{\min}, \\ \int \Phi_\pm(\mathbf{p})\, d\mathbf{p} < \infty, \\ \Phi_\pm(\mathbf{p}) \quad \text{is continuous.} \end{array}\right\} \qquad (5.2)$$

The inverse transforms of $\Phi_+(\mathbf{p})$ and $\Phi_-(\mathbf{p})$ will be denoted by $\varphi_+(\mathbf{x})$ and $\varphi_-(\mathbf{x})$ and we shall require also that $\varphi_+(\mathbf{x})$ and $\varphi_-(\mathbf{x})$ satisfy (1.21). These conditions can be satisfied (provided $d\Phi/dp_n > -\infty$ at $p = 0$) by taking

$$\Phi_+(\mathbf{p}) = \begin{cases} [\Phi(0) - \Phi_{\min}](1 - p/p')^n & \text{if } p \leq p' \\ 0 & \text{if } p \geq p' \end{cases} \qquad (5.3)$$

where p' is the smallest value of p at which $\Phi(\mathbf{p})$ attains its minimum value $\Phi_{\min}$ and n is a positive integer (≥ 2) chosen sufficiently large to make $(1 - p/p')^n \leq [\Phi(\mathbf{p}) - \Phi_{\min}]/[\Phi(0) - \Phi_{\min}]$, for $p \leq p'$. When $\Phi_+(\mathbf{p})$ is given by (3) $\varphi_+(\mathbf{x})$ can be computed explicitly and shown to satisfy (1.21), and hence so will $\varphi_-(\mathbf{x})$. [Even when $d\Phi/dp = -\infty$ at $p = 0$ there is every reason to believe that the division of Φ into Φ_+ and Φ_- can be made in a way to satisfy all our conditions, provided $\varphi(\mathbf{x})$ satisfies (1.21).]

Since $\varphi_+(\mathbf{x})$ is a nonnegative-definite function its contribution to W, which may be written W_+, can be estimated by the method used in Sec. III. In analogy with (3.11), the result has the form $W_+ \geq W_{+,\min}$ where

$$\lim_{\Omega\to\infty} W_{+,\min}/\Omega = -\tfrac{1}{2}\rho\gamma^\nu\varphi_+(0) + \tfrac{1}{2}\Phi_+(0)\rho^2. \qquad (5.4)$$

The other contribution W_- has a lower bound analogous to (4.15)

$$W_- \geq \tfrac{1}{2} \sum_i \sum_j N_i N_j w_{-,\min}(\mathbf{k}_{ij}) \qquad (5.5)$$

where as in (4.16) $\mathbf{k}_{ij}$ is the position vector from the center of the cell ω_i to that of ω_j, and

$$\begin{aligned} w_{-,\min}(\mathbf{k}_{ij}) &\equiv \operatorname{Min}_{\substack{x\in\omega_i \\ y\in\omega_j}} w_-(\mathbf{x} - \mathbf{y}, \gamma) \\ &= \operatorname{Min}_{r\in\omega_0} w_-(\mathbf{k}_{ij} + 2\mathbf{r}, \gamma), \end{aligned} \qquad (5.6)$$

where $w_-(\mathbf{x}, \gamma) = \gamma^\nu\varphi_-(\gamma\mathbf{x})$ and ω_0 is a cube of side $s + t$ centered at the origin.

The quadratic form in (5) can be diagonalized using a Fourier transformation. We define the function

$$W(\mathbf{p}) \equiv \sum_{\mathbf{k}} w_{-,\min}(\mathbf{k}) \exp(2\pi i\mathbf{p}\cdot\mathbf{k}) \qquad (5.7)$$

where the sum goes over all the different values taken by the vector $\mathbf{k}_{ij}$ as both i and j range over the values 1, 2, $\cdots$ M. These values lie on a simple cubic lattice of spacing $s + t$ with a lattice point at the origin and are inside (not on the surface of) a cube of volume $2^\nu\Omega$ centered at the origin. The sum in (7) therefore covers just $[2\Omega^{1/\nu}(s + t)^{-1} - 1] = [2M^{1/\nu} - 1]^\nu \equiv M'$ lattice points. The function $W(\mathbf{p})$ is periodic in $\mathbf{p}$, the unit cell of the periodicity being a cube of side $(s + t)^{-1}$. Although the notation does not show it, $W(\mathbf{p})$ depends on $s + t$, γ, and Ω as well as on $\mathbf{p}$.

The inverse of (7) is

$$w_{-,\min}(\mathbf{k}) = (M')^{-1} \sum_{\mathbf{p}} W(\mathbf{p}) \exp(-2\pi i\mathbf{p}\cdot\mathbf{k}) \qquad (5.8)$$

where the sum goes over M' values of $\mathbf{p}$ lying inside a cube of side $(s + t)^{-1}$ and on a lattice of spacing $(M')^{-1/\nu}(s + t)^{-1} = [2\Omega^{1/\nu} - (s + t)^{-1}.]$ Substituting (8) into (5) and rearranging, we obtain

$$W_- \geq \tfrac{1}{2}(M')^{-1} \sum_{\mathbf{p}} W(\mathbf{p})\, |n(\mathbf{p})|^2 \qquad (5.9)$$

where

$$n(\mathbf{p}) \equiv \sum_{i=1}^{M} N_i \exp(2\pi i \mathbf{p}\cdot\mathbf{k}_i) \quad (5.10)$$

and $\mathbf{k}_i$ is the position vector of the center of ω_i. For a lower bound on the sum (9) we may replace $W(\mathbf{p})$ by its minimum value, obtaining

$$\begin{aligned} W_- &\geq \tfrac{1}{2}(M')^{-1} \sum_{\mathbf{p}} |n(\mathbf{p})|^2 \operatorname*{Min}_{\mathbf{p}} W(\mathbf{p}) \\ &= \frac{1}{2} \sum_{i=1}^{M} N_i^2 \operatorname*{Min}_{\mathbf{p}} W(\mathbf{p}). \end{aligned} \quad (5.11)$$

Substituting this result into (4.6) and using (4.3) and (1.10) we obtain

$$\begin{aligned} A(N, \Omega, \gamma) &\geq kT \log [N!\,(M-1)!/(N+M-1)!] \\ &+ \operatorname*{Min}_{N_1\cdots N_M} \sum_{i=1}^{M} \{A^0(N_i, \omega) \\ &+ \tfrac{1}{2}N_i^2 \operatorname*{Min}_{\mathbf{p}} W(\mathbf{p})\} + W_{+,\min} + \bar{Q}_{\min} \end{aligned} \quad (5.12)$$

where $W_{+,\min}$ satisfies (4).

The inequality (12) is very similar to (4.19) and the effect on it of the triple limit process can be studied by the method used for (4.19). The result, analogous to (4.25), is

$$a(\rho, 0+) \geq \text{CE}\,\{a^0(\rho) + \tfrac{1}{2}\alpha_-\rho^2\} + \tfrac{1}{2}\alpha_+\rho^2, \quad (5.13)$$

where

$$\begin{aligned} \alpha_+ &\equiv (2/\rho^2) \lim_{s\to\infty} \lim_{\gamma\to 0} \lim_{\Omega\to\infty} W_{+,\min}/\Omega \\ &= \Phi_+(0) \end{aligned} \quad (5.14)$$

by (4), and

$$\alpha_- \equiv \lim_{s\to\infty} \lim_{\gamma\to 0} \lim_{\Omega\to\infty} (s+t)^\nu \operatorname*{Min}_{\mathbf{p}} W(\mathbf{p}). \quad (5.15)$$

The first step is to carry out the limit process $\Omega \to \infty$ in (15). It follows from the definition (7) that

$$|W(\mathbf{p}) - \lim_{\Omega\to\infty} W(\mathbf{p})| \leq \sum |w_{-,\min}(\mathbf{k})| \quad (5.16)$$

where the sum is over all lattice points $\mathbf{k}$ outside a cube of volume just less than $2^\nu\Omega$ centered at the origin. This sum is independent of $\mathbf{p}$, and because of (1.21) it tends to zero as $\Omega \to \infty$. Therefore $W(\mathbf{p})$ approaches its limit, as $\Omega \to \infty$ uniformly in p, so that

$$\begin{aligned} \lim_{\Omega\to\infty} \operatorname*{Min}_{\mathbf{p}} W(\mathbf{p}) &= \operatorname*{Min}_{\mathbf{p}} \lim_{\Omega\to\infty} W(\mathbf{p}) \\ &= \operatorname*{Min}_{\mathbf{p}} \Sigma(\mathbf{p}, \gamma) \end{aligned} \quad (5.17)$$

where

$$\Sigma(\mathbf{p}, \gamma) \equiv \sum_{\mathbf{k}} w_{-,\min}(\mathbf{k}) \exp(2\pi i \mathbf{p}\cdot\mathbf{k}), \quad (5.18)$$

the sum going over an infinite lattice with spacing $s + t$.

To study the effect of the next limit process, $\gamma \to 0$, we consider separately the cases where $\mathbf{p}$ is inside and outside a cube Γ of side $2p_0$ centered at the origin. The dependence of p_0 on γ will be chosen so that in the limit $\gamma \to 0$ the sum (18) may be replaced by an integral if $\mathbf{p} \in \Gamma$ and by zero otherwise.

By writing $\gamma\mathbf{q}$ for $\mathbf{p}$ and $\mathbf{x}$ for $\gamma\mathbf{k}$ in (18), and using an argument similar to that which gave (2.20), we find that

$$\begin{aligned} (s+t)^\nu &\lim_{\gamma\to 0} \Sigma(\gamma\mathbf{q}, \gamma) \\ &= \int \varphi_-(\mathbf{x}) \exp(2\pi i \mathbf{q}\cdot\mathbf{x})\, d\mathbf{x} \\ &= \Phi_-(\mathbf{q}). \end{aligned} \quad (5.19)$$

In order to use this result in (15), we must show also that the convergence to the limit is uniform provided $\gamma\mathbf{q} \in \Gamma$. To do this we use the definitions (3.1), (1.6), (18), and (6), to obtain the estimate

$$\begin{aligned} &|\Phi(\mathbf{p}/\gamma) - (s+t)^\nu \Sigma(\mathbf{p}, \gamma)| \\ &= \Big|\sum_{\mathbf{k}} \int_{\omega_0} e^{2\pi i \mathbf{p}\cdot\mathbf{k}}[e^{2\pi i \mathbf{p}\cdot\mathbf{y}} w_-(\mathbf{k}+\mathbf{y}, \gamma) \\ &\quad - \operatorname*{Min}_{\mathbf{z}\in\omega_0} w_-(\mathbf{k}+2\mathbf{z}, \gamma)]\, d\mathbf{y}\Big| \\ &\leq \sum_{\mathbf{k}} \int_{\omega_0} \{|e^{2\pi i \mathbf{p}\cdot\mathbf{y}} - 1|\, |w_-(\mathbf{k}+\mathbf{y}, \gamma)| \\ &\quad + |w_-(\mathbf{k}+\mathbf{y}, \gamma) - \operatorname*{Min}_{\mathbf{z}\in\omega_0} w_-(\mathbf{k}+2\mathbf{z}, \gamma)|\}\, d\mathbf{y} \\ &\leq \pi\nu p_0(s+t) \int |w_-(\mathbf{x}, \gamma)|\, d\mathbf{x} \\ &\quad + \Big|\int w_-(\mathbf{x}, \gamma)\, d\mathbf{x} - (s+t)^\nu \sum_k w_{-,\min}(\mathbf{k})\Big|. \end{aligned} \quad (5.20)$$

Since we have required that $\int |\varphi_-(\mathbf{y})|\, d\mathbf{y} < \infty$, the first part of the last member of (20) tends to zero with γ provided that

$$\lim_{\gamma\to 0} p_0 = 0. \quad (5.21)$$

The second part also tends to zero with γ, as in the argument leading to (2.20), because of the Riemann integrability of φ_-. Since both parts are independent of $\mathbf{p}$, the convergence to the limit in (19) is indeed uniform for $\gamma\mathbf{q} \in \Gamma$, and it follows that

$$(s+t)^{\nu} \lim_{\gamma\to 0} \min_{\mathbf{p}\in\Gamma} \sum(\mathbf{p},\gamma) = \lim_{\gamma\to 0} \operatorname*{Min}_{\mathbf{p}\in\Gamma} \Phi_-(\mathbf{p}/\gamma)$$
$$= \operatorname*{Min}_{\mathbf{q}} \Phi_-(\mathbf{q}) \equiv \Phi_-(0) \quad (5.22)$$

by (2), provided $\lim_{\gamma\to 0} p_0/\gamma = \infty$.

To complete the estimates of Min $\sum(\mathbf{p},\gamma)$ for (18) we must also find a lower bound for $\sum(\mathbf{p},\gamma)$ with $\mathbf{p}$ outside Γ. Whenever $\mathbf{p}$ is outside Γ, we can find a direction parallel to an edge of the cube Γ such that the component of $\mathbf{p}$ along this direction exceeds p_0. If $\mathbf{l}$ denotes a vector of length $s+t$ along this direction we therefore have

$$p_0(s+t) \leq \mathbf{p}\cdot\mathbf{l} \leq \tfrac{1}{2}, \quad (5.23)$$

the second inequality being a consequence of the fact [see (8)] that all the allowed values of $\mathbf{p}$ lie inside a cube of side $(s+t)^{-1}$ centered at the origin. On multiplying both sides of (18) by $1 - \exp(2\pi i\mathbf{p}\cdot\mathbf{l})$ and taking absolute values, we obtain, since $|\mathbf{l}| = s+t$,

$$2\sin(\pi\mathbf{p}\cdot\mathbf{l})\,\left|\sum(\mathbf{p}\cdot\gamma)\right|$$
$$= \left|\sum_{\mathbf{k}} [w_{-,\min}(\mathbf{k}) - w_{-,\min}(\mathbf{k}-\mathbf{l})]\exp(2\pi i\mathbf{p}\cdot\mathbf{k})\right|$$
$$\leq \sum_{\mathbf{k}} |w_{-,\min}(\mathbf{k}) - w_{-,\min}(\mathbf{k}-\mathbf{l})|$$
$$\leq \sum_{\mathbf{k}} \operatorname*{Max}_{\mathbf{x}\in\omega_0} w_-(\mathbf{k}+4\mathbf{x},\gamma)$$
$$- \sum_{\mathbf{k}} \operatorname*{Min}_{\mathbf{x}\in\omega_0} w_-(\mathbf{k}+4\mathbf{x},\gamma) \quad (5.24)$$

where ω_0 denotes, as in (6), a cube of side $(s+t)$ centered at the origin. Applying once more an argument of the type which led to (2.20), we find that both sums in the last member of (24) tend to the same integral in the limit $\gamma \to 0$. Thus (24) can be written in the form

$$\left|\sum(\mathbf{p},\gamma)\right| \leq \tfrac{1}{2}\,|\csc(\pi\mathbf{p}\cdot\mathbf{l})|\,\delta(\gamma) \quad (5.25)$$

where δ does not depend and on p and tends to 0 with γ. Combining (23) and (25) we obtain

$$\left|\sum(\mathbf{p},\gamma)\right| \leq \delta(\gamma)/4p_0(s+t) \quad \text{for } \mathbf{p} \text{ outside } \Gamma \quad (5.26)$$

since $\sin x \geq 2x/\pi$ for $0 \leq x \leq \pi/2$. Let us now choose p_0 in such a way that

$$\lim_{\gamma\to 0} \delta(\gamma)/p_0 = 0. \quad (5.27)$$

This is compatible with the previous requirements (5.21) and $\lim p_0/\gamma = \infty$; for example, the choice $p_0 \propto [\gamma + \delta(\gamma)]^{\frac{1}{2}}$ satisfies all three requirements. Combining (27) and (26) we obtain

$$\lim_{\gamma\to 0} \operatorname*{Min}_{\mathbf{p}\text{ outside }\Gamma} \sum(\mathbf{p},\gamma) = 0. \quad (5.28)$$

Combining (15) and (17) and then using (22) and (28), we obtain

$$\alpha_- = \lim_{s\to\infty} \lim_{\gamma\to 0} (s+t)^{\nu} \operatorname*{Min}_{\mathbf{p}} \sum(\mathbf{p},\gamma)$$
$$= \lim_{s\to\infty} \Phi_-(0) = \Phi_-(0) = \Phi_{\min}. \quad (5.29)$$

This completes the evaluation of α_-.

Our main results (2.25) and (13) can be summarized in the formula

$$\text{CE}\,\{a^0(\rho) + \tfrac{1}{2}\alpha_-\rho^2\} + \tfrac{1}{2}\alpha_+\rho^2 \leq a(\rho, 0+)$$
$$\leq \text{CE}\,\{a^0(\rho) + \tfrac{1}{2}\alpha\rho^2\}. \quad (5.30)$$

From (30) we can find $a(\rho, 0+)$ exactly provided the upper and lower bounds coincide. This can happen in two ways:

(i) Where $a^0(\rho) + \frac{1}{2}\alpha_-\rho^2$ coincides with its convex envelope, (e.g., if $\alpha_- = 0$), the function $a^0\rho + \frac{1}{2}\alpha\rho^2$ does so too since $\alpha = \alpha_+ + \alpha_- \geq \alpha_-$; thus (30) implies

$$a(\rho, 0+) = \text{CE}\,\{a^0(\rho) + \tfrac{1}{2}\alpha\rho^2\} = a^0(\rho) + \tfrac{1}{2}\alpha\rho^2$$
$$\text{if} \quad a^0(\rho) + \tfrac{1}{2}\alpha_-\rho^2 = \text{CE}\,\{a^0(\rho) + \tfrac{1}{2}\alpha_-\rho^2\}. \quad (5.31)$$

In the special case $\alpha_- = 0$ this reduces to (3.14).

(ii) If $\alpha_+ = 0$ then we have $\alpha = \alpha_-$ so that (30) reduces to

$$a(\rho, 0+) = \text{CE}\,\{a^0(\rho) + \tfrac{1}{2}\alpha\rho^2\} \quad \text{if} \quad \alpha_+ = 0. \quad (5.32)$$

The result (4.26) is a special case of this, since (4.1) implies $\Phi(0) \leq \Phi(\mathbf{p})$ for all $\mathbf{p}$.

VI. EQUATION OF STATE

For a system with finite γ, the thermodynamic pressure $p(\rho,\gamma)$ is given[7,9] by

$$p(\rho,\gamma) = -\partial(\rho^{-1}a(\rho,\gamma))/\partial(\rho^{-1})$$
$$= \rho\,\partial a(\rho,\gamma)/\partial\rho - a(\rho,\gamma), \quad (6.1)$$

the differentiation being at constant γ and also at constant T, although the dependence on T is not explicit in the notation. Since $a(\rho,\gamma)$ is[7,9] a convex function of ρ at constant γ the derivative on the right of (1) exists[16] except on a countable set of values of ρ; it seems likely that this countable set is in fact empty but no proof is known.

Taking the limit $\gamma \to 0$ on both sides of (6.1) we obtain

$$p(\rho, 0+) = \rho \lim_{\gamma\to 0} \partial a(\rho,\gamma)/\partial\rho - a(\rho, 0+). \quad (6.2)$$

To evaluate the right side of (2) we must show that the order of the operations $\gamma \to 0$ and $\partial/\partial\rho$

[16] Ref. 10, p. 94.

can be reversed. To this end we use the inequality[17] for the right- and left-hand derivatives of a convex function

$$\begin{aligned}[a(\rho-\epsilon,\gamma)-a(\rho)]/(-\epsilon) &\le D_-a(\rho,\gamma)\\ &\le D_+a(\rho,\gamma)\\ &\le [a(\rho+\epsilon,\gamma)-a(\rho)]/\epsilon\end{aligned} \tag{6.3}$$

where

$$D_\pm a(\rho,\gamma) \equiv \lim_{\epsilon\to 0}\,[a(\rho\pm\epsilon,\gamma)-a(\rho,\gamma)]/(\pm\epsilon), \tag{6.4}$$

with ϵ an arbitrary positive number.

Taking the limit $\gamma \to 0$ on both sides of (3), followed by the limit $\epsilon \to 0$, we obtain

$$\begin{aligned}D_-a(\rho,0+) &\le \lim_{\gamma\to 0} D_-a(\rho,\gamma)\\ &\le \lim_{\gamma\to 0} D_+a(\rho,\gamma) \le D_+a(\rho,0+).\end{aligned} \tag{6.5}$$

Hence if $a(\rho, 0+)$ is differentiable all four of these expressions are equal, so that $\lim_{\gamma\to 0} D_\pm a(\rho,\gamma)$ exists and is equal to $da(\rho, 0+)/d\rho$. Substituting this result into (2) we obtain

$$p(\rho, 0+) = (\rho d/d\rho - 1)a(\rho, 0+) \tag{6.6}$$

at all points where $a(\rho, 0+)$ is differentiable. Equation (6) can be interpreted geometrically by the statement that $-p(\rho, 0+)$ is the place where the tangent at ρ to the graph of $a(\rho, 0+)$ intercepts the a-axis.

Combining (6) with (5.31) or (5.32), we obtain

$$p(\rho, 0+) = (\rho d/d\rho - 1)\ \mathrm{CE}\ \{a^0(\rho) + \tfrac{1}{2}\alpha\rho^2\} \tag{6.7}$$

for all Kac potentials to which (5.31) or (5.32) applies, including nonnegative definite and nonpositive potentials.

If the graph of $a(\rho, 0+) = \mathrm{CE}\ \{a^0(\rho) + \frac{1}{2}\alpha\rho^2\}$ has a straight segment (see Fig. 3) then the geometrical interpretation of (6) shows that $p(\rho, 0+)$ is constant along this straight segment; and the chemical potential $[a(\rho, 0+) + p(\rho, 0+)]/\rho = da(\rho, 0+)/d\rho$ is also constant. This behavior of the thermodynamic functions characterizes a first-order phase transition.

Since the straight portion of the graph CE $\{a^0(\rho) + \frac{1}{2}\alpha\rho^2\}$ touches the graph $a^0(\rho) + \frac{1}{2}\alpha\rho^2$ at both ends, the quantity

$$\bar{p}_{vdw}(\rho) \equiv (\rho d/d\rho - 1)(a^0(\rho) + \tfrac{1}{2}\alpha\rho^2) \tag{6.8}$$

takes the same value, call it p_{sat}, at the two ends of the phase transition region. Moreover, if ρ_l and ρ_v are the abscissas of the two points of contact, then we have

$$\begin{aligned}\int_{\rho_v}^{\rho_l} [\bar{p}_{vdw}(\rho) - p_{sat}]\, d(\rho^{-1}) &= \{-[a^0(\rho) + \tfrac{1}{2}\alpha\rho^2 + p_{sat}]\rho^{-1}\}_{\rho_v}^{\rho_l}\\ &= \{(d/d\rho)[a^0(\rho) + \tfrac{1}{2}\alpha\rho^2]\}_{\rho_v}^{\rho_l}\end{aligned} \tag{6.9}$$

using (8) first as a definition of $\bar{p}_{vdw}(\rho)$ and then to evaluate p_{sat}. Since the slopes of the graph $a^0(\rho) + \frac{1}{2}\alpha\rho^2$ at the two points of contact are equal, the expression (9) vanishes. This shows that ρ_l and ρ_v are related to $\bar{p}_{vdw}(\rho)$ by Maxwell's equal-area construction (Fig. 1). The replacement of $a^0(\rho) + \frac{1}{2}\alpha\rho^2$ by its convex envelope is precisely equivalent to making Maxwell's modification (1.2) in the function $\bar{p}_{vdw}(\rho)$ defined by (8).

For values of ρ where the upper and lower bounds of $a(\rho, 0+)$, in (5.30) do not coincide we still obtain bounds on $\lim_{\gamma\to 0} \partial a(\rho,\gamma)/\partial\rho$ and hence on $p(\rho, 0+)$ by an argument due to Fisher[18]: owing to the convexity of the graph of $a(\rho, 0+)$ its slope at $\rho = \rho_0$ must lie between the slopes of tangents to its upper bound crossing its lower bound at $\rho = \rho_0$.

A result similar to that obtained for the pressure holds also for the internal energy density

$$u(\beta, \rho, \gamma) = (\partial/\partial\beta)[\beta a(\beta, \rho, \gamma)]. \tag{6.10}$$

Since $-\beta a(\beta, \rho, \gamma)$ is convex in β we obtain, as in the derivation of (6),

$$u(\beta, \rho, 0+) = \partial/\partial\beta[\beta a(\beta, \rho, 0+)]. \tag{6.11}$$

Unfortunately we have been unable to prove any similar general statements about the specific heat and compressibility which correspond to second derivatives of the free-energy density. At sufficiently low densities, however, one can show that all the density derivatives of $p(\rho, \gamma)$ approach the corresponding derivatives of $p(\rho, 0+)$ by using Vitali's theorem together with the results of Lebowitz and Penrose[19] for the convergence of virial expansions.

VII. THE PAIR DISTRIBUTION FUNCTION

In order to understand better the effect of the Kac potential on possible phase transitions in the reference system and in the van der Waals system, it is useful to study the two-body distribution function. (In this section we do not aim at such a high standard of rigor as in the earlier sections.)

[17] Obtained by making $\delta \to 0$ in Eq. (3.18.3) of Hardy, Littlewood and Polya (Ref. 10).

[18] M. E. Fisher, "Bounds for the Derivatives of the Free Energy and the Pressure of a Hard Core System near Close Packing" (to be published).

[19] J. L. Lebowitz and O. Penrose, J. Math. Phys. 5, 841 (1964).

Instead of the usual two-body distribution function[20] $n_2(\mathbf{x}_1, \mathbf{x}_2)$, we shall study its space average

$$\bar{n}_2(\mathbf{r}) \equiv \Omega^{-1} \int n_2(\mathbf{x}, \mathbf{x} + \mathbf{r})\, d\mathbf{x}. \qquad (7.1)$$

This distribution function when the system has N particles in a box Ω and a Kac parameter γ will be denoted by $\bar{n}_2(\mathbf{r}; N, \Omega, \gamma)$. This function can be determined from the formula[21]

$$\tfrac{1}{2} \int \bar{n}_2(\mathbf{r}; N, \Omega, \gamma) q'(\mathbf{r})\, d\mathbf{r} = (\partial/\partial\eta)\Omega^{-1} A[N, \Omega, \gamma, \eta q'(\mathbf{r})]_{\eta=0} \qquad (7.2)$$

which holds for arbitrary bounded functions $q'(\mathbf{r})$. The integration may be taken over all space, since $\bar{n}_2(\mathbf{r}; N, \Omega, \gamma) = 0$ for large r. The symbol $A[N, \Omega, \gamma, \eta q'(\mathbf{r})]$ denotes the free energy analogous to (1.10) when the short-range interaction potential is not $q(\mathbf{r})$ but $q(\mathbf{r}) + \eta q'(\mathbf{r})$, and η is a parameter. Taking the thermodynamic limit and then the van der Waals limit in (2) we obtain, provided all the relevant limits and the derivative exist,

$$\tfrac{1}{2} \int \bar{n}_2(\mathbf{r}; \rho, 0+) q'(\mathbf{r})\, d\mathbf{r} = (\partial/\partial\eta) a[\rho, 0+, \eta q'(\mathbf{r})]_{\eta=0} \qquad (7.3)$$

where

$$\bar{n}_2(\mathbf{r}; \rho, 0+) \equiv \lim_{\gamma\to 0} \lim_{\Omega\to\infty} \bar{n}_2(\mathbf{r}; \rho\Omega, \Omega, \gamma) \qquad (7.4)$$

and

$$a[\rho, 0+, \eta q'(\mathbf{r})] \equiv \lim_{\gamma\to 0} \lim_{\Omega\to\infty} A[\rho\Omega, \Omega, \gamma, \eta q'(\mathbf{r})]/\Omega. \qquad (7.5)$$

The permutation of the limit operations $\gamma \to 0$ and $\Omega \to \infty$ with integration and differentiation in deriving (3) is justified on the left by Lebesgue's theorem.[22] On the right it is justified [as in the argument based on (6.3) and (6.5)], by the convexity of $A[N, \Omega, \gamma, \eta q'(\mathbf{r})]$ as a function of η. This convexity can be verified by calculating the second derivative of A with respect to η. In a similar way we can obtain

$$\tfrac{1}{2} \int \bar{n}_2^0(\mathbf{r}; \rho) q'(\mathbf{r})\, d\mathbf{r} = (\partial/\partial\eta) a^0[\rho, \eta q'(\mathbf{r})]_{\eta=0} \qquad (7.6)$$

where the superscript zero denotes quantities belonging to the reference system with a modified short-range potential $q(\mathbf{r}) + \eta q'(\mathbf{r})$. When the conditions of validity for our main result (5.31) or (5.32) are satisfied, (3) reduces to

$$\tfrac{1}{2} \int \bar{n}_2(\mathbf{r}; \rho, 0+) q'(\mathbf{r})\, d\mathbf{r} = (\partial/\partial\eta)\, \mathrm{CE}\, \{a^0[\rho, \eta q'(\mathbf{r})] + \tfrac{1}{2}\alpha\rho^2\}_{\eta=0}. \qquad (7.7)$$

Two cases may be distinguished. In the first of these the graph of $a^0[\rho, \eta q'(\mathbf{r})] + \frac{1}{2}\alpha\rho^2$ coincides with that of its convex envelope and so (7) combines with (6) to give

$$\tfrac{1}{2} \int \bar{n}_2(\mathbf{r}; \rho, 0+) q'(\mathbf{r})\, d\mathbf{r} = \tfrac{1}{2} \int \bar{n}_2^0(\mathbf{r}, \rho) q'(\mathbf{r})\, d\mathbf{r} \qquad (7.8)$$

which implies

$$\bar{n}_2(\mathbf{r}; \rho, 0+) = \bar{n}_2^0(\mathbf{r}, \rho) \qquad (7.9)$$

since $q'(\mathbf{r})$ is arbitrary. In this case, therefore, the Kac potential does not affect the distribution of pairs of particles, as one might expect from (1.8).

In the alternative case, where the graph of $a^0[\rho, \eta q'(\mathbf{r})] + \frac{1}{2}\alpha\rho^2$ does not coincide with that of its convex envelope, the latter is a straight line touching the former at two places, say ρ_1 and ρ_2. Both ρ_1 and ρ_2 depend, in general, on η. The equation of this straight line may be written

$$\mathrm{CE}\, \{a^0[\rho, \eta q'(\mathbf{r})] + \tfrac{1}{2}\alpha\rho^2\} = [(\rho - \rho_1)a_2 + (\rho_2 - \rho)a_1](\rho_2 - \rho_1)^{-1} (\rho_1 < \rho < \rho_2) \qquad (7.10)$$

where

$$a_h \equiv a^0[\rho_h, \eta q'(\mathbf{r})] + \tfrac{1}{2}\alpha\rho_h^2 \qquad (h = 1, 2). \qquad (7.11)$$

On substituting this into (7) and using (9), which applies when $\rho = \rho_1$ or $\rho = \rho_2$, we obtain

$$\tfrac{1}{2} \int \bar{n}_2(\mathbf{r}; \rho, 0+) q'(\mathbf{r})\, d\mathbf{r} = \tfrac{1}{2} \int \left[\frac{\rho - \rho_1}{\rho_2 - \rho_1} \bar{n}_2^0(\mathbf{r}; \rho_2) + \frac{\rho_2 - \rho}{\rho_2 - \rho_1} \bar{n}_2^0(\mathbf{r}, \rho_1) \right] q'(\mathbf{r})\, d\mathbf{r} \qquad (7.12)$$

since

$$\frac{\partial}{\partial\rho_1} \left[\frac{(\rho - \rho_1)a_2 + (\rho_2 - \rho)a_1}{\rho_2 - \rho_1} \right] = \frac{\rho_2 - \rho}{(\rho_2 - \rho_1)^2} \left[a_1 - a_2 + (\rho_2 - \rho_1) \frac{\partial}{\partial\rho_1} a_1 \right] = 0, \text{ etc.} \qquad (7.13)$$

by virtue of the double tangent construction. [In Eq. (13) we are treating $\eta, \rho_1, \rho_2, \rho$ as independent

[20] T. L. Hill, *Statistical Mechanics* (McGraw-Hill Book Company, Inc., New York, 1956), Sec. 29.

[21] Equation (7.2) is the classical analog of a quantum formula given by Bogolyubov and Zubarev, Zh. Eskperim. i Teor. Fiz. **28**, 129 (1955) [English transl.: Soviet Phys.—JETP **1**, 83 (1955)]. This was first used in this type of work by M. E. Fisher.

[22] F. Riesz and B. Sz.-Nagy, *Functional Analysis* (Frederick Ungar Publishing Company, New York, 1955), p. 37.

variables.] Since $q'(\mathbf{r})$ is arbitrary, it follows from (12) that

$$\bar{n}_2(\mathbf{r};\rho,0+) = \frac{\rho-\rho_1}{\rho_2-\rho_1}\bar{n}_2^0(\mathbf{r};\rho_2) + \frac{\rho_2-\rho}{\rho_2-\rho_1}\bar{n}_2^0(\mathbf{r};\rho_1), \qquad (\rho_1<\rho<\rho_2). \tag{7.14}$$

This equation can be interpreted[23] to mean that two phases coexist whenever CE $\{a^0(\rho)+\frac{1}{2}\alpha\rho^2\} < a^0(\rho)+\frac{1}{2}\alpha\rho^2$; their pair distribution functions are $\bar{n}_2^0(\mathbf{r};\rho_1)$ and $\bar{n}_2^0(\mathbf{r};\rho_2)$, and their densities ρ_1 and ρ_2, respectively. This is consistent with the interpretation of a straight portion of the graph $a(\rho, 0+)$ as a first-order transition.

In interpreting the results (9) and (14) it should be remembered that their derivation involves the limit process $\gamma\to 0$; they therefore yield information only about values of r small compared with the range γ^{-1} of the Kac potential. For example, (14) implies a two-phase structure on a length scale $\ll\gamma^{-1}$ but not necessarily on a scale $\gg\gamma^{-1}$.

VIII. SIMPLE UPPER BOUND ON $a(\rho, \gamma)$

Of the various upper and lower bounds on $a(\rho,\gamma)$ obtained in Sec. II to V, the only one simple enough to be useful for finite values of γ is the lower bound (3.12) for nonnegative-definite Kac potentials. The other bounds are too unwieldy because they involve the network of cells $\omega_1 \cdots \omega_M$.

A simple upper bound on $a(\rho,\gamma)$ can be obtained by a method due in essence to Gibbs.[24] We rewrite (1.11) in the form

$$Z(N,\Omega,\gamma) = Z^0(N,\Omega)\langle e^{-\beta W}\rangle^0 \tag{8.1}$$

where W is the long-range contribution to the potential energy and $\langle\ \rangle^0$ indicates a canonical average over the reference system. Since $e^{-\beta W}$ is convex we have by (1.6) and (7.1)

$$\langle e^{-\beta W}\rangle^0 \geq \exp\langle -\beta W\rangle^0 = \exp\left[-\tfrac{1}{2}\beta\iint \gamma^\nu\phi(\gamma\mathbf{r})\bar{n}_2^0(\mathbf{x},\mathbf{x}+\mathbf{r})\,d\mathbf{x}\,d\mathbf{r}\right]. \tag{8.2}$$

Combining (1) and (2) and then taking the thermodynamic limit, we obtain

$$a(\rho,\gamma) \leq a^0(\rho) + \tfrac{1}{2}\gamma^\nu \int \varphi(\gamma\mathbf{r})\bar{n}_2^0(\mathbf{r};\rho)\,d\mathbf{r} \tag{8.3}$$

where $\bar{n}_2^0$ is defined as in Sec. VII. This inequality may be strengthened in the manner used in going from (2.24) to (2.25), to give

$$a(\rho,\gamma) \leq \mathrm{CE}\left\{a^0(\rho) + \tfrac{1}{2}\gamma^\nu\int\varphi(\gamma\mathbf{r})\bar{n}_2^0(\mathbf{r};\rho)\,d\mathbf{r}\right\}. \tag{8.4}$$

A system for which the right side of (8.4) can be evaluated is the one-dimensional system considered by Kac, Uhlenbeck, and Hemmer.[6] In our notation it is defined by (1.4) and (1.5). For this system, the last term in (8.3) is essentially the Laplace transform of $\bar{n}_2^0$, and (8.4) becomes

$$a(\rho,\gamma) \leq \mathrm{CE}\,\{a^0(\rho) + \tfrac{1}{2}\alpha\rho\gamma[(1+\gamma\rho^{-1}-\gamma r_0)e^{\gamma r_0}-1]^{-1}\} \tag{8.5}$$

with

$$a^0(\rho) = \rho kT\{\log[\rho(1-\rho r_0)^{-1}(2\pi\hbar^2/mkT)^{\frac{1}{2}}]-1\}. \tag{8.6}$$

In the van der Waals limit (5) reduces to (2.25).

The argument which led to (4) also applies to lattice gases. For example, if the short-range potential is taken as

$$q(\mathbf{r}) = \begin{cases}+\infty & \text{if } \mathbf{r}=0,\\ 0 & \text{if } \mathbf{r}\neq 0,\end{cases} \tag{8.7}$$

the part of the interaction potential which prevents more than one particle occupying any site, then $\bar{n}_2^0(\mathbf{r};\rho)$ vanishes for $\mathbf{r}=0$ and takes the value ρ^2 for $\mathbf{r}\neq 0$; consequently the lattice-gas analog of (4) leads to

$$a(\rho,\gamma) \leq \mathrm{CE}\,\{a^0(\rho)+\tfrac{1}{2}\rho^2\textstyle\sum' w(\mathbf{r},\gamma)\} \tag{8.8}$$

with

$$a^0(\rho) = kT[\rho\ln\rho + (1-\rho)\ln(1-\rho)]. \tag{8.9}$$

The sum $\sum'$ goes over the infinite lattice excluding $\mathbf{r}=0$.

IX. DISCUSSION

We have shown that for a class of Kac potentials, including nonnegative-definite and nonpositive potentials, the thermodynamic free-energy density is given in the van der Waals limit by

$$a(\rho,0+) = \mathrm{CE}\,\{a^0(\rho)+\tfrac{1}{2}\alpha\rho^2\} \tag{9.1}$$

and the equation of state by Maxwell's modification of the corresponding van der Waals equation of state. If $\alpha<0$, the graph of $a(\rho, 0+)$ may have straight portions; these correspond to first-order phase transitions both in the thermodynamic prop-

[23] J. E. Mayer, J. Chem. Phys. 15, 187 (1947); G. E. Uhlenbeck, P. Hemmer, and M. Kac, J. Math. Phys. 4, 229 (1963).

[24] See M. Girardeau, J. Chem. Phys. 40, 899 (1964).

erties and in the behavior of the pair distribution function for $r \ll \gamma^{-1}$. If $\alpha > 0$ on the other hand, the graph of $a(\rho, 0+)$ cannot have a straight portion and thermodynamically the system can have no phase transition. A paradoxical situation arises if the reference system has a phase transition and $\alpha > 0$; then by (7.9) the pair distribution function has the form characterizing a phase transition, yet there is no phase transition in the thermodynamic sense. The explanation is that the result (7.9), which indicates the coexistence of two phases, was obtained using the limit process $\gamma \to 0$ and may therefore be relied on only when $r \ll \gamma^{-1}$. On the other hand, the term $\frac{1}{2}\alpha\rho^2$ in $a(\rho, 0+)$ indicates that on the length scale where the Kac potential operates (distances $\approx$ or $\gg \gamma^{-1}$) the system is uniform since there is no transition. It appears therefore, that the repulsive Kac potential causes the distinct liquid and gas phases of a normal first-order transition to break into droplets or froth whose chacteristic length is $\gg r_0$ but $\ll \gamma^{-1}$. This fact might possibly find a practical application.

Some of the results on which this paper is based can be generalized to quantum mechanics. The results of Sec. III generalize immediately to quantum mechanics, and so do those of Sec. II provided the boundary condition on the wavefunction is that it must vanish when the center of any particle touches the wall of the container. The main result of Sec. VIII also generalizes to quantum mechanics by virtue of Bogolyubov's inequality.[25] On the other hand Sec. IV and V do not generalize so readily. Unfortunately this means that we can at present evaluate $a(\rho, 0+)$ rigorously only for nonnegative-definite Kac potentials, which cannot produce a phase transition. Thus it remains to be shown that a van der Waals phase transition can occur in a quantum system.

The results of this paper can easily be generalized to classical lattice gases. The proofs require only minor modifications, and the main results (5.31) and (5.32) are the same. Our results apply also to Ising spin systems since these are isomorphic to lattice gases. In this way the Bragg–Williams and Weiss theories can also be dealt with in a rigorous fashion.[26]

Another direction in which our results might be generalized is to weaken the conditions (1.20) and (1.21). For example, for nonnegative-definite Kac potentials the hard-core condition (1.20a) can be replaced by the condition $q(r) >$ const $r^{-\nu-\epsilon}$ for small r, which is sufficient[7,9] to ensure the existence of $a^0(\rho)$ and $a(\rho, \gamma)$. For more general Kac potentials the hard-core condition plays no part in the upper bound on $a(\rho, 0+)$ but is used to restrict the number of particles in a cell when lower bound on $a(\rho, 0+)$ is obtained. Possibly a more refined argument could dispense with the hard-core condition altogether.

A more interesting extension of this work would be to study Kac potentials satisfying neither of the conditions (i) and (ii) given at the end of Sec. V. For these potentials upper and lower bounds on $a(\rho, 0+)$ do not coincide, and it is possible that the behavior of the system is more complicated than in the van der Waals theory. For example the Kac potential might bring about spatial ordering with a length scale γ^{-1}. Finally from the physical point of view the most important extension of this work is to study the properties of the system when γ is finite but small. This is now being carried out along several lines.[27]

ACKNOWLEDGMENTS

We are grateful to M. Austin, M. E. Fisher, E. Lieb, D. Newman, and N. van Kampen for useful advice and suggestions.

[25] Cited in Ref. 4 of V. V. Tolmachev, Dokl. Akad. Nauk SSSR **134**, 1324 (1960) [Engl. transl.: Soviet Phys.–Doklady **5**, 984 (1961)]. See also M. Girardeau, J. Chem. Phys. **41**, 2945 (1964).

[26] In this connection see also the work of R. Brout, Phys. Rev. **115**, 824 (1959) (and later publications) who considered expansions in the inverse number of interacting spins, and G. A. Baker, Phys. Rev. **126**, 2072 (1962) who first considered the limiting process $\gamma \to 0$ in his study of spin systems.

[27] J. L. Lebowitz, G. Stell, and S. Baer, J. Math. Phys. **6**, 1282 (1965).

Volume 22, Number 13 PHYSICAL REVIEW LETTERS 31 March 1969

EXISTENCE OF THERMODYNAMICS FOR REAL MATTER WITH COULOMB FORCES

J. L. Lebowitz*
Belfer Graduate School of Science, Yeshiva University, New York, New York 10033

and

Elliott H. Lieb†
Department of Mathematics, Massachusetts Institute of Technology, Cambridge, Massachusetts 02139
(Received 3 February 1969)

It is shown that a system made up of nuclei and electrons, the constituents of ordinary matter, has a well-defined statistical-mechanically computed free energy per unit volume in the thermodynamic (bulk) limit. This proves that statistical mechanics, as developed by Gibbs, really leads to a proper thermodynamics for macroscopic systems.

In this note we wish to report the solution to a classic problem lying at the foundations of statistical mechanics.

Ever since the daring hypothesis of Gibbs and others that the equilibrium properties of matter could be completely described in terms of a phase-space average, or partition function, $Z = \mathrm{Tr}e^{-\beta H}$, it was realized that there were grave difficulties in justifying this assumption in terms of basic microscopic dynamics and that such delicate matters as the ergodic conjecture stood in the way. These questions have still not been satisfactorily resolved, but more recently still another problem about Z began to receive attention: Assuming the validity of the partition function, is it true that the resulting properties of matter will be extensive and otherwise the same as those postulated in the science of thermodynamics? In particular, does the thermodynamic, or bulk, limit exist for the free energy derived from the partition function, and if so, does it have the appropriate convexity, i.e., stability properties?

To be precise, if N_j are an unbounded, increasing sequence of particle numbers, and Ω_j a sequence of reasonable domains (or boxes) of volume V_j such that $N_j/V_j \to \text{constant} = \rho$, does the free energy per unit volume

$$f_j = -kT(V_j)^{-1} \ln Z(\beta, N_j, \Omega_j) \tag{1}$$

approach a limit [called $f(\beta,\rho)$] as $j\to\infty$, and is this limit independent of the particular sequence and shape of the domains? If so, is f convex in the density ρ and concave in the temperature β^{-1}? Convexity is the same as thermodynamic stability (non-negative compressibility and specific heat).

Various authors have evolved a technique for proving the above,[1,2] but always with one severe drawback. It had to be assumed that the interparticle potentials were short range (in a manner to be described precisely later), thereby excluding the Coulomb potential which is the true potential relevant for real matter. In this note we will indicate the lines along which a proof for Coulomb forces can be and has been constructed. The proof itself, which is quite long, will be given elsewhere.[3] We will also list here some additional results for charged systems that go beyond the existence and convexity of the limiting free energy.

To begin with, a sine qua non for thermodynamics is the stability criterion on the N-body Hamiltonian $H=E_K+V$. It is that there exists a constant $B\geq 0$ such that for all N,

$$V(r_1,\cdots,r_N)>-BN \quad \text{(classical mechanics)}, \qquad (2)$$

$$E_0>-BN \quad \text{(quantum mechanics)}, \qquad (3)$$

where E_0 is the ground-state energy in infinite space. (Classical stability implies quantum-mechanical stability, but not conversely.) Heuristically, stability insures against collapse. From the mathematical point of view, it provides a lower bound to f_j in (1). We wish to emphasize that stability of the Hamiltonian (H stability), while necessary, is insufficient for assuring the existence of thermodynamics. For example, it is trivial to prove H stability for charged particles all of one sign, and it is equally obvious that the thermodynamic limit does not exist in this case.

It is not too difficult to prove classical and thus also quantum-mechanical H stability for a wide variety of short-range potentials or for charged particles having a hard core.[2,4] But real charged particles require quantum mechanics and the recent proof of H stability by Dyson and Lenard[5] is as difficult as it is elegant. They show that stability will hold for any set of charges and masses provided that the negative particles and/or the positive ones are fermions.

The second requirement in the canonical proofs[1] is that the potential be tempered, which is to say that there exist a fixed r_0 and constants $C\geq 0$ and $\epsilon>0$ such that if two groups of N_a and N_b particles are separated by a distance $r>r_0$, their interparticle energy is bounded by

$$V(N_a\oplus N_b)-V(N_a)-V(N_b) \leq Cr^{-(3+\epsilon)}N_aN_b. \qquad (4)$$

Tempering is roughly the antithesis of stability because the requirements that the forces are not too repulsive at infinity insures against "explosion." Coulomb forces are obviously not tempered and for this reason the canonical proofs have to be altered. Our proof, however, is valid for a mixture of Coulomb and tempered potentials and this will always be understood in the theorems below. It is not altogether useless to include tempered potentials along with the true Coulomb potentials because one might wish to consider model systems in which ionized molecules are the elementary particles.

Prior to explaining how to overcome the lack of tempering we list the main theorems we are able to prove. These are true classically as well as quantum mechanically. But first three definitions are needed:

(D1) We consider s species of particles with charges e_i, particle numbers $N^{(i)}$, and densities $\rho^{(i)}$. In the following N and ρ are a shorthand notation for s-fold multiplets of numbers. The conditions for H stability (see above) are assumed to hold.

(D2) A neutral system is one for which $\sum_1^s N^{(i)}\times e_i=0$, alternatively $\sum_1^s\rho^{(i)}e_i=0$.

(D3) The ordinary s-species grand canonical partition function is

$$\sum_{N^{(s)}=0}^{\infty}\cdots\sum_{N^{(1)}=0}^{\infty}\prod_1^s z_i^{N^{(i)}}Z(N,\Omega). \qquad (5)$$

The neutral grand canonical partition function is the same as (5) except that only neutral systems enter the sum.

The theorems are the following:

(T1) The canonical, thermodynamic limiting free energy per unit volume $f(\beta,\rho)$ exists for a neutral system and is independent of the shape of the domain for reasonable domains. Furthermore, $f(\beta,\rho^{(1)},\rho^{(2)},\cdots)$ is concave in β^{-1} and jointly convex in the s variables $(\rho^{(1)},\cdots,\rho^{(s)})$.

(T2) The thermodynamic limiting microcanonical[6] entropy per unit volume exists for a neutral system and is a concave function of the energy per unit volume. It is also independent of domain shape for reasonable shapes and it is equal to the entropy computed from the canonical free energy.

(T3) The thermodynamic limiting free energy per unit volume exists for both the ordinary and the neutral grand canonical ensembles and are independent of domain shape for reasonable domains. Moreover, they are equal to each other and to the neutral canonical free energy per unit volume.

Theorem 3 states that systems which are not charge neutral make a vanishingly small contribution to the grand canonical free energy. While this is quite reasonable physically, it does raise an interesting point about nonuniform convergence because the ordinary and neutral partition functions are definitely not equal if we switch off the charge before passing to the thermodynamic limit, whereas they are equal if the limits are taken in the reverse order.

An interesting question is how much can charge neutrality be nonconserved before the free energy per unit volume deviates appreciably from its neutral value? The answer is in theorem 4.

(T4) Consider the canonical free energy with a surplus (i.e., imbalance) of charge Q and take the thermodynamic limit in either of three ways: (a) $QV^{-2/3}\to 0$; (b) $QV^{-2/3}\to\infty$; (c) $QV^{-2/3}\to$ const. In case (a) the limit is the same as for the neutral system while in case (b) the limit does not exist, i.e., $f\to\infty$. In case (c) the free energy approaches a limit equal to the neutral-system free energy plus the energy of a surface layer of charge Q as given by elementary electrostatics.

We turn now to a sketch of the method of proof and will restrict ourselves here to the neutral canonical ensemble. As usual, one first proves the existence of the limit for a standard sequence of domains. The limit for an arbitrary domain is then easily arrived at by packing that domain with the standard ones. The basic inequality that is needed is that if a domain Ω containing N particles is partitioned into D domains $\Omega_1, \Omega_2, \cdots, \Omega_D$ containing $N_1, N_2, \cdots, N_D$ particles, respectively, and if the interdomain interaction be neglected, then

$$Z(N,\Omega) \geq \prod_1^D Z(N_i, \Omega_i). \tag{6}$$

If Ω is partitioned into subdomains, as above, plus "corridors" of thickness $>r_0$ which are devoid of particles, one can use (4) to obtain a useful bound on the tempered part of the omitted interdomain interaction energy. We will refer to these energies as surface terms.

The normal choice[1] for the standard domains are cubes C_j containing N_j particles, with C_{j+1} being composed of eight copies of C_j together with corridors, and with $N_{j+1}=8N_j$. Neglecting surface terms one would have from (6) and (1)

$$f_{j+1} \leq f_j. \tag{7}$$

Since f_j is bounded below by H stability, (7) implies the existence of a limit. To justify neglect of the surface terms one makes the corridors increase in thickness with increasing j; although V_j^C, the corridor volume, approaches ∞ one makes $V_j^C/V_j\to 0$ in order that the limiting density not vanish. The positive ϵ of (4) allows one to accomplish these desiderata.

Obviously, such a strategy will fail with Coulomb forces, but fortunately there is another way to bound the interdomain energy. The essential point is that it is not necessary to bound this energy for all possible states of the systems in the subdomains; it is only necessary to bound the "average" interaction between domains, which is much easier. This is expressed mathematically by using the Peierls-Bogoliubov inequality[7] to show that

$$Z(N,\Omega) \geq e^{-\beta U}\prod_1^D Z(N_i, \Omega_i), \tag{8}$$

where U is the average interdomain energy in an ensemble where each domain is independent. U consists of a Coulomb part, U_C, and a tempered part, U_t, which can be readily bounded.[1]

We now make the observation, which is one of the crucial steps in our proof, that independently of charge symmetry U_C will vanish if the subdomains are spheres and are overall neutral. The rotation invariance of the Hamiltonian will produce a spherically symmetric charge distribution in each sphere and, as Newton[8] observed, two such spheres would then interact as though their total charges (which are zero) were concentrated at their centers.

With this in mind we choose spheres for our standard domains. Sphere S_j will have radius $R_j = p^j$ with p an integer. The price we pay for using spheres instead of cubes is that a given one, S_k, cannot be packed arbitrarily full with spheres S_{k-1} only. We prove, however, that it can be packed arbitrarily closely (as $k\to\infty$) if we use all the previous spheres $S_{k-1}, S_{k-2}, \cdots S_0$. Indeed for the sequence of integers $n_1, n_2, \cdots, n_j = (p-1)^{j-1}p^{2j}$ we can show that we can simultaneously pack n_j spheres S_{k-j} into S_k for $1\leq j\leq k$. The fractional volume of S_k occupied by the S_{k-j} spheres is $\varphi_j = p^{-3j}n_j$, and from (8) we then have

$$f_k \leq \varphi_1 f_{k-1} + \varphi_2 f_{k-2} + \cdots + \varphi_k f_0, \tag{9}$$

and

$$\sum_1^\infty \varphi_j = 1. \tag{10}$$

[Note that the inequality (6) is correct as it stands for pure Coulomb forces because U_C in (8) is identically zero. If short-range potentials are included there will also be surface terms, as in the cube construction, but these present only a technical complication that can be handled in the same manner as before.[1]] While Eq. (9) is more complicated than (7), it is readily proven explicitly that f_k approaches a limit as $k \to \infty$. [Indeed, it follows from the theory of the renewal equation[9] that (9) will have a limit if $\sum_1^\infty j\varphi_j < \infty$.]

The possibility of packing spheres this way is provided by the following geometrical theorem which plays the key role in our analysis. We state it without proof, but we do so in d dimensions generally and use the following notation: σ_d = volume of a unit d-dimensional sphere = $\frac{4}{3}\pi$ in three dimensions and $\alpha_d = (2^d - 1)2d^{\frac{1}{2}}$.

(T5) Let $p \geq \alpha_d + 2^d \sigma_d^{-1}$ be a positive integer. For all positive integers j, define radii $r_j = p^{-j}$ and integers $n_j = (p-1)^{j-1} p^{j(d-1)}$. Then it is possible to place simultaneously $\bigcup_j (n_j$ spheres of radius $r_j)$ into a unit d-dimensional sphere so that none of them overlap.

The minimum value of p required by the theorem in three dimensions is 27.

Many of the ideas presented here had their genesis at the Symposium on Exact Results in Statistical Mechanics at Irvine, California, in 1968, and we should like to thank our colleagues for their encouragement and stimulation: M. E. Fisher, R. Griffiths, O. Lanford, M. Mayer, D. Ruelle, and especially A. Lenard.

*Work supported by Air Force Office of Scientific Research, U. S. Air Force under Grant No. AFOSR 68-1416.

†Work supported by National Science Foundation Grant No. GP-9414.

[1]These developments are clearly expounded in M. E. Fisher, Arch. Ratl. Mech. Anal. 17, 377 (1964); D. Ruelle, Statistical Mechanics (W. A. Benjamin, Inc., New York, 1969). For a synopsis, see also J. L. Lebowitz, Ann. Rev. Phys. Chem. 19, 389 (1968).

[2]R. B. Griffiths, Phys. Rev. 176, 655 (1968), and footnote 6a in A. Lenard and F. J. Dyson [J. Math. Phys. 9, 698 (1968)]; O. Penrose, in Statistical Mechanics, Foundations and Applications, edited by T. Bak (W. A. Benjamin, Inc., New York, 1967), p. 98.

[3]E. H. Lieb and J. L. Lebowitz, "The Constitution of Matter," to be published.

[4]L. Onsager, J. Phys. Chem. 43, 189 (1939); M. E. Fisher and D. Ruelle, J. Math. Phys. 7, 260 (1966).

[5]F. J. Dyson and A. Lenard, J. Math. Phys. 8, 423 (1967); A. Lenard and F. J. Dyson, J. Math. Phys. 9, 698 (1968); F. J. Dyson, J. Math. Phys. 8, 1538 (1967).

[6]R. B. Griffiths, J. Math. Phys. 6, 1447 (1965).

[7]K. Symanzik, J. Math. Phys. 6, 1155 (1965).

[8]I. Newton, in Mathematical Principles, translated by A. Motte, revised by F. Cajori (University of California Press, Berkeley, Calif., 1934), Book 1, p. 193, propositions 71, 76.

[9]W. Feller, An Introduction to Probability Theory and Its Applications (J. Wiley & Sons, New York, 1957), 2nd ed. Vol. 1, p. 290.

Commun. math. Phys. 35, 87—92 (1974)

GHS and other Inequalities*

Joel L. Lebowitz

Belfer Graduate School of Science, Yeshiva University, New York, N.Y., USA

Received September 16, 1973

Abstract. We use a transformation due to Percus to give a simple derivation of the Griffiths, Hurst, and Sherman, and some other new inequalities, for Ising ferromagnets with pair interactions. The proof makes use of the Griffiths, Kelly, and Sherman and the Fortuin, Kasteleyn, and Ginibre inequalities.

1. Introduction

We consider an Ising spin system with ferromagnetic pair interactions; $\sigma_i = \pm 1$, $i \in \Lambda$, $i = 1, \ldots, |\Lambda|$,

$$H(\boldsymbol{\sigma}) = -1/2 \sum_{i \neq j} J_{ij}\sigma_i\sigma_j - \sum_i h_i\sigma_i, \qquad J_{ij} \geqq 0, \tag{1.1}$$

where H is the energy of the system and h_i are external magnetic fields.

The Ursell, or cluster function, $u_l(i_1, \ldots, i_l)$ play a central role in statistical mechanics. They are given for spin systems, by the relations [1]

$$u_l(i_1, \ldots, i_l) = \frac{\partial^l}{\partial h_{i_1} \ldots \partial h_{i_l}} \ln Z(\Lambda; \boldsymbol{h}, \boldsymbol{J}), \qquad l = 1, \ldots, |\Lambda| \tag{1.2}$$

where we have written $\boldsymbol{h}$ and $\boldsymbol{J}$ for the collections $\{h_i\}$, $\{J_{ij}\}$ and

$$Z(\Lambda; \boldsymbol{h}, \boldsymbol{J}) = \mathrm{Tr}_{\{\sigma\}} \exp[-H(\boldsymbol{\sigma})]. \tag{1.3}$$

Thus

$$\begin{aligned} u_1(i) &= \langle\sigma_i\rangle, \qquad u_2(i,j) = \langle\sigma_i\sigma_j\rangle - \langle\sigma_i\rangle\langle\sigma_j\rangle, \\ u_3(i,j,k) &= \langle\sigma_i\sigma_j\sigma_k\rangle - \langle\sigma_i\rangle\langle\sigma_j\sigma_k\rangle - \langle\sigma_j\rangle\langle\sigma_i\sigma_k\rangle \\ &\quad - \langle\sigma_k\rangle\langle\sigma_i\sigma_j\rangle + 2\langle\sigma_i\rangle\langle\sigma_j\rangle\langle\sigma_k\rangle, \ldots \end{aligned} \tag{1.4}$$

where $\langle\ \rangle$ denotes expectations with respect to the measure $\mu(\boldsymbol{\sigma}) = Z^{-1}\exp[-H]$; (we have set the temperature $\beta^{-1} = 1$).

The Griffiths, Kelly, and Sherman [2] (GKS) inequalities for this system apply when the h_i have the same sign for all i; say $h_i \geqq 0$, $\boldsymbol{h} \geqq 0$, (similar results hold by symmetry when $h_i < 0$, all i). They state

$$\langle\sigma_A\sigma_B\rangle \geqq \langle\sigma_A\rangle\langle\sigma_B\rangle, \qquad \boldsymbol{h} \geqq 0 \tag{1.5}$$

---

* Work supported in part by USAFOSR-73-2430.

where $A, B \subset \Lambda$ and $\sigma_A = \prod_{i \in A} \sigma_i$. The more recent Fortuin, Kasteleyn, and Ginibre [3] (FKG) inequalities apply whatever the signs of the h_i (assuming always of course that $J_{ij} \geqq 0$). FKG consider general increasing functions $f(\sigma_{i_1}, \ldots, \sigma_{i_n})$, $g(\sigma_{j_1}, \ldots, \sigma_{j_m})$, i.e. $f(\sigma_{i_1}, \ldots, \sigma_{i_l} = 1, \ldots, \sigma_{i_n}) \geqq f(\sigma_{i_1}, \ldots, \sigma_{i_l} = -1, \ldots, \sigma_{i_n})$. For such functions FKG prove that

$$\langle fg \rangle \geqq \langle f \rangle \langle g \rangle . \tag{1.6}$$

Note that if g is an increasing function $-g$ is a decreasing function and so the inequality in (1.6) is reversed if f is increasing and g is decreasing.

The more general form of the FKG inequalities, which we shall need later, state that if X is a "lattice", e.g. the subsets of Λ ordered by inclusion, and if $P(x)$, $x \in X$ is a normalized measure satisfying

$$P(x)\, P(y) \leqq P(x \vee y)\, P(x \wedge y), \quad \text{for all} \quad x, y \in X \tag{1.7}$$

then $\langle f g \rangle \geqq \langle f \rangle \langle g \rangle$ whenever $f(x)$ and $g(x)$ are increasing functions, $f(x) \geqq f(y)$ if $x \geqq y$. The expectations are now with respect to the measure P.

These inequalities which apply to more general (than pair) ferromagnetic type interactions have found many uses in statistical mechanics [1] (and more recently also in field theory [4]). They can be applied directly to the first two Ursell functions, $u_1(i) \geqq 0$, if $\boldsymbol{h} \geqq 0$, by GKS $u_2(i, j) \geqq 0$, for arbitrary fields by FKG. For simple proofs of GKS and FKG see [5] and [6] respectively.

There exists also another extremely useful inequality due to Griffiths, Hurst, and Sherman [7] (GHS) which *is* restricted to ferromagnetic pair interactions,

$$u_3(i, j, k) \leqq 0, \quad \text{for} \quad \boldsymbol{h} \geqq 0 . \tag{1.8}$$

(By symmetry $u_3 \geqq 0$ for $\boldsymbol{h} \leqq 0$.) The derivation of this inequality which is much more specialized than either the GKS or the FKG inequalities, is in many ways also more complicated involving combinatorial analysis, etc. It is the purpose of this note to derive the GHS inequality as part of a whole set of inequalities which are in turn derived, in a fairly simple way from the GKS and FKG inequalities.

2. Proof of GHS and other Inequalities

Let $\{s_i\}$ be a "duplicate" set of variables to the $\{\sigma_i\}$ with the same energy as in (1) and let $\langle \ \rangle'$ denote expectations with respect to the measure

$$\mu(\boldsymbol{\sigma}, \boldsymbol{s}) = Z^{-2}(\Lambda, \boldsymbol{h}, \boldsymbol{J}) \exp[H(\boldsymbol{\sigma}) + H(\boldsymbol{s})] . \tag{2.1}$$

Define $q_i=\frac{1}{2}(\sigma_i-s_i)$, $t_i=\frac{1}{2}(\sigma_i+s_i)$, $q_A=\prod_{i\in A} q_i$, $t_A=\prod_{i\in A} t_i$. q_i and t_i can take on the values, $-1, 0, 1$ with the constraint that $q_i=0\Rightarrow t_i=\pm 1$, $t_i=0\Rightarrow q_i=\pm 1$. We then readily find that (dropping the primes on $\langle\ \rangle'$)

$$\begin{gathered} u_1(i)=\langle t_i\rangle, \quad u_2(i,j)=2\langle q_i q_j\rangle \\ u_3(i,j,k)=4[\langle q_i q_j t_k\rangle-\langle q_i q_j\rangle\langle t_k\rangle], \end{gathered} \tag{2.2}$$

$$\begin{gathered} u_4(i,j,k,l)=8[\langle q_i q_j t_k t_l\rangle-\langle q_i q_j\rangle\langle t_k t_l\rangle] \\ -2[u_3(i,j,k)\langle t_l\rangle+u_3(i,j,l)\langle t_k\rangle]. \end{gathered} \tag{2.3}$$

Theorem. *For Ising system with ferromagnetic pair interactions*

$$\langle q_C\rangle \geqq 0, \qquad C\subset\Lambda, \tag{2.4}$$

$$\langle q_C t_D\rangle \leqq \langle q_C\rangle\langle t_D\rangle, \qquad \text{if} \quad \boldsymbol{h}\geqq 0,\ C, D\subset\Lambda, \tag{2.5a}$$

$$\langle q_C q_D\rangle \geqq \langle q_C\rangle\langle q_D\rangle. \tag{2.5b}$$

Remarks. (i) $\langle t_A\rangle\geqq 0$, and $\langle t_A t_B\rangle\geqq\langle t_A\rangle\langle t_B\rangle$, if $\boldsymbol{h}\geqq 0$, is a direct consequence of the GKS inequalities.

(ii) The proof of (2.4) is based on the GKS inequalities alone and is due to Percus [8] who introduced the "duplicate" set of $\boldsymbol{s}$ variables. (2.4) and (2.2) imply $u_2(i,j)\geqq 0$ for arbitrary h_i without the use of the FKG inequalities [8]. The proof of (2.5) is new and makes use of the GKS and the general form of the FKG inequalities discussed in Section 1. (2.5) together with (2.3) imply the GHS inequality.

Proof. The function $H(\boldsymbol{\sigma})+H(\boldsymbol{s})$ appearing in the measure $\mu(\boldsymbol{\sigma},\boldsymbol{s})$ in (2.1) becomes in terms of the variables $\boldsymbol{q}$ and $\boldsymbol{t}$,

$$\begin{aligned} H_\Lambda(\boldsymbol{\sigma})+H_\Lambda(\boldsymbol{s}) &= -\Big[1/2\sum_{i\neq j} 2J_{ij}q_i q_j\Big]-\Big[1/2\sum_{i\neq j}2J_{ij}t_i t_j+\Sigma\, 2h_i t_i\Big] \\ &= \hat{H}_\Lambda(\boldsymbol{q};0,2\boldsymbol{J})+\hat{H}_\Lambda(\boldsymbol{t};2\boldsymbol{h},2\boldsymbol{J}) \end{aligned} \tag{2.6}$$

where we have indicated explicitly the dependences of the energy on $\boldsymbol{h}$ and $\boldsymbol{J}$ and use the cap to remind us that q_i and t_i in (2.6) are *not* Ising variables since they take the values $-1, 0, 1$ with the restrictions mentioned earlier. Using (2.6) it is easy to see that for any functions $\phi(\boldsymbol{q})$ and $\Psi(\boldsymbol{t})$,

$$\langle\phi(\boldsymbol{q})\,\Psi(\boldsymbol{t})\rangle=\sum_{A\subset\Lambda} P(A)\,f(A)\,g(A), \tag{2.7}$$

where

$$\begin{aligned} P(A)=&\{\mathrm{Tr}_{q_I}\exp[-H_A(\boldsymbol{q};0,2\boldsymbol{J})]\} \\ &\cdot\{\mathrm{Tr}_{t_I}\exp[-H_{\tilde{A}}(\boldsymbol{t};2\boldsymbol{h},2\boldsymbol{J})]\}/Z^2(\Lambda;\boldsymbol{h},\boldsymbol{J}), \end{aligned} \tag{2.8}$$

$$f(A)=\{\mathrm{Tr}_{q_I}\phi(\boldsymbol{q})\exp[-H_A(q;0,2\boldsymbol{J})]\}/Z(A;0,2\boldsymbol{J}), \tag{2.9}$$

$$g(A)=\{\mathrm{Tr}_{t_I}\Psi(\boldsymbol{t})\exp[-H_{\tilde{A}}(\boldsymbol{t};2\boldsymbol{h},2\boldsymbol{J})]\}/Z(\tilde{A};0,2\boldsymbol{J}). \tag{2.10}$$

Here $\tilde{\Lambda}$ denotes the complement of A in Λ and the subscripts I as well as the absence of a cap on H_A and $H_{\tilde{A}}$ indicate that in taking traces in (2.8)–(2.10) over the q_i, $i \in A$, and t_i, $i \in \tilde{A}$, the q_i and t_i are to be treated as Ising variables taking on the values ± 1 only, while $q_i = 0$ for $i \in \tilde{A}$, $t_i = 0$ for $i \in A$. Clearly $P(A)$ is the probability with respect to the measure $\mu(\boldsymbol{\sigma}, \mathbf{s})$ that $q_i = \pm 1$, $t_i = 0$ for $i \in A$, while $f(A)$ is the expectations value of $\phi(q)$ in an Ising system in A with energy $H_A(q; 0, 2\boldsymbol{J})$. Similarly $g(A)$ is the expectation value of $\Psi(\boldsymbol{t})$ in an Ising system in $\tilde{A}$ with energy $H_{\tilde{A}}(\boldsymbol{t}; 2\boldsymbol{h}, 2\boldsymbol{J})$.

It is now readily seen that (2.4) holds since in this case $f(A) \geqq 0$ by GKS and $g(A) = 1$. To prove (2.5) we note first that it follows from the GKS inequalities that $\langle \sigma_B \rangle$, $B \subset \Lambda$ is non decreasing as Λ is increased, i.e. $\langle \sigma_B \rangle_{\Lambda'} \geqq \langle \sigma_B \rangle_\Lambda$ for $\Lambda' \supset \Lambda$ (and $\boldsymbol{h} \geqq 0$). Hence for $\phi(\boldsymbol{q}) = q_C$, $\Psi(\boldsymbol{t}) = t_D$, $f(A)$ is an increasing and $g(A)$ a decreasing function of the "size" of A, i.e. $f(A') \geqq f(A)$, $g(A') \leqq g(A)$ if $A' \supset A$. Similarly

$$\langle q_C q_B \rangle = \sum_{A \subset \Lambda} P(A)\, \chi(A) \geqq \sum_{A \subset \Lambda} P(A)\, \chi_1(A)\, \chi_2(A) \tag{2.11}$$

where

$$\chi(A) = \mathrm{Tr}_{q_I} \{q_C q_D \exp[-H_A(\boldsymbol{q}; 0, 2\boldsymbol{J})]\}/Z(A; 0, 2\boldsymbol{J}), \tag{2.12}$$

and $\chi_1(A)$, $\chi_2(A)$ are the corresponding Ising expectations of q_C and q_B respectively. The inequality in (2.11) follows from GKS which also shows that both $\chi_1(A)$ and $\chi_2(A)$ are increasing functions on the sets A.

Thus (2.5) will be a consequence of the FKG inequalities if we can show that the measure P satisfies (1.7), i.e.

$$P(A)\, P(B) \leqq P(A \cup B)\, P(A \cap B). \tag{2.13}$$

Using (2.8) we see that (2.13) is a consequence of the following lemma.

Lemma. *For an Ising spin system with general ferromagnetic interactions,*

$$H_\Lambda = -\sum_{Q \subset \Lambda} J_Q \sigma_Q, \quad J_Q \geqq 0 \tag{2.14}$$

$$Z(\Lambda_1)\, Z(\Lambda_2) \leqq Z(\Lambda_1 \cup \Lambda_2)\, Z(\Lambda_1 \cap \Lambda_2) \tag{2.15}$$

where $Z(\Lambda) = \mathrm{Tr}_\sigma \exp[H_\Lambda(\boldsymbol{\sigma})]$.

Remark. In applying (2.15) to prove (2.13), $\Lambda_1 = A$, $\Lambda_2 = B$ or $\Lambda_1 = \tilde{A}$, $\Lambda_2 = \tilde{B}$.

Proof. Setting $\Lambda_1 \cap \Lambda_2 = K_1$, $\Lambda_1 \backslash K_1 = K_2$, $\Lambda_2 \backslash K_1 = K_3$ we want to show that

$$F \equiv \ln Z(K_1 \cup K_2 \cup K_3) - \ln[Z(K_1 \cup K_2)\, Z(K_1 \cup K_3)/Z(K_1)] \geqq 0. \tag{2.16}$$

Now it follows readily from the GKS inequalities that (2.16) holds when there is *no* interaction between spins on sites in K_1 and spins on sites in K_3. Consider now the change in F when an interaction term of the form $J\sigma_{Q_1}\sigma_{Q_3}$, $J>0$, $Q_1\subset K_1$, $Q_3\subset K_3$ is added to the energy of the system. Taking the derivative of F with respect of J we obtain

$$\frac{\partial F}{\partial J} = \langle\sigma_{Q_1}\sigma_{Q_3}\rangle_{K_{123}} - \langle\sigma_{Q_1}\sigma_{Q_3}\rangle_{K_{13}} \tag{2.17}$$

where we have used the abbreviations $K_{12}=K_1\cup K_2$, etc. The right side of (2.17) is $\geqq 0$ by GKS and hence $F\geqq 0$. The lemma and the theorem are thus proven.

Remarks. (i) It follows from (2.3) that $u_4(i,j,k,l)\leqq 0$ whenever $\langle\sigma_k\rangle$ and $\langle\sigma_l\rangle$ (or, by symmetry the expectation values of any pair of spins in the set $\{i,j,k,l\}$) vanish, e.g. $\boldsymbol{h}=0$. Furthermore using the symmetry of u_4 and the fact that $\langle t_k t_l\rangle = \langle q_k q_l\rangle$ when $\boldsymbol{h}=0$, we have in this case

$$|u_4(i,j,k,l)|\leqq 2u_2(i,j)\,u_2(k,l)\,,$$
$$|u_4(i,j,k,l)|\leqq 2[u_2(i,j)\,u_2(k,l)\,u_2(i,k)\,u_2(j,l)]^{1/2}\,,$$
$$|u_4(i,j,k,l)|\leqq 2[u_2(i,j)\,u_2(k,l)\,u_2(i,k)\,u_2(j,l)\,u_2(i,l)\,u_2(j,k)]^{1/3}\,.$$

This may be of some relevance to the question of the decay rates of the different Ursell functions [9].

(ii) For $\boldsymbol{h}\geqq 0$, $\langle q_A\rangle$ is a decreasing and $\langle t_A\rangle$ is an increasing function of the external fields $\{h_i\}$.

(iii) By combining (2.4), (2.5) and the subsequence remark we find that for $\boldsymbol{h}\geqq 0$,

$$\langle q_A t_B\rangle\langle q_C\rangle \leqq \langle q_A\rangle\langle t_B\rangle\langle q_C\rangle \leqq \langle q_A q_C\rangle\langle t_B\rangle\,,$$
$$\langle q_A t_B\rangle\langle t_C\rangle \leqq \langle q_A\rangle\langle t_B\rangle\langle t_C\rangle \leqq \langle q_A\rangle\langle t_B t_C\rangle\,.$$

Acknowledgement. I am greatly indebted to J. Percus for communicating to me his method of using the "duplicate" set of variables s prior to publication. I also want to thank D. Ruelle for some useful comments.

References

1. See for example, Ruelle, D.: Statistical mechanics. New York: Benjamin 1969. — Lebowitz, J.L.: Commun. math. Phys. **28**, 313 (1972)
2. Griffiths, R.B.: J. Math. Phys. **8**, 478, 484 (1967). — Kelly, D.G., Sherman, S.: J. Math. Phys. **9**, 466 (1968)
3. Fortuin, C.M., Kasteleyn, P.W., Ginibre, J.: Commun. math. Phys. **22**, 89 (1971)
4. See for example, Simon, B.: Commun. math. Phys. **31**, 127 (1973)

5. Ginibre, J.: Commun. math. Phys. **16**, 310 (1970)
6. Cartier, P.: Seminaire Bourbaki, no. 431, 1972/73
7. Griffiths, R. B., Hurst, C. A., Sherman, S.: J. Math. Phys. **11**, 790 (1970)
8. Percus, J.: unpublished
9. Duneau, M., Souillard, B., Iagolnitzer, D.: Commun. math. Phys. **31**, 191 (1973) and Analyticity and strong cluster properties for classical gases with finite range interaction (preprint)

Communicated by D. Ruelle

J. L. Lebowitz
Belfer Graduate School of Science
Yeshiva University
500 West 185th Street
New York, N.Y. 10033, USA

JOURNAL OF COMPUTATIONAL PHYSICS 17, 10–18 (1975)

A New Algorithm for Monte Carlo Simulation of Ising Spin Systems*

A. B. BORTZ

Belfer Graduate School of Science, Yeshiva University, New York, New York 10033

M. H. KALOS

Courant Institute of Mathematical Sciences, New York University, New York, New York 10012

AND

J. L. LEBOWITZ

Belfer Graduate School of Science, Yeshiva University, New York, New York 10033

Received June 27, 1974; revised September 4, 1974

We describe a new algorithm for Monte Carlo simulation of Ising spin systems and present results of a study comparing the speed of the new technique to that of a standard technique applied to a square lattice of 6400 spins evolving via single spin flips. We find that at temperatures $T < T_c$, the critical temperature, the new technique is faster than the standard technique, being ten times faster at $T = 0.588\,T_c$. We expect that the new technique will be especially valuable in Monte Carlo simulation of the time evolution of binary alloy systems. The new algorithm is essentially a reorganization of the standard algorithm. It accounts for the a priori probability of changing spins before, rather than after, choosing the spin or spins to change.

INTRODUCTION

The Monte Carlo method has been used extensively to study spin one-half Ising systems with nearest neighbor interactions [1–8]. It has been used to study equilibrium properties [1] as well as the time evolution [2–8] of spin lattice systems having interactions described by the hamiltonian

$$\mathscr{H} = -J \sum_{i,j} S_i S_j - H \sum_i S_i\,; \qquad S_i = \pm 1, \tag{1}$$

* Research supported by AFOSR Grant #73-2430A and AEC Contract AT(11-1)-3077.

where the summation indices refer to sites on a d-dimensional lattice (usually with toroidal boundary conditions) and the first summation is restricted to nearest neighbor spins. New configurations of the lattice are generated from an initial configuration by reversing single spins (single spin flip studies [1–4]), or by interchanging pairs of unlike nearest neighbor spins (spin exchange studies [5–8]). Because a spin exchange leaves the number of up and down spins unchanged, it has been used to study binary alloys, identifying spin up with atomic species A and spin down with atomics species B.

The standard Monte Carlo algorithm generates a sequence of configurations as follows. (1) Given a configuration $\mathscr{C}$, the program selects at random with uniform probability a spin (or pair of nearest neighbor spins). (2) The program computes the probability P of reversing (interchanging) the spin(s) according to a thermodynamically reasonable formula. The formula reflects the fact that if the reversal (interchange) produces configuration $\mathscr{C}'$, the equilibrium ratio of the probability of $\mathscr{C}'$ to the probability of $\mathscr{C}$ is $\exp(-\Delta E/kT)$, where ΔE is the energy of $\mathscr{C}'$ less the energy of $\mathscr{C}$ and kT is Boltzmann's constant times the absolute temperature. (3) The program chooses at random a fraction R with uniform probability over the interval [0, 1). (4) Then it performs the reversal (interchange) if $R \leqslant P$. In time evolution problems, P is usually computed as

$$P = x/(1 + x); \qquad x = \exp(-\Delta E/kT). \tag{2}$$

An attempted reversal (interchange) at any lattice site (pair) represents the same amount of time as an attempted reversal (interchange) at any other site (pair). The number of attempts can then be used as a measure of time. In equilibrium studies, P is usually computed [9] as

$$P = \begin{cases} 1, & \text{if } \Delta E \leqslant 0, \\ \exp(-\Delta E/kT), & \text{otherwise,} \end{cases} \tag{3}$$

speeding up the program by eliminating step (3) if $\Delta E \leqslant 0$ and increasing the probability of generating a new configuration at the expense of losing the time variable. Both (2) and (3) describe a Markov process whose stationary state is the equilibrium distribution with fixed magnetic field (magnetization).

When the system is in or near equilibrium or some metastable state, the rate of generating new configurations using the standard algorithm becomes quite slow, since P is then usually very small. In ferromagnetic ($J > 0$) spin exchange problems, the rate of generating new configurations is even slower, since it becomes very unlikely that a pair of nearest neighbors, selected at random, will be different. Thus the usefulness of the standard algorithm is limited by the low probability of generating new configurations.

We describe here an algorithm which generates a new configuration with every choice of spin (neighbors) without changing the behavior of the process described by (2). It does so by accounting for the a priori probability of reversal (interchange) before, rather than after, choosing the site (pair) to change. In that sense it is a reorganization of the standard algorithm. Primarily because of extra bookkeeping, the computation time to generate a new configuration using this algorithm is longer than the computation time to attempt a new configuration using the standard algorithm. However, in situations in which the standard algorithm usually rejects the test configuration, the algorithm we present here generates new configurations much faster.

The n-Fold Way

The new algorithm, hereafter called the n-fold way, is based on the fact that there is a small number n of classes of sites (neighbor pairs) classifying sites (pairs) by their probability of reversal (interchange). For example, in the single spin flip square lattice with periodic boundary conditions there are only ten classes of sites, and we can number them as shown in Table I. The standard algorithm, as noted above, chooses among all sites (nearest neighbor pairs) with equal probability, determines that the selected site (pair) is of class i by scanning its neighbors, and finally decides whether to flip (interchange) by finding P, usually from a table of

TABLE I

Classifications of Spins in the Ten-Fold Way[a]

Class	Spin	Number of spin up nearest neighbors
1	Up	4
2	Up	3
3	Up	2
4	Up	1
5	Up	0
6	Down	4
7	Down	3
8	Down	2
9	Down	1
10	Down	0

[a] Lattice: square; Method: single spin flip; Periodic Boundary Conditions.

values P_i, and comparing it with a random fraction R. The n-fold way chooses among all sites (pairs) with a probability which is weighted in such a way that the probability of choosing a given site (pair) is proportional to its probability of flipping (interchanging). Thus once a site (pair) is selected, the flipping (interchanging) can be immediately performed.

The details of the n-fold way can be understood from the discussion that follows, a description of a test of the ten-fold way of Table I and the results of that test.

The Ten-Fold Way

We tested the ten-fold way (on a lattice of 80 × 80 spins) because many of the equilibrium properties of the (infinite) square Ising lattice, including the critical temperature T_c, are well known [10]. We describe the ten-fold way program in detail. Extensions to spin exchange models are obvious, but not trivial. Extensions to other lattices evolving via single spin flips are obvious and trivial.

The program was written for a CDC 6600 computer in Fortran Extended (FTN) language. An array LOC(6400) is partitioned into ten classes with moveable partitions. The value of an element of LOC specifies a spin's location in the 80 × 80 square. Its index and the partition keep track of the classes of the spins as follows. Let n_i be the number of spins in class i and let m_i be the number of spins whose class number is less than i. Then LOC($m_i + 1$) through LOC(m_{i+1}) contains the spins of class i (i.e., the partition is the m_i's and $n_i = m_{i+1} - m_i$). A second array LOOK(6400) is required in order to find the address in LOC and the class of a spin, given its position in the lattice. The position determines the index of an element of LOOK. The value of that element is equal to the address of the spin in LOC plus 2^{20} times the class of the spin. Thus the n-fold way requires an approximate doubling of core required for program variables (or tripling if one chooses to use two arrays, one for address in LOC and one for class, instead of packing the array LOOK).

Each spin flip requires the following calculations. First we calculate the ten numbers,

$$Q_i = \sum_{j=1}^{i} n_j P_j \,; \qquad i = 1, 2, \ldots, 10, \tag{4}$$

which change after each flip and therefore explicitly depend on time. Then we choose a random number R with uniform probability in the interval $[0, Q_{10})$. By letting the class of the selected spin be defined by i such that $Q_{i-1} \leqslant R < Q_i$ ($Q_0 \equiv 0$), we choose a spin in class i with probability $n_i P_i / Q_{10}$. We then choose a random integer l_i with uniform probability in the interval $[1, m_i]$ to find LCH, the particular spin to flip. Therefore the a priori probability of a given spin being chosen as LCH is P_i/Q_{10} which is proportional to P_i, its probability of flipping. The location in the lattice of LCH is stored in LOC ($m_i + l_i$).

To flip LCH, we rearrange LOC and its partitions in a relatively simple way, changing the class of LCH by ± 5 without changing the class of any other spins. But flipping LCH also changes the classes of its four nearest neighbors by ± 1. To find the neighbors' locations in LOC and their class (without checking against the partition of LOC, a time-consuming operation), we must look in LOOK. Once we find the neighbors in LOC, we change their classes by four more simple rearrangements of LOC. Whenever we rearrange LOC, we also make the corresponding changes in LOOK to preserve proper crossreferencing between the two arrays. The rearrangement of LOC can be made considerably simpler if we use a larger LOC array. We chose the size of LOC as we did in anticipation of memory limitations in three-dimensional simulations or in spin exchange simulations, both of which require a larger pair of arrays. (To have a reasonable crystal size in three dimensions, more lattice sites are needed. In spin exchange simulations, the n-fold way requires two arrays of pairs of neighbors, and there are twice as many pairs as spins in a square lattice, three times as many in a simple cubic.)

We increment the time variable t as follows. Q_{10} is the number of spins times the average probability that an attempt will produce a flip, given configuration $\mathscr{C}$. At each flip, the time is incremented by a stochastic variable Δt whose expectation value is proportional to Q_{10}^{-1}. Thus t is proportional to the number of attempts per site. We choose

$$\Delta t = -(\tau/Q_{10}) \ln R, \tag{5}$$

where R is a random fraction. This choice reflects properly the distribution of time intervals between flips, for a reasonable physical model. In that model, discussed in Appendix B, the cumulative time t is approximately proportional to real time.

Results

To discover circumstances under which the n-fold way is superior to the standard algorithm, we wrote a program for the standard algorithm which paralleled the ten-fold way program, using the same programming economies wherever possible. Using the same starting condition, we evolved both programs to steady state. We compared central processor time to produce all the flips needed to reach steady state in the two programs. We also calculated the ratio of times to produce a flip in steady state. The results are presented in Table II.

In zero magnetic field, the two programs perform about equally well at T_c, but as temperature decreases, the ten-fold way becomes markedly faster, being ten times as fast as the standard algorithm when $T/T_c = 0.588$. In nonzero field, the ten-fold way is even better. Almost certainly, previous authors [2–6] modified the standard algorithm using programming techniques which increased its speed without changing its basic structure. We could undoubtedly do the same to our ten-fold way program. Thus Table II should be regarded as simply illustrative.

TABLE II

Comparison of Two Algorithms: Standard and Ten-fold Way

				Approach to steady state				
				Calculation time (sec)				Time ratio in steady state
T/T_c	$4J/kT$	$2H/kT$	Initial configuration	Standard	Ten-fold Way	Flips	Tries	(standard/ten-fold)
0.588	3.0	0	Random	700	92	2.3×10^5	1.1×10^7	10
0.881	2.0	0	Random	1480	782	2.0×10^6	2.3×10^7	2.5
0.881	2.0	0.2	Random	19.0	13.0	3.0×10^4	3.0×10^5	3.6
0.881	2.0	0.2	All down	72.0	46.5	1.1×10^5	1.0×10^6	3.6
0.950	1.856	0	Random	553	415	1.0×10^6	8.6×10^6	1.4
1.000	1.763	0	Random	405	417	1.0×10^6	6.2×10^6	1.0
1.000	1.763	0	All up	16.5	12.3	3.0×10^4	2.6×10^5	1.1
1.000	1.763	0.1763	Random	25	21	5.0×10^4	3.9×10^5	1.8
1.000	1.763	0.1763	All up	2.56	1.75	2.5×10^3	3.6×10^4	2.0
1.000	1.763	0.1763	All down	49.3	41.4	1.0×10^5	7.3×10^5	2.1
1.100	1.602	0	Random	13.8	21.3	5.0×10^4	2.0×10^5	0.73
1.100	1.602	0.1602	Random	16.1	19.6	4.5×10^4	2.4×10^5	1.2

It illuminates the value of the novel organization of the n-fold way algorithm. From Table II, it is clear that even allowing for differences in programming technique, there are many situations in which the ten-fold way is definitely superior to the standard algorithm in terms of central processor time.

Extending the n-Fold Way

For ferromagnetic spin exchange studies, the n-fold way would certainly be a great improvement over the standard algorithm under most conditions. Since clustering of like spins occurs quickly, the standard algorithm would quickly waste a lot of time choosing like neighbors. Furthermore, even when an unlike pair is found, the energy change can be greater than in single spin-flip simulations (e.g., in the square lattice, there are six neighbor bonds that change in a spin exchange, but only four in a spin flip). Thus the probability of interchanging can be much less than the probability of flipping.

We expect the system to tend to a state in which the classes with the smallest probabilities of interchanging have the most pairs, especially at low temperatures. This expectation is confirmed by a previous study [7]. $P_{\min}$, the smallest probability of interchanging, therefore influences $\bar{P}$, the average probability of interchanging more than any other P_i does. Since $P_{\min}$ is smaller for interchanges than for flips, interchanges are, on the average, less likely than single spin flips at the same value of J/kT.

Based on the results presented in Table II and the above comments, we conclude that the n-fold way is a valuable alternate technique to the standard Monte Carlo algorithm for generating configurations of Ising spin systems. In many studies it may reduce computation time by an order of magnitude or more. It will be most valuable in studies involving the spin-exchange mechanism with ferromagnetic interactions. Its value in single spin-flip models, although clearly demonstrated, is limited since the greatest interest in those models is near $T = T_c$ where the standard technique appears to be as efficient as the more elaborate n-fold way.

It is interesting to contrast our algorithm with one proposed by Friedberg and Cameron [11] for the study of equilibrium Ising systems. Their method is a variant of the basic idea of Metropolis *et al.* [9] which generates the random spin flips in a way particularly well suited for modern digital computers. Our method was motivated primarily as a study of the time evolution of a system but it could be applied equally well at equilibrium. If so used it represents an entirely different approach which is or could be made complementary. Friedberg and Cameron stress carrying out the elementary operations in a very rapid way. We stress the need to emphasize sites likely to undergo transitions. But there is no reason why elements of both algorithms could not be combined.

APPENDIX A: AN INTERMEDIATE ALGORITHM

Previously we reported results of a Monte Carlo simulation of the time evolution of a two-dimensional binary alloy [7]. The algorithm used there was intermediate between the standard algorithm and the seven-fold way for spin exchanges on a ferromagnetic square Ising lattice. Using a redundant crossreferencing scheme (like the LOC/LOOK arrays), we selected an up spin (A atom) which had at least one down spin (B atom) neighbor; then we chose a neighbor at random. This made the spin exchange mechanism more like the single spin flip mechanism since each site chosen had a finite chance of being changed.

But even with this improvement, the algorithm still suffered from unlikely interchanges. Data from that study indicates that at $T/T_c = 0.588$, a system with equal concentrations of A and B atoms (equivalent to zero magnetization) required, on the average, about 40 choices of a pair of neighbors for each interchange once substantial clustering was evident. (In fact, that was nearly all of the time, for starting from a random initial state, this 40:1 ratio was established after only 3×10^4 interchanges in a study which ran to 2×10^6 interchanges.) For each interchange, there were, on the average, 13.8 AB pairs found, and for each AB pair found there were on the average 2.9 like pairs found. Thus under those circumstances, the more complicated seven-fold way would have saved considerable computation time.

APPENDIX B: COMPUTATION OF STOCHASTIC TIME VARIABLE

Suppose that our lattice system is immersed in a bath at fixed temperature T. Further suppose that this bath generates attempted flips randomly in space and time such that, on the average, there is one attempted flip per lattice site in time τ. We expect τ to depend on the temperature of the bath and the nature of the system, but only weakly on the state of the system. The probability of flipping a spin on a given random attempt is Q_{10}/N, where N is the number of lattice sites. Thus the probability of having a flip during the infinitesimal time interval dt is

$$p\,dt = (Q_{10}/\tau)\,dt. \tag{6}$$

We calculate the distribution of time intervals between flips by considering the probability that no flip occurs before time Δt has elapsed since the previous flip, $P(\Delta t)$.

The probability that no flip occurs before $\Delta t + dt$ has elapsed since the previous flip, $P(\Delta t + dt)$, must be less than $P(\Delta t)$ by dp, the probability that no flip occurs before Δt, but one occurs between Δt and $\Delta t + dt$. But dp is clearly $P(\Delta t)\, p\, dt$; thus,

$$P(\Delta t + dt) = P(\Delta t) - P(\Delta t)(Q_{10}/\tau)\, dt. \tag{7}$$

We rewrite (7) as a differential equation

$$P'(\Delta t) = (-Q_{10}/\tau)\, P(\Delta t), \tag{8}$$

which has solution

$$P(\Delta t) = \exp(-Q_{10}\, \Delta t/\tau), \tag{9}$$

where we have used the obvious boundary condition that $P(0) = 1$.

We can now calculate the stochastic time interval between flips. We choose a random fraction R uniformly over the interval (0, 1). We then set $P(\Delta t) = R$ and solve for Δt:

$$\Delta t = -(\tau/Q_{10}) \ln R. \tag{10}$$

Since we assume that τ is nearly independent of the state of the system, the cumulative time t is approximately proportional to real time.

Acknowledgment

We are grateful to Dr. K. Binder for valuable comments.

References

1. L. D. Fosdick, Monte Carlo Computations on the Ising Lattice, *in* "Methods in Computational Physics," Vol. 1, (B. Alder, Ed.) Academic Press, New York, 1963; and references therein.
2. H. Müller-Krumbhaar and K. Binder, *J. Stat. Phys.* **8** (1973), 1; and references therein.
3. K. Binder and E. Stoll, *Phys. Rev. Letters* **30** (1973), 47.
4. E. Stoll, K. Binder, and T. Schneider, *Phys. Rev. B* **8** (1973), 3266.
5. P. A. Flinn, *J. Stat. Phys.* **10** (1974), 89.
6. K. Binder, *Z. Physik*, to be published.
7. A. B. Bortz, M. H. Kalos, J. L. Lebowitz, and M. A. Zendejas, *Phys. Rev, B* **10** (1974), 535.
8. A. B. Bortz, *J. Stat. Phys.*, to be published.
9. N. C. Metropolis, A. W. Rosenbluth, M. N. Rosenbluth, and E. Teller, *J. Chem. Phys.* **21** (1953), 1087.
10. L. Onsager, *Phys. Rev.* **65** (1944), 117.
11. R. Friedberg and J. E. Cameron, *J. Chem. Phys.* **52** (1970), 6049.

Advances in Chemical Physics, Volume XXXVIII
Edited by Stuart A. Rice

IRREVERSIBLE THERMODYNAMICS FOR QUANTUM SYSTEMS WEAKLY COUPLED TO THERMAL RESERVOIRS

HERBERT SPOHN*

Belfer Graduate School of Science
Yeshiva University
New York, N.Y.

JOEL L. LEBOWITZ†

Service de Physique Théorique
CEN Saclay
Gif-sur-Yvette, France

CONTENTS

I. INTRODUCTION

Statistical mechanics is the link between the microscopic and macroscopic world. Put more poetically—since this article is dedicated to Ilya Prigogine, whose life is devoted to the discovery and exposition of the grand poetry of nature—statistical mechanics is the bridge between the

* On leave of absence from the Fachbereich Physik der Universität München. Work supported by a Max Kade Foundation Fellowship.

† John Simon Guggenheim Fellow. Permanent address: Belfer Graduate School of Science, Yeshiva University, New York, N.Y., where work is supported by AFOSR Grant.

world of the atom and the world of the object. Man lives in the latter world but wishes, for aesthetic as well as for practical reasons, to understand the relation between the two worlds. This is the task of statistical mechanics.

Traditionally, statistical mechanics is split into two parts: equilibrium and nonequilibrium. Our understanding of the former is, formally at least, essentially complete. The Gibbs formalism gives a well-specified prescription for computing all the equilibrium properties of a macroscopic system from a knowledge of its microscopic Hamiltonian. Compared to this, our understanding of nonequilibrium statistical mechanics is still very incomplete. There is no general prescription for computing from first principles the great variety of nonequilibrium behavior observed in macroscopic systems. Some progress has been made, however, in recent years, and Prigogine is one of the pioneers in this field. It is therefore appropriate that this article, which describes recent developments in one small corner of this vast field, should be dedicated to him on the occasion of his sixtieth birthday.

The nonequilibrium phenomena we shall deal with in this article are transport processes. We wish to find the "Gibbs ensemble" appropriate for the description of a system in which there are flows of energy or material. Unfortunately, no such prescription exists at the present time even if we restrict ourselves to the case of steady fluxes. It is only in the special case of the weak coupling or van Hove limit that a mathematically precise microscopic theory has been developed. This theory is applied here to the case of systems coupled weakly to several idealized heat reservoirs. Some new results concerning entropy production in such situations are also presented. This yields in some special cases a microscopic justification for the principle of minimal entropy production that Prigogine has used so effectively in the thermodynamics of irreversible processes.

The situation we consider, then, is that of a finite system (say, a metal bar) coupled to inexhaustible heat reservoirs (heating the metal bar at one end and cooling it at the other). The case, where the reservoirs are of a certain stochastic nature, has been studied in detail by Lebowitz et al.[1–4] To obtain a truly dynamical model we choose in the present paper as reservoirs (infinitely extended) ideal gases at equilibrium. Even with such a simplifying assumption very little can be said. Already the problem to show the existence of a steady heat transporting state can be solved only in very special cases (such as the harmonic chain); even worse, these examples all show anomalous transport properties (their heat conductivity is infinite). Only in the limit, where the coupling between the system and the reservoirs becomes weak (van Hove limit) does the situation improve. However, a price has to be paid: All quantities can be determined only to lowest nonvanishing order in the coupling constant λ. For example, the total amount of energy flowing through the system $\langle J\rangle_\lambda$ has an expansion of the form $\langle J\rangle_\lambda = \lambda^2\langle J_2\rangle + \cdots$. In the weak coupling limit $\langle J_2\rangle$ can be determined.

In the limit of weak coupling we can prove the approach to a steady state as $t\to\infty$, the Onsager relations and the principle of minimal entropy production. Our proofs are based on two fundamental properties of the time evolution (which is a quantum generalization of a classical Markov process) obtained in this limit: (*1*) the detailed balance property, which is a direct consequence of the fact that the reservoirs are at thermal equilibrium, and (*2*) the convexity of the entropy production.

Many things are omitted from this article. In particular we do not discuss at all the Kubo relations for computing transport coefficients, nor do we include recent work by the authors[5] on the heat flow in harmonic crystals or recent results about the heat flow in a random harmonic chain.[6] The reason is that we tried to cover the weak coupling situation as fully as possible at the moment.

II. THE MICROSCOPIC MODEL

We wish to describe steady-state nonequilibrium processes such as heat flow in a spatially confined system $\mathcal{S}$. To sustain a truly steady heat flow, diffusion, or chemical reactions throughout the system, we have to couple $\mathcal{S}$ to inexhaustible energy and particle reservoirs $\mathcal{R}$. In fact, we treat here only heat conduction, since the essential features of the problem appear already in this simplest case. Indeed, in the model we use, diffusion and chemical reactions would be treated in an analogous fashion and their inclusion would lead only to a "multiplication of indices" without giving any more physical insight. $\mathcal{R}$ is therefore chosen to be a collection of heat reservoirs at various temperatures.

Microscopically, the system $\mathcal{S}$ is described in the usual way. We associate with $\mathcal{S}$ a Hilbert space $\mathcal{H}$. The states of $\mathcal{S}$ are described by statistical operators or density matrices, which are positive operators on $\mathcal{H}$ normalized to trace one. Observables are self-adjoint operators on $\mathcal{H}$. The dynamics of the isolated system is governed by the unitary group $\exp(-iHt)$ generated by the Hamiltonian H. Since we want $\mathcal{S}$ to be a thermodynamically stable system, we have to assume that $Z = \mathrm{tr}(\exp(-\beta H))$ exists for all positive β. H must therefore have a purely discrete spectrum, which is bounded below and has no finite limit points; that is, there are only a finite number of energy levels in any finite interval ΔE.

An inexhaustible heat reservoir will be modeled microscopically here by an infinitely extended quantum mechanical system with certain additional structures. The necessity of an infinitely extended system can be seen from simple energy arguments: A finite reservoir can supply heat to (or extract heat from) the system only over a certain period of time. On a more precise level, we will later see that, in the "Markovian approximation," the influence of the reservoir on the system manifests itself through certain time correlation functions of the reservoir. To have an approach to stationarity it is necessary that these correlation functions decay faster than $1/t$. Such a decay is impossible for finite quantum systems, where because of the discreteness of the energy spectrum all correlation functions are almost periodic functions of t. Of course, if $\mathcal{R}$ is a very large quantum system, the correlation functions can be expected to have in general the right decay properties over long time spans. This expresses simply the fact that a large bucket of water "acts" for some purposes as a good heat reservoir. But clearly a piece of metal placed between two isolated buckets of water (or champagne) at different temperatures will never approach a truly stationary state in which heat is flowing through the metal bar (our system $\mathcal{S}$). It is therefore sensible to go over to the idealization of making $\mathcal{R}$ an infinitely extended system for which the faster than $1/t$ decay is expected to be true in general (and can be proved for certain simple systems like ideal gases).

The reservoir $\mathcal{R}$ is composed of r independent parts $\mathcal{R}^1, \ldots, \mathcal{R}^r$, corresponding to heat baths with different temperatures. With each one of them we associate a Hilbert space $\mathcal{H}_k^R$, a time evolution $\exp(-iH_k^R t)$, and an equilibrium state (density matrix) ω^k for the temperature β_k^{-1}. Then the Hilbert space of the reservoir is the product $\mathcal{H}^R = \mathcal{H}_1^R \otimes \cdots \otimes \mathcal{H}_r^R$. Since the various parts of the reservoir are isolated from each other the Hamiltonian H^R of $\mathcal{R}$ is the sum

$$H^R = H_1^R + \cdots + H_r^R \tag{II.1}$$

and the state of $\mathcal{R}$ with the kth part at temperature β_k^{-1} is

$$\omega^R = \omega^1 \otimes \cdots \otimes \omega^r \tag{II.2}$$

[The thoughtful reader might wonder how we incorporated the fact that the reservoir is infinitely extended and how we can define a Hamiltonian for a system with infinite energy. The reservoir enters in the description of the open system $\mathcal{S}$ only through certain time correlation functions $\langle A(t)B \rangle$. They are defined as the limit of the finite volume correlation functions $\langle A(t)B \rangle_V \to \langle A(t)B \rangle$ as $V \to \infty$, where $\langle \, \rangle_V$ denotes the average over the grand canonical ensemble. The crucial point is that we can always find a Hamiltonian H and a density matrix ω such that $\langle A(t)B \rangle =$ $\mathrm{tr}\,(\exp(iHt)A \exp(-iHt)B\omega)$. Of course, since H and ω are defined indirectly only through the limiting process $V \to \infty$, they will not have the form familiar from the finite system (e.g., H usually has the whole real line as continuous spectrum and an eigenvalue zero). See Refs. 7, 8.]

We couple $\mathcal{S}$ and $\mathcal{R}$ by some bounded interaction operator λH^{SR}, where $\lambda \geq 0$ is the coupling strength. Then the self-adjoint Hamiltonian of the coupled system $\mathcal{S}+\mathcal{R}$ is given by

$$H_\lambda = H \otimes 1^R + 1 \otimes H^R + \lambda H^{SR} \tag{II.3}$$

acting on the Hilbert space $\mathcal{H} \otimes \mathcal{H}^R$. The time evolution of the statistical operator W is given by

$$U_t^\lambda W = \exp(-iH_\lambda t) W \exp(iH_\lambda t) \tag{II.4}$$

It is convenient to think of U_t^λ as acting in the space of all density matrices, that is, in the so-called trace class $T(\mathcal{H} \otimes \mathcal{H}^R)$. (This is the space of all operators on $\mathcal{H} \otimes \mathcal{H}^R$ such that $\operatorname{tr}|W| < \infty$. It is a Banach space under the norm $\|W\|_1 = \operatorname{tr}|W|$.)

The reservoir by itself is of no interest to us. It forms merely a convenient microscopic device to induce steady flows throughout the system. Therefore we derive, following well-known methods,[9–18] a generalized master equation describing the reduced dynamics of the system. We assume that the initial state of $\mathcal{S}+\mathcal{R}$ is uncorrelated and of the form $\rho \otimes \omega^R$, where ρ is an arbitrary density matrix of $\mathcal{S}$. Then, in the Schrödinger picture, the reduced dynamics Λ_t^λ of the system is defined by taking the partial trace over all degrees of freedom of the reservoir

$$\Lambda_t^\lambda \rho = \operatorname{tr}_R [U_t^\lambda (\rho \otimes \omega^R)] \tag{II.5}$$

Λ_t^λ describes the time evolution of $\mathcal{S}$ coupled to the reservoirs. To derive the master equation, one thinks of (II.5) as a projection on $T(\mathcal{H} \otimes \mathcal{H}^R)$. Let us define an amplification $\mathcal{A}$: $T(\mathcal{H}) \to T(\mathcal{H} \otimes \mathcal{H}^R)$ by

$$\mathcal{A}\rho = \rho \otimes \omega^R \tag{II.6}$$

Then

$$P = \mathcal{A} \operatorname{tr}_R \tag{II.7}$$

projects $T(\mathcal{H} \otimes \mathcal{H}^R)$ to the subspace $T(\mathcal{H}) \otimes \omega^R$. We can now write (II.5) as $(\Lambda_t^\lambda \rho) \otimes \omega^R = PU_t^\lambda(\rho \otimes \omega^R)$. Let V_t^λ be the time evolution on $T(\mathcal{H} \otimes \mathcal{H}^R)$ generated by $P[H_\lambda, \cdot\,]P + (1-P)[H_\lambda, \cdot\,](1-P)$ and let us assume $\operatorname{tr}_R \omega^R H^{SR} = 0$. We consider U_t^λ as a perturbation of V_t^λ. Then, by a well-known formula (which can be verified directly by differentiation),

$$U_t^\lambda = V_t^\lambda - i\lambda \int_0^t ds\, V_{t-s}^\lambda (P[H^{SR}, \cdot\,](1-P) + (1-P)[H^{SR}, \cdot\,]P) U_s^\lambda \tag{II.8}$$

Projecting on the right and on the left, we obtain

$$PU_t^\lambda P = V_t^\lambda P - i\lambda \int_0^t ds\, V_{t-s}^\lambda P[H^{SR}, \cdot\,](1-P)U_s^\lambda P \tag{II.9}$$

and

$$(1-P)U_t^\lambda P = -i\lambda \int_0^t ds\, V_{t-s}^\lambda (1-P)[H^{SR}, \cdot\,]PU_s^\lambda P \tag{II.10}$$

and by substituting (II.10) into (II.9)

$$PU_t^\lambda P = V_t^\lambda P + \lambda^2 \int_0^t ds \int_0^S du\, V_{t-s}^\lambda P[H^{SR}, \cdot\,](1-P) \times V_{s-u}^\lambda (1-P)[H^{SR}, \cdot\,]PU_u^\lambda P \tag{II.11}$$

If we identify $PT(\mathcal{H}\otimes\mathcal{H}^R) = T(\mathcal{H})\otimes\omega^R$ with $T(\mathcal{H})$, then (II.11) can be written as

$$\Lambda_t^\lambda \rho = \Lambda_t^0 \rho + \lambda^2 \int_0^t ds \int_0^s du\, \Lambda_{t-s}^0 K^\lambda (s-u)\Lambda_u^\lambda \rho \tag{II.12}$$

which is the desired integral equation for the reduced dynamics. On differentiating, we obtain the generalized master equation of Nakajima, Prigogine, Résibois, and Zwanzig:[10–12]

$$\frac{d}{dt}\rho(t) = -i[H, \rho(t)] + \lambda^2 \int_0^t ds\, K^\lambda(t-s)\rho(s) \tag{II.13}$$

The integral kernel $K^\lambda(s)$ is given by

$$K^\lambda(s) = -\mathrm{tr}_R\,([H^{SR}, \cdot\,](1-P)V_s^\lambda(1-P)[H^{SR}, \cdot\,]\mathcal{A}) \tag{II.14}$$

$K^\lambda(s)$ admits a power series expansion in the coupling constant λ given by

$$K^\lambda(s) = \Lambda_s^0\left[K_0(s) + \sum_{n=1}^{\infty} (-i)^n\lambda^n \int_{0\le t_1\le\cdots\le t_n\le s} dt_n \ldots dt_1 K_n(s, t_n, \ldots, t_1)\right] \tag{II.15}$$

where

$$K_0(s) = -\mathrm{tr}_R\,([H^{SR}(s), \cdot\,](1-P)[H^{SR}, \cdot\,]\mathcal{A}) \tag{II.16}$$

$$K_n(s, t_n, \ldots, t_1) = \mathrm{tr}_R\,([H^{SR}(s), \cdot\,](1-P)[H^{SR}(t_n), \cdot\,](1-P) \ldots (1-P)[H^{SR}(t_1), \cdot\,](1-P)[H^{SR}, \cdot\,]\mathcal{A}) \tag{II.17}$$

and

$$[H^{SR}(t), \cdot\,] = U_{-t}^0[H^{SR}, \cdot\,]U_t^0 = [\exp(iH_0t)H^{SR}\exp(-iH_0t), \cdot\,] \tag{II.18}$$

If we choose for H^{SR} a linear coupling of the form $V \otimes V^R$ (or a sum of such linear couplings), then $K_n(s, t_n, \ldots, t_1)$ depends on the (equilibrium) multitime correlation functions $\mathrm{tr}_R(\omega^R V^R(t_n') \ldots V^R(t_1')V^R)$ with $V^R(t) = \exp(iH^R t)V^R \exp(-iH^R t)$.

The first step in studying nonequilibrium steady state properties of $\mathcal{S}$ is to obtain the stationary state ρ_{st}^λ. If the reservoir temperatures are not too far apart and if H^{SR} represents a "good coupling," it seems safe to assume that there exists a unique stationary state ρ_{st}^λ and that every initial state approaches this state as $t \to \infty$. Therefore we should define

$$\rho_{st}^\lambda = \lim_{t\to\infty} \Lambda_t^\lambda \rho \tag{II.19}$$

Unfortunately, our knowledge about the existence of this limit is at the present rather poor, which should come as no surprise, since (II.19) is related to establishing mixing (which is a stronger notion than ergodicity) for the infinite system $\mathcal{S} + \mathcal{R}$. Only in the case where $\mathcal{S} + \mathcal{R}$ together form either an infinite ideal solid (harmonic crystal) or an infinite ideal gas (or some other quasifree system) the existence of the limit (II.19) has been proved.[19–28] Of course, we are happy to admit any simplification on parts of the reservoir. But the system $\mathcal{S}$ should be something like a real physical system and the border cases ideal solid and ideal gas are of very limited interest; even worse, Fourier's law is not obeyed in such systems. They have an infinite thermal conductivity.[29,30] (This should be contrasted with the fact that an ideal gas obeys the laws of equilibrium thermodynamics.)

In view of this situation we look for a simplification of (II.13). If λ becomes small we might neglect all higher-order terms in (II.15). A further simplification is obtained if the memory terms present in $K_0(s)$ become negligible. From the structure of the integral kernel $K_0(s)$ we expect such a situation if the time scale on which $\rho(t)$ varies is much longer than the decay time of the kernel $K_0(s)$, that is, than the decay time of the correlation functions of the reservoir appearing in (II.16). For small λ and large t we then obtain the Born approximation to the master equation (II.13) as

$$\frac{d}{dt}\rho = -i[H, \rho] + \left(\lambda^2 \int_0^\infty ds\, K(s)\right)\rho \tag{II.20}$$

which is the generalized Pauli equation. The original Pauli equation considered only transitions between energy eigenstates. Equation II.20 is a Markovian evolution equation on the whole space $T(\mathcal{H})$ of statistical operators. A precise definition of this weak coupling limit involves two points. First, we have to go from the Schrödinger picture to the interaction picture in order to suppress oscillations of parts of the system that happen to be uncoupled. Second, to make up for the slowing of the decay as $\lambda \to 0$, we have to rescale the physical time t in such a way that $\lambda^2 t = \tau$ is kept fixed. This is the van Hove limit[31,32] that has been made rigorous under some conditions by Davies. The theorem of Davies, to be discussed in the next section, asserts the existence of

$$\lim_{\lambda \to 0} \Lambda^0_{-\lambda^{-2}\tau} \Lambda^\lambda_{\lambda^{-2}\tau} \rho = e^{L\tau} \rho \tag{II.21}$$

which is, roughly, the integrated form of (II.20). L is a linear operator on $T(\mathscr{H})$.

All our microscopic treatment of the thermodynamics of irreversible processes is in terms of the time evolution $\{e^{L\tau} | \tau \geq 0\}$ of $\mathscr{S}$ obtained in the weak coupling limit. It is therefore essential to discuss at this point precisely what information about the irreversible processes is preserved in the weak coupling limit. After all, we work in the interaction picture, we let $\lambda \to 0$ and the physical time $t \to \infty$, and a priori it is unclear whether these limits just destroy all that we are interested in.

The coupling of a real physical system to a heat reservoir is always at the surface via short-range forces. The interaction energy H^{SR} should therefore be roughly proportional to the surface, that is, should be small compared to the bulk energy H. This crude argument gives some hope that at least some relevant information is preserved in the weak coupling limit.

To get a somewhat more precise idea, let us assume that the limit (III.19) exists and is unique. Furthermore, let $\{e^{L\tau}\} | \tau \geq 0\}$ have a unique stationary state ρ_{st},

$$e^{L\tau} \rho_{st} = \rho_{st} \tag{II.22}$$

for all $\tau \geq 0$. Then we expect that

$$\lim_{\lambda \to 0} \rho^\lambda_{st} = \rho_{st} \tag{II.23}$$

The limit $t \to \infty$ does not enter, since we look at a sequence of stationary states. The interaction picture drops out, since, as we show later, the stationary state obtained in the weak coupling limit is invariant under the isolated time evolution, which means that ρ_{st} commutes with H. Thus in the weak coupling limit we should obtain the lowest (zeroth) order contribution in λ to the stationary state ρ^λ_{st}, that is, $\rho^\lambda_{st} = \rho_0 + \lambda \rho_1 + \cdots$ with $\rho_0 = \rho_{st}$. In the case of a single reservoir temperature β^{-1}, ρ_0 is just the canonical ensemble $Z^{-1} \exp(-\beta H)$.

Since ρ_0 commutes with H, ρ_0 does not describe the local heat flow inside the system. This phenomenon would be described by the first-order contribution ρ_1, which cannot be obtained by the methods of the weak coupling limit. However, we can still compute the total amount of energy flowing through the system to lowest (λ^2) order in λ. Let us assume that the reservoir consists of a left part $\mathcal{R}^l$ and a right part $\mathcal{R}^r$ at different temperatures. The steady-state heat flow $\langle J\rangle_\lambda$ can be defined as the change of energy of $\mathcal{S}$ due to the coupling to the left heat reservoir averaged over the stationary state

$$\langle J\rangle_\lambda = \left\langle\left(\frac{d}{dt}H(t)\right)_{\text{left}}\right\rangle_\lambda \tag{II.24}$$

where $\langle\cdot\rangle_\lambda$ denotes the average over the stationary state of $\mathcal{S}+\mathcal{R}$ and $(\cdot)_{\text{left}}$ denotes the time evolution coming from the coupling to the left reservoir. (The Hamiltonian generating this time evolution is given by $H\otimes 1 + 1\otimes H_l^R + \lambda H_l^{SR}$.) As $\lambda\to 0$, but $\lambda^2 t=\tau$ fixed, we should have the convergence

$$\langle (H(t))_{\text{left}}\rangle_\lambda \to \operatorname{tr}\,(\rho_{st}(H(\tau))_{\text{left}}) = \operatorname{tr}\,(\rho_{st}\exp\,(L^*_{\text{left}}\tau)(H)) \tag{II.25}$$

The interaction picture has again been "absorbed" by the invariance of ρ_{st} under the isolated time evolution of $\mathcal{S}$. If we assume that also the derivatives converge

$$\frac{1}{\lambda^2}\langle J\rangle_\lambda = \frac{1}{\lambda^2}\left\langle\left(\frac{d}{dt}H(t)\right)_{\text{left}}\right\rangle_\lambda \to \operatorname{tr}\left(\rho_{st}\left(\frac{d}{d\tau}H(\tau)\right)_{\text{left}}\right) = \operatorname{tr}\,(\rho_{st}L^*_{\text{left}}(H)) = \langle J\rangle \tag{II.26}$$

then it follows that the lowest-order contribution to the average heat flow is of order λ^2 and that $\lambda^{-2}\langle J\rangle_\lambda$ converges to a quantity $\langle J\rangle$, which is defined in Section V as the heat flow in the weak coupling limit.

Of course, these are plausibility arguments only. Their proof would require as prerequisite the proof of the existence of the limit (II.19). We investigated[5] these questions at a model system, where $\mathcal{S}$ is a classical harmonic chain of N particles (arbitrary couplings and masses) and $\mathcal{R}$ consists of two semi-infinite harmonic chains of equal masses (=1) and nearest neighbor coupling of unit strength coupled to the first and the Nth particle of $\mathcal{S}$. The conditions on "good coupling" are that all eigenfrequencies ω of $\mathcal{S}$ are in the range $0<\omega<2$ and that the sum of the first and Nth component of every eigenmode does not vanish. If they are fulfilled, then one can prove the existence of the limits (II.19) to (II.26) and the interchange of the various limits involved.

Let us summarize: We consider a spatially confined quantum mechanical system $\mathcal{S}$ coupled to inexhaustible heat reservoirs at various temperatures. We perform the weak coupling limit and obtain a Markovian time evolution for the system $\mathcal{S}$. We study irreversible processes in $\mathcal{S}$ in terms of this time evolution. Then, as long as we consider a steady-state situation, we obtain results that are valid to the lowest nonvanishing order in the coupling constant λ.

III. THE WEAK COUPLING LIMIT

We investigate in more detail under what conditions the weak coupling limit (II.21) of the generalized master equation (II.12) exists and study some of its properties. We follow with minor extensions (see Ref. 33) the work of Davies.[18] References 34 and 35 contain an extension of the theory developed in Ref. 18. A study of the weak coupling limit for specific model systems can be found in Refs. 36 to 39 and in Ref. 5. A rigorous model close to van Hove's original considerations can be found in Ref. 40. Readers interested in the various physical applications of the van Hove or weak coupling limit are referred to the review articles.[15,16] Starting with (II.12), we rescale the time $\lambda^2 t = \tau$ and go over to the interaction picture

$$\Lambda^0_{-\lambda^{-2}\tau}\Lambda^\lambda_{\lambda^{-2}\tau}\rho$$

$$= \rho + \int_0^\tau d\sigma \left\{ \Lambda^0_{-\lambda^{-2}\sigma}\left[\int_0^{\lambda^{-2}(\tau-\sigma)} dx\, \Lambda^0_{-x} K^\lambda(x)\right] \Lambda^0_{\lambda^{-2}\sigma} \right\} \Lambda^0_{-\lambda^{-2}\sigma}\Lambda^\lambda_{\lambda^{-2}\sigma}\rho \tag{III.1}$$

where ρ is the density matrix of the system at time $t = 0$. We recall that Λ^0_t is simply the uncoupled time evolution of the system, which does not depend on λ, $\Lambda^0_t\rho = e^{-iHt}\rho\, e^{iHt}$. Expanding K^λ as in (II.15), we expect that in the limit as $\lambda \to 0$ only the K_0 term will contribute. The integration up to $\lambda^{-2}(\tau - \sigma)$ extends to infinity as $\lambda \to 0$. Therefore, with $\rho(\tau) = \lim_{\lambda\to 0} \Lambda^0_{-\lambda^{-2}\tau}\Lambda^\lambda_{\lambda^{-2}\tau}\rho$,

$$\rho(\tau) = \rho(0) + \lim_{\lambda\to 0} \int_0^\tau d\sigma\, \Lambda^0_{-\lambda^{-2}\sigma} K \Lambda^0_{\lambda^{-2}\sigma}\rho(\sigma) \tag{III.2}$$

where

$$K = \int_0^\infty dx\, K_0(x) \tag{III.3}$$

The remaining limit is an averaging of K with respect to the free time evolution Λ^0_t. Since the spectrum of H is purely discrete, this time average always exists and is given by

$$L = \lim_{T\to\infty} \frac{1}{2T} \int_{-T}^{T} dt\, \Lambda^0_{-t} K \Lambda^0_t = \sum_{\omega \in \mathrm{Sp}([H,\cdot])} Q_\omega K Q_\omega \tag{III.4}$$

Here Q_ω are the spectral projections of $[H, \cdot\,]$ considered as an operator on the trace class $T(\mathcal{H})$ corresponding to the eigenvalue ω, i.e.,

$$Q_\omega \rho = \sum_{\varepsilon_m - \varepsilon_n = \omega} P_m \rho P_n$$

where

$$H = \sum_j \varepsilon_j P_j \tag{III.5}$$

is the spectral representation of the Hamiltonian H. Since H^{SR} was assumed to be bounded, K and L are bounded operators on the Banach space $T(\mathcal{H})$. Inserting (III.4) in (III.2), we obtain

$$\rho(\tau) = \rho(0) + \int_0^\tau d\sigma\, L\rho(\sigma) \tag{III.6}$$

or, equivalently,

$$\frac{d}{d\tau}\rho(\tau) = L\rho(\tau), \qquad \rho(\tau) = e^{L\tau}\rho = \Lambda_\tau \rho \tag{III.7}$$

which is the desired Markovian limit of (II.13).

Davies proves that with suitable bounds on the K_n's, our heuristic argumentation is indeed valid.

Theorem 1 (Davies). Let

$$\int_0^\infty dt\, \|K_0(t)\| < \infty \tag{III.8}$$

$$\int_{0 \le t_1 \le \cdots \le t_n \le t} dt_n \ldots dt_1 \|K_n(t, t_n, \ldots, t_1)\| \le a_n(t)$$

and let the $a_n(t)$ be bounded by

$$a_n(t) \le c_n t^{n/2} \tag{III.9}$$

for all $t \ge 0$, where $\sum_{n=1}^\infty c_n z^n$ has an infinite radius of convergence, and by

$$a_n(t) \le d_n t^{n/2-\varepsilon} \tag{III.10}$$

for some $\varepsilon > 0$, $d_n \ge 0$, and all $t \ge 0$. Then in the norm topology of $T(\mathcal{H})$

$$\lim_{\lambda \to 0} \|\Lambda^0_{-\lambda^{-2}\tau} \Lambda^\lambda_{\lambda^{-2}\tau} \rho - e^{L\tau}\rho\|_1 = 0 \tag{III.11}$$

for all $\rho \in T(\mathcal{H})$ uniformly on every finite interval $[0, \tau_0]$.

If the reservoir is quasifree (e.g., an ideal gas), then all higher-order correlation functions are expressible as sums of two-point correlation functions. As Davies shows, the conditions (III.8) and (III.10) reduce then (at least for a quasifree Fermi system) to a single condition on the two-point function. If H^{SR} is of the form $V \otimes V^R$, then the bound

$$|\mathrm{tr}_R(\omega_R V^R(t) V^R)| \leq a(1+t)^{-(1+\varepsilon)} \tag{III.12}$$

for some $\varepsilon > 0$ and all $t \geq 0$ is sufficient for the theorem to hold. Equation III.12 means that the time-dependent correlation function of V^R evolving under the isolated reservoir time evolution decays faster than t^{-1}. (See Ref. 41 for an extension to reservoirs that are not quasifree.)

For our further investigations we need a more explicit form of the generator L. First, we treat the case of a single reservoir at temperature β^{-1}. We assume that the interaction operator H^{SR} has the form

$$H^{SR} = \sum_\alpha V_\alpha \otimes V_\alpha^R \tag{III.13}$$

where $V_\alpha = V_\alpha \in B(\mathcal{H})$, $V_\alpha^R = V_\alpha^R \in B(\mathcal{H}^R)$, $\sum_\alpha \|V_\alpha\|^2 < \infty$, and, as before, $\mathrm{tr}_R\, (\omega^R V_\alpha^R) = 0$. $B(\mathcal{H})$ denotes all bounded operators on $\mathcal{H}$. Since K contains the correlation functions $\mathrm{tr}_R\, (\omega^R V_\gamma^R V_\alpha^R(t))$ and since L is the time average of K, the Fourier transforms

$$h_{\alpha\gamma}(\omega) = \int_{-\infty}^{\infty} dt\, e^{-i\omega t}\, \mathrm{tr}_R\, (\omega^R V_\gamma^R V_\alpha^R(t)) \tag{III.14}$$

will enter the generator L. We now exploit the fact that ω^R is a thermal equilibrium state to derive an important relation for $h_{\alpha\gamma}$. Thermal equilibrium states of both finite and infinite quantum systems are defined by the Kubo–Martin–Schwinger (KMS) boundary condition.[42,43] We call a state, defined by the density matrix W, a KMS state (thermal equilibrium state) for inverse temperature β if

$$\mathrm{tr}\, (WAB(t)) = \mathrm{tr}\, (WB(t - i\beta)A) \tag{III.15}$$

for all observables A, B, and all t. As usual $B(t - i\beta)$ is defined as $\exp\, (i(t - i\beta)H)B \exp\, (-i(t - i\beta)H)$, where H is the Hamiltonian of the system under consideration. (We disregard domain questions that can be handled, however.) The reader may check that for a system in a box, (III.15) admits as only solution $e^{-\beta H}/Z(\beta)$. If the system is infinitely extended, there might be (and in some cases definitely is) more than one solution of (III.15) indicating the existence of phase transitions.

Using the KMS boundary condition (III.15) for ω^R and, as a direct consequence, the time invariance of ω^R, we obtain

$$\begin{aligned} h_{\alpha\gamma}(\omega) &= e^{\beta\omega} \int_{-\infty}^{\infty} dt\, e^{-i\omega t}\, \mathrm{tr}_R\, (\omega^R V_\alpha^R(t) V_\gamma^R) \\ &= e^{\beta\omega} \int_{-\infty}^{\infty} dt\, e^{i\omega t}\, \mathrm{tr}_R\, (\omega^R V_\alpha^R V_\gamma^R(t)) = h_{\gamma\alpha}(-\omega)\, e^{\beta\omega} \end{aligned} \tag{III.16}$$

that is,

$$h_{\gamma\alpha}(-\omega) = e^{-\beta\omega} h_{\alpha\gamma}(\omega)$$

Since $\{h_{\alpha\gamma}(\omega)\}$ is the Fourier transform of time-dependent correlation functions, it is a positive matrix for all ω. Equation III.16 is essential for deriving the Onsager relations, the principle of minimal entropy production, and other basic postulates of thermodynamics of irreversible processes. To state it in a somewhat oversimplified way: In our model irreversible thermodynamics follows from the fact that the reservoirs are at thermal equilibrium.

We define the Hilbert transform $s_{\alpha\gamma}$ of $h_{\alpha\gamma}$ as

$$\begin{aligned} s_{\alpha\gamma}(\omega) &= i\int_0^\infty dt\, e^{-i\omega t}\, \mathrm{tr}_R\,(\omega^R V_\gamma^R V_\alpha^R(t)) - \tfrac{1}{2}h_{\alpha\gamma}(\omega) \\ &= \frac{1}{2\pi}\mathcal{P}\int_{-\infty}^{\infty} \frac{h_{\alpha\gamma}(\lambda)}{\lambda - \omega}\, d\lambda \end{aligned} \tag{III.17}$$

where $\mathcal{P}$ denotes the principal part, and introduce the operators

$$V_\alpha(\omega) = \sum_{\varepsilon_m - \varepsilon_n = \omega} P_n V_\alpha P_m = V_\alpha(-\omega) \tag{III.18}$$

If we now perform for the chosen H^{SR} the integration (III.3) and the average (III.4), we obtain for the generator $L: T(\mathcal{H}) \to T(\mathcal{H})$

$$\begin{aligned} L(\rho) = \sum_{\omega \in \mathrm{Sp}([H,\cdot])} \sum_{\alpha\gamma} \{ &-is_{\alpha\gamma}(\omega)[V_\gamma(\omega)^* V_\alpha(\omega), \rho] \\ &+ h_{\alpha\gamma}(\omega)([V_\alpha(\omega)\rho, V_\gamma(\omega)^*] + [V_\alpha(\omega), \rho V_\gamma(\omega)^*])\} \end{aligned} \tag{III.19}$$

Sp $([H, \cdot])$ denotes the spectrum of the Liouville–von Neumann operator $[H, \cdot]$. It is the set $\{\varepsilon_m - \varepsilon_n | \varepsilon_m, \varepsilon_n \in \mathrm{Sp}\,(H)\}$. The generator L is the sum of a Hamiltonian part L_a (the $s_{\alpha\gamma}$-term) and a dissipative part L_s (the $h_{\alpha\gamma}$-term). Here $e^{Lt}\rho$ describes the time evolution of states (Schrödinger picture). As in the quantum mechanics of isolated systems, one is often interested in the time evolution of observables A (Heisenberg picture), which is clearly to be defined by

$$\begin{aligned} \mathrm{tr}\,(A\, e^{Lt}(\rho)) &= \mathrm{tr}\,(e^{L^*t}(A)\rho) \\ \mathrm{tr}\,(AL(\rho)) &= \mathrm{tr}\,(L^*(A)\rho) \end{aligned} \tag{III.20}$$

The dual L^* of L is defined by (III.20) and acts now on observables. With the help of (III.19) and (III.20) we find

$$\begin{aligned} L^*(A) = \sum_{\omega \in \mathrm{Sp}([H,\cdot])} \sum_{\alpha\gamma} \{ &is_{\alpha\gamma}(\omega)[V_\gamma(\omega)^* V_\alpha(\omega), A] \\ &+ h_{\alpha\beta}(\omega)([V_\gamma^*(\omega), A] V_\alpha(\omega) + V_\gamma^*(\omega)[A, V_\alpha(\omega)]\} \end{aligned} \tag{III.21}$$

Three points should be noted: (*1*) The reservoir temperature enters via $h_{\alpha\gamma}$ through the KMS condition (III.16). (*2*) The "reservoir part" of the interaction H^{SR} is reflected by the numerical coefficients $h_{\alpha\gamma}$ and $s_{\alpha\gamma}$, the "system part" of H^{SR} by the $V_\alpha(\omega)$'s. (*3*) The Hamiltonian H of $\mathcal{S}$ enters indirectly through the $V_\alpha(\omega)$'s by definition (III.18).

We list a few properties of the generator L: .

Property 1. The canonical ensemble $\rho_\beta = Z^{-1}\, e^{-\beta H}$ is stationary,

$$L(\rho_\beta)=0, \qquad e^{Lt}\rho_\beta=\rho_\beta \tag{III.22}$$

for all $t\geq 0$. To prove (III.22), we insert ρ_β in (III.19) and use (III.16). In general, ρ_β is the only stationary state of $\{\Lambda_t | t\geq 0\}$ and every initial state approaches ρ_β as $t\to\infty$ (see Section VI, Theorem 3).

Property 2. Λ_t commutes with the free time evolution Λ_t^0,

$$e^{Lt}(e^{-iHt}\rho\, e^{iHt})=e^{-iHt}(e^{Lt}\rho)\, e^{iHt} \tag{III.23}$$

In particular, all functions of the energy in $T(\mathcal{H})$ form an invariant subspace for $\{\Lambda_t | t\geq 0\}$. Equation III.23 follows from the fact that L was defined as the time average over an operator K; see (III.4).

Property 3. Let the spectrum of H be nondegenerate. Then, by property 2, statistical operators diagonal in the energy basis are transformed among themselves. If we denote the diagonal elements of a density matrix in the energy representation by p_j, they satisfy the Pauli master equation

$$\frac{d}{dt}p_j(t)=\sum_k (W_{jk}p_k(t)-W_{kj}p_j(t)) \tag{III.24}$$

$p_j(t)\geq 0$, $\sum_j p_j(t)=1$. The coefficients W_{jk} satisfy the detailed balance condition

$$W_{jk}\, e^{-\beta\epsilon_k}=W_{kj}\, e^{-\beta\epsilon_k}\geq 0 \tag{III.25}$$

Equation III.25 is a consequence of (III.16). Since the spectrum of H is nondegenerate, the $s_{\alpha\gamma}$-term of L vanishes. In the dissipative part L_s of L $2V_\alpha(\omega)\rho V_\gamma(\omega)^*$ gives the first term and $-V_\gamma(\omega)^* V_\alpha(\omega)\rho-\rho V_\gamma(\omega)^* V_\alpha(\omega)$ gives the second term in (III.24).

Property 4. The generator L satisfies the (quantum generalization of the) detailed balance condition with respect to ρ_β

$$L_a(\rho_\beta)=0 \tag{III.26}$$

$$L_s(A\rho_\beta)=L_s^*(A)\rho_\beta \tag{III.27}$$

Equation III.26 is obvious; (III.27) requires some algebraic manipulations, where use is made of (III.16).

Let us define the scalar product $\langle A|B\rangle_\beta = \mathrm{tr}\,[\rho_\beta AB^*]$ on $B(\mathcal{H})$. Then we can rewrite (III.26) and (III.27) as

$$\langle L_a^*(A)|B\rangle_\beta = -\langle A|L_a^*(B)\rangle_\beta$$
$$\langle L_s^*(A)|B\rangle_\beta = \langle A|L_s^*(B)\rangle_\beta \tag{III.28}$$

Therefore L_a^* is antisymmetric and L_s^* is symmetric with respect to the scalar product $\langle\cdot|\cdot\rangle_\beta$. Here (III.28) is the quantum generalization of the detailed balance (III.25) as proposed and studied by Alicki[44] and Kossakowski, Frigerio, Gorini, and Verri.[45] Equation III.28 is characteristic for a system weakly coupled to a single heat reservoir. If $\mathcal{S}$ is coupled in a nontrivial way to two reservoirs at different temperatures, detailed balance is not fulfilled.

Although at first sight the generator L has a rather complicated structure, properties 1 to 3 allow one to visualize the time evolution $\{e^{Lt}|t\geq 0\}$ fairly well. We expand $\rho(t)=e^{Lt}\rho$ in an energy basis. Then, generally, the off-diagonal elements $\langle n|\rho(t)m\rangle$, $n\neq m$, decay exponentially, whereas the diagonal elements $\langle n|\rho(t)n\rangle$, by transitions according to (III.24), finally reach their canonical equilibrium value. Certainly this picture has to be modified in some cases. If the spectrum of H is degenerate, then property 3 is no longer true. One may have then transitions between diagonal and off-diagonal elements. Clearly, also if the coupling is trivial, say $V_\alpha=1$ for $\alpha=1$ and $V_\alpha=0$ otherwise, then $L=0$ (since the weak coupling limit is obtained in the interaction picture). In Theorem 3 we give a criterion for what is the exception and what the rule.

Let $\mathcal{S}$ be coupled now to the heat reservoirs $\mathcal{R}^1,\ldots,\mathcal{R}^r$ in equilibrium at inverse temperatures $\beta_1,\ldots,\beta_r$. Since the reservoir parts are independent from each other, the generator L of the weak coupling limit (III.7) is the sum of the generators $L_1,\ldots,L_r$

$$L = L_1 + \cdots + L_r \tag{III.29}$$

L_k generates the time evolution of $\mathcal{S}$ weakly coupled only to $\mathcal{R}^k$, $k=1,\ldots,r$. Each L_k has the form given in (III.19) with $h^k_{\alpha\gamma}(\omega)$ satisfying the KMS condition (III.16) with reciprocal temperature β_k.

IV. DETAILED BALANCE AND THE KMS CONDITION

In the preceding section we showed that for a system coupled to a single reservoir at temperature β^{-1}: KMS (of the reservoir)+weak coupling→ detailed balance with respect to the canonical ensemble ρ_β (of the system). The importance of the detailed balance property derives from the fact that, as is seen later, together with the convexity of the entropy production, it implies essentially all postulates of thermodynamics of irreversible processes. In this section we want to argue the converse: detailed balance of the system + weak coupling → KMS (of the reservoir), which may be viewed as a stability property[46,47] that singles out the equilibrium states. This indicates the deep connection between detailed balance (and therefore irreversible thermodynamics) and the KMS boundary condition. The results of this section are due to Kossakowski, Frigerio, Gorini, and Verri,[45] where also the proofs may be found. (The notions used in this section are slightly more abstract than those used otherwise, and it may be omitted without preventing understanding of Sections V to VIII.)

To begin with, we abandon our particular microscopic model, retaining only the assumption that the time evolution of our open quantum-mechanical system $\mathcal{S}$ is Markovian. This leads us to the definition of a dynamical semigroup.[48] A dynamical semigroup is a one-parameter family of linear mappings

$$\Lambda_t : T(\mathcal{H}) \to T(\mathcal{H}), \qquad t \geq 0 \tag{IV.1}$$

which are positive, $\rho \geq 0$ implies $\Lambda_t\rho \geq 0$, are trace-preserving, $\text{tr}\,(\Lambda_t\rho) = \text{tr}\,(\rho)$ form a semigroup

$$\Lambda_{t+s} = \Lambda_t \Lambda_s \tag{IV.2}$$

and are continuous in the trace norm,

$$\lim_{t\to 0_+} \|\Lambda_t\rho - \rho\|_1 = 0 \tag{IV.3}$$

for all $\rho \in T(\mathcal{H})$. One can show that Λ_t is a contraction, $\|\Lambda_t\| = 1$. Positivity and preservation of the trace ensure that Λ_t maps density matrices on density matrices, (IV.2) is the Markovian property, and (IV.3) ensures, by the Hille–Yosida theorem,[49] the existence of a densely defined generator L such that $\Lambda_t = e^{Lt}$. In terms of the Heisenberg picture, defined by $\text{tr}\,(A\Lambda_t(\rho) = \text{tr}\,(\Lambda_t^*(A)\rho)$ for all $\rho \in T(\mathcal{H})$, $A \in B(\mathcal{H})$, we have the following properties:

(i′) $\Lambda_t^* B(\mathcal{H})_+ \subset B(\mathcal{H})_+$
(ii) $\Lambda_t^*(1) = 1$
(iii) $\Lambda_{t+s}^* = \Lambda_t^* \Lambda_s^*$
(iv′) $\Lambda_t^*(A) \to A$ ultraweakly when $t \to 0_+$
(v) Λ_t^* is ultraweakly continuous

$B(\mathcal{H})_+$ denotes the positive elements in $B(\mathcal{H})$ and a sequence $A_n \in B(\mathcal{H})$ converges ultraweakly to $A \in B(\mathcal{H})$, if $\text{tr}\,(A_n\rho) \to \text{tr}\,(A\rho)$ for all $\rho \in T(\mathcal{H})$.

The reduced dynamics $\Lambda_t^{\lambda *}$ [cf. (II.5)] has another important property, as first pointed out by Kraus[50] in the context of state changes caused by quantum measurements and subsequently used by several authors.[51,52] To state this property, we recall a definition. Let $\mathscr{A}$ and $\mathscr{B}$ be C^*-algebras, and denote by $M(n)$ (n integer ≥ 1) the algebra of $n \times n$ complex matrices. A linear map $\phi: \mathscr{A} \to \mathscr{B}$ is called n-positive, if the map

$$\begin{aligned} &\phi_n : \mathscr{A} \otimes M(n) \to \mathscr{B} \otimes M(n) \\ &\phi_n(A \otimes M) = \phi(A) \otimes M, \qquad A \in \mathscr{A}, \quad M \in M(n) \end{aligned} \tag{IV.4}$$

is positive (i.e., maps a positive operator to a positive operator). ϕ is called completely positive, if it is n-positive for all integers n. The reduced dynamics $\Lambda_t^{\lambda^*}$ is completely positive for all t, since it is the composition of a *-automorphism and a conditional expectation. Therefore it seems physically well motivated to replace the positivity (i′) by the stronger property

(i) Λ_t^* is completely positive for all $t \geq 0$

For technical reasons it is convenient to require the continuity property

(iv) $$\lim_{t \to 0_+} \|\Lambda_t^* - 1\| = 0 \tag{IV.5}$$

which is stronger than (iv′). The norm continuity of Λ_t^* implies the existence of a bounded, ultraweakly continuous generator $L^*: B(\mathscr{H}) \to B(\mathscr{H})$ of Λ_t^*.

A family of mappings $\Lambda_t^*: B(\mathscr{H}) \to B(\mathscr{H})$, $t \geq 0$, with properties (i) to (v) is called a quantum-dynamical semigroup. (Readers interested in the theory of quantum-dynamical semigroups are referred to the book by Davies[53] and the review article by Gorini et al.[33] The review article by Ingarden and Kossakowski[54] may serve as introduction.) Recently, Lindblad[51] proved that the general form of the generator of a quantum-dynamical semigroup is given by

$$L^*(\mathscr{A}) = i[H, A] + \frac{1}{2} \sum_{j \in I} \{[V_j^*, A] V_j + V_j^* [A, V_j]\} \tag{IV.6}$$

where H is a bounded self-adjoint operator, $V_j \in B(\mathscr{H})$ such that $\sum_{j \in I} V_j^* V_j$ converges ultraweakly, and the sum on the right-hand side converges ultraweakly. We note that the generator (III.21) obtained in the weak coupling limit is of the form (IV.6).

In analogy to (III.28) we define detailed balance for a quantum-dynamical semigroup.

Definition 1:

Let $\{\Lambda_t^* | t \geq 0\}$ be a quantum-dynamical semigroup with generator L^*. Let $\rho_0 \in T(\mathscr{H})$ be a density matrix with no zero eigenvalue and invariant under

Λ_t and let a product on $B(\mathscr{H})$ be defined by

$$\langle A|B\rangle_0 = \mathrm{tr}\,(AB^*\rho_0) \tag{IV.7}$$

Then $\{\Lambda_t^*|t\geq 0\}$ satisfies the detailed balance condition with respect to ρ_0 if L^* can be written as a sum

$$L^* = L_a^* + L_s^* \tag{IV.8}$$

such that

$$L_a^* = i[H,\cdot\,] \text{ with } H \text{ self-adjoint and } [H,\rho_0]=0$$

(then L_a^* is antisymmetric $\langle L_a^*(A)|B\rangle_0 = -\langle A|L_a^*(B)\rangle_0$)

$$L_s^* \text{ is symmetric, } \langle L_s^*(A)|B\rangle_0 = \langle A|L_s^*(B)\rangle_0$$

If Λ_t^* satisfies detailed balance with respect to ρ_0 and if the spectrum of ρ_0 is nondegenerate, then Λ_t induces on the states diagonal in the ρ_0-basis a classical Markov process that satisfies the well-known detailed balance of the form (III.25). In this sense Definition 1 generalizes the classical case. The generators of quantum-dynamical semigroups satisfying the detailed balance condition have been completely classified.[44,45]

Let us assume now that the reservoir $\mathscr{R}$ is in an arbitrary state ω^R invariant under the uncoupled time evolution. As long as Davies' conditions (III.8) to (III.10) are satisfied, we can perform the weak coupling limit and obtain again a semigroup with a generator of the form (III.21). However, h_{ij} will not satisfy (III.16), in general. Let us fix for a moment the reservoir part of the interaction $V_1^R = A$, $V_2^R = B$, $V_j^R = 0$, $j\neq 1, 2$, and let us denote h_{12} by h_{AB}. If we assume in addition that the semigroup obtained in the weak coupling limit satisfies the detailed balance condition with respect to some stationary ρ_0, then this imposes certain restrictions on h_{AB} and on h_{BA}. If we assume detailed balance to be satisfied for a whole class of system Hamiltonians (in fact six-level systems suffice), then we obtain a whole class of restrictions on h_{AB} and on h_{BA}. The interesting point is that all these restrictions together admit only the single solution

$$h_{AB}(-\omega) = e^{-\beta\omega} h_{BA}(\omega) \tag{IV.9}$$

for some β independent of A and B. By Fourier transformations we obtain the KMS condition (III.15) for A and B. We repeat the argument for other observables, for which Davies' conditions (Theorem 1) are satisfied. This shows then that ω^R has to be a KMS state on all these observables. Therefore if the state ω^R of the reservoir $\mathscr{R}$ is such that all systems weakly coupled to $\mathscr{R}$ in the state ω^R satisfy the detailed balance condition, then ω^R has to be a thermal equilibrium (KMS) state.

V. ENTROPY PRODUCTION

Entropy production plays a central role in nonequilibrium thermodynamics.[55,56] The entropy production σ is the source term in the (local) balance equation for the entropy density

$$\frac{\partial S}{\partial t} = -\text{div}\,\mathbf{J}_S + \sigma \tag{V.1}$$

S being the local entropy density and $\mathbf{J}_S$ being the vector field of entropy flow per unit area per unit time. Here σ is assumed to be always nonnegative, since entropy can only be created, never destroyed; for reversible processes $\sigma = 0$. This is the (local formulation) of the second law of thermodynamics. It turns out that in the phenomenological theory σ can be written as a bilinear form in the fluxes characterizing the irreversible process and in the conjugate thermodynamic forces. If only heat conduction can occur in the system, then the entropy production is given by

$$\sigma = -\frac{1}{T^2}\mathbf{J} \,.\, (\text{grad } T) \tag{V.2}$$

where T is the temperature field and $\mathbf{J}$ is the vector field for the heat flow. (Here $\mathbf{J}$ is the flux and $-1/T^2$ grad T is the conjugate thermodynamic force.)

In our attempt to give a microscopic foundation to the thermodynamics or irreversible processes we should, therefore, in the first place, give a microscopic expression for the entropy production. To begin with, we treat the case of the system $\mathcal{S}$ weakly coupled to a reservoir at a single temperature β^{-1}. The time evolution of $\mathcal{S}$ is given by $\{e^{Lt} | t \geq 0\}$ in the weak coupling approximation (III.7). The entropy S of $\mathcal{S}$ in the state ρ is defined by the well-known expression of von Neumann

$$S(\rho) = -\text{tr}\,(\rho \log \rho) \tag{V.3}$$

The entropy flow is $-\beta\, đQ$ (Boltzmann's constant k is set equal to one), where $đQ$ is the heat transfer from the reservoir to the system (i.e., dQ is the change of energy of S that occurs at a reservoir temperature β^{-1} and is reversible as far as the reservoir is concerned).

$$\begin{gathered} đQ = \text{tr}\,(\rho(t)L^*(H)) = \text{tr}\left(\rho \frac{d}{dt} H(t)\right) \\ \mathbf{J}_S = -\beta\, đQ = \frac{d}{dt}\,\text{tr}\,\rho(t) \log \rho_\beta \end{gathered} \tag{V.4}$$

where $\rho_\beta = Z^{-1} e^{-\beta H}$. Therefore, using the balance equation (V.1) in the integrated form, we obtain for the (total) entropy production

$$\sigma(\rho)=\frac{d}{dt}(\operatorname{tr}(\rho(t)\log\rho_\beta)-\operatorname{tr}(\rho(t)\log\rho(t)))|_{t=0} \tag{V.5}$$

$$\sigma(\rho)=-\frac{d}{dt}S(e^{Lt}\rho|\rho_\beta)|_{t=0} \tag{V.6}$$

$S(\rho|\rho_\beta)$ is the relative entropy[57] of ρ with respect to ρ_β. Equation (V.6) is our basic definition.

Equation (V.5) is only a formal expression for the relative entropy, since both summands may be infinite. One way to define the relative entropy properly is

$$\begin{aligned} S(A|B)&=\sum_{i,j}|\langle a_i|b_j\rangle^2(a_i\log a_i-a_i\log b_j+b_j-a_i)\\ &(=\operatorname{tr}A\log A-\operatorname{tr}A\log B+\operatorname{tr}B-\operatorname{tr}A)\end{aligned} \tag{V.7}$$

where A, B are statistical operators with eigenvalues $\{a_i\}$, $\{b_j\}$ and eigenvectors $\{|a_i\rangle\}$, $\{|b_j\rangle\}$. Since the second factor is nonnegative, $S(A|B)\geq 0$, possible $S(A|B)=\infty$. We note that $S(A|B)=\infty$, if the range of A is not contained in the range of B. (The range of A is the linear span of all eigenvectors to nonzero eigenvalues of A.)

We collect the essential properties of the entropy production in

Theorem 2. Let $\dim\mathcal{H}=N<\infty$ and let L be given by (III.19). Then the entropy production

$$\sigma(\rho)=-\frac{d}{dt}S(\rho|\rho_\beta)|_{t=0}$$

is given by

$$\begin{aligned}\sigma(\rho)&=\operatorname{tr}(L(\rho)(\log\rho_\beta-\log\rho))\\ &=\sum_{\omega\in\operatorname{Sp}[H,\cdot\,]}\sum_{j,k=1}^{N}\left\{\sum_{\alpha,\gamma}h_{\alpha\gamma}(\omega)\langle\Psi_j|V_\alpha(\omega)\Psi_k\rangle\langle\Psi_k|V_\gamma(\omega)^*\Psi_j\rangle\right\}\\ &\quad\times\{(\rho_k-\rho_je^{-\beta\omega})(\log\rho_k-\log\rho_j+\beta\omega)\}\end{aligned} \tag{V.8}$$

where $\rho=\sum_{j=1}^N\rho_j|\Psi_j\rangle\langle\Psi_j|$ is the spectral decomposition of ρ and where $\operatorname{tr}(L(\rho)\log\rho)$ is defined by

$$\begin{aligned}\operatorname{tr}(L(\rho)\log\rho)&=\sum_{j=1}^{N}\langle\Psi_j|L(\rho)\Psi_j\rangle\log\rho_j\\ \langle\Psi_j|L(\rho)\Psi_j\rangle\log\rho_j&=\begin{cases}-\infty & \text{if } \langle\Psi_j|L(\rho)\Psi_j\rangle\neq 0 \text{ and } \rho_j=0\\ 0 & \text{if } \langle\Psi_j|L(\rho)\Psi_j\rangle=0\end{cases}\end{aligned} \tag{V.9}$$

Here σ has the properties:

(i) $\sigma(\rho)\geq 0$, possibly $\sigma(\rho)=\infty$. If $\{h_{\alpha\gamma}(\omega)\}>0$ (as a matrix) for all $\omega\in \mathrm{Sp}([H,\cdot])$, then $\sigma(\rho)=0$ if and only if $L(\rho)=0$.

(ii) σ is a convex function, that is,

$$\sigma(\lambda\rho_1+(1-\lambda)\rho_2)\leq\lambda\sigma(\rho_1)+(1-\lambda)\sigma(\rho_2) \tag{V.10}$$

$0\leq\lambda\leq 1$, with statistical operators ρ_1, ρ_2.

Proof: Since $\dim\mathcal{H}<\infty$ and since $\log\rho_\beta=-\beta H-Z$, we have

$$S(\rho(t)|\rho_\beta)=\mathrm{tr}\,(\rho(t)\log\rho(t))+\beta\,\mathrm{tr}\,(\rho(t)H)+\beta Z \tag{V.11}$$

The derivative of the second term is $\mathrm{tr}\,(L(\rho)\log\rho_\beta)$. Using the spectral decomposition of $\rho(t)$, $\rho(t)=\sum_j\rho_j(t)|\Psi_j(t)\rangle\langle\Psi_j(t)|$, we obtain

$$\frac{d}{dt}\mathrm{tr}\,(\rho(t)\log\rho(t))|_{t=0}=\sum_k\{\rho_k\log\rho_k(\langle\dot\Psi_k(0)|\Psi_k\rangle+\langle\Psi_k|\dot\Psi_k(0)\rangle)+\dot\rho_k(0)\log\rho_k\} \tag{V.12}$$

If $\rho_k=0$ and $\dot\rho_k(0)>0$, then $\dot\rho_k(0)\log\rho_k=-\infty$. If $\rho_k=0$ and $\rho_k(0)=0$, then using a Taylor expansion of $\rho_k(t)$ ($\rho_k(t)$ is an analytic function) and performing the limit as $t\to 0_+$, one obtains $\dot\rho_k(0)\log\rho_k=0$. Since

$$\langle\Psi_k|L(\rho)\Psi_k\rangle=\rho_k(\langle\dot\Psi_k(0)|\Psi_k\rangle+\langle\Psi_k|\dot\Psi_k(0)\rangle)+\dot\rho_k(0) \tag{V.13}$$

we have

$$\frac{d}{dt}\mathrm{tr}\,(\rho(t)\log\rho(t))|_{t=0}=\mathrm{tr}\,(L(\rho)\log\rho) \tag{V.14}$$

with the convention (V.9). We note that for $t>0$, $\rho_k(t)=0$ implies $\dot\rho_k(t)=0$, since $\rho_k(t_-)$, $\rho_k(t_+)\geq 0$.

We compute the entropy production $\sigma(\rho)=-\mathrm{tr}\,(L(\rho)\log\rho)-\mathrm{tr}\,(L(\rho)\beta H)$. We have $\mathrm{tr}\,(L_a(\rho)\log\rho)=0$, since $[\rho,\log\rho]=0$, and $\mathrm{tr}\,(L_a(\rho)H)=0$, since $[H, V_\gamma^*(\omega)V_\alpha(\omega)]=0$. The contribution of the symmetric part to the entropy production is

$$\sigma(\rho)=2\sum_{\omega\in\mathrm{Sp}([H,\cdot])}\sum_{j,k=1}^{N}\sum_{\alpha,\gamma}h_{\alpha\gamma}(\omega)$$
$$\times\langle\Psi_j|V_\alpha(\omega)\Psi_k\rangle\langle\Psi_k|V_\gamma(\omega)\Psi_j\rangle\rho_k(\log\rho_k-\log\rho_j+\beta\omega) \tag{V.15}$$

$$=2\sum_{\omega\in\mathrm{Sp}([H,\cdot])}\sum_{j,k=1}^{N}\sum_{\alpha,\gamma}h_{\alpha\gamma}(\omega)$$
$$\times\langle\Psi_j|V_\alpha(\omega)\Psi_k\rangle\langle\Psi_k|V_\beta(\omega)\Psi_j\rangle e^{-\beta\omega}\rho_j(\log\rho_j-\log\rho_k-\beta\omega) \tag{V.16}$$

where we used (III.16) and (III.18). Adding (V.15) and (V.16) and dividing by 2, one obtains (V.8).

From (V.8) it is obvious that $\sigma(\rho)\geq 0$. If $L(\rho)=0$, then $\sigma(\rho)=0$ by (V.8). Let $\sigma(\rho)=0$. Since $\{h_{\alpha,\gamma}(\omega)\}$ is a positive matrix we can diagonalize it. Let $h_\gamma(\omega)\geq 0$ be its eigenvalues and let $\{\tilde{V}_\gamma(\omega)\}$ be $\{V_\alpha(\omega)\}$ expressed in the eigenbasis of $\{h_{\alpha,\gamma}(\omega)\}$. Then

$$\sigma(\rho)=\sum_{\omega\in \mathrm{Sp}([H,\cdot\,])}\sum_{j,k=1}^{N}\sum_{\gamma} h_\gamma(\omega)|\langle\Psi_j|\tilde{V}_\gamma\Psi_k\rangle|^2 \times(\rho_k-\rho_j)\,e^{-\beta\omega}(\log\rho_k-\log\rho_j+\beta\omega)=0 \qquad \text{(V.17)}$$

implies

$$h_\gamma(\omega)\langle\Psi_j|\tilde{V}_\gamma(\omega)\Psi_k\rangle(\rho_k-\rho_j\,e^{-\beta\omega})=0 \qquad \text{(V.18)}$$

for all γ, $\omega\in\mathrm{Sp}\,([H,\cdot\,])$ and $j, k=1,\ldots, N$. Therefore

$$[\tilde{V}_\gamma(\omega), \rho\, e^{\beta H}]=0 \qquad \text{(V.19)}$$

for all γ and $\omega\in\mathrm{Sp}\,([H,\cdot\,])$. Let $\mathcal{N}$ be the set of all operators commuting with all $\tilde{V}_\gamma(\omega)$. Then

$$\rho=A\,e^{-\beta H} \qquad \text{(V.20)}$$

with $A\in\mathcal{N}$. We write the generator (III.19) in terms of the $\tilde{V}_\gamma(\omega)$. Then, since $A\in\mathcal{N}$,

$$[\tilde{V}_\gamma(\omega)^*\tilde{V}_\gamma(\omega), A\,e^{-\beta H}]=0 \qquad \text{(V.21)}$$

which implies $L_a(\rho)=0$. By the detailed balance (III.27)

$$L_s(A\,e^{-\beta H})=L_s^*(A)\,e^{-\beta H}=0 \qquad \text{(V.22)}$$

since $A\in\mathcal{N}$. Therefore $L(\rho)=0$.

To prove the convexity of σ we note that $\mathrm{tr}\,(L(\rho)\log\rho_\beta)$ is linear. Therefore we have to show only the convexity of $-\mathrm{tr}\,(L_s(\rho)\log\rho)$. Since $\{h_{\alpha\gamma}(\omega)\}$ is a positive matrix we can diagonalize it for every fixed ω. Then (III.19) takes the form

$$L_s(\rho)=\sum_j [V_j, \rho V_j^*]+[V_j\rho, V_j^*] \qquad \text{(V.23)}$$

where the V_j's are certain linear combinations of the $V_\alpha(\omega)$'s. Therefore

$$-\mathrm{tr}\,(L_s(\rho)\log\rho)=2\sum_j\{\mathrm{tr}\,(V_j^*V_j\rho\,\log\rho)-\mathrm{tr}\,(V_j\rho V_j^*\,\log\rho)\} \qquad \text{(V.24)}$$

A theorem due to Lieb[58] asserts that the function

$$\rho\to-\mathrm{tr}\,(\rho^q V_j\rho^{1-q}V_j^*) \qquad \text{(V.25)}$$

$0 \le q \le 1$, is convex. For $q = 0$ the function is affine. Therefore the derivative at $q = 0$

$$\rho \to \operatorname{tr}(V_j\rho \log \rho V_j^*) - \operatorname{tr}(\log \rho V_j \rho V_j^*) \tag{V.26}$$

is convex. This implies that $-\operatorname{tr} L_s(\rho) \log \rho$ is convex. ■

COMMENTS

1. The technical restriction $\dim \mathcal{H} < \infty$ is annoying and is not overcome easily. Since $S(\rho|\rho_\beta)$ is quite often infinite, it is difficult to prove that $-S(\Lambda_t\rho|\rho_\beta)$ is differentiable and that its derivative is given by (V.8). Of course, we can define the entropy production by (V.8) and then all properties stated are still valid. Our theorem holds also for the classical Markov process with finite state space defined by the Pauli equation (III.24) with detailed balance (III.25). For a general classical Markov process with detailed balance[1,2] one encounters the same difficulties as in the case of an infinite dimensional Hilbert space.

2. Property (ii) is the nonnegativity of the entropy production as required by the phenomenological theory. In general, ρ_β will be the only stationary state. Then $\sigma(\rho) > 0$ for all $\rho \ne \rho_\beta$. This means that $-S(\rho(t)|\rho_\beta)$ is strictly increasing reaching in the limit as $t \to \infty$ the value $-S(\rho_\beta|\rho_\beta) = 0$. At $t = 0$, $-S(\rho(t)|\rho_\beta)$ may have an infinite slope, but at any later time its slope is finite. Typically, this happens if the initial state has, for example, a sharp energy or is otherwise rather pure.

3. Property (iii) is the essential stability property of the entropy production. It guarantees that, as the system is coupled to several reservoirs, there is always a unique state of minimal entropy production. (This statement is too strong. It would be true, if one could prove strict convexity of σ. In general, there is a convex set of states with minimal entropy production.) This situation is analogous to the one in thermodynamics, where the concavity of the entropy ensures stability.

Let us now consider the case of the system $\mathcal{S}$ coupled to reservoirs $\mathcal{R}^1, \ldots, \mathcal{R}^r$ at inverse temperatures $\beta_1, \ldots, \beta_r$. The time evolution of $\mathcal{S}$ is generated by $L = L_1 + \cdots + L_r$. To find a microscopic expression for the entropy production we start again from the balance equation (V.1). The entropy S is still defined by (V.3). The entropy flow is given by

$$J_s = -\sum_{k=1}^{r} \beta_k \, đQ_k \tag{V.27}$$

where dQ_k is the heat transfer from the kth reservoir to the system (i.e., the change of energy of S due to the coupling to the kth reservoir). Therefore

$$đQ_k = \operatorname{tr}(\rho \, L_k^*(H)) \tag{V.28}$$

and, by (V.14),

$$\begin{aligned}\sigma(\rho) &= \sum_{k=1}^{r} -\beta_k \operatorname{tr}(\rho L_k^*(H)) - \operatorname{tr}(L(\rho)\log\rho)\\ &= \sum_{k=1}^{r} \{-\beta_k \operatorname{tr}(L_k(\rho)H) - \operatorname{tr}(L_k(\rho)\log\rho)\}\\ &= \sum_{k=1}^{r} \sigma_k(\rho) \qquad \text{(V.29)}\end{aligned}$$

The entropy production σ (for a specified state of the system) is the sum of the entropy productions σ_k, where σ_k is the entropy production of $\mathcal{S}$ coupled only to the kth reservoir. As a direct consequence of Theorem 2, we note

Corollary 1. Let $\sigma = \sum_{k=1}^{r} \sigma_k$. Then $\sigma \geq 0$. Here $\sigma(\rho)=0$ if and only if $\sigma_k(\rho)=0$ for $k=1,\ldots,r$. If for every L_k the canonical ensemble ρ_β is the only stationary state (in some temperature region), then $\sigma(\rho)=0$ if and only if $\beta_1=\beta_2=\cdots=\beta_r=\beta$ and $\rho=\rho_\beta$.

The entropy production σ is convex.

In the steady state ρ_s, defined by

$$L(\rho_s)=0, \qquad L=L_1+\cdots+L_r \qquad \text{(V.30)}$$

the entropy production is a bilinear form $\sigma(\rho_s)=\sum_{k=1}^{r} \beta_k J_k$, where $J_k = \operatorname{tr}(\rho_s L_k^*(H))$ is the steady-state heat flow from the kth reservoir to the system. Since ρ_s depends on $\beta_1,\ldots,\beta_r$, J_k does too. We choose some reference temperature β and define the thermodynamic forces by $X_k = -\beta_k+\beta$. Then, using the conservation law $\sum_{k=1}^{r} J_k = \operatorname{tr}(L(\rho_s)H)=0$, σ becomes a bilinear form in the fluxes and the thermodynamic forces

$$\sigma(\rho_s)=\sum_{r=1}^{k} X_k J_k(X_1,\ldots,X_r) \qquad \text{(V.31)}$$

In the *linear* approximation

$$J_k(X_1,\ldots,X_r)=\sum_{j=1}^{r} L_{kj}(\beta)X_j \qquad \text{(V.32)}$$

and the entropy production is a quadratic form

$$\sigma(\rho_s)=\sum_{k,j=1}^{r} L_{kj}(\beta)X_k X_j \qquad \text{(V.33)}$$

The $L_{kj}(\beta)$ are the kinetic coefficients related to the heat conductivities.

From the convexity of σ it follows immediately that

$$\{L_{kj}(\beta)\} \geq 0 \tag{V.34}$$

Since $\{L_{kj}(\beta)\}$ is a matrix on a real vector space, (V.34) does not imply its symmetry $L_{kj} = L_{jk}$ (Onsager relations). As we shall see, however, in Section VII the symmetry essentially follows from the detailed balance property.

VI. APPROACH TO STATIONARITY

On physical grounds one expects that, in general, the reservoir will drive the system to a unique stationary state. Of course, the coupling between the system and the reservoir might be poor (say, $H^{SR} = 0$). Then, since we work in the interaction picture, several stationary states should appear, and, depending on the initial state ρ, $\rho(t)$ should approach one of them as $t \to \infty$.

Definition 2:
A generator L (or the dynamics $\{e^{Lt}|t \geq 0\}$) is called relaxing, if there exists a unique stationary state ρ_s, $L(\rho_s) = 0$, and if

$$\lim_{t\to\infty} e^{Lt}\rho = \rho_s \tag{VI.1}$$

for all initial states ρ.

For a single temperature reservoir, the relative entropy is an efficient tool to study the approach to equilibrium. By Theorem 1 $-d/dt\, S(\rho(t)|\rho_\beta) > 0$ unless $L(\rho) = 0$. Therefore every initial state ρ tends to a limit as $t \to \infty$,

$$\lim_{t\to\infty} e^{Lt}\rho = A_\rho e^{-\beta H} \tag{VI.2}$$

where $A_\rho \in \mathcal{N}$ [cf. (V.19)]. L is relaxing if and only if $\mathcal{N}$ consists only of multiples of the unit operator. In that case the unique stationary state is the canonical ensemble.

One would like to have a condition for relaxation that is more directly related to H and H^{SR}. If we assume that $\{h_{\alpha\gamma}(\omega)\} > 0$ for all $\omega \in \mathrm{Sp}\,([H, \cdot\,])$ and all inverse temperatures β, then $\mathcal{N} = \{V_\alpha(\omega)|\alpha,\ \omega \in \mathrm{Sp}\,([H, \cdot\,])\}'$. $'$ denotes the set of all operators commuting with the ones in $\{\cdot\}$ [i.e., $\mathcal{N}$ is the set of all operators commuting with all $V_\alpha(\omega)$'s]. If, furthermore, $\{H, V_\alpha|\alpha\}' = \{\mathbb{C}1\}$, then, as can be shown,[57] $\mathcal{N} = \{\mathbb{C}1\}$. Hence $\{h_{\alpha\gamma}(\omega)\} > 0$ and $\{H, V_\alpha|\alpha\}' = \{\mathbb{C}1\}$ is a sufficient condition for relaxation. In the case of several reservoirs at different temperatures one can use the fact that, if $L_1, \ldots, L_r$ are relaxing, then also $L = L_1 + \cdots + L_r$ is relaxing ($\dim \mathcal{H} < \infty$).

For the case of an infinite dimensional Hilbert space one has to use much more powerful methods than we introduced so far to prove[60] (cf. also Refs. 59, 61–63).

Theorem 3. Let the dynamics of $\mathcal{S}$ coupled to the reservoirs $\mathcal{R}^1, \ldots, \mathcal{R}^r$ be given by the generator $L = L_1 + \cdots + L_r$ as in (III.29). If for given inverse temperatures $\beta_1, \ldots, \beta_r$ $\{h_{\alpha\gamma}^{(k)}(\omega)\} > 0$ for all $\omega \in \mathrm{Sp}([H, \cdot\,])$, $k = 1, \ldots, r$, and if

$$\{H, V_\alpha^{(k)} | \alpha, k = 1, \ldots, r\}' = \{\mathbb{C}1\} \qquad \text{(VI.3)}$$

(i.e., only multiples of the unit operator commute with H and the $V_\alpha^{(k)}$'s), then L is relaxing.

Remark: For readers who did not omit Section IV. Theorem 3 is the special case of a very fine theorem.[60] Let $\{\Lambda_t^* | t \geq 0\}$ be a quantum-dynamical semigroup with generator L^*. Then we define the dissipation function[51] $D: B(\mathcal{H}) \times B(\mathcal{H}) \to B(\mathcal{H})$ by

$$D(A, B) = L^*(A^*B) - L^*(A^*)B - A^*L^*(B) \qquad \text{(VI.4)}$$

One shows that $D(A, A) \geq 0$. If $D(A, A) = 0$ for all A, then $L^*(A) = i[H, A]$, which means that there is no dissipation at all. Obviously, $D(1, 1) = 0$. The theorem of Frigerio states that if $D(A, A) = 0$ implies $A = c1$, then $\{\Lambda_t | t \geq 0\}$ is relaxing. In this case we have complete dissipation. Therefore the dissipation function D characterizes the lack of reversibility of L^*. Using the form (IV.6) of the generator L^*, the preceding condition is equivalent to $\{V_j | j \in I\}' = \{\mathbb{C}1\}$.

Theorem 3 gives us a physically very intuitive criterion for an effective coupling, causing the system to approach a unique steady state. $\{h_{\alpha\gamma}^{(k)}(\omega)\} > 0$ means that the reservoir couples to all relevant frequencies of $\mathcal{S}$ and (VI.3) expresses that, besides the unit operator, there is no other operator that commutes with the Hamiltonian H and the interaction operators $V_\alpha^{(k)}$. To put it somewhat differently: If we form arbitrary linear combinations and products of H and the $V_\alpha^{(k)}$'s (and take weak limits), then we obtain in this way any bounded operator, which means that H and the $V_\alpha^{(k)}$'s together have to be sufficiently incompatible.

VII. ONSAGER RELATIONS

We show the validity of the Onsager relations[64]

$$L_{kj}(\beta) = L_{jk}(\beta) \qquad \text{(VII.1)}$$

[cf. (V.33)], for the system $\mathscr{S}$ weakly coupled to several reservoirs. In the special case of the Pauli master equation (III.24), Hepp[65,66] proved (VII.1) using the derivation of the Onsager relations for a classical system coupled to stochastic reservoirs as given by Bergmann, Lebowitz et al.[1,67]

Let $\rho_s = \rho_s(\beta_1, \ldots, \beta_r)$ be the unique stationary state of the generator $L = L_{1,\beta_1} + \cdots + L_{r,\beta_r}$, where we indicated explicitly the temperature dependence of the kth generator. Then the kinetic coefficients $L_{kj}(\beta)$ are defined by

$$L_{kj}(\beta) = \frac{\partial}{\partial \beta_i} \operatorname{tr}\,[L_{k,\beta_k}(H)\rho_s(\beta_1, \ldots, \beta_r)]|_{(\beta_1,\ldots,\beta_r)=(\beta,\ldots,\beta)} \tag{VII.2}$$

$k \neq j$. By the conservation law $\sum_k \operatorname{tr}\,(L_k^*(H)\rho_s) = 0$, $L_{jj}(\beta) = -\sum_{k\neq j} L_{kj}(\beta)$. Our proof of (VII.1) consists of two steps: First, we derive a convenient expression for the kinetic coefficients $L_{kj}(\beta)$ and then show their symmetry. The essential ingredient of the proof is the detailed balance property (III.27), which was a direct consequence of the KMS condition for the thermal equilibrium state of the reservoirs.

Lemma 1. Let $\dim \mathscr{H} < \infty$. Let the operator for the heat flow from the kth reservoir at inverse temperature β_k to the system be $J_{k,\beta_k} = L_{k,\beta_k}^*(H)$ and let $L_{1,\beta_1} + \cdots + L_{r,\beta_r}$ have a unique stationary state $\rho_s(\beta_1, \ldots, \beta_r)$ in some neighborhood of $(\beta, \ldots, \beta)$. Then the kinetic coefficients $L_{kj}(\beta)$ are given by

$$L_{kj}(\beta) = -\operatorname{tr}\left[\frac{1}{L^*}(J_{k,\beta})J_{j,\beta}\rho_\beta\right] \tag{VII.3}$$

$$= \int_0^\infty dt\, \operatorname{tr}\,[J_{k,\beta}(t)J_{j,\beta}\rho_\beta] \tag{VII.4}$$

with $J_{k,\beta}(t) = e^{L^* t} J_{k,\beta}$ and $L = L_{1,\beta} + \cdots + L_{r,\beta}$.

Proof: Since the canonical ensemble ρ_β is invariant for $L_{j,\beta}$, we have

$$L'_{j,\beta}(\rho_\beta) = L_{j,\beta}(H\rho_\beta) = J_{j,\beta}\rho_\beta \tag{VII.5}$$

where $'$ denotes the derivative with respect to β and where we used $L_{a,k}(H) = 0$ and the detailed balance (III.27) for $L_{s,k}$. Since $\rho_s(\beta, \ldots, \beta) = \rho_\beta$

$$(L_{1,\beta} + \cdots + L_{j,\beta+\delta} + \cdots + L_{r,\beta})(\rho_s(\beta, \ldots, \beta+\delta, \ldots, \beta)) - L(\rho_\beta) = 0 \tag{VII.6}$$

which implies

$$L'_{j,\beta}(\rho_\beta) = -L(\rho'_s(\beta)) \tag{VII.7}$$

and, by (VII.5),

$$L(\rho_s'(\beta)) = -J_{i,\beta}\rho_\beta \tag{VII.8}$$

Since, by assumption, the kernel of L is one-dimensional (i.e., the eigenvalue zero is nondegenerate), (VII.8) admits the one-parameter family of solutions

$$\rho_s'(\beta) = c\rho_\beta - \frac{1}{L}(J_{i,\beta}\rho_\beta) \tag{VII.9}$$

$c \in \mathbb{R}$, which defines $(1/L)(J_{j,\beta}\rho_\beta)$. Since $\mathrm{tr}\,(J_{k,\beta}\rho_\beta) = 0$,

$$L_{kj}(\beta) = -\mathrm{tr}\left[J_{k,\beta}\frac{1}{L}(J_{j,\beta}\rho_\beta)\right] \tag{VII.10}$$

ρ_β is the only stationary state of L. By the discussion in Section VI this implies that $\lim_{t\to\infty} e^{Lt}(J_{j,\beta}\rho_\beta) = c\rho_\beta$ and that, except zero, all eigenvalues of L lie in the open left-hand plane. Therefore we can express the inverse in (VII.10) as Laplace transform

$$L_{kj}(\beta) = \int_0^\infty dt\; \mathrm{tr}\,[J_{k,\beta}\, e^{Lt}(J_{j,\beta}\rho_\beta)] \tag{VII.11}$$

where we used again $\mathrm{tr}\,(J_{k,\beta}\rho_\beta) = 0$. The integral (VII.11) is finite. ■

It is no coincidence that (VII.4) looks like the well-known Kubo formula for transport coefficients. One can derive (VII.4) in the framework of linear response theory adapted to our model.[4,68]

Theorem 4. Let $\dim \mathcal{H} < \infty$ and let L_a and L_s be the antisymmetric and the symmetric part of $L = L_{1,\beta_1} + \cdots + L_{r,\beta_r}$. If in some neighborhood of $(\beta, \ldots, \beta)$ the unique stationary state ρ_s of L is also the unique stationary state of L_s, then the Onsager relations (VII.1) are valid.

Proof: Since the stationary state is already defined by $L_s(\rho_s) = 0$, we can repeat the derivation of (VII.4) with L_s instead of L and obtain (VII.4) with L replaced by L_s. By the detailed balance property (III.27)

$$e^{L_s t}(A\rho_\beta) = e^{L_s^* t}(A)\rho_\beta \tag{VII.12}$$

Since $[J_{k,\beta}, \rho_\beta] = 0$

$$\mathrm{tr}\,[e^{L_s^* t}(J_{k,\beta})J_{j,\beta}\rho_\beta] = \mathrm{tr}\,[e^{L_s^* t}(J_{j,\beta})J_{k,\beta}\rho_\beta] \tag{VII.13}$$

which implies (VII.1). ■

The conditions of the theorem are certainly met, when $L_a = 0$. They are also fulfilled, whenever the spectrum of H is nondegenerate. Since the free time evolution commutes with e^{Lt} and since ρ_s is unique

$$[H, \rho_s] = 0 \tag{VII.14}$$

In general, this does not imply $L_a(\rho_s) = 0$. However, if the spectrum of H is nondegenerate, then ρ_s is a function of H and therefore $L_a(\rho_s) = 0$. In general, one obtains

$$\text{tr}\,[e^{(L_s^*+L_a^*)t}(J_{k,\beta})J_{j,\beta}\rho_\beta] = \text{tr}\,[e^{(L_s^*-L_a^*)t}(J_{j,\beta})J_{k,\beta}\rho_\beta] \tag{VII.15}$$

Without additional assumptions (VII.15) does not imply the Onsager relations (VII.1).

At first sight it is somewhat surprising that the Onsager relations should not be valid in full generality. The reason is to be sought in our construction of the generator L. If we consider L as a function of the Hamiltonian H, then L is discontinuous whenever two eigenvalues meet. (This is so, since the projection P_j on the eigenspace for the energy ε_j changes discontinuously.) There is, however, a unique way to extend L continuously through the points of discontinuity. If we choose this (at finitely many points) modified $\tilde{L}$ as generator of the time evolution, then all our results remain valid and the Onsager relations hold without any restriction.

Remark: The reader will have noticed that we used the assumption $\dim \mathcal{H} < \infty$, which is, in view of physical applications, rather restrictive. The present theorem generalizes to infinite dimensional Hilbert spaces. Since we assumed the partition function to be finite, H is necessarily unbounded. Therefore the heat flow $L^*(H)$ (properly defined) becomes unbounded and one has to be somewhat careful with the integral (VII.4). For clarity and brevity of the argument, we thought it best to avoid such technicalities.

VIII. PRINCIPLE OF MINIMAL ENTROPY PRODUCTION

In the phenomenological theory the steady state may be characterized, under certain conditions, by a minimum of the entropy production compatible with the external constraints imposed on the system.[69–71] This property is valid only if the fluxes are proportional to the thermodynamic forces (i.e., only in the linear regime). As an illustration we consider the case of thermal conduction, where the entropy production as given by irreversible thermodynamics is [cf. (V.2)]

$$\sigma = \int \mathbf{J} \cdot \text{grad}\,\frac{1}{T}\,dV = \int L_{qq}\left(\text{grad}\,\frac{1}{T}\right)^2 dV \tag{VIII.1}$$

in the linear approximation with $\mathbf{J} = L_{qq}$ grad $1/T$. For fixed boundary conditions the minimum of σ is given by

$$\Delta \frac{1}{T} = 0 \tag{VIII.2}$$

which implies

$$\frac{\partial}{\partial t} T = 0 \tag{VIII.3}$$

by the heat equation $(\partial/\partial t)T = \alpha\, \Delta T$.

In our model the stationary states are defined by $L(\rho) = 0$. They form a convex set (any convex combination of stationary states is stationary). The states of minimal entropy production are defined by $\sigma(\rho)$ minimal. By the convexity of σ they also form a convex set. The external constraints are already specified by the reservoir temperatures appearing in L and σ. In analogy to the phenomenological theory it is tempting to conjecture

$$L(\rho) = 0 \Leftrightarrow \sigma(\rho) = \text{minimal} \tag{VIII.4}$$

If all reservoir temperatures are the same, (VIII.4) has been proved in Theorem 2. If the reservoir temperatures are different, one can show by explicit examples (e.g., dim $\mathscr{H} = 2$) that (VIII.4) is not true in general. From our preceding discussion this should come as no surprise, since for any finite temperature difference we are already in the nonlinear regime (the steady-state heat flow is not proportional to the temperature difference, in general). To derive the principle of minimal entropy production we have to proceed in a somewhat different way.

Let $L = L_{1,\beta_1} + \cdots + L_{r,\beta_r}$ have a unique stationary state $\rho_s(\beta_1, \ldots, \beta_r)$ in some neighborhood of $(\beta, \ldots, \beta)$. We then expand

$$\rho_s(\beta_1, \ldots, \beta_r) = \rho_\beta + \sum_{k=1}^{r} \rho_{s,k}(\beta)(\beta_k - \beta) + \cdots \tag{VIII.5}$$

Similarly, let $\sigma = \sigma_{1,\beta_1} + \cdots + \sigma_{r,\beta_r}$ have a unique minimum $\rho_m(\beta_1, \ldots, \beta_r)$. By convexity and the smooth dependence of σ_{k,β_k} on β_k, $\rho_m(\beta_1, \ldots, \beta_r)$ depends smoothly on the inverse temperatures. Furthermore, by Theorem 1, $\rho_m(\beta, \ldots, \beta) = \rho_\beta$. We expand

$$\rho_m(\beta_1, \ldots, \beta_r) = \rho_\beta + \sum_{k=1}^{r} \rho_{m,k}(\beta)(\beta_k - \beta) + \cdots \tag{VIII.6}$$

Then the principle of minimal entropy production asserts that

$$\rho_{m,k}(\beta) = \rho_{s,k}(\beta) \tag{VIII.7}$$

for $k = 1, \ldots, r$ [i.e., (VIII.4) is valid in the linear approximation].

Theorem 5. Let dim $\mathscr{H}<\infty$ and let the spectrum of H be nondegenerate. Let $L_{1,\beta_1}+\cdots+L_{r,\beta_r}$ have a unique stationary state $\rho_s(\beta_1,\ldots,\beta_r)$ and σ_{r,β_r} have a unique minimum $\rho_m(\beta_1,\ldots,\beta_r)$ in some neighborhood of $(\beta_1,\ldots,\beta)$. Then

$$\rho_{m,k}(\beta)=\rho_{s,k}(\beta) \tag{VIII.7}$$

Proof: In Lemma 1 we showed that $\rho_{s,k}(\beta)$ is defined by

$$L(\rho_{s,k}(\beta))=-L^*_{k,\beta}(H)\rho_\beta, \tag{VIII.8}$$

where $L=L_{1,\beta}+\cdots+L_{r,\beta}$ (VII.8).

Let $U(t)=e^{-iHt}$. Then

$$\begin{aligned}\sigma_{k,\beta_k}(U(t)\rho U(t)^*)&=-\text{tr}\,[U(t)L_{k,\beta_k}(\rho)U(t)^*U(t)(\log\rho+\beta_kH)U(t)^*]\\&=\sigma_{k,\beta_k}(\rho)\end{aligned} \tag{VIII.9}$$

Therefore, since the minimal state is unique by assumption,

$$[\rho_m(\beta_1,\ldots,\beta_r),H]=0 \tag{VIII.10}$$

Since the spectrum of H is nondegenerate, ρ_m is a function of H. Therefore it suffices to study the minimum of σ amongst the states diagonal in the energy basis $\{\Psi_j|j=1,\ldots,\dim\mathscr{H}\}$.

We consider σ as a function of the eigenvalues ρ_j of ρ. Then the minimum of σ is defined by

$$\frac{\partial}{\partial\rho_j}\sigma(\rho)=0=-\sum_{l=1}^{r}\left\{\text{tr}\,(L_{l,\beta_l}(\rho_j)(\log\rho+\beta_lH)+\langle\Psi_j|L_{l,\beta_l}(\rho)\Psi_j\rangle\frac{1}{\rho_j}\right\} \tag{VIII.11}$$

where $L_{l,\beta_l}(\rho_j)=L_{l,\beta_l}(\rho_j|\Psi_j\rangle\langle\Psi_j|)$ and $\rho=\rho_m(\beta_1,\ldots,\beta_r)$. By Theorem 1, $\sigma_{k,\beta+\delta}(\rho_{\beta+\delta})=0$. Taking the derivative at $\delta=0$

$$\langle\Psi_j|L'_{k,\beta}(\rho_\beta)\Psi_j\rangle-\langle\Psi_j|L'_{k,\beta}(H\rho_\beta)\Psi_j\rangle=0 \tag{VIII.12}$$

for all j, where $'$ denotes the derivative with respect to δ at $\delta=0$. In (VIII.11) we choose now $(\beta_1,\ldots,\beta_k,\ldots,\beta_r)=(\beta,\ldots,\beta+\delta,\ldots,\beta)$ and we obtain by taking the derivative at $\delta=0$

$$\begin{aligned}&\sum_{l\neq k}\text{tr}\left[L_{l,\beta}(\rho_{\beta,j})\frac{1}{\rho_\beta}\rho_{m,k}(\beta)\right]+\text{tr}\left[L_{k,\beta}(\rho_{\beta,j})\left(\frac{1}{\rho_\beta}\rho_{m,k}(\beta)+H\right)\right]\\&\quad+\sum_{l=1}^{r}\langle\Psi_j|L_{l,\beta}(\rho_{m,k}(\beta))\Psi_j\rangle\frac{1}{\rho_{\beta,j}}+\langle\Psi_j|L'_{k,\beta}(\rho_\beta)\Psi_j\rangle\frac{1}{\rho_{\beta,j}}=0\end{aligned} \tag{VIII.13}$$

for all j, where $\rho_{\beta,j}$ is the jth eigenvalue of ρ_β. Inserting (VIII.12), we obtain

$$\left\langle \Psi_j \middle| \left\{ \sum_{l=1}^{r} L_{l,\beta}(\rho_{m,k}(\beta)) + L_{k,\beta}^*(H)\rho_\beta \right\} \Psi_j \right\rangle + \rho_{\beta,j}^2 \left\langle \Psi_j \middle| \left\{ \sum_{l\neq k} L_{l,\beta}^*\left(\frac{1}{\rho_\beta}\rho_{m,k}(\beta)\right) \right.\right.$$

$$\left.\left. + L_{k,\beta}^*\left(\frac{1}{\rho_\beta}\rho_{m,n}(\beta)+H\right)\right\} \Psi_j \right\rangle = 0 \qquad \text{(VIII.14)}$$

The second term can be rewritten as

$$L^*\left(\frac{1}{\rho_\beta}\rho_{m,k}(\beta)\right) + L_{k,\beta}^*(H) = \frac{1}{\rho_\beta} L(\rho_{m,k}(\beta)) + L_{k,\beta}^*(H) \qquad \text{(VIII.15)}$$

where we used the detailed balance property (III.27) and the fact that $L_a(f(H)) = 0$ for any function f of H. Therefore (VIII.14) vanishes for

$$L(\rho_{m,k}(\beta)) = -L_{k,\beta}^*(H)\rho_\beta \quad \blacksquare \qquad \text{(VIII.16)}$$

Concerning the technical restrictions in Theorem 5, the same remarks as before apply. If we define the entropy production by (V.8), then the theorem is also valid for an infinite dimensional Hilbert space.

References

1. P. G. Bergmann and J. L. Lebowitz, *Phys. Rev.*, **99,** 578 (1955).
2. J. L. Lebowitz, *Ann. Phys.*, **1,** 1 (1956).
3. J. L. Lebowitz, *Phys. Rev.*, **114,** 1192 (1959).
4. J. L. Lebowitz and A. Shimony, *Phys. Rev.*, **128,** 391 (1962).
5. H. Spohn and J. L. Lebowitz, *Commun. Math. Phys.*, **54,** 97 (1977).
6. J. B. Keller, G. C. Papanicolaou, and J. Weilemann, "Heat Conductance in a One-dimensional Random Medium," to appear in *Comm. Pure Appl. Math.*
7. D. Ruelle, *Statistical Mechanics; Rigorous Results*, W. A. Benjamin, New York, 1969.
8. G. G. Emch, *Algebraic Methods in Statistical Mechanics and in Quantum Field Theory*, Wiley, New York, 1972.
9. L. van Hove, *Physica*, **23,** 441 (1957).
10. S. Nakajima, *Prog. Theor. Phys.*, **20,** 948 (1958).
11. I. Prigogine and P. Résibois, *Physica*, **27,** 629 (1961).
12. R. Zwanzig, *J. Chem. Phys.*, **33,** 1338 (1960).
13. R. Zwanzig, *Lect. Theor. Phys. (Boulder)*, **3,** 106 (1960).
14. P. N. Argyres and P. L. Kelley, *Phys. Rev.*, **134,** A98 (1964).
15. F. Haake, *Statistical Treatment of Open Systems by Generalized Master Equations*, Springer Tracts Mod. Phys., **66** (1973).
16. G. S. Agarwal, Progress in Optics, **XI,** 1 (1973).
17. C. Favre and P. Martin, *Helv. Phys. Acta*, **41,** 333 (1968).
18. E. B. Davies, *Commun. Math.* **39,** 91 (1974).
19. H. Narnhofer, *Acta Phys. Austr.*, **36,** 217 (1972).
20. H. Narnhofer, *Acta Phys. Austr.*, Suppl. XI, 527 (1973).
21. D. Robinson and O. E. Lanford III, *Commun. Math. Phys.*, **24,** 193 (1972).
22. D. Robinson, *Commun. Math. Phys.*, **31,** 171 (1973).

23. G. L. Sewell, *Ann. Phys.*, **85**, 336 (1974).
24. C. Radin, *J. Math. Phys.*, **11**, 2945 (1970).
25. G. G. Emch and C. Radin, *J. Math. Phys.*, **12**, 2043 (1971).
26. O. E. Lanford III and J. L. Lebowitz, "Time Evolution and Ergodic Properties of Harmonic Systems," in LNP, Vol. 38, Springer, Berlin, 1975, p. 144.
27. L. Van Hemmen, *Dynamics and Ergodicity of the Infinite Harmonic Crystal*, thesis, University of Croningen, 1976.
28. S. Goldstein, J. L. Lebowitz, and M. Aizenman, "Ergodic Properties of Infinite Systems," in LNP Vol. 38, Springer, Berlin, 1975, p. 112.
29. Z. Rieder, J. L. Lebowitz, and E. Lieb, *J. Math. Phys.*, **8**, 1073 (1967).
30. A. Casher and J. L. Lebowitz, *J. Math. Phys.*, **12**, 1701 (1971).
31. L. van Hove, *Physica*, **21**, 517 (1955).
32. L. van Hove, *Physica*, **23**, 441 (1957).
33. V. Gorini, A. Frigerio, M. Verri, and E. C. G. Sudarshan, *Properties of Quantum Markovian Master Equations*, University of Milan, preprint, 1976.
34. E. B. Davies, *Math. Ann.*, **219**, 147 (1976).
35. E. B. Davies, *Ann. Inst. Poinc.*, **11**, 265 (1975).
36. E. B. Davies, *Commun. Math. Phys.*, **33**, 171 (1973).
37. E. B. Davies, *J. Math. Phys.*, **15**, 2036 (1974).
38. E. B. Davies, *Lett. Math. Phys.*, **1**, 31 (1976).
39. J. V. Pulé, *Commun. Math. Phys.*, **38**, 241 (1974).
40. P. Martin and G. G. Emch, *Helv. Phys. Acta*, **48**, 59 (1975).
41. E. B. Davies and J. P. Eckmann, *Helv. Phys. Acta*, **48**, 731 (1975).
42. R. Haag, N. M. Hugenholtz, and M. Winnink, *Commun. Math. Phys.*, **5**, 215 (1967).
43. N. M. Hugenholtz, in R. F. Streater, Ed., *Mathematics for Contemporary Physics*, Academic, London, 1971.
44. R. Alicki, *Rep. Math. Phys.*, **10**, 249 (1976).
45. A. Kossakowski, A. Frigerio, V. Gorini, and M. Verri, *Commun. Math. Phys.*, **57**, 97 (1977).
46. R. Haag, D. Kastler, and E. Trych-Pohlmeyer, *Commun. Math. Phys.*, **38**, 173 (1974).
47. M. Aizenman, G. Gallavotti, S. Goldstein, and J. L. Lebowitz, *Commun. Math. Phys.*, **48**, 1 (1976).
48. A. Kossakowski, *Rep. Math. Phys.*, **3**, 247 (1972).
49. K. Yosida, *Functional Analysis*, Springer, Berlin, 1972.
50. K. Kraus, *Ann. Phys.*, **64**, 311 (1971).
51. G. Lindblad, *Commun. Math. Phys.*, **48**, 119 (1976).
52. V. Gorini, A. Kossakowski, and E. C. G. Sudarshan, *J. Math. Phys.*, **17**, 821 (1976).
53. E. B. Davies, *Quantum Theory of Open Systems*, Academic, London, 1976.
54. R. S. Ingarden and A. Kossakowski, *Ann. Phys.*, **89**, 451 (1975).
55. S. R. deGroot and P. Mazur, *Non-equilibrium Thermodynamics*, North-Holland, Amsterdam, 1962.
56. I. Prigogine, *Introduction to Thermodynamics of Irreversible Processes*, Wiley, New York, 1967.
57. G. Lindblad, *Commun. Math. Phys.*, **33**, 305 (1973).
58. E. H. Lieb, *Adv. Math.*, **11**, 267 (1973).
59. H. Spohn, *Lett. Math. Phys.*, **2**, 33 (1977).
60. A. Frigerio, *Quantum dynamical semigroups and approach to equilibrium*, University of Milan, preprint.
61. E. B. Davies, *Commun. Math. Phys.*, **19**, 83 (1970).
62. H. Spohn, *Rep. Math. Phys.*, **10**, 139 (1976).

63. D. E. Evans, *Commun. Math. Phys.*, **54,** 293 (1977).
64. L. Onsager, *Phys. Rev.*, **37,** 405; **38,** 2265 (1931).
65. K. Hepp, *Z. Phys.*, **B20,** 53 (1975).
66. K. Hepp, "Results and Problems in Irreversible Statistical Mechanics of Open Systems," in H. Araki, Ed., LNP, Vol. 39, Springer, Berlin, 1975.
67. C. R. Willis and P. G. Bergmann, *Phys. Rev.*, **128,** 391 (1962).
68. H. Spohn, *Linear Response Theory for Weakly Coupled Systems*, in preparation.
69. I. Prigogine, *Etude thermodynamique des phénomènes irreversible*, Desoer, Liege, 1947.
70. P. Mazur, *Bull. Acad. Roy Belg. Cl. Sc.*, **38,** 182 (1952).
71. R. Haase, *Z. Naturforsch.*, **6a,** 522 (1951).

Journal of Statistical Physics, Vol. 34, Nos. 3/4, 1984

Nonequilibrium Steady States of Stochastic Lattice Gas Models of Fast Ionic Conductors

Sheldon Katz,[1,2] Joel L. Lebowitz,[1] and Herbert Spohn[1,3]

Received September 6, 1983

We investigate theoretically and via computer simulation the stationary nonequilibrium states of a stochastic lattice gas under the influence of a uniform external field E. The effect of the field is to bias jumps in the field direction and thus produce a current carrying steady state. Simulations on a periodic 30×30 square lattice with attractive nearest-neighbor interactions suggest a nonequilibrium phase transition from a disordered phase to an ordered one, similar to the para-to-ferromagnetic transition in equilibrium $E = 0$. At low temperatures and large E the system segregates into two phases with an interface oriented parallel to the field. The critical temperature is larger than the equilibrium Onsager value at $E = 0$ and increases with the field. For repulsive interactions the usual equilibrium phase transition (ordering on sublattices) is suppressed. We report on conductivity, bulk diffusivity, structure function, etc. in the steady state over a wide range of temperature and electric field. We also present rigorous proofs of the Kubo formula for bulk diffusivity and electrical conductivity and show the positivity of the entropy production for a general class of stochastic lattice gases in a uniform electric field.

KEY WORDS: Steady states; stochastic lattice gas models; fast ionic conductors.

Supported in part by National Science Foundation Grant DMR81-14726 and NATO Grant 040.82.

[1] Department of Mathematics and Physics, Rutgers University, New Brunswick, New Jersey 08903.

[2] Present and permanent address: Department of Physics, Villanova University, Villanova, Pennsylvania. Work supported in part by a Lafayette College Junior Faculty Leave Grant.

[3] Present and permanent address: Theoretische Physik, Theresienstrasse 37, 8 Munchen 2, Germany. Work supported in part by a Heisenberg fellowship of the Deutsche Forschungsgemeinschaft.

0022-4715/84/0200-0497$03.50/0 © 1984 Plenum Publishing Corporation

INTRODUCTION

The equilibrium properties of macroscopic systems, with a given microscopic Hamiltonian, can be obtained as suitable averages in well-defined Gibbs ensembles. While this only solves the problem in principle—the evaluation of integrals involving approximately 10^{23} variables is not simple—it does provide a starting point for both qualitative understanding and quantitative approximations to equilibrium behavior. In addition many interesting equilibrium phenomena, such as phase transitions, can be studied explicitly in simplified lattice model systems.

Our understanding of the statistical mechanics of nonequilibrium phenomena is much less satisfactory at the present time. We do not have any general prescription for choosing appropriate ensembles even for the simplest nonequilibrium systems, e.g., those in a stationary state in which steady particle or heat currents are maintained. Attempts to model such systems have so far yielded more or less explicit stationary ensembles only for noninteracting systems, e.g., ideal gases and harmonic crystals.[1]

In this note we describe theoretical investigations and computer simulations of the stationary nonequilibrium states of an interacting system often used to model a fast ionic conductor.[2] The nonequilibrium state is produced by a uniform external electric field E, which causes a steady current to flow in the system: the system may be thought of as a closed loop, represented by periodic boundary conditions. The system also interacts, stochastically, with a heat reservoir at a fixed temperature T. The reservoir absorbs the heat generated by the current and maintains the steady state.

In this idealized model system the particles or ions are located at lattice sites and interact with a nearest-neighbor pair potential J. Their stochastic dynamics consists of jumps to neighboring unoccupied sites (with periodic boundary conditions), i.e., Kawasaki-type dynamics. In the absence of an electric field, $E = 0$, the transition rates are such that the stationary state of the system is the Gibbs equilibrium state at the reservoir temperature T. The uniform field E, which is *not* the gradient of a potential, biases the jumps in the direction of E. (We may think of the field as produced by a uniformly increasing magnetic flux through a loop.)

Our rigorous results for this system are—despite the simplicity of the model—confined to rather general properties. We prove that, as expected, the total entropy production is positive and show that to *linear* order in E the steady state and the state of minimum entropy production coincide. We prove the validity of the Kubo formula and the Einstein relation for the electrical conductivity and bulk diffusivity at zero field. The Kubo formula for the bulk diffusivity also holds in the steady state with finite field E.

(This is, as far as we know, the first rigorous proof of the Kubo relation for a steady-state transport coefficient in an interacting system.)

The computer simulations on the other hand give more detailed if less reliable (certainly less rigorous) results. They were done on a half-filled one-dimensional lattice with 200 sites and on a two-dimensional square lattice of 900 sites—both with periodic boundary conditions. The results in two dimensions suggest the existence of a phase transition at low temperatures which is strongly influenced, perhaps even dominated, by the electric field. This is indicated by the apparent segregation of the system in a direction perpendicular to the field into a "fluid" and "gas" phase. For the equilibrium system, $E = 0$, a segregation into "randomly oriented" regions takes place (in a macroscopic system) below the Onsager critical temperature $T_c \cong 0.44|J|/K_B$.[3] Defining the corresponding $T_c(E)$ as the apparent critical temperature in presence of the field we find that it increases with E for $J > 0$ (attractive) and decreases, possibly to zero, when E is very large, for $J < 0$ (repulsive). A rough phase diagram for $J > 0$, $E = \infty$, can thus be obtained. A preliminary account of our results has been given in Ref. 4.

In Section 1 we describe our model in more detail. In Sections 2 and 3 we provide some theoretical background for the numerical results presented in the following two sections. We report results on the steady-state current, the zero-frequency conductivity and bulk diffusivity, the structure function, the nearest-neighbor correlations, and the specific heat. The density of the system is always $1/2$ but we cover a wide range of temperatures and electric fields. The last three sections describe the exact results mentioned earlier and also include a discussion on the various ways used to determine numerically the bulk diffusivity. Some detailed calculations are left for the appendices.

1. DESCRIPTION OF THE MODEL AND DYNAMICS

We consider a stochastic lattice gas in the presence of a constant external electric field. Particles move on a simple (hyper) cubic lattice of dimension d. There is at most one particle per lattice site (hard-core exclusion). The system is confined to a box $\Lambda \subset \mathbb{Z}^d$ with periodic boundary conditions. The state space of the system is the set of all possible configurations. We denote the occupation variables by η_x,

$$\eta_x = \begin{cases} 1 & \text{if site } x \in \Lambda \text{ is occupied} \\ 0 & \text{is site } x \in \Lambda \text{ is empty} \end{cases}$$

and the full particle configuration by $\eta = \{\eta_x\}_{x \in \Lambda}$.

The dynamics is completely specified by prescribing the jump rates of the particles. To incorporate the hard-core exclusion it is convenient to think of them as exchange rates and to define

$$c(y,x,\eta) \equiv c(x,y,\eta) \geqslant 0 \tag{1.1}$$

as the rate at which the occupation at sites x and y are exchanged when the configuration is η. If $\eta_x = \eta_y$, then the exchange does not alter the configuration of the particles, considered identical here, and we may fix the value of $c(x,y,\eta)$ arbitrarily, e.g., equal to zero. (The situation would be different if we were to consider the motion of test particles, i.e., self-diffusion.[5,6]) If $\eta_x \neq \eta_y$, e.g., if $\eta_x = 1$ and $\eta_y = 0$, then $c(x,y,\eta)$ is the rate at which the particle at x jumps to the empty site at y. Letting η^{xy} denote the configuration η with occupation at sites x and y interchanged, i.e.,

$$(\eta^{xy})_z = \begin{cases} \eta_y & \text{for} \quad z = x \\ \eta_x & \text{for} \quad z = y \\ \eta_z & \text{for} \quad z \neq x, y \end{cases}$$

the time evolution of a probability distribution $\rho_t(\eta)$ is given by the master equation

$$\begin{aligned} \frac{d}{dt}\rho_t(\eta) &= \frac{1}{2}\sum_{x,y\in\Lambda}\{c(x,y,\eta^{xy})\rho_t(\eta^{xy}) - c(x,y,\eta)\rho_t(\eta)\} \\ &= L\rho_t(\eta) \end{aligned} \tag{1.2}$$

To proceed we have to specify the jump rates. General considerations provide us with certain constraints, but as will be seen later some important properties of the model may depend on the detailed choice of the rates. We list and motivate our conditions on the jump rates $c_E(x,y,\eta)$. (We have added the subscript E to make the dependence on the electric field explicit.)

(i) For simplicity we allow only nearest-neighbor jumps, i.e., $c_E(x,y,\eta) = 0$ for $|x-y| > 1$. To avoid degeneracies it is also useful to assume that $c_E(x,y,\eta) > 0$ for $|x-y| = 1$ and $\eta_x \neq \eta_y$. This condition assures that the number of particles is the only conserved quantity. We will further assume, as is usually done, that $c_E(x,y,\eta)$ depends only on the $4d-2$ nearest neighbors of the bond (x,y).

(ii) Our model should describe correctly thermal equilibrium. Therefore for zero electric field we impose the condition of detailed balance

$$c_0(x,y,\eta) = c_0(x,y,\eta^{xy})e^{-\beta[H(\eta^{xy}) - H(\eta)]} \tag{1.3}$$

Here β is the inverse temperature and $H(\eta)$ is the energy of the configuration η. This we choose to be given by the usual nearest-neighbor Hamiltonian

$$H(\eta) = -\tfrac{1}{2}J \sum_{x,y\in\Lambda,\ |x-y|=1} \eta_x\eta_y \tag{1.4}$$

$J > 0$ corresponds to an attractive and $J < 0$ to a repulsive interaction.

To model the electric field one assumes that jumps in the direction of the field are favored and jumps opposite to the direction of the field are suppressed. Physically it seems reasonable still to assume a local detailed balance but now including the work done by the electric field in the jump. This yields

$$c_E(x,y,\eta) = c_E(x,y,\eta^{xy})e^{-\beta[H(\eta^{xy})-H(\eta)]}e^{E\cdot(x-y)(\eta_x-\eta_y)} \tag{1.5}$$

Physically E stands for the combination $q\beta E$, where q is the charge and E the applied external electric field.

Remark. Notice that because of the periodic boundary conditions the electric field energy cannot be included in the Hamiltonian. In particular

$$Z^{-1}\exp\left[-\beta H(\eta) + \sum_x E\cdot x\eta_x\right] \tag{1.6}$$

is *not* invariant under the dynamics with rates $c_E(x,y,\eta)$, i.e., it is not a stationary solution of (1.2). For boundary conditions corresponding to a closed box (1.6) would be invariant. This state would describe then a gas in a uniform field in thermal equilibrium with no transport, which is clearly of no interest to us here.

(iii) Finally, it seems natural, although not necessary, to assume certain symmetry and homogeneity properties of the jump rates. We require that the rates are invariant under translations, i.e., $c_E(x+a, y+a, \tau_a\eta) = c_E(x,y,\eta)$ where $\tau_a\eta$ denotes the configuration η shifted by a, modulo the periodicity of Λ. We also require that the rates are left invariant under a reflection of both the configuration and the E-field. Lastly if we think of particles as positive charges and empty sites as negative charges, then it is physically natural to assume symmetry under charge conjugation. This means, in lattice gas language, that if we change η_x to $1-\eta_x$ for all x and E to $-E$ then the rates remain unchanged.

There is a simple way to implement all these symmetries through setting

$$c(x,y,\eta) = \phi(\beta H(\eta^{xy}) - \beta H(\eta) - E\cdot(x-y)(\eta_x-\eta_y)) \tag{1.7}$$

with some function ϕ which, by detailed balance (1.5) has to satisfy $\phi(h) = \phi(-h)e^{-h}$. In principle ϕ could still depend on $|E|$ and β; in fact the rate of hopping is strongly temperature dependent. However, it seems reasonable to assume that this dependence enters ϕ only multiplicatively and that otherwise ϕ is independent of β and E. Since a change in time scale does not affect the steady state, our prime interest here, we may as well assume that ϕ is independent of β and E. In particular we may adopt then one of the well-known choices for ϕ at $E = 0$: for the Metropolis dynamics[7] $\phi(h) = 1$ for $h \leqslant 0$ and $\phi(h) = e^{-h}$ for $h \geqslant 0$ and for the Kawasaki dynamics[8] $\phi(h) = 1/(1 + e^h)$.

For simplicity we shall absorb the coupling constant into the inverse temperature and set βJ equal to 4β. Then $\beta > 0$ corresponds to an attractive interaction, $\beta < 0$ to a repulsive interaction, and $\beta = 0$ to a noninteracting system.

2. STATIONARY STATES

We are interested in the steady nonequilibrium situation and investigate here general properties of stationary states. A stationary probability distribution $\rho_{E,N}$ is defined by

$$L_E \rho_{E,N}(\eta) = 0 \tag{2.1}$$

Because the number N of particles is conserved we have a stationary distribution for each $N = 0, \ldots, |\Lambda|$ and since the jump rates are nondegenerate these stationary distributions are unique. Expectations with respect to $\rho_{E,N}$ are denoted by $\langle \cdot \rangle_{E,\rho}$, where $\rho = N/|\Lambda|$. $\rho_{0,N}$ is the canonical equilibrium state.

The theory of Markov chains ensures an exponentially fast approach to stationarity for finite Λ. However, because of the conservation law, the approach to stationarity is diffusive and therefore slow. The time needed to approach stationarity is estimated to be of the same order as that required for a random walk in Λ to reach its equilibrium. Thus the first nonzero eigenvalue of L_E is expected to be of the order $|\Lambda|^{-2}$. At low temperatures the approach to stationarity may be even slower.

In contrast to the thermal equilibrium situation, $E = 0$, $\rho_{E,N}$ is defined only through (2.1) and not known explicitly, except for the few cases mentioned below. We outline briefly our information about $\rho_{E,N}$:

$\rho_{E,N}$ has to inherit the symmetries of the rates. Therefore $\rho_{E,N}$ is invariant under translations, i.e.,

$$\rho_{E,N}(\eta) = \rho_{E,N}(\tau_a \eta) \tag{2.2}$$

If for some reflection we denote by $R\eta$ the reflected configuration η and by RE the correspondingly reflected field, then

$$\rho_{E,N}(\eta) = \rho_{RE,N}(R\eta) \tag{2.3}$$

If we denote by $C\eta$ the configuration η with particles and empty sites interchanged, then

$$\rho_{E,N}(\eta) = \rho_{-E,|\Lambda|-N}(C\eta) \tag{2.4}$$

The symmetry of the probability distribution implies of course the corresponding symmetry of expectations. For example, (2.3) implies that the average current is odd in E, and together with (2.4) we conclude that the average current is even in the density around $\rho = 1/2$.

We note that in general $\rho_{E,N}$ is *not* even in E. Higher-order correlations which cannot be related to each other by reflection distinguish between the states $\rho_{E,N}$ and $\rho_{-E,N}$.

In a few cases (2.1) can be solved explicitly. For the particular choice

$$c_E(x, y, \eta) = e^{E\cdot(x-y)(\eta_x - \eta_y)/2} \tag{2.5}$$

which corresponds to $\beta = 0$ and $\phi(h) = e^{-h/2}$ in (1.7), the invariant distribution $\rho_{E,N}$ is obtained for arbitrary dimension by randomly placing N particles in Λ. The invariant distribution does not depend on E and corresponds to the equilibrium state at infinite temperature. Notice, however, that merely setting $\beta = 0$ in (1.5) does not ensure that the stationary state is of that simple form.

In one dimension there is a whole class of nontrivial jump rates, cf. Appendix 4, for which the canonical equilibrium state $Z^{-1}e^{-\beta H(\eta)}$ with N particles is invariant under L_E independently of the magnitude of E. Therefore all quantities of interest such as steady-state current, structure function, can be computed explicitly. In two or more dimensions, with the exception of (2.5), it is impossible to choose the jump rates in such a way that the thermal equilibrium state remains invariant as the E field is turned on.

All explicit solutions have the feature that the steady state is independent of E. Except for a few particles on a ring[9] we have no example with a nontrivial E dependence.

3. STEADY STATE CURRENT, ZERO FIELD CONDUCTIVITY

A quantity of basic physical interest is the average current in the steady state. We give first some theoretical background and describe in the following section our numerical results.

Let us define the current function $j_E(x, y)$ for the bond (x, y) by

$$j_E(x, y)(\eta) = c_E(x, y, \eta)(\eta_x - \eta_y) \tag{3.1}$$

This is the expected jump rate from x to y in a configuration η. The average current in the mth coordinate direction is then given by

$$j_m(\beta, E) = \langle j_E(x, x + e_m)\rangle_{E,\rho} \tag{3.2}$$

Here e_m is the unit vector pointing along the positive m axis. By translation invariance the average does not depend on x and by reflection symmetry it is odd in E.

The zero field conductivity is given by

$$\sigma_{mn} = \frac{\partial}{\partial E_n} \langle j_E(x, x + e_m)\rangle_{E,\rho}|_{E=0} \tag{3.3}$$

Let us denote the differentiation in (3.3) by a prime. Then

$$\begin{aligned}\sigma_{mn} &= \sum_\eta \rho_{0,N}(\eta) c_0'(x, x + e_m, \eta)(\eta_x - \eta_{x+e_m}) \\ &+ \sum_\eta \rho_{0,N}'(\eta) c_0(x, x + e_m, \eta)(\eta_x - \eta_{x+e_m})\end{aligned} \tag{3.4}$$

Here $\rho_{0,N}$ is the canonical equilibrium state. Note that we have two contributions: the first one results because of the E dependence of the current function and the second one because of the E dependence of the steady state. Using the detailed balance condition (1.5), cf. Appendix 1 for the computational details, these two terms correspond, respectively, to the static and dynamic part of the conductivity

$$\begin{aligned}\sigma_{mn} &= \tfrac{1}{2}\delta_{mn}\langle c_0(0, e_n, \eta)(\eta_0 - \eta_{e_n})^2\rangle_{0,\rho} \\ &- \int_0^\infty dt \sum_x \langle j_0(x, x + e_n) e^{L_0^* t} j_0(0, e_m)\rangle_{0,\rho}\end{aligned} \tag{3.5}$$

The expectations and rates are the zero field equilibrium expectations and rates. L_0^* is the adjoint of L_0 defined in (1.2). By reflection symmetry the second contribution in (3.5) and therefore the conductivity tensor are diagonal.

Using the same reasoning it can be shown that the frequency-dependent conductivity $\sigma_{mn}(\omega)$ is also of the form (3.5) with the addition of the factor $e^{i\omega t}$ under the time integral. Thus only the change in the steady state, which results from applying the electric field, leads to a frequency dependence of the conductivity. In particular, in all exactly soluble examples mentioned above, the steady state does not depend on E and therefore $\sigma_{mn}(\omega)$ is frequency independent.

We will show in Section 6 that (3.5) is really the Kubo formula for the conductivity. It is not immediately obvious that this is in fact what (3.5) is (note in particular that the second term is negative), but this is indeed the case. Just as for Hamiltonian systems, the Kubo formula here expresses the transport coefficient as the space-time integral over the corresponding current–current correlation function. It is the definition of the appropriate current function which requires some care in the stochastic case.

Since by detailed balance (1.3) $e^{L_0^* t}$ is symmetric with respect to the equilibrium weight $Z^{-1}e^{-\beta H(\eta)}$, and L_0^* is negative, we have the spectral representation

$$\int_0^\infty dt \sum_x \langle j_0(x, x+e_n) e^{L_0^* t} j_0(0, e_m)\rangle_{0,\rho}$$
$$= \delta_{mn} \int_0^\infty dt \int_0^\infty \mu(d\lambda) e^{-\lambda t} = \delta_{mn} \int_0^\infty \lambda^{-1} \mu(d\lambda) \tag{3.6}$$

with some positive measure $\mu(d\lambda)$. This yields immediately the upper bound

$$\sigma_{mn} \leqslant \tfrac{1}{2} \delta_{mn} \langle c_0(0, e_n, \eta)(\eta_0 - \eta_{0_n})^2 \rangle_{0,\rho} \tag{3.7}$$

As noticed by Halperin,[10] (3.7) can be improved by Jensen's inequality to give

$$\sigma_{mn} \leqslant \tfrac{1}{2} \delta_{mn} \langle c_0(0, e_n, \eta)(\eta_0 - \eta_{e_n})^2 \rangle_{0,\rho} - \left[\int_0^\infty \mu(d\lambda) \right]^2 \Big/ \int_0^\infty \mu(d\lambda)\lambda$$
$$= \tfrac{1}{2} \delta_{mn} \langle c_0(0, e_n, \eta)(\eta_0 - \eta_{e_n})^2 \rangle_{0,\rho}$$
$$- \left[\sum_x \langle j_0(x, x+e_n) j_0(0, e_n)\rangle_{0,\rho} \right]^2 \Big/ \sum_x \langle j_0(x, x+e_n) L_0^* j_0(0, e_n)\rangle_{0,\rho} \tag{3.8}$$

The Mori expansion (continued fraction expansion)[11] goes beyond (3.8). One assumes that some moments of the spectral measure $\mu(d\lambda)$, i.e.,

$$\int_0^\infty \mu(d\lambda)\lambda^k = \sum_x \langle j_0(x, x+e_n)(L_0^*)^k j_0(0, e_n)\rangle_{0,\rho} \tag{3.9}$$

have been computed for $k = 0, 1, \ldots, 2M-1$. Then $\mu(d\lambda)$ is taken to be the sum of M delta functions fitted to yield the moments (3.9) and one evaluates on this basis $\int_0^\infty \mu(d\lambda) 1/\lambda$. The Mori expansion is based on the assumption that the spectral distribution is dominated by a few frequencies. As will be discussed in Section 4, we checked this assumption for the one-dimensional model and found it not to be valid in this case.

4. NUMERICAL RESULTS I: STEADY STATE CURRENT, CONDUCTIVITY

For our numerical study we used in one dimension a system of 200 lattice sites with 100 particles and in two dimensions a 30×30 square lattice with 450 particles. Since we always kept the density equal to $1/2$, we do not indicate the density dependence. We use standard Monte Carlo procedure. A bond (x, y) is randomly selected. If $\eta_x = \eta_y$ a new bond is randomly selected. If $\eta_x \neq \eta_y$ then in two dimensions if in the attempted exchange the energy difference including the work done by the electric field, ΔU, is negative the exchange is performed, and if ΔU is positive, then the exchange is performed with probability $e^{-\beta\Delta U}$. This choice corresponds to setting $\phi(h) = 1$ for $h \leqslant 0$ and $\phi(h) = e^{-h}$ for $h \geqslant 0$ in (1.7). In one dimension we modified somewhat the Metropolis dynamics in order to still have a nontrivial E dependence of the steady state even at $\beta = 0$: we used the rates

$$\begin{aligned}
&c_E(x,x+1;0,1,0,0) = 1, \qquad c_E(x,x+1;0,0,1,0) = e^{-E} \\
&c_E(x,x+1;1,1,0,1) = 1, \qquad c_E(x,x+1;1,0,1,1) = e^{-E} \\
&c_E(x,x+1;1,1,0,0) = \tfrac{1}{2}\min(1,e^{-4\beta}), \\
&c_E(x,x+1;0,0,1,1) = \tfrac{1}{2}e^{-E}\min(1,e^{-4\beta}) \\
&c_E(x,x+1;0,1,0,1) = \tfrac{1}{2}\min(1,e^{4\beta}), \\
&c_E(x,x+1;1,0,1,0) = \tfrac{1}{2}e^{-E}\min(1,e^{4\beta})
\end{aligned} \tag{4.1}$$

They are still of the form (1.7) but with ϕ depending on E.

The natural time scale is calibrated in Monte Carlo step per site (MCS) units. One MCS is 900 (200) attempted exchanges in two (one) dimensions. The time scale of the master equation (1.2) is in MCS units. The particles start in a random configuration and are then quenched to the desired temperature at $E = 0$. The system evolves for 15000 MCS to ensure a typical equilibrium configuration. This configuration is saved for the following runs. Then the E field is gradually turned on. We allow for another 15000 MCS to ensure a typical steady-state configuration. Quantities such as the current and correlation functions were calculated by averaging over 25000 MCS in a given run. The data quoted are an average over at least six runs for $d = 1$ (two to five runs for $d = 2$) at a given temperature and electric field.

To obtain the steady-state current in the mth direction one counts the total number, $N_{+,m}(t)$, of jumps in the direction of e_m and the total number, $N_{-,m}(t)$, of jumps in the direction of $-e_m$ during a time interval of length t. The current is then

$$j_m(\beta, E) = \frac{1}{|\Lambda| t}\left[N_{+,m}(t) - N_{-,m}(t)\right], \qquad t \gg 1 \tag{4.2}$$

Also of interest is the number of exchanges per bond per unit time numerically defined by

$$w_m(\beta, E) = \frac{1}{|\Lambda| t}\left[N_{+,m}(t)_+ N_{-,m}(t)\right], \qquad t \gg 1 \tag{4.3}$$

This quantity is given by

$$w_m(\beta, E) = \langle c_E(x, x + e_m, \eta)(\eta_x - \eta_{x+e_m})^2 \rangle_{E,\rho} \tag{4.4}$$

$w_m(\beta, E)$ measures how often an exchange is actually performed. In thermal equilibrium $j_m(\beta, 0) = 0$. If the E field points in the mth direction, then for large fields $N_{+,m}(t) \gg N_{-,m}(t)$ and $j_m(\beta, E) \approx w_m(\beta, E)$. This defines then the region of saturation. Any further increase in the E field no longer changes the steady state. In Fig. 1 we show an example with $d = 1$ and $\beta = 0.6$. Since E enters in the exponential, $E = 5$ means that jumps opposite to the field are suppressed by a factor 7×10^{-3}. Numerically this corresponds to $E = \infty$.

For given β the current increases first linearly in E and then reaches saturation as $E \to \infty$. For low temperatures and attractive interaction the current is strongly suppressed because of clustering. For repulsive interaction the current is also suppressed at low temperatures, but the phenomenon sets in more slowly. There is no symmetry under $\beta \to -\beta$. The maximum zero-field conductivity and maximum saturation current occur at small negative β. We fit our data to the empirical guess $a(1 - e^{-bE})$,

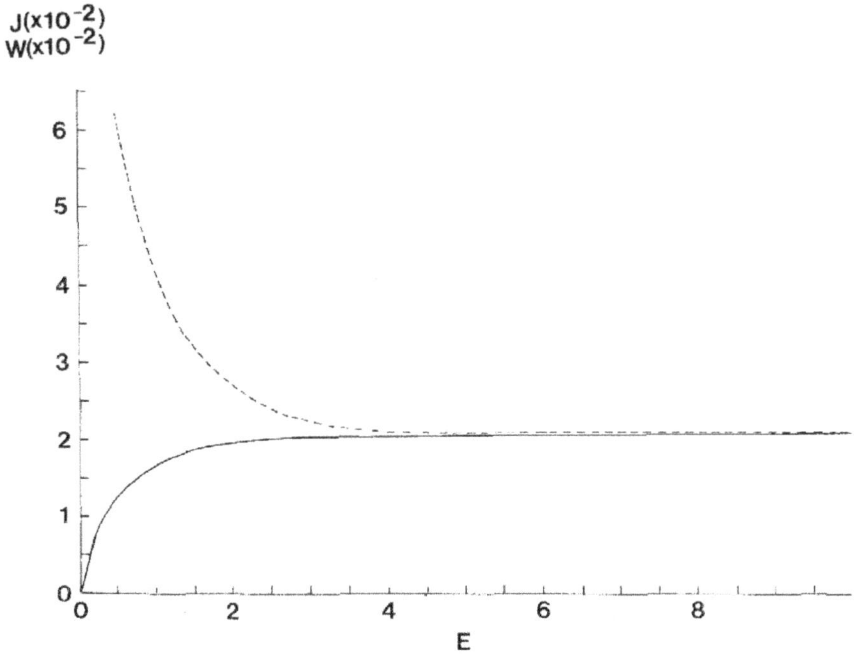

Fig. 1. Current (solid line) and average exchange rate (dotted line) as a function of electric field in one dimension at $\beta = 0.6$.

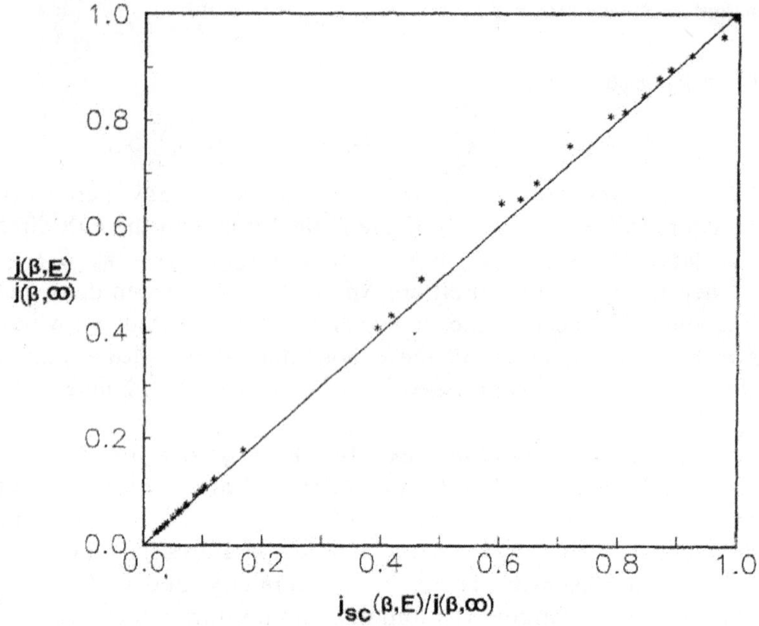

(a)

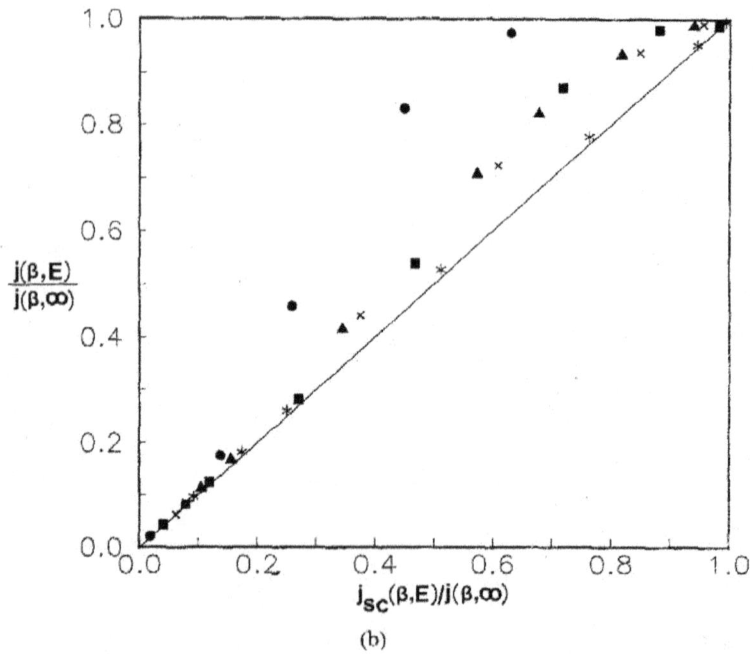

(b)

Fig. 2. (a) Scaling of the current in one dimension. Included are data for $\beta = -0.4, -0.2, 0.0, 0.2, 0.4, 0.6$ and several values of electric field from zero to saturation. (b) Scaling of the current in two dimensions. The asterisks correspond to $\beta = 0.0$, the full triangles to $\beta = 0.2$, the full squares to $\beta = 0.4$, the crosses to $\beta = -0.2$ and the full circles to $\beta = -0.4$.

$E \geqslant 0$, such that the zero-field conductivity, $(\partial/\partial E_m)j_m(\beta,0)=\sigma(\beta,0)$, and the saturation current are exact. This yields the scaling function

$$j_{sc}(\beta,E)=j(\beta,\infty)(1-e^{-E\sigma(\beta,0)/j(\beta,\infty)}) \tag{4.5}$$

Figures 2a and 2b show that the data fit reasonably well in one dimension and in two dimensions at high temperatures. In all cases $j(\beta,E)\geqslant j_{sc}(\beta,E)$.

Physically one would expect that a strong E field should override any kind of thermal dependence of the steady state. The detailed balance condition by itself does not imply such a behavior for the simple reason that it gives no connection between different (β,E) values. But if we choose the rates to be of the form (1.7) with ϕ independent of β and E and with $\phi(h)\to$ const as $h\to-\infty$, then in one dimension as $|E|\to\infty$ the rates are up to a multiplicative factor of the form (2.5) and therefore the steady state is the infinite temperature equilibrium state *independently* of the choice of β. This is why we choose rates (4.1) which still contain a temperature dependence even at $E=\infty$. In two or more dimensions if the field is directed along one of the coordinate axes, but not otherwise, even rates of the form (1.7) still retain a nontrivial temperature dependence at $E=\infty$ because of the jumps orthogonal to the field which are still governed by the thermal rates. These facts show the strong dependence of the stationary states on the dynamics, which in a way reflects the rigidity of our model: particles can only be on lattice sites, the lattice is never distorted, jumps can take place only to nearest neighbors.

From the steady-state current we obtain directly the zero-field conductivity $\sigma(\beta,0)\equiv\sigma(0)$. We list our numerical values in Table I.

According to (3.5) the zero-field conductivity has two contributions: a first one which is just one half of the average exchange rate, and a second one of dynamical origin. We follow the literature and denote the static contribution by $\sigma(\infty)$. As a measure for the dynamical contribution to the conductivity one defines the correlation factor

$$f_c=\frac{\sigma(0)}{\sigma(\infty)} \tag{4.6}$$

$f_c\leqslant 1$ by (3.7). By (4.2) and (4.3) f_c is numerically obtained as

$$f_c=\lim_{E\to 0}\frac{2}{E}\frac{N_+(t)-N_-(t)}{N_+(t)+N_-(t)} \tag{4.7}$$

with t large in order to have small statistical fluctuations. In Table I, we list the correlation factor at different temperatures in one and two dimensions, respectively. It is seen that the dynamical contribution to the zero-field conductivity is not negligible in general. In one dimension $f_c<1$ even at

Table I[a]

(a)

β	$\sigma(\infty)$ theory	$\sigma(\infty)$	$\sigma(0)$	$\sigma(0)$ upper bound	f_c	$\int dt\langle j(t)j(0)\rangle$	one-pole	error (%)	D
−0.4	0.1808	0.181	0.148	0.161	0.82	0.033	0.0196	41	1.318
−0.2	0.1920	0.192	0.165	0.175	0.84	0.027	0.0168	38	1.146
0.0	0.1875	0.188	0.174	0.179	0.93	0.014	0.0083	41	0.696
0.2	0.1288	0.129	0.111	0.118	0.86	0.017	0.0108	36	0.298
0.4	0.0812	0.081	0.067	0.073	0.83	0.014	0.0083	41	0.146
0.6	0.0475	0.047	0.038	0.043	0.81	0.009	0.0047	48	0.057

(b)

β	$\sigma(\infty)$	$\sigma(0)$	f_c	D
−0.4	0.037	0.029	0.78	0.504
−0.2	0.091	0.088	0.96	0.694
0.0	0.125	0.125	1	0.500
0.2	0.065	0.059	0.91	0.091
0.4	0.024	0.019	0.79	0.002

[a] Part (a) refers to one dimension, with columns (1) $\sigma(\infty)$ the infinite frequency part of the conductivity, analytically computed from the first term of (3.5). (2) The same quantity numerically determined through the average exchange rate. (3) The static conductivity $\sigma(0)$. (4) The theoretical upper bound (3.8). (5) $f_c = \sigma(0)/\sigma(\infty)$. (6) The time integral in (3.5). (7) The same quantity in the one-pole approximation. (8) The relative error of the one-pole approximation. (9) The bulk diffusion coefficient determined from the Einstein relation $D = \sigma/\chi$. Part (b) refers to the same quantities in two dimensions.

$\beta = 0$. This originates in our particular choice (4.1). If in (4.1) $\frac{1}{2}\min(1, e^{\pm 4\beta})$ is replaced by $\min(1, e^{\pm 4\beta})$, then $f_c = 1$ at $\beta = 0$.

For one-dimensional models with arbitrary rates Singer[12,13] computed analytically, with the aid of a computer algorithm, the first five moments of the spectral measure $\mu(d\lambda)$, cf. (3.9), i.e.,

$$\sum_x \langle j_0(x, x+1, \eta)(L_0^*)^k j_0(0, 1, \eta)\rangle_{0,\rho} \tag{4.8}$$

for $k = 0, 1, \ldots, 5$. This yields then a theoretical value for the upper bound (3.8) which is listed in Table I, column 7. Singer evaluated the one-, two-, and three-pole approximation according to the Mori scheme. He found that the one-pole approximation differs only little from the three-pole one. From Table I, column 8, we see that the one-pole approximation is about 40% off over the whole temperature range, which indicates that the spectral distribution $\mu(d\lambda)$ is not sharply peaked.

For one-dimensional models Zeyher[14] and Singer[12] argue that the current–current correlation function has a long time tail, i.e.,

$$\int_0^\infty \mu(d\lambda)e^{-\lambda t} \sim t^{-3} \tag{4.9}$$

for long times. In our computer simulation the current–current correlation function decayed within 2MCS to the noise level over the whole temperature range. This would seem to indicate that any long time tail—if present—must have a very small coefficient.

As will be discussed in Section 6, the bulk diffusion coefficient D and the zero-field conductivity are related by the Einstein relation

$$D = \sigma/\chi \tag{4.10}$$

where χ is the compressibility,

$$\chi(\rho) = \sum_x \left(\langle \eta_x \eta_0 \rangle_{0,\rho} - \rho^2\right) \tag{4.11}$$

We used series expansions for $\chi(\rho)$[15,16] and list in Table I, columns 9 and 13 our numerical values for the bulk diffusivity at density 1/2.

5. NUMERICAL RESULTS II: STRUCTURE FUNCTION, NEAREST-NEIGHBOR CORRELATIONS, LONG-RANGE ORDER, AND PHASE TRANSITION

A quantity of physical interest is the density–density correlation function in the steady state. Its Fourier transform defines the zero-frequency structure function $S(k)$ which is proportional to the differential scattering cross section. The small-k behavior of $S(k)$ yields information about the decay of correlations in the steady state and on the possible occurrence of long-range order.

We investigated numerically the structure function in one and two dimensions. The structure function is defined by

$$S_{\beta,E}(k) = \frac{1}{|\Lambda|} \sum_\eta \rho_{E,N}(\eta) \left| \sum_{x\in\Lambda} e^{ikx}\eta_x \right|^2 \tag{5.1}$$

k runs over the first Brillouin zone. In one dimension $N = |\Lambda|/2$ and $k = (2\pi/200)n$ with $n = 0, 1, \ldots, 199$. In two dimensions $k_1 = (2\pi/30)n_1$, $k_2 = (2\pi/30)n_2$ with $n_1, n_2 = 0, \ldots, 29$. For a repulsive interaction it is more instructive to study the staggered structure function $\tilde{S}_{\beta,E}(k)$ which is obtained by shifting in $S_{\beta,E}(k)$ each component of k by π. Numerically the steady-state average (5.1) is obtained by time averaging.

Since the results depend on the dimension, we discuss one and two dimensions separately.

$d = 1$, *Attractive Interaction.* For our choice of rates the E field acts like an increased ferromagnetic coupling. This is reflected in the structure function $S(k)$, cf. Fig. 3. The central peak increases with increasing field and the correlation length increases roughly by a factor of 4 from $E = 0$ to

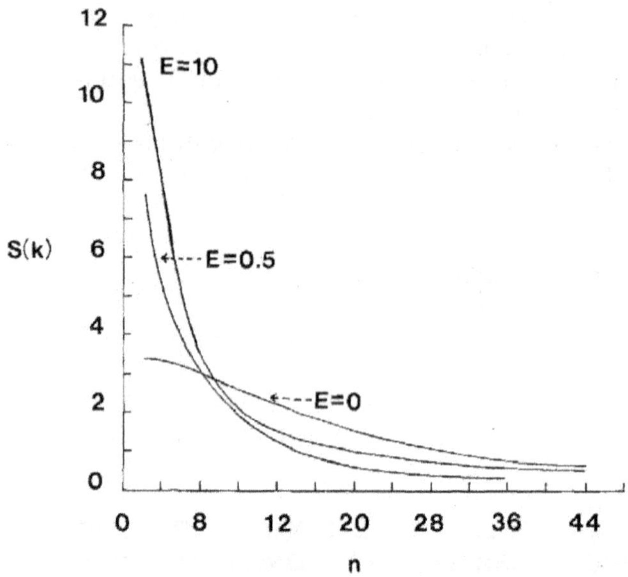

Fig. 3. The structure function $S(k)$ where $k = (2\pi/200)n$ for several values of electric field in one dimension at $\beta = 0.6$.

$E = \infty$. The "effective temperature," determined by fitting either to the equilibrium nearest-neighbor correlation or to the equilibrium correlation length, decreases roughly by a factor of $\frac{1}{2}$. The observed phenomenon cannot be of general nature. If we were to choose rates satisfying (A.26), then the correlation length would be independent of E, and if we were to choose the Metropolis dynamics, then for large fields the rates would become independent of β and the correlations would tend to zero as $E \to \infty$.

$d = 1$, *Repulsive Interaction*. Over the temperature range considered the structure function is practically independent of E.

$d = 2$, *Attractive Interaction*. In two dimensions we oriented the field always along the negative y axis. According to Onsager the equilibrium (infinite) system has a phase transition at $\beta_c \cong 0.44$. For $\beta > \beta_c$ there is a liquid (high density) and a gas (low density) phase. As $\beta \to \beta_c$, at $\rho = 1/2$, the correlation length diverges and the structure function becomes singular at $\mathbf{k} = 0$. This behavior is also reflected in the 30×30 system. Since the number of particles does not fluctuate, $S_{\beta,E}(\mathbf{k} = 0) = 0$ for all E and β. The maximum of $S_{\beta,0}(\mathbf{k})$ occurs at $\mathbf{k} = (2\pi/30, 0)$ and by symmetry also at $\mathbf{k} = (0, 2\pi/30)$. As we approach the region around 0.44 this maximum increases rapidly by an order of magnitude.

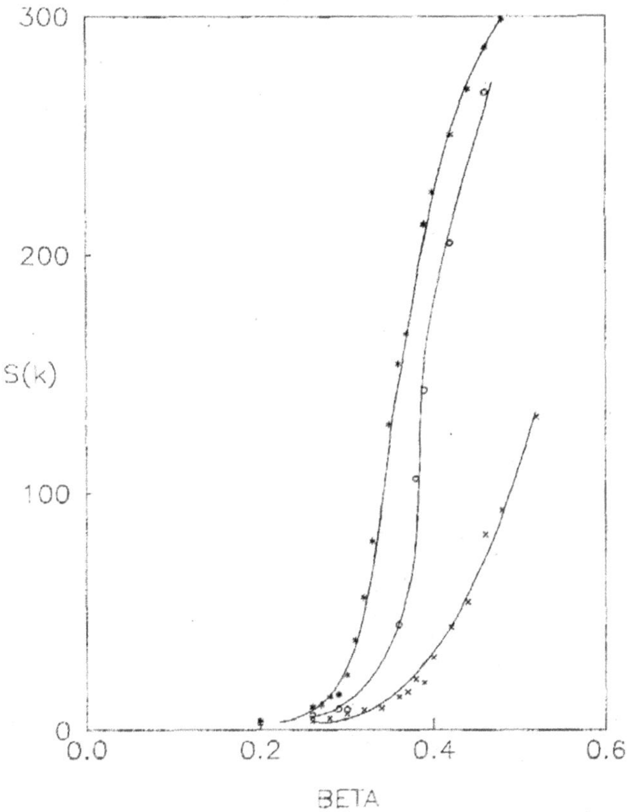

Fig. 4. The structure function $S(k)$ at the value $k_x = 2\pi/30$, $k_y = 0$, as a function of β, for various electric fields. The crosses correspond to $E = 0$, the open circles to $E = 0.75$, and the asterisks to $E = \infty$.

For finite E (always oriented along the negative y axis) we observe first that the region of rapid increase in the maximum of $S_{\beta,E}(\mathbf{k})$ occurs at smaller β, cf. Fig. 4. At $E = \infty$ the rapid increase occurs around $\beta = 0.32$. Secondly the structure function becomes highly anisotropic. While $S_{\beta,\infty}(0, 2\pi/30) \simeq 1$ over the whole temperature range considered, $S_{\beta,\infty}(2\pi/30, 0)$ increases from 1 to about 225. This suggests defining a critical temperature $\beta_c(E)$ depending on E, as the value for which $S_{\beta_c(E),E}(2\pi/30, 0) \cong 50$ (the choice 50 being somewhat arbitrary of course). Studying typical configurations the following picture emerges: for $\beta < \beta_c(E)$ we have a disordered configuration with some tendency to form clusters whereas for $\beta > \beta_c(E)$ and $E \neq 0$ typical configurations form striplike clusters oriented parallel to the field. In Fig. 5 we display some

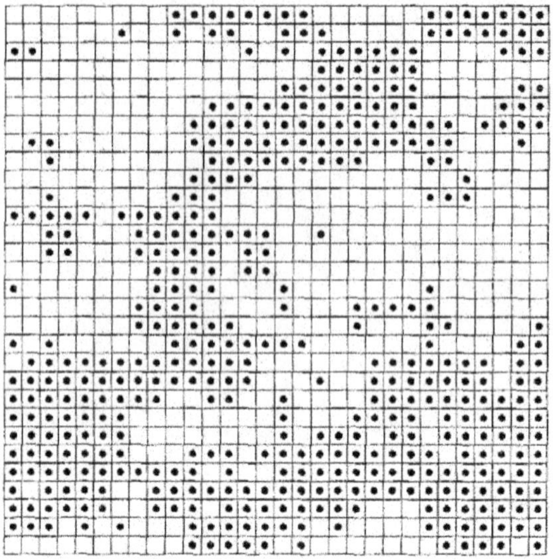

(a)

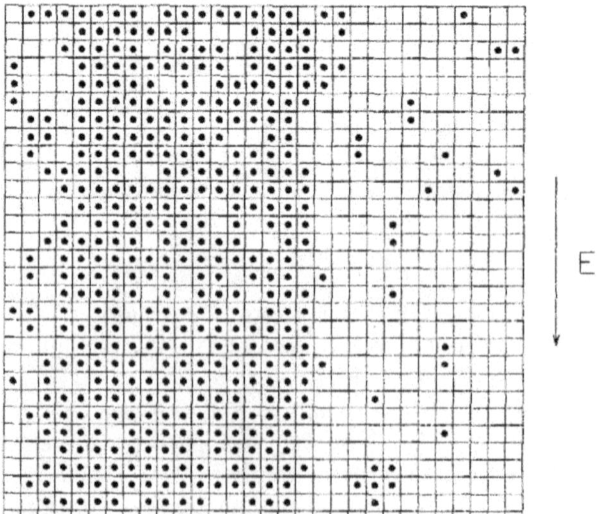

(b)

Fig. 5. (a) Equilibrium configuration at $\beta = 0.40$, $E = 0$. (b) A typical steady-state configuration, showing the formation of fully occupied columns at $\beta = 0.40$, and saturation field.

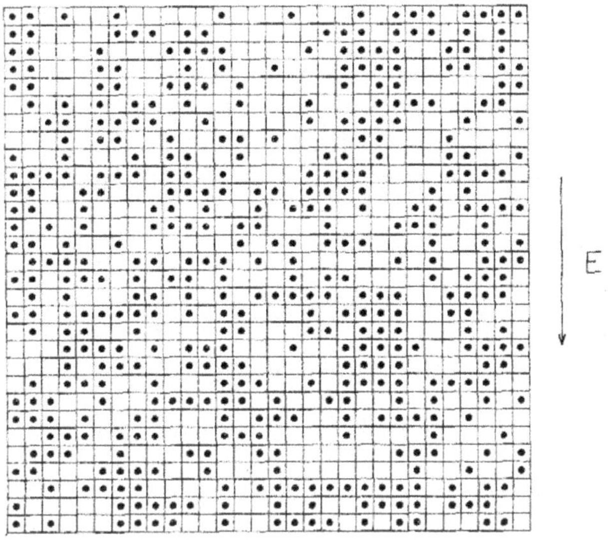

(c)

Fig. 5. (continued) (c) A typical steady-state configuration at $\beta = 0.20$ and saturation field.

Table II. Some Values of the Structure Function $S(K)$ for: (a) $E = 0$, $K = 0.4$, (b) $E = \infty$, $K = 0.4$, and (c) $E = \infty$, $K = 0.2$

n_1		0	1	2	3	4	5	
n_2	0	0.0	27.4	8.8	4.3	2.9	1.9	
	1	33.8	16.2	8.3	4.3	3.1	2.1	
	2	9.7	6.8	5.0	2.9	2.0	1.6	(a)
	3	4.5	4.4	2.9	2.2	1.6	1.4	
	4	2.9	2.7	2.3	1.5	1.3	1.2	
	5	1.9	1.5	1.4	1.4	1.3	0.9	
n_1		0	1	2	3	4	5	
n_2	0	0.0	226.2	1.5	18.6	1.6	4.4	
	1	0.7	0.7	1.0	0.9	1.2	1.1	
	2	0.5	0.5	0.7	0.7	0.7	0.9	(b)
	3	0.4	0.4	0.5	0.5	0.6	0.7	
	4	0.4	0.4	0.4	0.5	0.5	0.4	
	5	0.4	0.4	0.3	0.5	0.5	0.5	
n_1		0	1	2	3	4	5	
n_2	0	0.0	3.8	3.4	3.4	2.4	2.2	
	1	0.9	1.2	2.2	2.9	2.4	2.3	
	3	1.0	1.2	1.7	2.0	1.8	2.2	(c)
	4	1.0	1.0	1.1	1.4	1.8	1.5	
	5	0.7	0.9	0.9	1.1	1.2	1.4	

typical configurations contrasting the cases $\beta < \beta_c(\infty)$ with $\beta > \beta_c(\infty)$ and $E = 0$ with $E = \infty$ for $\beta > \beta_c(\infty)$. As a more quantitative measure we also show the structure function for small k in Table II.

To give an explanation which is at least qualitative for the occurrence of strip configurations it is useful to consider very low temperatures. If $E = 0$, then jumps which decrease the number of nearest neighbors are essentially forbidden. The ground state is a droplet which has the smallest surface, and at low temperatures the system tries to minimize surface energy. On the other hand if the work done by the field is larger than the binding energy, say $E = \infty$, then jumps in the direction of the field still

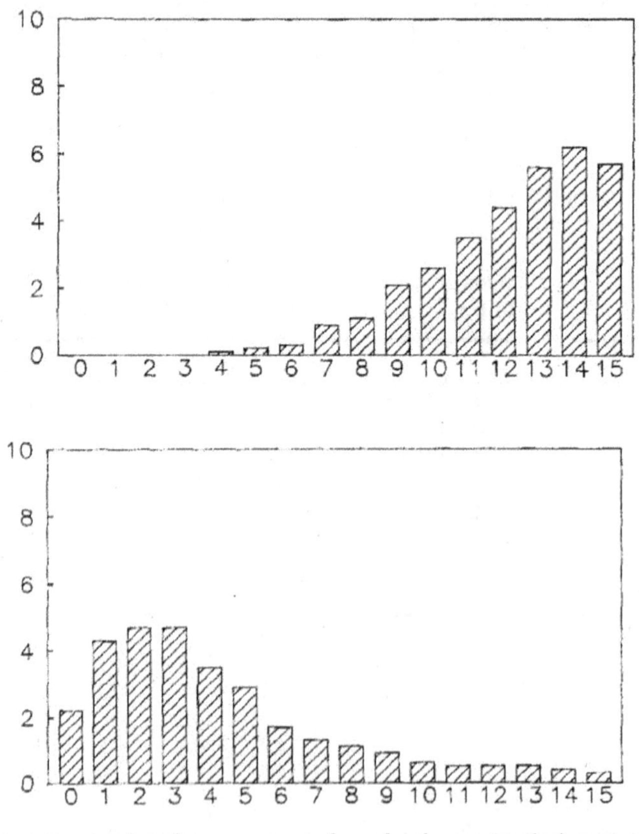

Fig. 6. Histograms showing the average number of columns (vertical axis) with a given number of occupied sites. Because of symmetry, columns with m occupied sites and $(30 - m)$ unoccupied sites are considered equivalent to columns with m unoccupied sites and $(30 - m)$ occupied sites. Because such symmetry does not exist for $m = 15$, contributions for $m = 15$ were multiplied by two in compiling the histogram. (a) Histogram from $\beta = 0.20$ and saturation field. (b) Histogram for $\beta = 0.40$ and saturation field.

have rate one independently of the number of nearest neighbors, provided the jump is not blocked. Therefore the most slowly changing configurations are strips parallel to the field. In analogy to thermal equilibrium we would expect that the transition from disordered to strip configurations becomes sharp for the infinite system.

We may learn something about the phase diagram in the β, ρ plane for large E ($E \cong \infty$) by looking at the densities of the "liquid" and "gas" phases as they appear in typical configurations at $\beta > \beta_c(\infty)$. Counting the number in each column we can make a histogram whose peaks, for very large systems, should occur at ρ_v and ρ_l, respectively. By the symmetries (2.3)–(2.4), $\rho_l - \rho_v = m^*$, where m^* is the spontaneous magnetization in spin language. Such histograms are shown in Fig. 6. To take account of the finite (small) size of our system we estimate m^* by computing the difference

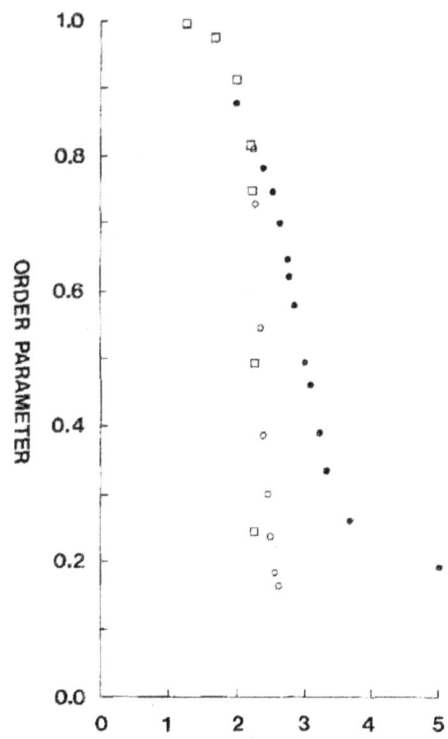

Fig. 7. Order parameter vs. $1/\beta$. The solid circles represent our Monte Carlo data in maximum electric field. The open squares represent the exact zero electric field magnetization and the open circles represent Landau's Monto Carlo results for a simulation of the two-dimensional Ising model on a 30×30 lattice.

between the average of the magnetization squared in the vertical columns and horizontal rows and dividing the result by $L^2 = 900$. (This would equal one when the strip formation is perfect and zero when the system is isotropic.) The $m^*(T)$ curve so obtained, for $E = \infty$, is shown in Fig. 7.

Also shown in Fig. 7 is the exact infinite volume equilibrium, $E = 0$, $m^*(T)$ curve as well as some $m^*(T)$ obtained[17] from computer simulations on a 30×30 lattice at $E = 0$. It appears that the nonequilibrium $m^*(T)$ is less steep than the equilibrium one. If this is really the case it would indicate a critical magnetization exponent larger than the equilibrium 1/8. This would then be similar to what is found for the phase separation of a binary fluid under a shear flow.[18]

The development of two phases at $\beta = \beta_c$ is accompanied at $E = 0$ by a logarithmic singularity in the specific heat as $\beta \to \beta_c$. To find out whether there is a similar behavior in the presence of the field we investigated the specific heat at $E = \infty$ and the saturation current as a function of β. The specific heat, cf. Fig. 8, has a maximum around 0.37 which coincides

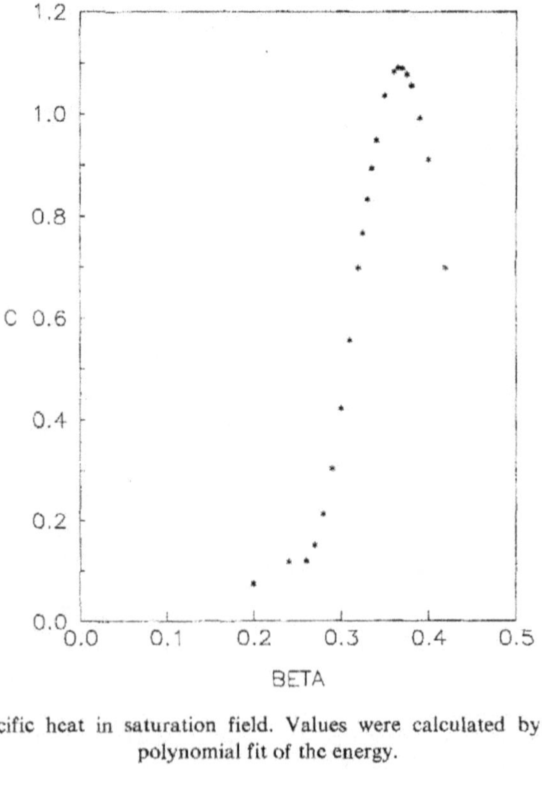

Fig. 8. The specific heat in saturation field. Values were calculated by differentiating a polynomial fit of the energy.

roughly with the critical temperature obtained from the structure function. Compared with the thermal equilibrium, $E = 0$, specific heat, the maximum is not as pronounced. This may also be due to the nature of the singularity in the thermodynamic limit being changed by the field.

We have also looked for any apparent nonsmooth behavior of the current at $\beta_c(E)$. We found that the saturation current seems to have a break around $\beta = 0.3$. In Fig. 9 we plot the saturation current with respect to β. It seems possible that in the thermodynamic limit the slope develops a jump discontinuity.

We also investigated the truncated nearest-neighbor correlations in the vertical and horizontal direction; cf. Fig. 10. Since above $\beta_c(E)$ the column density typically differs from 1/2, cf. Fig. 6, we truncate the vertical nearest-neighbor correlation in each column separately by the column

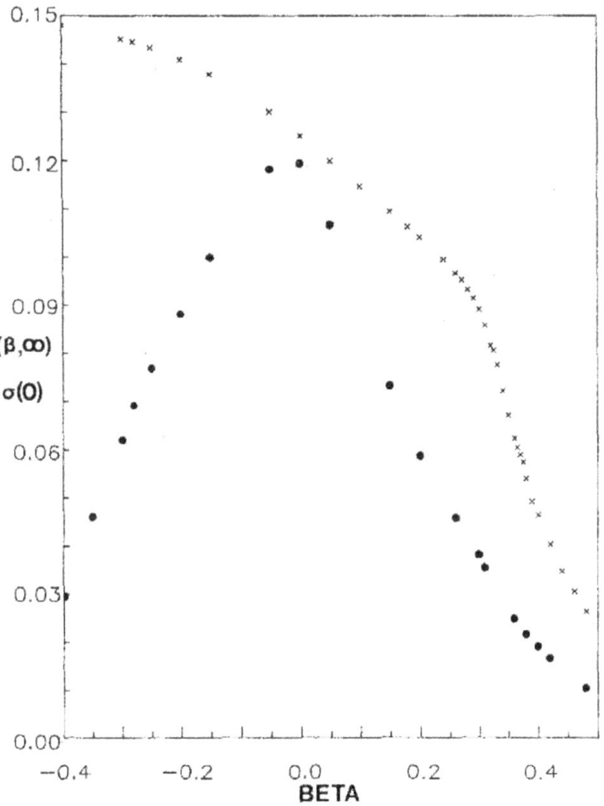

Fig. 9. The crosses represent the saturation current. The solid circles represent the zero field conductivity, $\sigma(0)$.

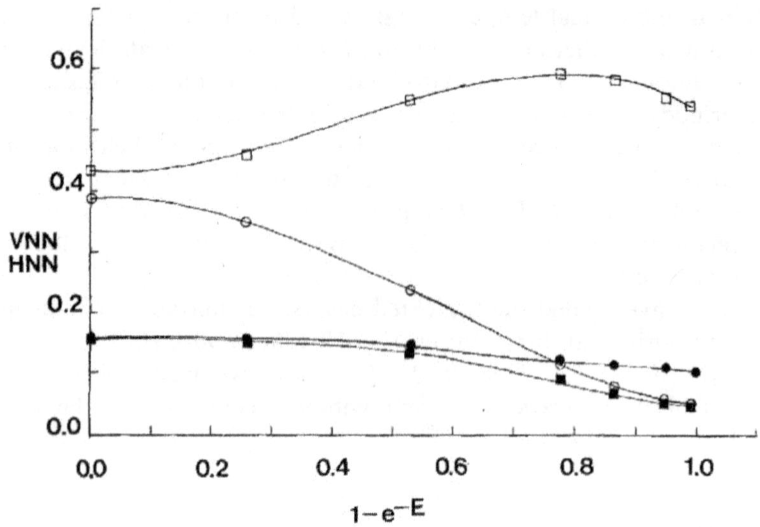

Fig. 10. Truncated vertical nearest-neighbor correlation functions (VNN) and horizontal nearest-neighbor correlation functions (HNN). The open (closed) circles correspond to VNN for $\beta = 0.4$ (0.2). The open (closed) squares correspond to HNN for $\beta = 0.4$ (0.2). The difference between VNN and HNN at $E = 0$ is an indication of statistical fluctuation.

density squared. For consistency we use the same truncation in all other cases too. We see that the correlations become anisotropic for $E \neq 0$. The qualitative behavior of the nearest-neighbor correlations may be explained as resulting from two partially opposing tendencies of the field: (i) to produce strips and thereby to increase correlations and (ii) to decrease the correlations within each column (recall that in one dimension with Metropolis dynamics the $E = \infty$ state is completely random). At small β the ordering tendency dominates and the correlations increase with increasing E. At large β the truncated correlations still increase for small E whereas at the critical field strength, $E \cong 1.5$ for $\beta = 0.4$, strips are formed and the randomization tendency within each column dominates. More study is clearly required for a full understanding of the correlations in this nonequilibrium steady state.

$d = 2$, *Repulsive Interaction*. For $E = 0$ at sufficiently low temperatures the particles try to arrange themselves in a chessboard configuration. For large field the particles jump with rate one to the unoccupied site ahead of them in the $-y$ direction. Therefore the field breaks the antiferromagnetic structure. This mechanism is supported numerically. The average current is larger than the one for the ferromagnetic coupling of the same magnitude; see Ref. 4. The nearest-neighbor correlations increase with increasing field, i.e., particles become less negatively correlated. The horizontal nearest-neighbor correlation is always less negative than the vertical one. The staggered structure function is only slightly anisotropic. Its maximum, still at either $\mathbf{k} = (2\pi/30, 0)$ or $\mathbf{k} = (0, 2\pi/30)$ is substantially decreased when going from $E = 0$ to $E = \infty$.

6. KUBO'S FORMULA AND BULK DIFFUSIVITY

Kubo's formula relates transport coefficients to the space-time integral over the corresponding current–current correlation function in equilibrium. First we discuss how this is defined for stochastic lattice gases. We then prove the formula for $\sigma(0)$, the zero field conductivity, and show that $\sigma(0)$ is related to the bulk diffusivity for $E = 0$ by an Einstein relation. We then comment on various methods to determine the latter numerically. We close by explaining the corresponding situation for nonzero E.

We define the integrated current through the bond (x, y) by

$\mathscr{J}(x, y; dt)$ = number of particles which cross the bond from x to y minus number of particles which cross the bond from y to x during the time interval $[t, t + dt]$

$\mathscr{J}(x, y; dt)$is defined for every "history" (i.e., stochastic trajectory) of the particle system. The density $\mathscr{J}(x, y; t)$ of the integrated current, defined by $\mathscr{J}(x, y; t)\, dt = \mathscr{J}(x, y; dt)$, is the *actual* current through the bond (x, y). It has a δ function with weight 1 at times when a particle jumps from x to y and a δ function with weight -1 at times when a particle jumps from y to x. The actual current should not be confused with the current function (3.1): $j_E(x, y)(\eta)$ is the *average* current *when* the particle configuration is η. For an arbitrary initial probability distribution and an aribitrary dynamics L the average current at time t is given by the average of $\mathscr{J}$ over trajectories followed by an average over configurations. In particular

$$\langle \mathscr{J}(x, y; dt)\rangle = \langle e^{L^* t} j(x, y)\rangle\, dt \tag{6.1}$$

In thermal equilibrium the average current is zero,

$$\langle \mathscr{J}(x, y; dt)\rangle_{0,\rho} = \langle j_0(x, y)\rangle_{0,\rho}\, dt = 0 \tag{6.2}$$

The current–current correlation function in equilibrium is then defined by

$$\langle \mathscr{J}(x, x + e_m; dt)\, \mathscr{J}(y, y + e_n; ds)\rangle \tag{6.3}$$

By translation invariance it depends only on $x - y$ modulo the periodicity of Λ and by time invariance only on $t - s$. In order to compute (6.3) we replace the integrated current by its short time approximation

$$\mathscr{J}(x, y; [t, t+\tau]) = \eta_{x,t}(1-\eta_{y,t})(1-\eta_{x,t+\tau})\eta_{y,t+\tau} - (1-\eta_{x,t})\eta_{y,t}\eta_{x,t+\tau}(1-\eta_{y,t+\tau}) \qquad (6.4)$$

for small τ. Here $\eta_{x,t} = 1$, if site x is occupied at time t, and $\eta_{x,t} = 0$ if site x is empty at time t. We insert the short time approximation in (6.3) with the short time intervals $[0, ds]$ and $[t + ds, t + ds + dt]$. We have to compute then an expectation depending on four different times. At this stage we exploit the Markov property of the stochastic time evolution. Using detailed balance the average equals then

$$-ds\, dt\, \langle j_0(x, x+e_m) e^{L_0^* t} j(y, y+e_n)\rangle \qquad (6.5)$$

To obtain the equal time contribution we set in (6.3) both time intervals equal to $[0, \tau]$ and use the short time approximation. For small τ the average equals then

$$\tau\delta_{xy}\delta_{mn}\langle c_0(x, x+e_n, \eta)(\eta_x - \eta_{x+e_n})^2\rangle_{0,\rho} \qquad (6.6)$$

Therefore

$$\begin{aligned}\langle \mathscr{J}(x, x+e_m; dt)\, \mathscr{J}(y, y+e_n; ds)\rangle_{0,\rho} &= \delta(t-s)\, dt\, ds\, \delta_{xy}\delta_{mn}\langle c(0, e_n, \eta)(\eta_0 - \eta_{e_n})^2\rangle_{0,\rho} \\ &\quad - dt\, ds\, \langle j_0(x, x+e_m) e^{L_0^*|t-s|} j_0(y, y+e_n)\rangle_{0,\rho} \end{aligned} \qquad (6.7)$$

Comparing with (3.5) we obtain for the conductivity

$$\sigma_{mn} = \frac{1}{2}\int_{-\infty}^{\infty} \sum_x \langle \mathscr{J}(x, x+e_m; dt)\, \mathscr{J}(0, e_n; 0)\rangle_{0,\rho} \qquad (6.8)$$

The zero-field conductivity equals the space-time integral over the equilibrium current–current correlation function. This is Kubo's famous result derived in another context.[19] The integral over an autocorrelation function is non-negative and therefore $\sigma \geqslant 0$ which is not immediately obvious from (3.5).

In the infinite volume limit the conductivity is still defined by (6.8) with the average now over the infinite volume equilibrium state at inverse temperature β and density ρ. If the equilibrium state has good cluster properties, then the integrand of (6.8) converges in the limit $\Lambda \to \mathbb{Z}^d$, $N \to \infty$, $N/|\Lambda| \to \rho$. However, we have no proof that the finite volume conductivity converges in the same limit to the infinite volume one.

Bulk Diffusion

We proceed now to the connection between the zero-field conductivity and the bulk diffusivity in thermal equilibrium. The bulk diffusivity is unambiguously defined only for the infinitely extended system. We consider the truncated density–density correlation function

$$\langle \eta_{x,t}\eta_{0,0}\rangle_{0,\rho} - \rho^2 \tag{6.9}$$

in equilibrium at given inverse temperature β and density ρ. One expects that for large x and t the density–density correlation function behaves diffusively as

$$\langle \eta_{x,t}\eta_{0,0}\rangle_{0,\rho} - \rho^2 \cong \chi(\rho)e^{At}(x,0) \tag{6.10}$$

where $e^{At}(q,0)$ is the transition probability for a Brownian particle with diffusion matrix $D(\rho)$. In other words $e^{At}(q,0)$ is the fundamental solution of the diffusion equation

$$\frac{\partial}{\partial t} f(q,t) = \sum_{m,n=1}^{d} \frac{\partial}{\partial q_m} D_{nm}(\rho) \frac{\partial}{\partial q_n} f(q,t) \tag{6.11}$$

By definition $D(\rho)$ is the bulk diffusion coefficient at density ρ. $\chi(\rho)$ is determined by summing both sides of (6.10) over x. Since $\int dq\, e^{At}(q,0) = 1$ by conservation of mass, we obtain

$$\chi(\rho) = \sum_x \left(\langle \eta_x \eta_0\rangle_{0,\rho} - \rho^2\right) \tag{6.12}$$

i.e., $\chi(\rho)$ is the static compressibility. By the assumed reflection symmetry $D(\rho)$ is diagonal. Therefore given (6.10) it is reasonable to adopt the following definition of the bulk diffusion coefficient:

$$D_{nn}(\rho) \equiv \lim_{t\to\infty} \frac{1}{2\chi(\rho)} \frac{1}{t} \sum_x (x_n)^2\left(\langle \eta_{x,t}\eta_{0,0}\rangle_{0,\rho} - \rho^2\right) \tag{6.13}$$

For high temperature, using the conservation of mass and the good cluster properties of the equilibrium state, we prove in Appendix 2 that the limit (6.13) exists and that

$$D_{mn}(\rho) = \frac{1}{2\chi(\rho)} \int_{-\infty}^{\infty} \sum_{x\in\mathbb{Z}^d} \langle \mathscr{J}(x, x+e_m; dt)\mathscr{J}(0, e_n; 0)\rangle_{0,\rho} \tag{6.14}$$

Comparing with (6.8) we obtain the Einstein relation

$$D_{mn}(\rho) = \frac{1}{\chi(\rho)} \sigma_{mn} \tag{6.15}$$

with σ_{mn} the infinite volume conductivity. By measuring the conductivity one can therefore determine the bulk diffusivity.

There have been various numerical investigations of the bulk diffusivity. We want to briefly compare here the various methods. In principle there are at least five different ways for determining the bulk diffusion constant. In three of them one uses the relationship (6.15). Therefore with these methods one has to determine the zero-field conductivity and, independently, the static compressibility, which can be done by equilibrium techniques.

(1) One measures the zero field conductivity by measuring the steady-state current for small field. This is the method used here. In practice just a single value of the field suffices. Since the current is an average over all bonds, statistical fluctuations are rather small.

(2) One determines numerically the total current–current correlation function in thermal equilibrium as

$$\frac{1}{|\Lambda|t}\left\langle\left(\sum_{x\in\Lambda}\mathscr{J}(x,x+e_m;[0,t])\right)\left(\sum_{y\in\Lambda}\mathscr{J}(y,y+e_n;[0,t])\right)\right\rangle_{0,N} \tag{6.16}$$

For large times it has to tend to the constant $2\sigma_{mn}$. In this way one evaluates numerically directly the integral (6.14). This method looks rather promising to us. But it has not been employed systematically yet. Sadiq[20] used this method in the case of a two-dimensional system with two conserved quantities near the critical point. The results are modest, in the sense that there are large statistical fluctuations. In our opinion, the reason lies more in the ambitious goal pursued in Ref. 20 than in the inadequacy of the method.

(3) Murch and Thorn[21,22] measured the conductivity for a simple cubic and a two-dimensional honeycomb lattice with both repulsive and attractive interactions. They use a time-dependent method and compute the average drift of an initially sharp mass distribution. The error bars are rather large. It is not clear to us how the precise density dependence of the bulk diffusivity can be accessed by this method.

(4) One prepares the system initially in a slowly, sinusoidally varying density. This density decays then to uniform equilibrium and the rate of decay is a measure of the bulk diffusion coefficient. Kutner, Binder, and Kehr[23] used this method in the case of a three-dimensional system on a simple cubic lattice with attractive interactions. They cover the whole density range at various temperatures. To our knowledge, this is the most systematic measurement of a bulk diffusivity. It is not so clear to us how well this method can handle statistical errors and finite wavelength and frequency effects.

(5) One establishes the system in a nonequilibrium steady state by imposing different densities at the boundaries. This is achieved by creation and destruction of particles at the boundary. In the steady state the mass current is proportional to the bulk diffusivity for small density difference. Murch[24] used this method for a three-dimensional simple cubic lattice with attractive and repulsive interactions. Since the current is an average over all bonds, statistical errors are presumably small. However, finite size

effects and nonlinear dependence on the density difference may be important.

Certainly, all five methods work in principle: which one of the methods gives the must reliable numerical value for the bulk diffusivity still waits for clarification. In our opinion, stationary methods are closest to the theoretical definition.

So far we have focused on the zero-field conductivity and diffusivity. The finite field diffusivity is still given by the space-time integral over the current–current correlation function. Since to our knowledge this has not been given in the literature, we explain the formal derivation in Appendix 3. There cannot be any simple relationship like (6.15) between diffusivity and conductivity at nonzero field. To see this we define the rates

$$\tilde{c}_E(x, y, \eta) = \alpha(E) c_E(x, y, \eta) \tag{6.17}$$

with $\alpha(E) > 0$ and independent of x, y, and η and c_E satisfying the detailed balance condition (1.5). $\tilde{c}_E$ then also satisfies the detailed balance condition. Since (6.17) just amounts to a change of the time scale, the stationary states of the dynamics with rates $\tilde{c}_E$ are the same as those with c_E. Therefore

$$\tilde{j}_n(\beta, E) \equiv \langle \tilde{j}_n(0, e_n) \rangle_{E,\rho} = \alpha(E) j_n(\beta, E) \tag{6.18}$$

and

$$\tilde{\sigma}_{mn}(\beta, E) = \left[\frac{\partial}{\partial E_n} \alpha(E) \right] j_m(\beta, E) + \alpha(E) \sigma_{mn}(\beta, E) \tag{6.19}$$

By an appropriate choice of $\alpha(E)$ we can produce essentially any current–field characteristic we want. In particular, the finite field conductivity can be negative, which is not possible for the bulk diffusivity.

Of course, we can still follow the computation in Appendix 1 and express the conductivity as

$$\sigma_{mn}(E) = \left\langle \frac{\partial}{\partial E_n} j_E(0, e_m) \right\rangle_{E,\rho} + \int_0^\infty dt \sum_\eta \rho_{E,N}(\eta) \left(\frac{\partial}{\partial E_n} L_E^* \right) e^{L_E^* t} j_E(0, e_m)(\eta) \tag{6.20}$$

7. ENTROPY PRODUCTION

In linear irreversible thermodynamics the steady state can be obtained as the minimum of the entropy production.[25] We take this as our motivation to investigate here whether a corresponding principle could be valid on the level of stochastic particle dynamics. We comment at the end of this section on the entropy production for a classical particle system in contact with heat reservoirs.

We define the entropy production $\sigma(\rho)$ in the state ρ as

$$\sigma(\rho) = \text{change in system entropy} + \text{flow of entropy from the system to the surroundings}$$

The change in the system entropy in the state ρ is

$$-\frac{d}{dt}\sum_{\eta}\rho_t(\eta)\log\rho_t(\eta)\Big|_{\rho_t=\rho} \tag{7.1}$$

with ρ_t governed by the master equation (1.2). The entropy flow to the surroundings has two contributions: (a) there is a change in the energy of the system when a transition changes the number of nearest neighbors. The corresponding entropy flow is

$$-\beta\sum_{\eta}\rho(\eta)L^*H(\eta) \tag{7.2}$$

(b) There is work done by the electric field which is absorbed by the surroundings as heat at the temperature β^{-1}. This is equal to E times the average current

$$-\tfrac{1}{2}\sum_{\eta}\rho(\eta)\sum_{\substack{x,y\\|x-y|=1}} c_E(x,y,\eta)E\cdot(x-y)(\eta_x-\eta_y) \tag{7.3}$$

Remember that E corresponds to βE. Therefore, adding (7.1) to (7.3), we obtain

$$\begin{aligned}\sigma(\rho) = -\tfrac{1}{2}\sum_{\substack{x,y\\|x-y|=1}}\sum_{\eta}\{&[c_E(x,y,\eta^{xy})\rho(\eta^{xy}) - c_E(x,y,\eta)\rho(\eta)]\log\rho(\eta)\\ &+\rho(\eta)c_E(x,y,\eta)\beta(H(\eta^{xy})-H(\eta))\\ &+\rho(\eta)c_E(x,y,\eta)E\cdot(x-y)(\eta_x-\eta_y)\}\end{aligned} \tag{7.4}$$

Note that the first two terms are the time derivative of the "free energy," $E-TS$, whereas the third term cannot be written as a time derivative.

Let $\nu(\rho) = e^{\beta H(\eta)}\rho(\eta)$. Then using the detailed balance condition (1.5) we can rewrite (7.4) as

$$\begin{aligned}\sigma(\nu e^{-\beta H}) = \tfrac{1}{4}\sum_{\substack{x,y\\|x-y|=1}}\sum_{\eta} c_E(x,y,\eta)e^{-\beta H(\eta)}&[\nu(\eta) - e^{E\cdot(x-y)(\eta_x-\eta_y)}\nu(\eta^{xy})]\\ \times&[\log\nu(\eta) - \log e^{E\cdot(x-y)(\eta_x-\eta_y)}\nu(\eta^{xy})]\end{aligned} \tag{7.5}$$

The function $f(x, y) = (x - y)(\log x - \log y)$ is non-negative for $x, y > 0$ and we conclude that

$$\sigma(\rho) \geqslant 0 \tag{7.6}$$

In fact, for $E \neq 0$, $\sigma(\rho) > 0$ because $\sigma(\rho) = 0$ if and only if $\nu(\eta) = e^{E \cdot (x-y)(\eta_x - \eta_y)}\nu(\eta^{xy})$, which cannot be satisfied because of the periodic boundary conditions. Since for $0 \leqslant \lambda \leqslant 1$, $f((1-\lambda)x_1, (1-\lambda)y_1) + f(\lambda x_2, \lambda y_2) \leqslant (1-\lambda)f(x_1, y_1) + \lambda f(x_2, y_2)$ with strict inequality if $(x_1, y_1) \neq \alpha(x_2, y_2)$, σ is *strictly* convex, i.e.,

$$\sigma((1-\lambda)\rho_1 + \lambda\rho_2) < (1-\lambda)\sigma(\rho_1) + \lambda\sigma(\rho_2) \tag{7.7}$$

for $\rho_1 \neq \rho_2$. Therefore σ has a unique minimum $\rho_{E,\min}$, the state of minimal entropy production at given field E.

How is $\rho_{E,\min}$ linked to the steady state $\rho_{E,N}$ defined by (2.1)? From simple examples we learn that $\rho_{E,\min} \neq \rho_{E,N}$ in general. However to first order in E they agree, i.e.,

$$\frac{\partial}{\partial E_n} \rho_{E,N}\bigg|_{E=0} = \frac{\partial}{\partial E_n} \rho_{E,\min}\bigg|_{E=0} \tag{7.8}$$

We computed already the left-hand side of (7.8), see (3.3)–(3.5), as

$$-\left(\int_0^\infty dt \sum_x e^{L_0^* t} j_0(x, x+e_n)\right) Z^{-1} e^{-\beta H} \tag{7.9}$$

To obtain the right-hand side we take the variational derivative of σ at $\nu_{E,\min}$. This gives

$$\begin{aligned} \sum_{\substack{x,y \\ |x-y|=1}} c_E(x, y, \eta)\Big[1 - e^{E \cdot (x-y)(\eta_x - \eta_y)} \nu_{E,\min}(\eta^{xy}) \nu_{E,\min}(\eta)^{-1} \\ + \log \nu_{E,\min}(\eta) - \log \nu_{E,\min}(\eta^{xy}) - E \cdot (x-y)(\eta_x - \eta_y)\Big] \\ = 0 \end{aligned} \tag{7.10}$$

We solve (7.10) to first order in E and obtain

$$\begin{aligned} \sum_{\substack{x,y \\ |x-y|=1}} c_0(x, y, \eta)\left\{\frac{\partial}{\partial E_n}\left[\nu_{E,\min}(\eta^{xy}) - \nu_{E,\min}(\eta)\right]\right\}\Bigg|_{E=0} \\ = \sum_x c_0(x, x+e_n, \eta)(\eta_x - \eta_{x+e_n}) \end{aligned} \tag{7.11}$$

in agreement with (7.8) and (7.9).

We remark that the agreement between steady state and state of minimal entropy production to linear order in E relies in an essential way on the stochastic nature of the dynamics. If we consider a system of classical particles in contact with stochastic heat reservoirs[26] then even to linear order in the temperature difference the steady state does not agree with the state of minimal entropy production. The reason is that, since the free energy is invariant under the Hamiltonian time evolution, the entropy production for a given measure is independent of the Hamiltonian. On the other hand the first-order correction to the steady state depends on the particular form of the Hamiltonian.

8. THE INFINITE-VOLUME LIMIT

Physically, one is typically interested in the bulk properties of the system. Therfore, as in thermal equilibrium, surface effects should be eliminated by taking the thermodynamic limit where the size of the system tends to infinity keeping the density fixed. We choose then a sequence Λ_j of periodic boxes, $\Lambda_j \nearrow \mathbb{Z}^d$ as $j \to \infty$, and a sequence N_j of number of particles such that $\lim_{j\to\infty} N_j/|\Lambda_j| = \rho$ and study the limit of the correlation functions

$$\lim_{j\to\infty} \left\langle \prod_{x\in A} \eta_x \right\rangle_{E, N_j/|\Lambda_j|} \equiv \int \mu_{E,\rho}(d\eta) \prod_{x\in A} \eta_x \tag{8.1}$$

where A is an arbitrary finite set of sites. By compactness the limit taken along suitable subsequences exists and defines the infinite volume state $\mu_{E,\rho}$. By construction $\mu_{E,\rho}$ is stationary and translation invariant.

By just using compactness one loses any specific information about the infinite volume states. But we conjecture that they have the following properties: For sufficiently small β for any density ρ and field E the limit state $\mu_{E,\rho}$ should cluster in the sense that truncated correlation functions decay at infinity. On the other hand for given E and $\beta > \beta_c(E)$ one should have two density ranges $0 \leqslant \rho \leqslant \rho_g$ and $\rho_l \leqslant \rho \leqslant 1$ such that for every ρ within that range $\mu_{E,\rho}$ has good clustering whereas for $\rho_g < \rho < \rho_l$ the limit (8.1) will be a suitable superposition of the states μ_{E,ρ_g} and μ_{E,ρ_l}. In particular the limit correlation functions in (8.1) do not cluster and the system exhibits long-range order. As already noted the segregation curve has to be symmetric around $\rho = 1/2$. A typical configuration for μ_{E,ρ_l} would have an infinite cluster of particles with holes stretched along the field direction. If in this configuration we interchange particles and empty sites, then we obtain a typical configuration for μ_{E,ρ_g}.

Another approach to the infinite volume limit is to study directly the stationary states of the infinite volume dynamics. We consider the space Ω of all configurations on $\mathbb{Z}^d$, i.e., $\Omega = \{0, 1\}^{\mathbb{Z}^d}$. It has been shown[25] that the evoluation equation

$$\frac{d}{dt} f_t(\eta) = (L_E^* f_t)(\eta) \tag{8.2}$$

with

$$(L_E^* f)(\eta) = \frac{1}{2} \sum_{x,y \in \mathbb{Z}^d} c_E(x, y, \eta)[f(\eta^{xy}) - f(\eta)] \tag{8.3}$$

acting on local functions has a unique solution $e^{L_E^* t}$ on $C(\Omega)$, the space of bounded and continuous functions on Ω. The kernel $e^{L_E^* t}(\eta \mid d\eta')$ defines the transition probability for the infinite system. Invariant probability measures are characterized by

$$\int_\Omega \mu(d\eta)(L_E^* f)(\eta) = 0$$

for all local functions f. The problem is then to classify all invariant states for given parameters E and β. One class of stationary states is provided by the construction (8.1). These steady states are translation invariant and carry a current. The dynamics in these states is not reversible. Another class of invariant states is obtained by the infinite volume of the Gibbs state (1.6). Their density tends to unity in the direction of the field and to zero opposite to the direction of the field. In these states the dynamics is reversible. If we restrict out attention to steady states which are translation invariant, then we would hope that they coincide exactly with the ones obtained through the limit (8.1).

ACKNOWLEDGMENTS

We thank H. van Beijeren and O. Penrose for valuable comments. J. Marro has started computer simulations in $d = 3$.

APPENDIX 1

We fill in the details in deriving (3.5). Differentiating the detailed balance condition (1.5) at $E_n = 0$ one obtains

$$\begin{aligned} c_0'(x, x + e_m, \eta) &= c_0'(x, x + e_m, \eta^{x\,x+e_m})e^{-\beta(H(\eta^{x\,x+e_m}) - H(\eta))} \\ &\quad + c_0(x, x + e_m, \eta)e^{-\beta(H(\eta^{x\,x+e_m}) - H(\eta))}(\eta_x - \eta_{x+e_m})\delta_{mn} \end{aligned} \tag{A.1}$$

Inserting (A.1) in the first term of (3.4) yields

$$\tfrac{1}{2}\langle c(x, x + e_n)(\eta_x - \eta_{x+e_n})^2\rangle\delta_{mn} \tag{A.2}$$

Differentiating the invariance condition at $E_n = 0$ one obtains

$$L\rho_{0,N}' + L'\rho_{0,N} = 0 \tag{A.3}$$

By (A.1)

$$L'\rho_{0,N} = -\Big[\sum_x j(x, x+e_m)\Big]\rho_{0,N} \tag{A.4}$$

Using again detailed balance the first-order correction is then

$$\rho'_{0,N} = -\Big[\int_0^\infty dt \sum_x e^{L^*t} j(x, x+e_m)\Big]\rho_{0,N} \tag{A.5}$$

Inserting in the second term of (3.4) gives the desired result.

APPENDIX 2

We prove the existence of the limit (6.13) under the condition of sufficiently high temperatures. Our argument does not exclude the physically absurd possibility $D(\rho) \equiv 0$. In the following manipulations we invoke the following lemma.

Lemma. Let f and g be local functions and let τ_x denote the shift by $x \in \mathbb{Z}^d$. If ρ and β satisfy the high-temperature condition $|\beta\mu(\rho)| + 2d|\beta| < \pi/4$ with $\mu(\rho)$ the chemical potential as a function of ρ, then

$$|\langle g\tau_x e^{L^*t} f\rangle_{\rho,\beta}| \leqslant c_1 e^{-c_2|x|} \tag{A.6}$$

with positive constants c_1, c_2 depending on g, f, and t. Here $\langle\cdot\rangle_{\rho,\beta}$ denotes the average over the infinite volume equilibrium state at inverse temperature β and density ρ.

Proof. Since f is local, the existence of the dynamics implies that $e^{L^*t}f$ is quasilocal in the sense that, for fixed t, it depends only exponentially little on the occupation variables far out. The explicit dependence follows from estimates similar to those of Holley and Stroock.[28,29] Our high-temperature condition guarantees exponential clustering.[29,30] Both results together imply (A.6). ■

The constant c_1 grows exponentially with t. (A.6) is therefore useful only for finite t.

From the definition of the integrated current we deduce the conservation law

$$\eta_{x,t} - \eta_{x,0} = \sum_{m=1}^{d} \{\mathscr{J}(x-e_m, x; [0,t]) - \mathscr{J}(x, x+e_m; [0,t])\} \tag{A.7}$$

for every history. Because of the bound (A.6) the following identities are valid:

$$\frac{1}{t}\sum_x (x_n)^2(\langle\eta_{x,t}\eta_{0,0}\rangle_{\rho,\beta}-\rho^2)$$

$$=-\frac{1}{2t}\sum_x (x_n)^2\langle(\eta_{x,t}-\eta_{x,0})(\eta_{0,t}-\eta_{0,0})\rangle_{\rho,\beta}+\frac{1}{t}\sum_x (x_n)^2(\langle\eta_x\eta_0\rangle_{\rho,\beta}-\rho^2)$$

$$=\frac{1}{t}\sum_x\langle\mathscr{J}(x,x+e_n;[0,t])\mathscr{J}(0,e_n;[0,t])\rangle_{\rho,\beta}$$

$$+\frac{1}{t}\sum_x(\langle\eta_x\eta_0\rangle_{\rho,\beta}-\rho^2) \tag{A.8}$$

We used here detailed balance and partially integrated twice. As $t\to\infty$ the second term vanishes. For the first term we use (6.7). Then

$$(\text{A.8})=\langle c(0,e_n,\eta)(\eta_0-\eta_{e_n})^2\rangle_{\rho,\beta}$$

$$-\frac{1}{t}\int_0^t ds\int_0^t ds'\,h(|s-s'|)+O(1/t) \tag{A.9}$$

with

$$h(t)=\sum_x (x,x+e_n)e^{L_0^\dagger|t|}j(0,e_n)\rangle_{0,\rho}$$

The integrand of the second term is positive. Therefore its limit as $t\to\infty$ exists and is either finite or infinite. If it would be infinite, then the sum of the first two terms would be negative, which contradicts the positivity of the total current–current correlation function. We conclude then that the limit (6.13) exists and is given by (6.14).

The argument given yields no information about the decay of $h(t)$. In principle the two terms of (3.5) could cancel each other in the limit $t\to\infty$.

APPENDIX 3

We want to show here that the bulk diffusivity at finite E is still given by the space-time integral over the truncated current–current correlation function. We have too little control over the dynamics and over the stationary state to make our argument rigorous. We assume the existence of an infinite volume state $\langle\cdot\rangle_\rho$ with density ρ which is stationary in time, invariant under spatial translation, and has some reasonable clustering in space and time. The detailed balance condition (1.5) is not used. In the case of a constant electric field $\langle\ \rangle_\rho\equiv\langle\ \rangle_{E,\rho}$.

The basic assumption is that the truncated density–density correlation function,

$$\langle \eta_{x,t}\eta_{0,0}\rangle_\rho - \rho^2 \tag{A.10}$$

behaves for large x and t as the fundamental solution of the linearized macroscopic equation

$$\frac{\partial}{\partial t}\rho(q,t) = -\sum_{m=1}^{d}\left(\frac{\partial}{\partial\rho} j_m(\rho)\right)\frac{\partial}{\partial q_m}\rho(q,t)$$
$$+\sum_{m,n=1}^{d}\frac{\partial}{\partial q_m} D_{mn}(\rho)\frac{\partial}{\partial q_n}\rho(q,t) \tag{A.11}$$

Here, by definition, $D(\rho)$ is the bulk diffusion matrix and $j(\rho)$ is the stationary current, i.e., $j_m(\rho) = \langle j(0,e_m)\rangle_\rho$. This allows us to identify the macroscopic quantities with microscopic expressions. The constant drift term appears because we no longer assume reversibility.

The first moment of (A.10) should grow linearly in t and identify then the correct drift. The normalization is

$$\chi(\rho) = \sum_x \left(\langle \eta_x\eta_0\rangle_\rho - \rho^2\right) \tag{A.12}$$

Using the conservation law we obtain

$$\frac{1}{\chi(\rho)}\sum_x x_m\left(\langle \eta_{x,t}\eta_{0,0}\rangle_\rho - \rho^2\right)$$
$$= \frac{1}{\chi(\rho)}\sum_x x_m\left(\langle(\eta_{x,t}-\eta_{x,0})\eta_{0,0}\rangle_\rho + \langle\eta_x\eta_0\rangle_\rho - \rho^2\right)$$
$$= \frac{1}{\chi(\rho)}\sum_x x_m\left(\sum_{n=1}^{d}\langle(\mathscr{J}(x-e_n, x;[0,t]) - \mathscr{J}(x, x+e_n;[0,t]))\eta_{0,0}\rangle_\rho\right)$$
$$= \frac{1}{\chi(\rho)}\int_0^t ds \sum_x \left[\langle \eta_x e^{L^*s} j(0,e_m)\rangle_\rho - \langle\eta_0\rangle_\rho\langle j(0,e_m)\rangle_\rho\right]$$
$$= t\,\frac{1}{\chi(\rho)}\sum_x\left(\langle\eta_x j(0,e_m)\rangle_\rho - \langle\eta_0\rangle_\rho\langle j(0,e_m)\rangle_\rho\right) \tag{A.13}$$

Let

$$\langle\cdot\rangle_{\rho(\lambda)} = Z(\lambda)^{-1}\left\langle\cdot\exp\left(\lambda\sum_x\eta_x\right)\right\rangle_\rho \tag{A.14}$$

Then

$$\sum_x (\langle \eta_x j(0,e_m)\rangle_\rho - \langle \eta_0\rangle_\rho \langle j(0,e_m)\rangle_\rho)$$

$$= \frac{\partial}{\partial\lambda} \langle j(0,e_m)\rangle_{\rho(\lambda)}\Big|_{\lambda=0} = \left[\frac{\partial}{\partial\rho} j_m(\rho)\right] \frac{\partial\rho(\lambda)}{\partial\lambda}\Big|_{\lambda=0} = \chi(\rho)\frac{\partial}{\partial\rho} j_m(\rho) \tag{A.15}$$

which shows that

$$\frac{1}{\chi(\rho)} \sum_x x_m (\langle \eta_{x,t}\eta_{0,0}\rangle_\rho - \rho^2) = t\frac{\partial}{\partial\rho} j_m(\rho)$$

as claimed above.

To identify the bulk diffusion matrix we define

$$\sum_{m,n=1}^{d} a_m^* a_n D_{mn}(\rho)$$

$$= \lim_{t\to\infty} \frac{1}{2t} \left[\frac{1}{\chi(\rho)} \sum_x \left| \sum_{m=1}^{d} a_m x_m \right|^2 (\langle \eta_{x,t}\eta_{0,0}\rangle_\rho - \rho^2) - \left| \frac{1}{\chi(\rho)} \sum_x \sum_{m=1}^{d} a_m x_m (\langle \eta_{x,t}\eta_{0,0}\rangle_\rho - \rho^2) \right|^2 \right] \tag{A.16}$$

We follow the same steps as in (A.8) and obtain

$$\frac{1}{2t}\frac{1}{\chi(\rho)} \sum_x \left| \sum_{m=1}^{d} a_m x_m \right|^2 (\langle \eta_{x,t}\eta_{0,0}\rangle_\rho - \rho^2)$$

$$= \sum_x \sum_{m,n=1}^{d} a_m^* a_n (\langle \mathscr{J}(x,x+e_m;[0,t])\mathscr{J}(0,e_n;[0,t])\rangle_\rho - t\langle j(0,e_m)\rangle_\rho t\langle j(0,e_n)\rangle_\rho) \tag{A.17}$$

The correct spatial truncation appears through the partial integration. To compute the current–current correlation function we use again the short-time approximation (6.4) and follow the argument given there. Then

$$\langle \mathscr{J}(0,e_m;ds)\mathscr{J}(x,x+e_n;dt)\rangle_\rho$$

$$= \delta(t-s)\,dt\,ds\,\delta_{mn}\delta_{x0}\langle c(0,e_n)(\eta_0 - \eta_{e_n})^2\rangle_\rho$$

$$+ dt\,ds\,\langle j(0,e_m)\partial_{0e_m} e^{L^*|t-s|} j(x,x+e_n)\rangle_\rho \tag{A.18}$$

Here ∂_{xy} interchanges the occupations at x and y, i.e., $(\partial_{xy}f)(\eta) = f(\eta^{xy})$.

Note that in thermal equilibrium detailed balance implies $\partial_{0e_m}^* j(0, e_m) = -j(0, e_m)$ with the definition $\langle \partial_{0e_m}^* j(0, e_m) f \rangle_\rho = \langle j(0, e_m) \partial_{0e_m} f \rangle_\rho$ for all local functions f. Inserting (A.18) in (A.17) we obtain

$$
\begin{aligned}
&\sum_{m,n=1}^{d} a_m^* a_n D_{mn}(\rho) \\
&\quad = \frac{1}{2\chi(\rho)} \sum_{n=1}^{d} |a_n|^2 \langle c(0, e_n)(\eta_0 - \eta_{e_n})^2 \rangle_\rho \\
&\quad\quad + \sum_{m,n=1}^{d} a_m^* a_n \left\{ \lim_{t \to \infty} \frac{1}{2t} \int_0^t ds \int_0^t ds' \right. \\
&\quad\quad\quad \times \left[\frac{1}{\chi(\rho)} \sum_x \left(\langle j(0, e_m) \partial_{0e_m} e^{L^*|s-s'|} j(x, x+e_n) \rangle_\rho \right. \right. \\
&\quad\quad\quad\quad \left. - \langle j(0, e_m) \rangle_\rho \langle j(0, e_n) \rangle_\rho \right) \\
&\quad\quad\quad\quad \left. \left. - \left(\frac{\partial}{\partial \rho} j_m(\rho) \right) \left(\frac{\partial}{\partial \rho} j_n(\rho) \right) \right] \right\}
\end{aligned}
\tag{A.19}
$$

We have to check that (A.19) contains the correct truncation in time. For this we have to study

$$
\lim_{t \to \infty} \frac{1}{\chi(\rho)} \sum_x \left(\langle \tau_x g e^{L^* t} f \rangle_\rho - \langle g \rangle_\rho \langle f \rangle_\rho \right) \tag{A.20}
$$

where $g = \partial_{0e_m}^* j(0, e_m)$ and $f = j(0, e_n)$. We write (A.20) as

$$
\lim_{t \to \infty} \frac{1}{\chi(\rho)} \frac{\partial}{\partial \lambda} Z(\lambda)^{-1} \left\langle (E^{L^* t} f) \exp\left(\lambda \sum_x \tau_x g \right) \right\rangle_\rho \Big|_{\lambda = 0}
$$

and interchange the differentiation and the limit. Then

$$
\begin{aligned}
&(\text{A.20}) \\
&\quad = \frac{\partial}{\partial \lambda} \lim_{t \to \infty} Z(\lambda)^{-1} \left\langle (e^{L^* t} f) \exp\left(\lambda \sum_x \tau_x g \right) \right\rangle_\rho \Big|_{\lambda = 0} \frac{\partial}{\partial \lambda} \langle f \rangle_{\rho(\lambda)} \Big|_{\lambda = 0}
\end{aligned}
\tag{A.21}
$$

where $\langle \cdot \rangle_{\rho(\lambda)}$ is the stationary state with density $\rho(\lambda)$ obtained in the limit as $t \to \infty$ of $Z(\lambda)^{-1}\langle \cdot \exp(\lambda \sum_x \tau_x g)\rangle_\rho$. Therefore

$$(\mathrm{A.20}) = \frac{1}{\chi(\rho)}\left(\frac{\partial}{\partial \rho}\langle f\rangle_\rho\right)\frac{\partial \rho(\lambda)}{\partial \lambda}\bigg|_{\lambda=0}$$

$$= \left[\frac{1}{\chi(\rho)}\sum_x (\langle \eta_x f\rangle_\rho - \langle \eta_0\rangle_\rho \langle f\rangle_\rho)\right]\left[\frac{1}{\chi(\rho)}\sum_x (\langle \eta_x g\rangle_\rho - \langle \eta_0\rangle_\rho \langle g\rangle_\rho)\right] \tag{A.22}$$

where the same argument as in (A.15) is used. Inserting $g = \partial^*_{0e_m} j(0, e_m)$ and $f = j(0, e_n)$ and using (A.15) and

$$\sum_x (\langle \eta_x \partial^*_{0e_m} j(0, e_m)\rangle_\rho - \langle \eta_0\rangle_\rho \langle j(0, e_m)\rangle_\rho)$$

$$= \sum_x (\langle j(0, e_m)\partial_{0e_m}\eta_x\rangle_\rho - \langle \eta_0\rangle_\rho \langle j(0, e_m)\rangle_\rho)$$

$$= \sum_x (\langle \eta_x j(0, e_m)\rangle_\rho - \langle \eta_0\rangle_\rho \langle j(0, e_m)\rangle_\rho) \tag{A.23}$$

we find that the truncation is indeed as given in (A.19).

Assuming the integrable decay of the total current correlation function

$$D_{mn}(\rho) = \frac{1}{2\chi(\rho)}\langle c(0, e_n)(\eta_0 - \eta_{e_n})^2\rangle_\rho \delta_{mn}$$

$$+ \int_{-\infty}^{\infty} dt \left\{ \frac{1}{\chi(\rho)} \sum_x \left[\langle j(0, e_m)\partial_{0e_m} e^{L^*|t|} j(x, x + e_n)\rangle_\rho \right.\right.$$

$$\left. - \langle j(0, e_m)\rangle_\rho \langle j(0, e_n)\rangle_\rho \right]$$

$$\left. - \left[\frac{\partial}{\partial \rho} j_m(\rho)\right]\left[\frac{\partial}{\partial \rho} j_n(\rho)\right]\right\} \tag{A.24}$$

Note that the spatial sum and the time integration cannot be interchanged. First one has to perform the spatial truncation and the spatial sum. The total current–current correlation function has then to be truncated in time and integrated over all times. The same structure was found for the one-dimensional system of hard rods.[31]

APPENDIX 4

We want the steady state to be independent of E. (2.1) results then in

$$\sum_{\substack{x,y \\ |x-y|=1}} c_E(x, y, \eta)(e^{E\cdot(x-y)(\eta_x - \eta_y)} - 1) = 0 \tag{A.25}$$

which should be read as an equation for the jump rates. To solve it in one dimension we use that the rates depend only on the nearest neighbors of the bond $(x, x+1)$ and set

$$\begin{aligned} c_E(x, x+1; 1,1,0,1) &= \alpha_1(E) \\ c_E(x, x+1; 0,1,0,1) &= \alpha_2(E) \\ c_E(x, x+1; 1,1,0,0) &= \alpha_3(E) \\ c_E(x, x+1; 0,1,0,0) &= \alpha_4(E) \end{aligned} \tag{A.26}$$

with the rates in the opposite direction determined by the detailed balance condition (1.5). (A.25) is then satisfied only if the rates are related by

$$\begin{aligned} \alpha_2 &= e^{\beta}\alpha_3 \\ \alpha_1 - \alpha_2 - \alpha_3 + \alpha_4 &= 0 \end{aligned} \tag{A.27}$$

In linear response this result was noted in Ref. 13. If we require in addition to (A.27) that for given β (1.7) should be satisfied for all E with a function ϕ independent of E, then this is possible only for $\beta = 0$ and $\phi(h) = e^{-h/2}$.

In more than one dimension (A.25) has as only solution $c_E(x, y, \eta) = e^{-E\cdot(x-y)(\eta_x - \eta_y)/2}$.

REFERENCES

1. Cf. J. L. Lebowitz, *Prog. Theor. Phys.* **64**:35 (1978); and references therein.
2. W. Dietrich, P. Fulde, and I. Peschel, *Adv. Phys.* **29**:527 (1980); and references therein.
3. L. Onsager, *Phys. Rev.* **65**:117 (1944); *Nuovo Cimento Suppl.* **6**:261 (1949).
4. S. Katz, J. L. Lebowitz, and H. Spohn, *Phys. Rev. B* **28**:1655 (1983).
5. A. D. LeClaire, in *Physical Chemistry. An Advanced Treatise*, Vol. 10, p. 261, edited by H. Eyring, D. Henderson and W. Jost (Academic Press, New York, 1970).
6. C. Kipnis and S. R. S. Varadhan, to appear.
7. N. Metropolis, A. W. Rosenbluth, M. M. Rosenbluth, A. H. Teller, and E. Teller, *J. Chem. Phys.* **21**:1087 (1953).
8. Cf. *Monte Carlo Methods in Statistical Mechanics*, K. Binder, ed. (Springer, Berlin, 1979).
9. J. C. Kimball and L. W. Adams, *Phys. Rev. B* **18**:5851 (1978).
10. B. I. Halperin, *Phys. Rev. B* **8**:4437 (1973).
11. H. Mori, *Prog. Theor. Phys.* **33**:423 (1965).
12. H. Singer, Diploma thesis, Universität, Berlin, 1980, unpublished.
13. H. Singer and I. Peschel, *Z. Phys.* **B39**:333 (1980).
14. R. Zeyher, *Solid State Commun.* **36**:33 (1980).
15. M. F. Sykes and M. E. Fisher, *Physica* **28**:919 (1962).
16. C. Domb and M. F. Sykes, *Proc. R. Soc. (London)* **A235**:247 (1956).
17. D. P. Landau, *Phys. Rev. B* **13**:2997 (1976).
18. A. Onuki and K. Kawasaki, *Ann. Phys. (N.Y.)* **121**:456 (1979); and **131**:217 (1981). D. Beysens and M. Gbadamassi, *Phys. Rev. A* **22**:2250 (1980).
19. R. Kubo, *J. Phys. Soc. Jpn.* **12**:570 (1957).
20. A. Sadiq, *Phys. Rev. B* **9**:2299 (1974).
21. G. E. Murch and R. Thorn, *J. Phys. Chem. Solids* **39**:1301 (1978).

22. G. E. Murch and R. Thorn, *Phil. Mag.* **37**:85 (1977).
23. R. Kutner, K. Binder, and K. W. Kehr, Diffusion in concentrated lattice gases II, preprint, KFA Jülich.
24. G. E. Murch, *Phil. Mag.* **A41**:159 (1980).
25. Cf. S. R. de Groot and P. Mazur, *Non Equilibrium Thermodynamics* (North-Holland, Amsterdam, 1982).
26. J. L. Lebowitz and P. G. Bergmann, *Phys. Rev.* **99**:578 (1955); *Ann. Phys. (N.Y.)* **1**:1 (1957).
27. T. M. Liggett, The stochastic evolution of infinite systems of interacting particles, *Lecture Notes Math.* **598**:249 (1977).
28. R. A. Holley and D. W. Strook, *Commun. Math. Phys.* **48**:249 (1976).
29. R. A. Holley and D. W. Strook, *Z. Wahrscheinlichkeitstheorie verw. Gebiete* **35**:87 (1976).
30. R. B. Israel, *Commun. Math. Phys.* **50**:245 (1976).
31. H. Spohn, *Ann. Phys. (N.Y.)* **141**:353 (1982).

VOLUME 55, NUMBER 19 PHYSICAL REVIEW LETTERS 4 NOVEMBER 1985

Rigorous Derivation of Reaction-Diffusion Equations with Fluctuations

A. De Masi,[a] P. A. Ferrari,[b] and J. L. Lebowitz
Department of Mathematics and Physics, Rutgers University, New Jersey 08903
(Received 1 August 1985)

We study interacting particle (spin) systems on a lattice under the combined influence of Glauber (spin flip) and simple exchange (Kawasaki) dynamics. We prove that when the conserving exchanges occur on a microscopically fast scale the macroscopic density (magnetization) evolves according to an autonomous nonlinear diffusion-reaction equation. Microscopic fluctuations about the deterministic macroscopic evolution are found explicitly. They grow, with time, to become infinite, when the deterministic solution is unstable.

PACS numbers: 05.60.+w, 05.40.+j

In this note we show how to derive rigorous nonlinear diffusion-reaction–type equations[1] for the macroscopic evolution of microscopically simple model systems particles on a lattice with stochastic microscopic dynamics.[2] These types of equations are often used to describe chemically reacting systems, population genetics, etc.[1] Our analysis contains some of the essential elements involved in the transition from microscopic to macroscopic evolution equations in general systems: the suitable rescaling of space and time.[3] By such rescalings one takes account of the central fact that there are a very large number of atoms in each drop of macroscopic fluid and that there is a big spread between microscopic and macroscopic time scales. This brings in the "law of large numbers" which is crucial for obtaining deterministic autonomous macroscopic equations for quantities which fluctuate on the microscopic scale.

The simplicity of our model systems permits us to obtain also these fluctuations around the deterministic solutions directly from the microscopic dynamics. These fluctuations form a Gaussian field with an amplitude of $O(\epsilon^{d/2})$ around stable solutions of the macroscopic equations but grow exponentially or like a power around unstable solutions.

In the simplest case these solutions are spatially uniform stationary solutions of the nonlinear equation, some of which are stable and others unstable. They can be interpreted (in some cases) as coexisting stationary states of our microscopic lattice system, even before we take the limit $\epsilon \to 0$. N.B.: These stationary states are generally *not* equilibrium Gibbs states with *any* finite-range (or rapidly decaying) potential. They are nonequilibrium stationary states which can have phase transition even in one dimension.

The study of such stationary nonequilibrium microscopic states is in itself a problem of great interest and was one motivation for undertaking the present work.[4] Another motivation is our hope that the analysis of the microscopic fluctuations about the deterministic macroscopic equations will add to our understanding of the instabilities and pattern formation associated with nonlinear evolution equations.[5] This may be particularly so in cases where stochastic lattice models are invented to simulate physically important and mathematically intractable nonlinear equations, e.g., those describing the motion of an interface between two fluids.[6]

Our system is a simple cubic lattice in d dimensions, at each site of which there is a spin $\sigma(\mathbf{x}) = \pm 1$, $\sigma = \{\sigma(\mathbf{x}) | \mathbf{x} \in Z^d\}$. (Equivalently we can think of occupied and empty sites or of two chemical species.) We consider two mechanisms by which a configuration of the lattice σ changes with time: a Glauber dynamics in which a spin flips at a site $\mathbf{x}$, $\sigma \to \sigma^{\mathbf{x}}$, with a rate $c(\mathbf{x};\sigma)$, and a Kawasaki dynamics in which unequal spins at neighboring sites exchange, $\sigma + \sigma^{\mathbf{x},\mathbf{y}}$, with a constant rate.

There are very few restrictions on $c(\mathbf{x};\sigma)$. For concreteness it is useful to think of a Glauber dynamics which satisfies detailed balance for the Gibbs state of a one-dimensional Ising model with nearest-neighbor interactions J at reciprocal temperature β,

$$c(\mathbf{x};\sigma) = \{1 - \gamma\sigma(x)[\sigma(x+1) + \sigma(x-1)] + \gamma^2\sigma(x+1)\sigma(x-1), \tag{1}$$

where $\gamma = \tanh\beta J$.

The exchange process, on the other hand, being independent of the spin configuration on neighboring sites, acts *as if* the system were at an infinite temperature, $\beta = 0$. We shall furthermore assume that this exchange rate is very fast compared to the flip rate, i.e., the magnetization-conserving dynamics will take place, in the language of the introductory paragraphs, on a microscopic time scale. This change in scales will be denoted by ϵ^{-2}, with $\epsilon << 1$.

The "equation of motion" for $f(\sigma,t)$, the expected value at time t of a function f, when the spin configuration at $t=0$ is σ, is a sum of a Glauber and a simple exchange Kawasaki term,

$$\frac{df(\sigma,t)}{dt}=\sum_{\mathbf{x}\in Z^d} c(\mathbf{x};\sigma)[f(\sigma^{\mathbf{x}},t)-f(\sigma,t)]+\epsilon^{-2}\sum_{|\mathbf{x}-\mathbf{y}|=1}[f(\sigma^{\mathbf{x},\mathbf{y}},t)-f(\sigma,t)]=L_G f+\epsilon^{-2}L_K f. \tag{2}$$

We shall now define the macroscopic magnetization density by rescaling space by ϵ^{-1}. (The relation between space and time rescalings depends on the microscopic dynamics.[3]) Let Λ^δ be a cubical box with sides of length σ, $\Lambda_\mathbf{r}^\delta$, centered on $\mathbf{r}\in R^d$. The magnetization density on this scale is

$$m^\epsilon(\mathbf{r},t;\sigma)=S_t^\epsilon[\epsilon^d\sum_{\epsilon\mathbf{x}\in\Lambda_\mathbf{r}^\delta}\sigma(\mathbf{x})], \tag{3}$$

where $S_t^\epsilon=\exp[L_G+\epsilon^{-2}L_K]t$. $m^\epsilon(\mathbf{r},t;\ )$ is a random variable whose probability distribution depends on the initial probability distribution of the system. We shall assume the latter to have good cluster properties and that as $\epsilon\to 0$, $\epsilon^{-d}|\langle\sigma(\mathbf{x})\rangle_0-m_0(\epsilon\mathbf{x})|\to 0$, where $m_0(\mathbf{r})$ is a smooth function of $\mathbf{r}$, $\mathbf{r}\in R^d$, $|m_0(\mathbf{r})|\leq 1$, and $\langle\ \rangle_0$ denotes expectations with respect to the initial distribution.

In the limit $\epsilon\to 0$ there is a true separation between the microscopic and macroscopic scales and $m_0(\mathbf{r})$ becomes the macroscopic magnetization at $t=0$. We now expect that the exchanges which become infinitely fast on the macroscopic time scale when $\epsilon\to 0$ will cause the spins in the box $\Lambda_\mathbf{r}^\delta$ to be distributed independently (with a product measure) at the instantaneous value of the magnetization $m^\epsilon(\mathbf{r},t;\sigma)$. This magnetization will change because of fluxes through boundaries of $\Lambda_\mathbf{r}^\delta$ and spin flips which try to make the spins correlated. On the other hand $\Lambda_\mathbf{r}^\delta\to\infty$ so $m^\epsilon(\mathbf{r},t;\ )$ should become a deterministic variable $m(\mathbf{r},t)$ (fluctuations should go to zero), and its time evolution be determined by use of this product measure to evaluate the changes it undergoes. This is exactly what happens:

Our results form two theorms.

Theorem 1.—In the limit $\epsilon\to 0$, $m^\epsilon(\mathbf{r},t;\sigma)\to\int_{\Lambda_\mathbf{r}^\delta}m(\mathbf{r}',t)d^3r^1$, a deterministic (nonfluctuating) function of $\mathbf{r}$ and t, with the density $m(r,t)$ satisfying the equation

$$\partial m(\mathbf{r},t)/\partial t=\nabla^2 m+F(m(\mathbf{r},t)) \tag{4}$$

with the initial condition $m(\mathbf{r},0)=m_0(\mathbf{r})$.

$F(m)=-\langle 2\sigma(0)c(0;\sigma)\rangle_{\nu_m}$ is a polynomial in m—the average being taken with respect to the to the Bernoulli (product) measure ν_m, $\langle\sigma(\mathbf{x})\rangle_{\nu_m}=m$ for all $\mathbf{x}$. This is the same as the infinite-temperature equilibrium state with magnetization per site m. $F(m(\mathbf{r},t))$ is in fact equal to the almost sure value of $\lim_{\epsilon\to 0}L_G m^\epsilon(\mathbf{r},t;\sigma)$ in the Bernoulli measure $\nu_{m(\mathbf{r},t)}$ which is what we expected. For example in (1), we have

$$\frac{\partial m(r,t)}{\partial t}=\frac{\partial^2 m}{\partial r^2}+2(2\gamma-1)m-2\gamma^2 m^3. \tag{5}$$

To see the microscopic fluctuations in the magnetization—corresponding to the deviations of the probability distribution at time t from a product measure—we must magnify them in an appropriate way. This is given by the next theorem.

Theorem 2.—Let

$$\phi^\epsilon(\mathbf{r},t;\sigma)=\epsilon^{-d/2}[m^\epsilon(\mathbf{r},t;\sigma)-m(\mathbf{r},t)];$$

then

$$\phi^\epsilon(\mathbf{r},t;\sigma)\underset{\epsilon\to 0}{\to}\int\phi(\mathbf{r},t),$$

a random Gaussian field satisfying the following Ornstein-Uhlenbeck–type stochastic equation:

$$\frac{\partial\phi(\mathbf{r},t)}{\partial t}=\nabla^2\phi+2F'(m(\mathbf{r},t))\phi+\frac{dW(\mathbf{r},t)}{dt}, \tag{6}$$

where $W(\mathbf{r},t)$ is a Brownian motion or Wiener process (dW/dt is "white" noise) with the covariance

$$\langle W(\mathbf{r},t)W(\mathbf{r}',t')\rangle=\min(t,t')[-(1-m^2)\nabla^2\delta(\mathbf{r}-\mathbf{r}')-\nabla(1-m^2)\nabla\delta(\mathbf{r}-\mathbf{r}')+4f(m)\delta(\mathbf{r}-\mathbf{r}')], \tag{7}$$

where $f(m)=\langle c(0;\sigma)\rangle_{\nu_m}[=1-\gamma(2-\gamma)m^2$, for example (1)].

The equal-time correlations of the fluctuation field ϕ,

$$c(\mathbf{r},\mathbf{r}';t)=\langle\phi(\mathbf{r},t)\phi(\mathbf{r}',t)\rangle,$$

satisfy the following equations:

$$c(\mathbf{r},\mathbf{r}';t)=[1-m^2(\mathbf{r},t)]\delta(\mathbf{r}-\mathbf{r}')+\tilde{c}(\mathbf{r},\mathbf{r}';t), \tag{8}$$

$$\partial\tilde{c}(\mathbf{r},r';t)/\partial t=\Delta\tilde{c}_t+2F'(m)\tilde{c}_t-2\delta(r-r')[(\nabla m)^2-F'(m)(1-m^2)+mF(m)-2f(m)], \tag{9}$$

$$\tilde{c}(\mathbf{r},\mathbf{r}',0)=0.$$

The roof of these theorems uses a dual branching process; cf. Liggett,[2] Sect. 3, for a clear presentation of duality. This reduces the study of a Markov process on an uncountable state space ($\{-1,1\}^Z$ in our case) to the study of its simpler dual process, a Markov chain on a countable state space. We can then obtain estimates on the correlations built up by the Glauber dynamics in a macroscopic region, scale ϵ^{-1}, in a macroscopic time, scale ϵ^{-2}. The details of the proofs are rather long and will be presented elsewhere.[7] For the case when there are no flips our results reduce to those of De Masi *et al.*[3]

We shall discuss the behavior of the macroscopic magnetization and its fluctuations in terms of the one-dimensional example in (5). The general case should be very similar although less is known about the solutions of the diffusion-reaction equation in $d > 1$.

Equation (5) has for a given $m(r,0) = m_0(r)$, $|m_0| \leq 1$, a unique solution $m(r,t)$ for $t \geq 0$, with $|m(q,t)| \leq 1$.[1] The general time-dependent solutions can be rather complicated. The analysis of their stability, particularly for the "propagating" ones, is a subject of great current interest in terms of "pattern selection."[4] We shall only discuss here the simplest case, when $m(r,t) = \bar{m}(\gamma)$ is independent of r and t.

We note first that for $\gamma < 0$, corresponding to antiferromagnetic interactions, $\bar{m}(\gamma) = 0$ is the unique translationally invariant stationary solution—and it is stable. For the more interesting ferromagnetic case, $\gamma > 0$, the solution $\bar{m} = 0$ is unique for $\gamma < \gamma_c = \frac{1}{2}$. For $\gamma > \gamma_c$ there are three solutions: $\bar{m} = 0$, and $\bar{m} = \pm m^* = \pm(2\gamma - 1)^{1/2}/\gamma$ corresponding to the magnetized state. The latter solutions are stable while the $\bar{m} = 0$ solution is unstable. The value γ_c clearly corresponds to a mean-field–type critical point—its origin lying in the decorrelations among the spins induced by the fast exchanges.

The instability of the $\bar{m} = 0$ state, for $\gamma > \gamma_c$, is reflected in the unbounded growth of the fluctuations about this state obtained from the solution of (9). For the example (1) the equation for the covariance with initial condition $m_0 = 0$ is

$$\partial \tilde{c}_t/\partial t = \Delta \tilde{c}_t - 4(1-2\gamma)\tilde{c}_t + 8\gamma\delta(r - r'),$$

$$\tilde{c}(r,r';0) = 0,$$

and the solution is

$$\tilde{c}_t(r,r') = 8\gamma \int_0^t ds\,(4\pi s)^{-1/2} \exp[(r-r')^2/4s] \exp[-4(1-2\gamma)s].$$

For $\gamma > \gamma_c$, $\tilde{c}_t \to \infty$ as $t \to \infty$, the growth being like $\sqrt{t}$ for γ_c and exponential for $\gamma > \gamma_c$, while for $\gamma < \gamma_c$,

$$\tilde{c}_t(r,r') \to \frac{\gamma/2}{\gamma_c - \gamma} \exp[-2(1-2\gamma)^{1/2}|r - r'|]$$

so that the Gaussian field ϕ approaches a stationary state with exponentially decaying covariances (on the macroscopic scale).

We thank Sheldon Goldstein, Errico Presutti, Herman Rost, and Herbert Spohn for some very useful discussions. This work was supported by the National Science Foundation under Grant No. DMR81-14726. One of us (A.D.M.) acknowledges partial support from the Consiglio Nazionale delle Ricerche, and another of us (P.A.F.) from the Conselho Nacional de Desenvolvimento Cientifico e Technologico, under Grant No. 201682-83.

[a]Permanent address: Dipartamento di Matematica, Università dell'Aquila, 67100 l'Aquila, Italy.

[b]Permanent address: Instituto de Matematica e Estatistica, Universidade de São Paulo, São Paulo, Brasil.

[1]J. Smoller, *Shock Waves and Reaction-Diffusion Equations*, (Springer-Verlag, New York, 1983), and references therein; D. Aronson and H. Weinberger, *Nonlinear Diffusion in Population Genetics, Combustion, and Nerve Propagation*, Springer Lecture Notes in Mathematics, Vol. 446 (Springer-Verlag, New York, 1975), pp. 5–49.

[2]T. M. Liggett, *Interacting Particle Systems* (Springer-Verlag, New York, 1985), and references therein.

[3]A. De Masi, N. Ianiro, A. Pellegrinotti, and E. Presutti, in *Nonequilibrium Phenomena II, from Stochastics to Hydrodynamics*, edited by J. L. Lebowitz and E. W. Montroll (Elsevier, New York, 1984), and references therein.

[4]S. Katz, J. L. Lebowitz, and H. Spohn, Phys. Rev. B **28**, (1983), and J. Stat. Phys. **34**, 497 (1984); H. van Beijeren and L. S. Schulman, Phys. Rev. Lett. **53**, 806 (1984).

[5]E. Ben Jacob, H. Brand, G. Dee, L. Kramer, and J. S. Langer, Physica (Utrecht) **14**, 348 (1985).

[6]L. Patterson, Phys. Rev. Lett. **52**, 2281 (1985); L. Kadanoff, J. Stat. Phys. **39**, 267 (1985).

[7]A. De Masi, P. A. Ferrari, and J. L. Lebowitz (to be published). A preliminary account of this work was presented at the International Conference on Mathematical Problems from the Physics of Fluids, Rome, June 1985.

VOLUME 56, NUMBER 12 PHYSICAL REVIEW LETTERS 24 MARCH 1986

Rigorous Derivation of Reaction-Diffusion Equations with Fluctuations. A. DE MASI, P. A. FERRARI, and J. L. LEBOWITZ [Phys. Rev. Lett. 55, 1947 (1985)].

On page 1947, in the right-hand column, line 25 should read, ". . . exchange, $\sigma \rightarrow \sigma^{x,y}$, with . . . ." The opening curly parenthesis in Eq. (1) should be deleted.

On page 1948, in the left-hand column, the sentence beginning in line 6 should read, "Let Λ_r^δ be a cubical box with sides of length δ, centered on $\mathbf{r} \in R^d$." Line 20 should read, ". . . magnetization density at . . . ." In line 27, Λ_r^δ should read $\epsilon^{-d}\Lambda_r^\delta$. In line 29, $m(\mathbf{r},t)$ should be deleted.

On page 1948, in the right-hand column, line 1 should read, "$\rightarrow \int_{\Lambda_r^\delta} m(\mathbf{r}',t)d^dr'$, a . . . ." The sentence containing Eq. (5) should begin, "For the example in (1) we have . . . ."

We alter the discussion of Theorem 2 for clarity. The text should be replaced by the following:

Theorem 2.—Let

$$\phi^\epsilon(\mathbf{r},t;\sigma) = \epsilon^{-d/2}[m^\epsilon(\mathbf{r}',t;\sigma) - \int_{\Lambda_r^\delta} m(\mathbf{r}',t)d^dr'];$$

then

$$\phi^\epsilon(\mathbf{r},t;\sigma) \underset{\epsilon\to 0}{\rightarrow} \int \phi(\mathbf{r}',t)d^dr',$$

a random Gaussian field satisfying the following Ornstein-Uhlenbeck–type stochastic equation:

$$\frac{\partial\phi(\mathbf{r},t)}{\partial t} = \nabla^2\phi + F'(m(\mathbf{r},t))\phi + H(\mathbf{r},t), \tag{6}$$

where $H(\mathbf{r},t)$ is "white" noise with the covariance

$$\langle H(\mathbf{r},t)H(\mathbf{r}',t')\rangle = \delta(t-t')\{2\nabla_r \cdot \nabla_{r'}[(1-m^2)\delta(\mathbf{r}-\mathbf{r}')] + 4f(m)\delta(\mathbf{r}-\mathbf{r}')\}, \tag{7}$$

where $f(m) = \langle c(0;\sigma)\rangle_{\nu_m} [= 1-\gamma(2-\gamma)m^2$, for example (1)].

The equal-time correlations of the fluctuation field ϕ,

$$c(\mathbf{r},\mathbf{r}';t) = \langle\phi(\mathbf{r},t)\phi(\mathbf{r}',t)\rangle,$$

satisfy the following equations:

$$c(\mathbf{r},\mathbf{r}';t) = [1-m^2(\mathbf{r},t)]\delta(\mathbf{r}-\mathbf{r}') + \tilde{c}(\mathbf{r},\mathbf{r}';t), \quad \tilde{c}(\mathbf{r},\mathbf{r}',) = 0, \tag{8}$$

$$\begin{aligned}\partial\tilde{c}(\mathbf{r},\mathbf{r}';t)/\partial t = {} & [\nabla_r^2 + \nabla_{r'}^2 + F'(m(\mathbf{r},t)) + F'(m(\mathbf{r}',t))]\tilde{c}(\mathbf{r},\mathbf{r}';t) \\ & -2\delta(\mathbf{r}-\mathbf{r}')[(\nabla m)^2 - F'(m)(1-m^2) + mF(m) - 2f(m)],\end{aligned} \tag{9}$$

The proof of these theorems uses a dual branching process; cf. Liggett,[2] Sect. 3, for a clear presentation of duality. This reduces. . . .

On page 1949, in line 17 of the first column, $|m(q,t)|$ should read $|m(\mathbf{r},t)|$. The equations on page 1949 should be replaced by

$$\partial\tilde{c}/\partial t = -4(1-2\gamma)\tilde{c} + 8\gamma\delta(\mathbf{r}-\mathbf{r}') + 2\nabla^2\tilde{c}, \quad \tilde{c}(r,r';0) = 0, \tag{10}$$

and the solution is

$$\tilde{c}(\mathbf{r},\mathbf{r}';t) = 8\gamma\int_0^t ds\,(8\pi s)^{-1/2}\exp[-(\mathbf{r}-\mathbf{r}')^2/8s]\exp[-4(1-2\gamma)s]. \tag{11}$$

For $\gamma > \gamma_c$, $\tilde{c}_t \rightarrow \infty$ as $t \rightarrow \infty$, the growth being like $\sqrt{t}$ for γ_c and exponential for $\gamma > \gamma_c$, while for $\gamma < \gamma_c$,

$$\tilde{c}(\mathbf{r},\mathbf{r}';t) \rightarrow \frac{\gamma}{(\gamma_c-\gamma)^{1/2}}\exp[-2(\gamma_c-\gamma)^{1/2}|\mathbf{r}-\mathbf{r}'|] \tag{12}$$

Journal of Statistical Physics, Vol. 50, Nos. 3/4, 1988

Statistical Mechanics of the Nonlinear Schrödinger Equation

Joel L. Lebowitz,[1] Harvey A. Rose,[2] and Eugene R. Speer[3]

Received September 21, 1987

We investigate the statistical mechanics of a complex field ϕ whose dynamics is governed by the nonlinear Schrödinger equation. Such fields describe, in suitable idealizations, Langmuir waves in a plasma, a propagating laser field in a nonlinear medium, and other phenomena. Their Hamiltonian

$$H(\phi) = \int_\Omega [\tfrac{1}{2}|\nabla\phi|^2 - (1/p)\,|\phi|^p]\,dx$$

is unbounded below and the system will, under certain conditions, develop (self-focusing) singularities in a finite time. We show that, when Ω is the circle and the L^2 norm of the field (which is conserved by the dynamics) is bounded by N, the Gibbs measure ν obtained is absolutely continuous with respect to Wiener measure and normalizable if and only if p and N are such that classical solutions exist for all time—no collapse of the solitons. This measure is essentially the same as that of a one-dimensional version of the more realisitc Zakharov model of coupled Langmuir and ion acoustic waves in a plasma. We also obtain some properties of the Gibbs state, by both analytic and numerical methods, as N and the temperature are varied.

KEY WORDS: Nonlinear Schrödinger equation; statistical mechanics; unbounded Hamiltonians; singularities; Gibbs measures.

1. INTRODUCTION

1.1. General Background

The essence of statistical mechanics is the replacement of the study of the microscopic dynamical trajectory of an individual macroscopic system by

[1] Departments of Mathematics and Physics, Rutgers University, New Brunswick, New Jersey 08903.

[2] Los Alamos National Laboratory, Los Alamos, New Mexico 87545.

[3] Department of Mathematics, Rutgers University, New Brunswick, New Jersey 08903.

0022-4715/88/0200-0657$06.00/0

the study of appropriate ensembles or probability measures on the phase space of the system. The success of this program, especially for systems in equilibrium, has made a virtue of necessity. Instead of trying to solve the initial value problem for a system containing a very large (say 10^{23}) number of particles, which is clearly an impossible task even in principle, we obtain information about values of macroscopic observables by taking averages over Gibbs probability distributions containing only a few parameters (particle density, temperature, etc.).

While the rigorous justification of the theory is still not fully understood, its success leaves no doubt about its utility. In fact, the results obtained from a suitable probability measure, which includes information about both typical behavior and fluctuations, are generally more relevant than the solution of a specific initial value problem for understanding the behavior of real systems.[1] With this in mind, we present here an extension of the statistical mechanical formalism to a continuum field with a Hamiltonian that is unbounded below. The system models certain aspects of the behavior of a plasma excited by an electromagnetic field, say that of a laser[2–4] (this is our primary motivating example), as well as the propagation of a laser beam in a nonlinear medium and other phenomena (see Ref. 5 for references).

To put our work in context, we recall first the structure of Gibbs distributions of particle systems. In classical mechanics (the quantum case is similar) the Gibbs probability distribution for finding a system consisting of N particles in a compact spatial region Ω is a set of microscopic states dX_N is given by

$$\mu(dX_N) = Z_N^{-1} \exp[-\beta H(X_N)]\, dX_N$$

where H is the Hamiltonian of the system, β is the reciprocal temperature, and Z_N is a normalization constant (also the partition function). Provided now that $H(X_N)$ is suitably bounded from below (H stability) and satisfies other reasonable conditions,[6] one can take the thermodynamic limit $N \to \infty$, $|\Omega| \to \infty$, $N/|\Omega| \to \rho$ (where $|\Omega|$ is the volume of Ω), the correct idealization of a macroscopic system, and obtain a well-defined measure on the resulting infinite-dimensional phase space.

The situation becomes more complicated when the appropriate microscopic description of the system is in terms of a continuum field, say $\phi(\mathbf{x})$, $\mathbf{x} \in \Omega$. Such a description arises either when ϕ represents a fundamental field of nature, e.g., the electromagnetic or Yang–Mills field, or when ϕ is itself a coarse-grained (reduced) description of the microscopic system, e.g., the density and/or velocity field of a fluid. While the statistical mechanical treatment of the latter "derivative" fields, which endows them with fluctuations, may appear at first sight artificial (first coarsening, then refining), there are often very good reasons for wanting to focus attention on the collective variables of the field rather than on the atomic degrees of freedom—e.g., in fluid turbulence it is in the hydrodynamic modes that the interesting action takes place.

The difficulties encountered in constructing statistical mechanical theories of fields are associated with the infinite number of degrees of freedom present already in a finite volume—the ultraviolet problem. This causes no real trouble in the linear case, i.e., when the Hamiltonian is quadratic in the fields, since the degrees of freedom essentially decouple—say, in a Fourier representation—to give the free field. (One of the theory's first applications, the statistical treatment of blackbody radiation, is an example of such a field.) In the nonlinear case, comonly treated in Euclidean quantum field theory, the difficulties are indeed serious when the spatial dimension d is greater than three. For $d \leqslant 3$, on the other hand, all can be made well for the type of polynomial Hamiltonian usually considered in field theory. These Hamiltonians are bounded from below and permit extensions of the usual tricks.[7] The situation is quite different when the Hamiltonian appropriate to the continuum field is not bounded below, the case we consider there.

It is *a priori* clear that this situation can arise only when the system has, in the coarse-grained description provided by the collective modes, the potentiality for some kind of instability on the appropriate scale. This is indeed the case for many systems, including suitable plasmas irradiated with laser light. Here the collective modes represent Langmuir waves and the instability is the soliton or caviton collapse, which gives rise to plasma turbulence. The physics of the problem—including the reasons for expecting an equilibrium treatment of the collective modes to be of any relevance, even though the system is clearly not in thermodynamic equilibrium—are given in Section 3; there we also discuss the Zakharov model, which contains the ion acoustic waves in addition to Langmuir waves. Here we go on to describe our simpler model and some features of our approach to its statistical mechanics.

1.2. Formulation of the Problem

We construct and study equilibrium Gibbs measures of systems whose microscopic state is modeled by a complex continuum field $\phi(x)$ depending on one space dimension. We take the domain Ω to be the interval $[0, L]$, with periodic boundary conditions, and consider Hamiltonians of the form

$$H(\phi) = \frac{1}{2}\int_0^L |\phi'(x)|^2\, dx - \frac{1}{p}\int_0^L |\phi(x)|^p\, dx \tag{1.1}$$

where the real and imaginary parts of ϕ are the conjugate canonical variables and p will be assumed to satisfy $p \geqslant 3$. The corresponding equation of motion for a time-dependent field $u(x, t)$ ($= [\phi(x)](t)$) is

$$iu_t = -u_{xx} - |u|^{p-2}u \tag{1.2}$$

with $x \in [0, L]$, which is to be supplemented by boundary conditions

$$u(0, t) = u(L, t), \qquad t \in \mathbb{R} \tag{1.3}$$

and an initial condition

$$u(x, 0) = \phi_0(x), \qquad 0 \leqslant x \leqslant L \tag{1.4}$$

Equation (1.2) will be referred to as the nonlinear Schrödinger equation (NLSE)[8]; it is a generalization of the usual case, for which $p = 4$. A related vector NLSE in $d = 3$ space dimensions and with $p = 4$ describes in certain regimes the Langmuir waves in a plasma, i.e., the time evolution of the envelope of the propagating electric field. For the problem of a laser field propagating in a nonlinear medium, t corresponds to the space coordinate in the direction of propagation and $d = 2$, $p = 4$ is the physically relevant case.

Starting from smooth initial data, solutions of the NLSE with $p = 4$ and $d \geqslant 2$ can develop singularities in a finite time. On the other hand, in one dimension (1.2) is well known to be integrable for $p = 4$.[9] To get a singular behavior in one dimension similar to that encountered in the physically interesting cases, one must consider (1.2) with $p \geqslant 6$. [Generally the critical value of p in dimension d is $p_c = (2d + 4)/d$.] This suggests, and there is in fact reason to believe,[10] that by studying (1.1) and (1.2) for $p > 4$ one may obtain insight into the behavior of the $p = 4$ system in high dimensions.

Our basic problems then will be to investigate the existence and properties of Gibbs measures with formal densities $Z^{-1} \exp[-\beta H(\phi)]$, with H given in (1.1). A little thought shows that the unboundedness of $H(\phi)$ from below makes this totally impossible without some additional restrictions. Fortunately there is a natural way to restrict the (infinite-dimensional) phase space of this system: we make use of the fact that (1.2) conserves, in addition to the energy, the L^2 norm (analogous to the particle number) of the field:

$$N(\phi) = \int_0^L |\phi(x)|^2 \, dx \tag{1.5}$$

We can therefore try to construct Gibbs measures with specified values of N or with a bound on N; this is exactly what we do in this paper.

We remark that these measures live on rather ragged field configurations $\phi(x)$ for which the classical energy is infinite; in fact, on Brownian paths. Physical and mathematical problems raised by these irregular configurations will be discussed in Sections 3 and 5.

We do not attempt to go to an infinite-volume (thermodynamic) limit, however, as we would in the traditional case with a Hamiltonian bounded from below. The reason is that, due to the unboundedness of H, our system has a tendency to concentrate its field into local packets, driving the kinetic energy up and the potential energy down; the ground state of the system is a soliton (or caviton in plasma language) whose strength grows with N. Now, if we tried to let $N \to \infty$ as $L \to \infty$, keeping N/L fixed, the system would concentrate the available N and make the field blow up locally. This is just what happens for $p > p_c$ even for finite N and is in fact what makes this model interesting for the study of plasma turbulence.

The remainder of the paper is organized as follows. In Section 2 we describe some aspects of the behavior of the NLSE as a classical evolution equation (for the case $d = 1$ with periodic boundary conditions) and state our results on the existence of the statistical ensemble; we find that the conditions under which the ensemble is well defined correspond closely to conditions guaranteeing a well-defined classical evolution. In Section 3 we describe briefly some of the plasma physics underlying our model and related properties of the statistical ensemble. Section 4 is devoted to proofs of results given in preceding sections. Concluding in Section 5, we survey some open problems.

2. SUMMARY OF RESULTS

2.1. The Initial Value Problem

In this section we study the nonlinear Schrödinger equation on a finite interval as an initial value problem. The conclusions we reach are slight modifications of known results, and will not in fact be used subsequently. Nevertheless, they form an important motivational background for the construction of the statistical ensemble to be given in Section 2.2: in a technical sense, because both sets of results depend critically on a standard interpolation inequality, and in a physical sense, because the conditions for the existence of a solution to the initial value problem are essentially the same as those for the existence of the ensemble.

The problem we wish to study is given in (1.2)–(1.4). The two terms in the Hamiltonian (1.1) will be called, respectively, the kinetic and potential energy of the field, and we will assume that both these energies are finite for the initial field configuration $\phi_0(x)$.

Solutions of the initial value problem for (1.2) have been extensively studied in the case when the domain is the entire real line $\mathbb{R}$ (see in particular Refs. 8 and 11). Many of these results extend readily to the problem on an interval; the most relevant will be summarized in Theorem 2.1 below. In brief, solutions always exist for a time period which may in general depend on the initial condition. Under further conditions on the degree of nonlinearity and on the L^2 norm (1.5) of the initial data, solutions will exist for all time. As will be discussed in Section 2.2, these conditions also guarantee the existence of the statistical ensemble.

Before stating the theorem, we must establish some notation that will be used throughout this paper. We will in general be considering complex-valued functions defined either on the real line $\mathbb{R}$ or an interval $I \subset \mathbb{R}$ of length L, usually taken to be $[0, L]$. We use standard L^q norms on I,

$$\|\phi\|_q \equiv \left[\int_{x \in I} |\phi(x)|^q \, dx\right]^{1/q}$$

and write $\|\phi\|_{0,q}$ for the corresponding norm on $\mathbb{R}$. Finally, we let $H^1 \equiv H^1(I)$ be the Sobolev space of functions with finite kinitic energy, that is, the space of functions ϕ on I satisfying periodic boundary conditions and having finite norm

$$\|\phi\|_{H^1}^2 \equiv \|\phi'\|_2^2 + \|\phi\|_2^2$$

We can now state precisely the relevant results for the initial value problem. In doing so, it is convenient to reformulate (1.2) and (1.4) as an integral equation:

$$u(\cdot, t) = U(t)\phi_0 + i\int_0^t U(t-s)(|u(\cdot, s)|^{p-2} u(\cdot, s)) \, ds \tag{2.1}$$

where $U(t) = \exp(-itH_0)$ with $H_0 = -d^2/dx^2$. Our theorem will deal with functions in H^1, so that that boundary condition (1.3) will be satisfied automatically.

Theorem 2.1. Suppose that $p \geqslant 3$ and that $\phi_0(x) \in H^1$. Then:

1. Equation (2.1) has a solution $u(x, t)$, defined in some interval $|t| < T$, which lies in H^1 (as a function of x).
2. If $u(x, t)$ is a solution of (2.1) with $u(\cdot, t) \in H^1$, then $H(u(\cdot, t))$ and $\|u(\cdot, t)\|_2^2$ are constant in time.
3. There exists a constant N_1, with $0 < N_1 < \infty$, such that:
   (a) If $p < 6$ or if $p = 6$ and $\|\phi_0\|_2^2 < N_1$, we may take $T = \infty$.

(b) If $p \geqslant 6$, there exists an initial value ϕ_0 and a finite T such that $\lim_{t \to T} \|u_x(\cdot, t)\|_2^2 = \infty$. If $p > 6$, $\|\phi_0\|_2^2$ may be chosen arbitrarily small; if $p = 6$, $\|\phi_0\|_2^2$ may be chosen arbitrarily close to N_1.

4. The constant N_1, which occurs in the critical case $p = 6$, satisfies $N_0 \leqslant N_1 \leqslant 2N_0$, where N_0 is given by

$$N_0 = (3/C_6)^{1/2} = \|\hat{\phi}\|_2^2 \tag{2.2}$$

Here C_6 is the best possible constant in a certain interpolation inequality (4.1).

Proof. See Section 4.

We remark that in the corresponding theorem on $\mathbb{R}$, or on I with Dirichlet of Neumann boundary conditions, one has $N_1 = N_0$; it seems likely that this is true for periodic boundary conditions also, although we do not have a proof. In the corresponding result for the existence of the statistical ensemble, Theorem 2.3, the critical value of the L^2 norm in the case $p = 6$ is also N_0.

2.2. The Statistical Ensemble

We now wish to define a statistical ensemble for the system considered in the previous section, with formal (unnormalized) Gibbs measure

$$\begin{aligned} &\exp[-\beta H(\phi)] \prod_{x \in [0, L]} d\phi(x) \\ &\quad = \exp\left(\frac{\beta}{p}\int_0^L |\phi|^p \, dx\right)\left[\exp\left(-\frac{\beta}{2}\int_0^L |\phi'|^2 \, dx\right) \prod_{x \in [0, L]} d\phi(x)\right] \end{aligned} \tag{2.3}$$

where $d\phi$ denotes Lebesgue measure in the complex plane. Our goal in this section is to give (2.3) a precise meaning as a normalizable measure; we begin with a discussion of two preliminary difficulties.

First, the "Lebesgue measure" $\prod d\phi(x)$ is ill defined. Note, however, that the quantity in brackets on the right-hand side of (2.3) is a formal version of the well-known measure for a massless free field in the interval $[0, L]$; this is a Wiener measure supported on continuous but not differentiable fields. (More precisely, the measure is a product of measures for the real and imaginary parts of ϕ; each of these is a superposition of Brownian bridge measures, since we impose periodic boundary conditions.)

Second, H fails badly to be bounded below. The problem is acute for large fields: if we scale $\phi \to \alpha\phi$, the total energy $H(\phi)$ rapidly approaches negative infinity as α increases, and thus (2.3) cannot be normalizable.

Since the dynamics conserves the L^2 norm of ϕ, however, it seems natural to consider either of two modified ensembles, in which we restrict to those ϕ that satisfy either

$$\|\phi\|_2^2 = N \tag{2.4}$$

or

$$\|\phi\|_2^2 \leqslant N \tag{2.5}$$

For technical simplicity, we wish to avoid the problem, inherent in (2.4), of restricting our Wiener measures to a sphere in L^2, and therefore adopt (2.5) in defining our ensemble.

Remark 2.1. More generally, one could consider measures of the form

$$e^{-\beta H(\phi)} G(\|\phi(x)\|_2^2) \prod_{x \in [0, L]} d\phi(x)$$

where G falls off sufficiently fast at infinity to ensure normalizability; (2.5) is of this form with G a characteristic function. The choice

$$G(X) = e^{-\beta \mu X}$$

yields the usual grand canonical ensemble, but a scaling argument as above shows that this will not be normalizable. For $p < 6$, a modified version, with

$$G(X) = e^{-\beta \mu X^q} \tag{2.6}$$

$q = (p+2)/(6-p)$ and sufficiently large μ avoids this difficulty, as discussed in Remark 4.1 below.

Remark 2.2. A lattice version of the ensemble (2.6) has been investigated numerically, in conjunction with J. O'Connell. In the spirit of the grand canonical ensemble, one would like to probe different values of $\|\phi\|_2^2$ by adjusting the "chemical potential" μ. In the simulations, however, we find a sharp transition in typical behavior at a critical value μ_c corresponding roughly to that necessary for normalizability in the continuum. Above μ_c, the measure is concentrated near uniform field configurations with rather low values of $\|\phi\|_2^2$; below μ_c, on configurations with almost all the field at one lattice site and with large values of $\|\phi\|_2^2$. (These typical states are metastable over a range of values near μ_c.) Thus, (2.6) does not appear suitable for investigating regions of moderate $\|\phi\|_2^2$. Numerical investigations of these regions using (2.4) are discussed in Section 3.3.

Thus, we will define our Gibbs measure by

$$\left[\exp\left(\frac{\beta}{p}\int_0^L |\phi|^p\, dx\right)\right] \chi_{\{\phi \mid \|\phi\|_2^2 \leqslant N\}}\, d\mu_\beta(\phi) \tag{2.7}$$

where μ_β is the appropriate version of the Wiener measure. We will show that (2.7) is well defined and normalizable when N and p are such that the dynamical problem, with arbitrary initial data ϕ satisfying (2.5), has solitions (with finite kinetic energy) for all time, that is, when $p<6$ or when $p=6$ and $N<N_0$.

There is a simple formal argument that indicates that this should be true. Under these restrictions on p and N, the interpolation inequalities of Section 4, (4.2) (for $p=6$) or (4.3) (for $p<6$), together with (2.5), immediately yield a lower bound for H on the space H^1:

$$H(\phi) \geqslant \tfrac{1}{2}\varepsilon \int_0^L |\phi'|^2\, dx - K_\varepsilon$$

Formally, this implies the normalizability of (2.7) via

$$\exp[-\beta H(\phi)] \prod_{x \in [0,L]} d\phi(x) \leqslant \tilde{K}\left[\exp\left(-\frac{\varepsilon\beta}{2}\int_0^L |\phi'|^2\, dx\right)\right] \prod_{x \in [0,L]} d\phi(x) \tag{2.8}$$

since the right-hand side of (2.8) is again a well-defined Wiener measure. The goal is to transform this formal argument (which deals with smooth fields) into a rigorous argument for Wiener paths; the basic technique is always to integrate over a family of Wiener paths that are close to some smooth field to which the interpolation inequalities can be applied.

We can now state our main result. Let C denote the set of continuous, complex-valued functions ϕ on $[0, L]$ that satisfy $\phi(0)=\phi(L)$, and let μ_β be the Wiener measure on C corresponding to the bracketed terms in (2.3). (The measure μ_β is defined formally in Section 4.) Then we have the following result:

Theorem 2.2. The function

$$F(\phi) \equiv \left[\exp\left(\frac{\beta}{p}\|\phi\|_p^p\right)\right] \chi_{\{\phi \mid \|\phi\|_2^2 \leqslant N\}}$$

(a) Is in $L^1(C; \mu_\beta)$ if $p<6$ and $N<N_0$.

(b) Is not in $L^1(C; \mu_\beta)$ if $p>6$ or if $p=6$ and $N>N_0$.

Proof. See Section 4.

Thus, when the parameters satisfy the hypotheses of case (a), we can define the partition function

$$Z(\beta, N) = \int_C F(\phi)\, d\mu_\beta(\phi) \tag{2.9}$$

and a probability measure $\nu_{\beta,N}$:

$$d\nu_{\beta,N}(\phi) = Z(\beta, N)^{-1} F(\phi)\, d\mu_\beta(\phi) \tag{2.10}$$

3. NATURE OF STATISTICAL STATES

Having constructed statistical states for the NLSE, we now discuss their significance for the study of plasmas and describe some of their properties. We begin by recalling a more sophisticated model for a plasma and its relation to the NLSE.

3.1. The Zakharov Model

The collective response of a nearly collisionless plasma when the electron temperature is much greater than the ion temperature is dominated by electron plasma waves (Langmuir waves) and ion acoustic waves. Under appropriate conditions (energy density of the waves small compared to particle thermal energy densities, and characteristic length scales large compared to a Debye length), Zakharov's model is a useful description of the nonlinear coupling of these collective modes. In one dimension, neglecting dissipation terms and using appropriate units, we can write the Zakharov equations (ZE) in the form

$$iu_t = -u_{xx} + nu \tag{3.1}$$

$$n_{tt} - c^2 n_{xx} = c^2(|u|^2)_{xx} \tag{3.2}$$

with $u(x, t)$ the electrostatic envelope field, $n(x, t)$ the ion density fluctuation field, and c the speed of sound in the plasma. In discussing these equations, we will again consider the case of periodic boundary conditions on the interval $[0, L]$. We note that solutions of the ZE preserve the L^2 norm of u and the mean ion density field $\int_0^L n\, dx$.

Equation (3.2) can also be written in terms of an auxiliary field $V(x, t)$ [a hydrodynamic flux for the transport of ions, constrained by $\int_0^L V(x, t)\, dx = 0$] as

$$n_t = -c^2 V_x \tag{3.3a}$$

and

$$V_t = -n_x - |u|_x^2 \tag{3.3b}$$

In the form (3.1), (3.3) these are Hamilton's equations for the Hamiltonian

$$H_Z = \frac{1}{2}\int_0^L \left[|u_x|^2 + \frac{1}{2}(n^2 + c^2V^2) + n|u|^2 \right] dx \tag{3.4}$$

with (Re u, Im u) and ($\tilde{n}$, $\tilde{V}$) as pairs of conjugate variables, where we write $\tilde{n} = 2^{-1/2}n$ and $\tilde{V} = 2^{-1/2}\int^x V\,dx$.

The nonlinear Schrödinger equation (with $p = 4$) is closely related to the Zakharov equations in several ways. First, in the adiabatic limit $c \to \infty$, (3.2) reduces to $n = -|u|^2 + K$ and (3.1) becomes equivalent to the NLSE (for rigorous results on this limit see Ref. 12). Second, any solution of the NLSE of the form $u(x, t) = e^{i\omega t}w(x)$ yields a solution of the ZE if n is taken to be $n(x) = -|w(x)|^2$; thus, the ZE and the NLSE have corresponding plane wave and soliton solutions. Finally, consider a formal Gibbs measure for the Hamiltonian (3.4):

$$e^{-\beta H_Z}\chi_{\{\int |u|^2 dx \leq N\}} \prod_x [d^2u(x)\, d\tilde{n}(x)\, d\tilde{V}(x)] \tag{3.5}$$

Because the n and V fields appear quadratically in H_Z, we can integrate over $\tilde{n}$ and $\tilde{V}$ explicitly to compute the marginal distribution of the u field; this yields precisely the formal NLSE Gibbs measure (2.3). In this light the results of Section 2.2 may be regarded as a construction of the probability distribution for the electrostatic envelope field and ion density field in the Zakharov model. On the other hand, while the NLSE is completely integrable in the case $p = 4$, the Zakharov equations (so far as we know) are not; as a result, much more is known about solutions of the former.

There have been other studies of the statistical mechanics of the NLSE and classical nonlinear field theories. In Ref. 13 (and references therein) attention is restricted to Hamiltonians that are bounded below. In Refs. 14 and 15 an attempt is made to study Langmuir wave turbulence as a critical phenomenon. Only Ref. 14 is directly related to our work in that the starting point is the grand canonical ensemble. Here an ultraviolet cutoff is assumed, but infrared questions for the existence of the measure (the system is treated in $\mathbb{R}^d$) are not discussed. No distinction is made between different dimensions, and the fact that the exponent $p = 4$ is respectively subcritical, critical, and supercritical in dimensions one through three does not influence the analysis. The significance of the cutoff measure in the supercritical case is not made clear.

In Ref. 16 the emphasis is different, in that the relationship between a statistical dynamical theory (the DIA of R. H. Kraichnan) and equilibrium statistical mechanics is explored. Also, attention was restricted to a Fourier truncation of the NLSE and only negative temperatures (so there is no tendency for soliton formation) are considered.

As regards other field theories, we have little to say, except to point out that there have been many studies of the equilibrium statistical mechanics of the Fourier-truncated Euler and incompressible ideal MHD equations (see Ref. 17). In these models there is the unusual circumstance that while the dynamics is nonlinear, the equilibrium measure can be Gaussian.

3.2. Physical Interpretation of the Ensembles $v_{\beta, N}$

We now discuss the extent to which the ensembles constructed in Section 2.2 are applicable to the physics of real systems and what light they can shed on that physics. Of course, the one-dimensional nature of our model implies that any application to two- or three-dimensional systems can be at best qualitative. On the positive side we would like to stress again that the conditions (on p and N) for the existence of the measures $v_{\beta,N}$ are essentially the same as those for the existence of a global time evolution for the NLSE; by itself this suggests that the standard picture of a measure-preserving dynamics makes sense under these conditions. There are several difficulties with this simple picture, however.

We first remark that, with probability one, field configurations in our ensembles have infinite kinetic energy $\frac{1}{2}\int |u_x|^2 \, dx$. This is closely connected with a problem in applying the standard picture of the relation between the ensemble and the dynamics: we have in fact no dynamics defined on these singular configurations. (The problem of defining such a dynamics is discussed further in Section 5.) This infinite energy also represents a problem for relating physical quantities calculated in our ensemble to those that might be measured in the real system; certainly we do not expect agreement for the kinetic energy or other quantities that depend on smoothness of the configurations. We argue, however, that microscopic processes in a physical system would provide an ultraviolet cutoff if included in a more realistic model and that, lacking this refined version, we expect that quantities with finite expectation in our model would accurately represent cutoff-independent quantities in a more realistic calculation. (In a plasma these processes can be dissipative, a point discussed briefly in Section 5.)

A second interpretational problem is that the usual justification for the introduction of equilibrium ensembles, ergodicity of the dynamics, certainly does not apply to the NLSE in the integrable case $p=4$. This comment does not hold for the NLSE with other values of p or for the ZE. On the other hand, one of us (H.A.R., in conjunction with G. Doolen) has studied numerically the time evolution of solutions of the one-dimensional ZE, and the results suggest a lack of ergodicity even here: equipartition of kinetic energy into sound modes of different wavelengths seems to fail. In Section 5 we discuss briefly a Langevin model which has formally the same invariant measure and may be ergodic, but also point out that this is not a realisitic model of dissipation in an excited plasma.

A third and more serious difficulty is that the conditions on p and N under which our ensembles exist do not correspond, even qualitatively, to those of the most interesting physical systems. In the case of a plasma, for example, the space dimension d is 3 and the exponent $p=4$ is greater than the corresponding critical exponent $p_c = 10/3$. Thus, there are classical initial conditions for which solutions do not exist for all time; the typical behavior is a self-focusing instability or collapse. Dissipative processes in a true plasma will eventually halt this collapse of the Langmuir wave packet, and the Zakharov model is expected to be relevant only during time intervals when these processes are not acting. To gain insight into the collapse in our statistical model, we would have to go beyond what we have done here. Some possible approaches are discussed in Section 5. For dimension two, the exponent $p=4$ for the laser propagation problem is the critical value, and the existence theory for smooth solutions of the NLSE, together with our results for $d=1$, $p=6$, would suggest the existence of an N_0 such that an ensemble exists for $N < N_0$.

Accepting for the moment some applicability of our ensembles to physical systems, or at the least to one-dimensional versions of them, we ask what information they can provide. In the numerical results mentioned above there appear to be qualtiatively different regimes of behavior for solutions of the ZE: at low energy the solution $u(x, t)$ is typically (i.e., at typical times) fairly uniform as a function of x, while at high energy the typical solution is a single, stable soliton. At intermediate conditions there is a regime in which several solitons may coexist, and solitons continually coalesce and are created. The question of whether the existence of these different regimes might be related to a phase transition in the usual statistical mechanical sense has been a prime motivating factor in this paper.

3.3. Phase Transitions

We now turn to a discussion of the properties of the measures $\nu_{\beta,N}$ defined in (2.10). In particular, we wish to address the question of whether such states exhibit any phase transition as β and N vary, corresponding to the qualitative differences noted in numerical studies. Such a transition could reveal itself in the context of a rigorous treatment of the ensemble through a lack of analyticity of the partition function or other physical quantities along certain curves in the β–N plane.

We have done Monte Carlo studies on a lattice version of the ensemble studied in this paper—more specifically, of the ensemble with restriction (2.4), which we do not expect to differ significantly from the actually constructed ensemble with restriction (2.5). Here we see roughly two regimes in the β–N plane: typical configurations show no spatial structure for low N or for low β, and soliton-like structures at higher N and β. One parameter used to distinguish these regimes is

$$S = \left\langle \int_0^L |\phi|^4 \, dx \right\rangle \Big/ \left\langle \int_0^L |\phi|^2 \, dx \right\rangle^2$$

where now $\langle \cdot \rangle$ refers to a time average in the Monte Carlo stochastic process. When β is varied at large N, S shows a relatively sharp increase

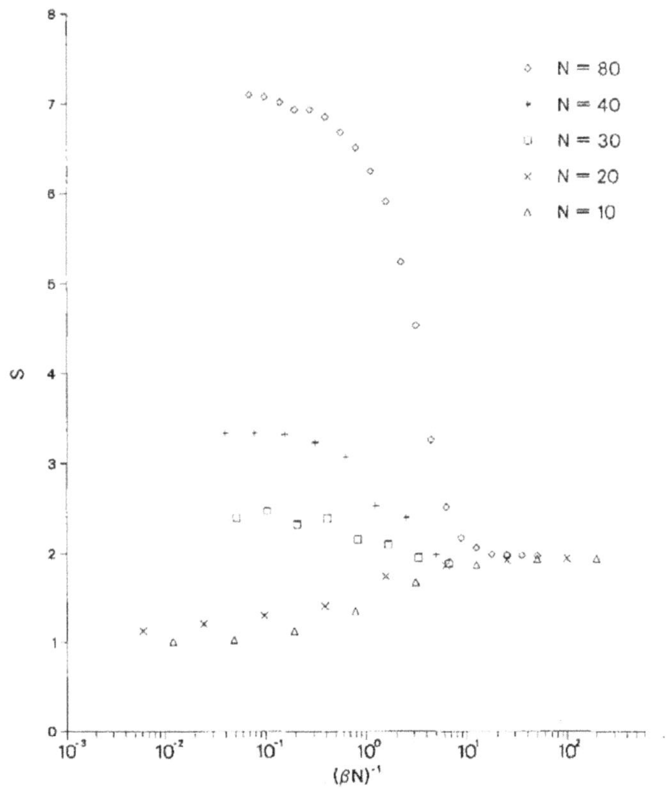

Fig. 1. Values of S for $0.0005 \leqslant \beta \leqslant 4.096$ and $10 \leqslant N \leqslant 80$.

from its high-temperature value (2 in the continuum, slightly less on the lattice) to the value S_N corresponding to the single-soliton solution of the time-independent NLSE with L^2 norm N. Figure 1 is a plot of values of S versus a scaled temperature $(\beta N)^{-1}$, with $p=4$, for several values of N; this scaling tends to bring the transitions into coincidence and is the one suggested by Proposition 3.1 below. We also seek to determine the general character of a typical field configuration by computing the quantity

$$F(x)=\langle|\tilde{\phi}(x)|^2\rangle$$

where $\tilde{\phi}$ is the translate of ϕ for which the maximum magnitude occurs at a fixed point, say $x_{\max}$. At high temperature, fluctuations produce an artificial spike in $F(x)$ at $x_{\max}$; to eliminate this effect, we look at the ratio $F(x)/F_0(x)$, where the subscript 0 identifies quantities computed in an ensemble with no potential energy term in the Hamiltonian. Figure 2 is a

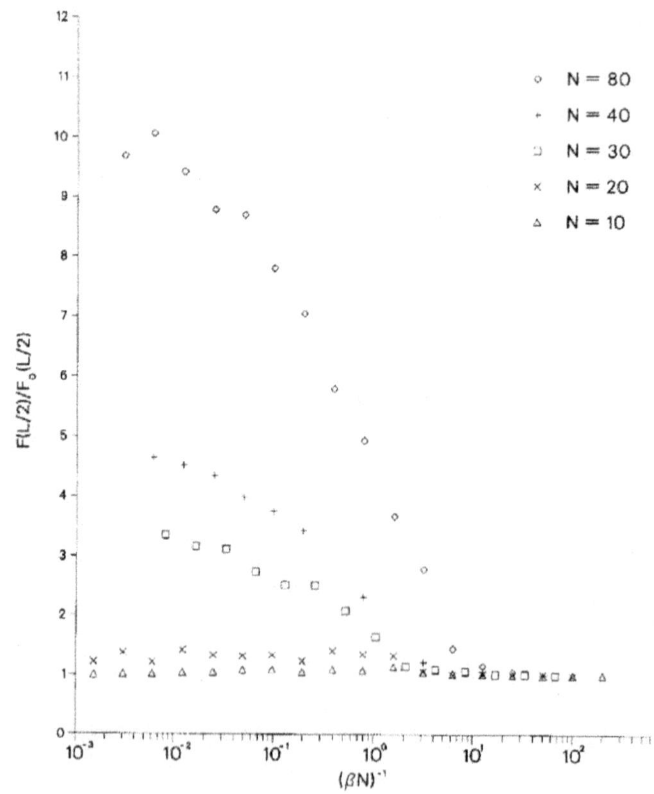

Fig. 2. Values of $F(L/2)/F_0(L/2)$ for $0.005 \leqslant \beta \leqslant 4.096$ and $10 \leqslant N \leqslant 80$.

plot of the quantity $F(x_{max})/F_0(x_{max})$, which represents a (normalized) height of the typical soliton, using the same horizontal scale as Fig. 1. Figures 3 and 4 present, for $N = 80$ and $N = 30$ respectively, the graphs of $F(x)/F_0(x)$ for 14 different values of β, ranging from 0.0005 to 4.096 and logarithmically equally spaced. For reference, we give in Fig. 5 the graphs of $F_0(x)$ for $N = 80$ at the same values of β; since the model with no potential energy is invariant under the scaling $\phi \to \rho\phi$, $N \to \rho^2 N$, $\beta \to \rho^{-2}\beta$, the graphs for other values of N are similar. (All data have been taken with $p = 4$ and $L = 1.0$ and with 40 lattice sites. In Figs. 3–5 we take I to be the interval $[0, L]$ and choose $x_{max} = L/2$.) A phase transition does not appear to be inconsistent with these data, but the numerical studies do not settle the point.

Given the nature of the observed typical field configurations at low temperatures or high values of N, one might suspect the existence of measures not satisfying translation invariance, corresponding to concentration of the measure on fields near a particular soliton-like structure. The

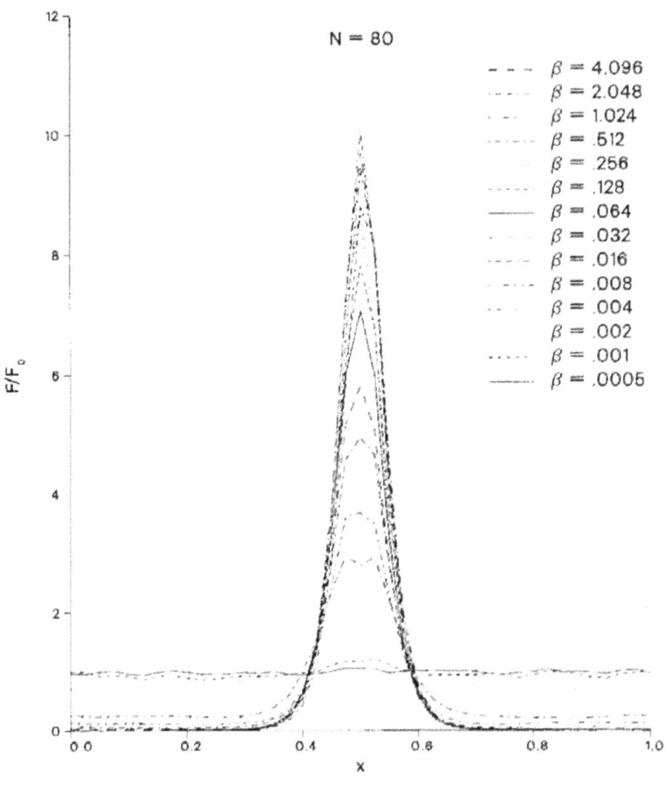

Fig. 3. Normalized field profiles at $N = 80$ for $0.0005 \leq \beta \leq 4.096$.

translation-invariant measure $\nu_{\beta,N}$ constructed in Section 2.2 could then be a convex superposition of non-translation-invariant measures. We see no mechanism for the generation of such measures, however. One might try to localize typical solitions by introducing an external potential into the Hamiltonian (1.1) via a term $\gamma \int_0^L V(x)\,\phi(x)\,dx$, yielding a modified measure $\nu_{\beta,N,\gamma}$, and then consider the limit $\gamma \to 0$. The measure $\nu_{\beta,N,\gamma}$ will be absolutely continuous with respect to $\nu_{\beta,N}$, however, and we will have

$$\lim_{\gamma \to 0} \nu_{\beta,N,\gamma} = \lim_{\gamma \to 0} \left\{ \exp\left[-\beta\gamma \int_0^L V(x)\,\phi(x)\,dx \right] \right\} \nu_{\beta,N} = \nu_{\beta,N}$$

(weakly) by the Lebesgue dominated convergence theorem.

We finally discuss the analyticity of the measure in the parameters β and N. Although we are unable to settle completely the question of analyticity, the following result limits the possible curves of critical points

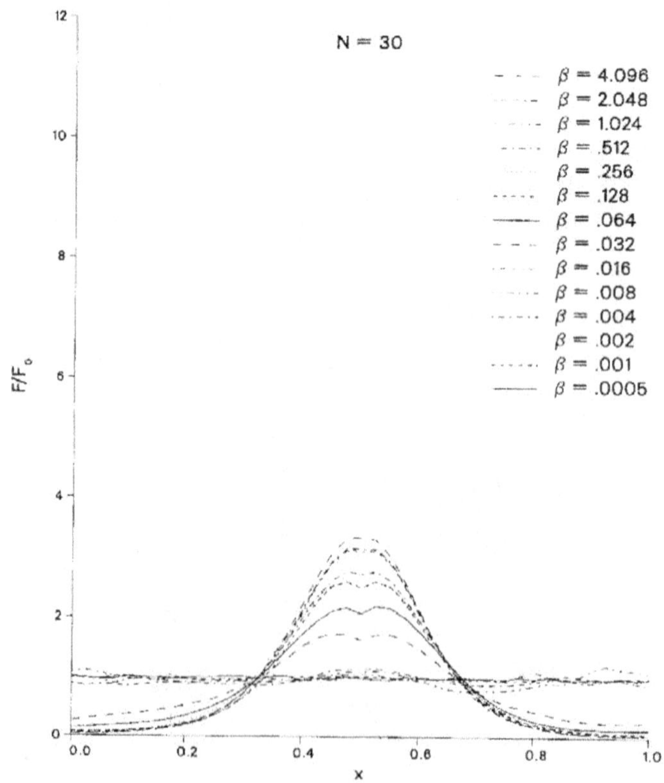

Fig. 4. Normalized field profiles at $N = 30$ for $0.0005 \leqslant \beta \leqslant 4.096$.

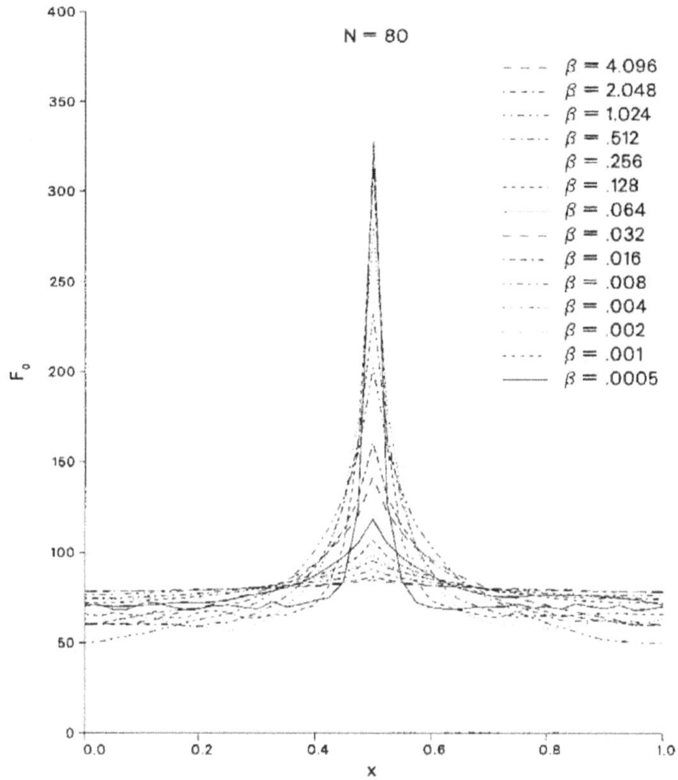

Fig. 5. Field profiles for Hamiltonian with no potential energy at $N=80$ for $0.0005 \leqslant \beta \leqslant 4.096$.

in the β–N plane to those of the form $\beta N = K$, with K a constant. (This is the scaling used in Figs. 1 and 2.) We will treat only the partition function $Z(\beta, N)$ defined in (2.9), although our conclusions would apply equally to the expectations of physical observables.

Proposition 3.1. Suppose that $p \leqslant 6$ and that $N > 0$. Then $Z(\alpha, \alpha^{-1}N)$ is (real) analytic in α for $\alpha > 0$ and, if $p = 6$, $\alpha^{-1}N < N_0$.

Proof. See Section 4.

4. PROOFS

In this section we give proofs of the theorems stated in Sections 2 and 3.

All the results depend criticaly on a standard interpolation inequality[18] relating various function norms. On $\mathbb{R}$ this has the form

$$\|\phi\|_{0,p}^{p} \leqslant C_p \|\phi'\|_{0,2}^{(p-2)/2} \|\phi\|_{0,2}^{(p+2)/2} \tag{4.1}$$

where $2 < p < \infty$. Weinstein[11] has shown that the best possible constant in (4.1) is

$$C_p = \frac{p}{2} \|\hat{\phi}\|_{0,2}^{2-p}$$

where $\hat{\phi}$ is a ground state for the equation

$$(p-2)\phi'' - (p+2)\phi + \phi^{p-1} = 0$$

that is, a nonzero solution with minimal L^2 norm. (4.1) implies a similar result on the interval I:

Lemma 4.1. (a) If $2 < p \leqslant 6$, then for any $\varepsilon > 0$ there is a constant $K_\varepsilon > 0$ such that, for $\phi \in H^1$,

$$\|\phi\|_p^p \leqslant (C_p + \varepsilon) \|\phi'\|_2^{(p-2)/2} \|\phi\|_2^{(p+2)/2} + K_\varepsilon \|\phi\|_2^p \tag{4.2}$$

(b) If $2 < p < 6$, then for any $\varepsilon > 0$ there is a constant $K'_\varepsilon > 0$ such that, for $\phi \in H^1$,

$$\|\phi\|_p^p \leqslant \varepsilon \|\phi'\|_2^2 \|\phi\|_2^{p-2} + K'_\varepsilon \|\phi\|_2^p \tag{4.3}$$

Proof. For $\phi \in H^1(I)$ we may be periodicity assume that $|\phi(0)|$ $[= |\phi(L)|] \leqslant L^{-1/2} \|\phi(x)\|_2$. Choose $\delta > 0$ and let $\hat{\phi}$ be the continuous functions on $\mathbb{R}$ which agrees with ϕ on $[0, L]$, vanishes on $(-\infty, -\delta]$ and $[L+\delta, \infty)$, and is affine on $[-\delta, 0]$ and $[L, L+\delta]$. Since

$$\|\phi\|_p^p \leqslant \|\hat{\phi}\|_{0,p}^p$$

$$\|\hat{\phi}\|_{0,2}^2 = \|\phi\|_2^2 + \frac{2\delta}{3} |\phi(0)|^2 \leqslant \left(1 + \frac{2\delta}{3L}\right) \|\phi\|_2^2$$

and

$$\|\hat{\phi}'\|_{0,2}^2 = \|\phi'\|_2^2 + \frac{2}{\delta} |\phi(0)|^2 \leqslant \|\phi'\|_2^2 + \frac{2}{\delta L} \|\phi\|_2^2$$

(4.2) follows from (4.1) and the inequality $(a+b)^q \leqslant a^q + b^q$, valid for $a, b \geqslant 0$ and $q = (p-2)/4$, by an appropriate choice of δ. Similarly, (4.3) follows from the observation that for any $\eta > 0$ there is a $\tilde{K}_\eta > 0$ such that $(a+b)^q \leqslant \eta a b^{q-1} + \tilde{K}_\eta b^q$, and a choice of η. ∎

Remark 4.1. The constant $\tilde{K}_\eta$ varies (for small η) as $\eta^{(p-2)/(p-6)}$, so that, by taking $\eta \sim \|\phi\|_2^{-(p-2)}$, we may derive the inequality

$$\frac{1}{p}\|\phi\|_p^p \leqslant \tfrac{1}{2}\|\phi'\|_2^2 + C_1\|\phi\|_2^{(2p+4)/(6-p)} + C_2\|\phi\|_2^p$$

valid for $p < 6$. This is the point of departure for the proof of normalizability of the modified grand canonical ensemble discussed in Remark 2.1.

We now sketch the proof of Theorem 2.1, emphasizing only those elements that differ from corresponding proofs on the entire line $\mathbb{R}$.

Proof of Theorem 2.1. Parts 1 and 2. The fixed-point argument used to prove existence in Ref. 8, and the regularitzation argument used to prove the conservation laws, go through directly in the space $H^1 = H^1(I)$.

Part 3(a), $N_1 \geqslant N_0$. The elegant discussion of Ref. 11 applies here, with N_0 given by (2.2) and with the use of (4.2) or (4.3) instead of (4.1). Since the constant N_0 will reoccur in our discussion of the statistical ensemble, and since the discussion is based on the important interpolation inequality, it seems worthwhile to repeat the argument here. We discuss only the critical case $p = 6$. The fixed-point argument of Ref. 8 will in fact prove existence globally in time if supplemented by an *a priori* bound on the H^1 norm of the solution. But from (4.2) and the conservation laws,

$$\begin{aligned} H(\phi_0) &= H(u(\cdot, t)) \\ &= \tfrac{1}{2}\|u_x(\cdot, t)\|_2^2 - \tfrac{1}{6}\|u(\cdot, t)\|_6^6 \\ &\geqslant \tfrac{1}{2}[1 - (N_0^{-2} + \varepsilon)\|\phi_0\|_2^4]\,\|u_x(\cdot, t)\|_2^2 - K_{3\varepsilon}\|\phi_0\|_2^6 \end{aligned}$$

If $\|\phi\|_2^2 < N_0$, we may choose ε so small that $(N_0^{-2} + \varepsilon)\|\phi_0\|_2^4 \equiv \kappa < 1$, yielding the bound

$$\|u_x(\cdot, t)\|_2^2 \leqslant (1-\kappa)^{-1}[H(\phi_0) + K_{3\varepsilon}\|\phi_0\|_2^6]$$

Part 3(b), $N_1 \leqslant 2N_0$. The argument of Glassey[19] shows that Part 3(b) holds, with $N_1 = N_0$, for the NLSE on $\mathbb{R}$, and has been extended by Kavian[20] to certain bounded domains. In particular, Kavian shows that 3(b) holds on the interval I with Dirichlet, Neumann, or periodic boundary conditions; his example gives $N_1 \leqslant N_0$ in the first two cases, but $N_1 \leqslant 2N_0$ in the periodic case of interest to us. It seems likely that $N_1 = N_0$ here also, but we do not have a proof. ∎

We next turn to the discussion of Theorem 2.2, and begin by establishing some notation and defining the Wiener measures we will use. Recall that C denotes the set of continuous, complex-valued functions ϕ on $[0, L]$ that satisfy $\phi(0)=\phi(L)$; for any $\psi \in C$ let $B_\varepsilon(\psi)$ denote the ball $\{\phi \in C \mid \|\phi-\psi\|_\infty < \varepsilon\}$. We also denote by $C^{a,b}$ the set of continuous functions on $[0, L]$ that satisfy $\phi(0)=a$, $\phi(L)=b$.

Now define the measure $\mu_\beta^{a,b:L}$ on $C^{a,b}$ as the natural restriction of the Wiener measure with diffusion constant β^{-1}, that is, as the measure with marginal distributions (see Ref. 21)

$$\operatorname{Prob}(\{\phi \mid \phi(x_i) \in A_i,\ 1 \leqslant i \leqslant n\})$$
$$= p_\beta^L(a, b)^{-1} \int_{y_i \in A_i} \prod_{i=1}^{n+1} p_\beta^{x_i - x_{i-1}}(y_i, y_{i-1})\, dy_1 \cdots dy_n$$

Here $0 = x_0 < x_1 < \cdots < x_n < x_{n+1} = L$, $A_1, \ldots, A_n$ are Borel subsets of $\mathbb{C}$, dy is Lebesgue measure on $\mathbb{C}$, $y_0 = a$, $y_{n+1} = b$, and for $y, y' \in \mathbb{C}$,

$$p_\beta^x(y, y') \equiv (\beta/2\pi x) \exp(-\beta |y - y'|^2/2x)$$

Finally, define the measure μ_β on $C = \bigcup_{\alpha \in \mathbb{C}} C^{a,a} \simeq \mathbb{C} \times C^{0,0}$ as $da \times d\mu_\beta^{0,0;L}$, with da Lebesgue measure on $\mathbb{C}$; equivalently, μ_β has marginal distributions

$$\operatorname{Prob}(\{\phi \mid \phi(x_i) \in A_i,\ 1 \leqslant i \leqslant n+1\})$$
$$= p_\beta^L(0, 0)^{-1} \int_{y_i \in A_i} \prod_{i=1}^{n+1} p_\beta^{x_i - x_{i-1}}(y_i, y_{i-1})|_{y_0 = y_{n+1}}\, dy_1 \cdots dy_{n+1} \tag{4.4}$$

We will occasionally need the corresponding spaces and measures defined for real-valued paths; these will be denoted $C_{\mathbb{R}}^{a,b}$, $\mu_{\mathbb{R},\beta}^{a,b;L}$, etc. In particular, for $y, y' \in \mathbb{R}$,

$$p_{\mathbb{R},\beta}^x(y, y') \equiv (\beta/2\pi x)^{1/2} \exp(-\beta |y - y'|^2/2x)$$

Before proving the theorem we give a minor technical result.

Lemma 4.2. For any $\eta > 0$ there is an $\varepsilon > 0$ such that, if $\phi_1, \phi_2 \in C$ satisfy $\|\phi_1 - \phi_2\|_\infty < \varepsilon$, then

$$|\,\|\phi_1\|_q^q - \|\phi_2\|_q^q| < \eta(\|\phi_1\|_q^q + 1)$$

for $q = 2, p$.

Proof. We omit the straightforward proof.

As a preliminary to the proof of the first part of Theorem 2.2, we must define some quantities associated with ordinary (real-valued) Brownian motion in one dimension with diffusion constant β^{-1}, and with paths $\omega(x)$ satisfying $\omega(0)=0$. Fix $\theta>0$ and let $T_\beta(\omega)$ and $S_\beta(\omega)$ be the times the path first hits $y=\theta$ and $|y|=\theta$, respectively. Elementary calculations (see, e.g., Ref. 22) show that the random variables T_β and S_β have densities

$$f_\beta(x)=\theta(\beta/2\pi x^3)^{1/2}\exp(-\beta\theta^2/2x) \tag{4.5}$$

and

$$g_\beta(x)=2\theta\,(\beta/2\pi x^3)^{1/2}\sum_{k=0}^{\infty}(-1)^k\exp\{-\beta[(2k+1)\theta]^2/2x\}$$

respectively, so that

$$2c(\beta)^{-1}f_\beta(x)\leqslant g_\beta(x)\leqslant 2f_\beta(x) \tag{4.6}$$

for $c(\beta)=1-\exp(-8\beta\theta^2/2L)$. Finally, for $|y|<\theta$ and $|y'|<\theta$ we let $q_\beta^x(y,y')$ be the transition density from y to y' in time x for Brownian motion with absorbing barriers at $\pm\theta$; that is, q satisfies

$$2\beta q_{\beta,x}=q_{\beta,y'y'}$$
$$q_\beta^0(y,y')=\delta(y-y')$$
$$q_\beta^x(y,\theta)=q_\beta^x(y,-\theta)=0$$

Now suppose that $\omega(x)$ is a real-valued continuous path satisfying $\omega(0)=a$, $\omega(L)=a+l\theta$ for $l\in\mathbb{Z}$ (i.e., $\omega\in C_{\mathbb{R}}^{a,a+l\theta}$); we introduce variables that keep track of the way in which ω attains values of the form $a+j\theta$, $j\in\mathbb{Z}$. Specifically, if $D_{\alpha,\theta}$ is the set of numbers of this form, we define $U_0(\omega), U_1(\omega),..., U_{M(\omega)+1}(\omega)$ inductively by

$$U_0(\omega)=0$$

and, if $X_k\equiv\{x>U_k\,|\,\omega(x)\in D_{\alpha,\theta},\ \omega(x)\neq\omega(U_k)\}$ is nonempty,

$$U_{k+1}(\omega)=\inf X_k$$

We let $M(\omega)$ be the smallest value of k for which X_k is empty, and set $U_{M(\omega)+1}(\omega)=L$. Finally, we define $\Gamma_1(\omega),..., \Gamma_{M(\omega)}(\omega)\in\{-1,1\}$ by the condition

$$\omega(U_k)=\omega(U_{k-1})+\Gamma_k\theta,\qquad k=1,..., M$$

As functions on the measure space $(C_{\mathbb{R}}^{a,a+l\theta}, \mu_{\mathbb{R},\beta}^{a,a+l\theta})$, M, Γ, and U are random variables with joint distribution given by

$$\mu_{\mathbb{R},\beta}^{a,a+l\theta;L}(\{M=m,\ \Gamma_j=\gamma_j,\ j=0,\dots,m, \text{ and } u_j \leqslant U_j \leqslant u_j+du_j,\ j=1,\dots,m\}) \equiv \delta_{\sum \gamma_i, l}\, p_{\mathbb{R},\beta}^{L}(a, a+l\theta)^{-1}\, h_\beta(m,u)\, du_1 \cdots du_m \tag{4.7}$$

where

$$h_\beta(m,u) = \prod_{j=1}^{m} \left[\tfrac{1}{2} g_\beta(u_j - u_{j-1})\right] q_\beta^{(L-u_m)}(0,0)\, du_1 \cdots du_m$$

Finally, we let $\xi(\omega) \in H_1([0,L])$ be the piecewise linear interpolant that agrees with ω at the points $x = U_j(\omega)$; note that $\|\omega - \xi(\omega)\|_\infty < 2\theta$.

Proof of Theorem 2.2(a). For a path $\phi \in C$ we let $A(\phi) \equiv \phi(0)$ and perform the construction of the preceding paragraph separately for $\phi^{\mathrm{R}} = \operatorname{Re}\phi$ and $\phi^{\mathrm{I}} = \operatorname{Im}\phi$, with $l=0$ in each case, producing random variables $M^{\mathrm{R}} \equiv M(\phi^{\mathrm{R}})$, etc. Let $\psi(\phi) = \xi(\phi^{\mathrm{R}}) + i\xi(\phi^{\mathrm{I}})$; ψ is a piecewise linear interpolant of ϕ with $\|\psi(\phi) - \phi\|_\infty < 2\sqrt{2}\theta$. Lemma 4.3 guarantees that we can, first, choose θ small enough that for all ϕ with $\|\phi\|_2^2 \leqslant N$,

$$\|\psi(\phi)\|_2^2 < \begin{cases} N_1, & \text{if } p=6 \\ N+1, & \text{if } p<6 \end{cases}$$

where $N < N_1 < N_0$, and second, by another application of Lemma 4.3 and use of Eq. (2.2) and the inequality (4.2) for $p=6$, or (4.3) for $p<6$, choose η and then θ so that

$$\frac{1}{p}\|\phi\|_p^p < \frac{1}{p}\left[(1+\eta)\,\|\psi(\phi)\|_p^p + \eta\right] < \frac{1}{2}(1-\lambda)\,\|\psi(\phi)'\|_2^2 + K$$

for constants λ, $K>0$ which are independent of ϕ. Finally, observe that if $\|\phi\|_2^2 \leqslant N$, then $|\phi(x)|^2 \leqslant N/L$ for some $x \in [0,L]$, and since for all $x \in 0, L]$,

$$|\phi(0) - \phi(x)| \leqslant \theta\left[(M^{\mathrm{R}}+1)^2 + (M^{\mathrm{I}}+1)^2\right]^{1/2}$$

M^{R} and M^{I} must satisfy

$$(M^{\mathrm{R}})^2 + (M^{\mathrm{I}})^2 \geqslant c_1 A^2 - c_2 \tag{4.8}$$

for appropriate constants c_1 and c_2.

Because $\psi(\phi)$ depends on ϕ only through the random variables A, M, U, and Γ, we have

$$\int_C F(\phi)\, d\mu_\beta(\phi)$$

$$\leqslant \int_{\|\phi\|_2^2 \leqslant N} \exp[\beta(1-\lambda)\,\|\psi(\phi)'\|_2^2/2 + K]\, d\mu_\beta$$

$$\leqslant (\exp K) \int_C da \sum_{m^R, m^I} \sum_{\gamma^R, \gamma^I} \int du^R\, du^I\, \delta_{\sum \gamma_i^R, 0}\, \delta_{\sum \gamma_i^I, 0}\, p_\beta^L(a, a)^{-1}$$

$$\times \{\exp[\beta(1-\lambda)\, \|\psi(\phi)'\|_2^2/2]\}\, h_\beta(m^R, u^R)\, h_\beta(m^I, u^I) \qquad (4.9)$$

Write

$$\|\psi(\phi)'\|_2^2 = \|\xi(\phi^R)'\|_2^2 + \|\xi(\phi^I)'\|_2^2$$

and note that

$$\|\xi(\phi^R)'\|_2^2 = \sum_{j=1}^{M^R} \frac{\theta^2}{U_j^R - U_{j-1}^R}$$

Then (4.5), (4.6), and the easy estimate $q_\beta(y) \leqslant C q_{\lambda\beta}(y)$ for $|y| < \theta$ imply that

$$\{\exp[\beta(1-\lambda)\, \|\xi(\phi^R)'\|_2^2/2]\}\, h_\beta(m^R, u^R)$$
$$\leqslant C[\lambda^{-1/2} c(\lambda\beta)]^{m^R}\, h_{\lambda\beta}(m^R, u^R)$$

with a similar inequality for the terms involving the imaginary part. Thus, with an obvious estimate of the sum over γ, (4.9) becomes

$$\int_C F(\phi)\, d\mu_\beta(\phi)$$

$$\leqslant C^2 e^K \int_C da \sum_{m^R, m^I} [2\lambda^{-1/2} c(\lambda\beta)]^{m^R + m^I}$$

$$\times \int du^R\, du^I\, p_\beta^L(a, a)^{-1}\, h_{\lambda\beta}(m^R, u^R)\, h_{\lambda\beta}(m^I, u^I) \qquad (4.10)$$

Now observe that the last line of (4.10) is, except for an incorrect normalization factor, one contribution from (4.7) to $\mu_{\lambda\beta}^{a,b;L}(C^{a,b}) = 1$, where $b = a + m^R\theta + im^I\theta$ [specifically, the contribution from terms for which $\gamma^I \equiv \gamma^R \equiv 0$ and $m^I = m^R = 0$ in (4.7)]. Hence

$$\int_C F(\phi)\, d\mu_\beta(\phi)$$

$$\leqslant C^2 e^K \int_C da \sum_{m^R, m^I} \int du\, [2\lambda^{-1/2} c(\lambda\beta)]^{m^R + m^I}\, p_\beta^L(a, a)^{-1}\, p_{\lambda\beta}^L(a, b) \quad (4.11)$$

Since

$$p^L_{\lambda\beta}(a, b) = (\lambda\beta/2\pi)\exp\{-\lambda\beta\theta^2[(m^R)^2 + (m^I)^2]/2L\}$$

the sum over m^R and m^I in (4.11) is convergent; moreover, the lower bound (4.8) implies that the resulting integral over a also converges. Thus

$$\int_C F(\phi)\, d\mu_\beta(\phi) < \infty \quad \blacksquare$$

For the proof of the second part of the theorem we will need an Onsager–Machlup formula for the measure μ_β (see Ref. 23 for further references).

Lemma 4.3. Suppose that $\psi \in C$ is absolutely continuous and that ψ' is of bounded variation on $[0, L]$; without loss of generality we redefine $\psi'(L)$ so that $\psi'(L) = \psi'(0)$. Then

$$\begin{aligned}
&\mu_\beta(B_\varepsilon(\psi)) \\
&= \left\{\exp\left[-\frac{\beta}{2}\int_0^L |\psi'(x)|^2\, dx\right]\right\} \int_{B_\varepsilon(0)} \exp\left(-\beta \operatorname{Re}\int_0^L \phi\, d\psi'\right) d\mu_\beta(\phi) \\
&\geqslant \left\{\exp\left[-\frac{\beta}{2}\int_0^L |\psi'(x)|^2\, dx\right] \exp[-\beta\varepsilon V(\psi')]\right\} \mu_\beta(B_\varepsilon(0))
\end{aligned}$$

where $V(\psi')$ is the total variation of ψ' on $[0, L]$.

Proof. For $n \geqslant 1$ we introduce mesh points $x_k = kL/n$, $0 \leqslant k \leqslant n$, and set

$$B_\varepsilon^{(n)}(\psi) = \{\phi \in C \mid |\phi(x_k) - \psi(x_k)| < \varepsilon,\ 1 \leqslant k \leqslant n\}$$

Clearly $\mu_\beta(B_\varepsilon(\psi)) = \lim_{n\to\infty} \mu_\beta(B_\varepsilon^{(n)}(\psi))$. On the other hand, from (4.4)

$$\begin{aligned}
&\mu_\beta(B_\varepsilon^{(n)}(\psi)) \\
&= p^L_\beta(0, 0)^{-1} \int_{|z_i| < \varepsilon} \prod_{i=1}^n p^{L/n}_\beta(\psi(x_i) + z_i, \psi(x_{i-1}) + z_{i-1})\, dz_1 \cdots dz_n \\
&= \exp\left[-(\beta n/2L) \sum_{i=1}^n |\psi(x_i) - \psi(x_{i-1})|^2\right] \\
&\quad \times \int_{B_\varepsilon^{(n)}(0)} \exp\left\{-(\beta n/L) \operatorname{Re} \sum_{i=1}^n [\psi(x_i) - \psi(x_{i-1})]\right. \\
&\quad \times \left.[\phi(x_i) - \phi(x_{i-1})]\right\} d\mu_\beta(\phi)
\end{aligned}$$

where $z_0 = z_n$ by convention. But since $\psi' \in L^2$,

$$\lim_{n \to \infty} \frac{n}{L} \sum_{i=1}^{n} |\psi(x_i) - \psi(x_{i-1})|^2 = \int_0^L |\psi'(x)|^2 \, dx$$

Moreover, if for $\phi \in C$ we let $\phi^{(n)}$ be the piecewise linear interpolant of ϕ on the mesh points, we have

$$\frac{n}{L} \sum_{i=1}^{n} [\psi(x_i) - \psi(x_{i-1})][\phi(x_i) - \phi(x_{i-1})] = \int_0^L \psi'(x)\, \phi^{(n)\prime}(x) \, dx$$

$$= \int_0^L \phi^{(n)} d\psi'$$

Since $\phi^{(n)} \to \phi$ uniformly as $n \to \infty$, the equality in the statement of the lemma follows from the dominated convergence theorem; the inequality is then immediate. ∎

Proof of Theorem 2.2(b). For notational simplicity we replace the interval $[0, L]$ by the interval $[-L/2, L/2]$. We may choose the ground state solution $\hat{\phi}$ of the NLS on $\mathbb{R}$ discussed in Section 2 so that it is positive and symmetric under reflection through the origin, lies in $H^1(\mathbb{R}) \cap C^\infty(\mathbb{R})$, and satisfies $\|\hat{\phi}\|_{0,p}^p = (p/2)\, \|\hat{\phi}'\|_{0,2}^0$, $\|\hat{\phi}\|_{0,2}^2 = N_0$. Let $\tilde{\phi} = \alpha\hat{\phi}$ for some $\alpha > 1$; in the case $p = 6$ we choose α such that $\|\tilde{\phi}\|_2^2 < N$, while for $p > 6$, α is arbitrary. Clearly, $H(\tilde{\phi}) < 0$.

Now for any ϕ defined on $\mathbb{R}$ and any $\rho > 0$ we define

$$(T_\rho \phi)(x) = \rho^{4/(p-2)}\, \tilde{\phi}(\rho^2 x)$$

For ϕ sufficiently regular that the corresponding norms are finite, both $\|\phi\|_{0,p}^p$ and $\|\phi'\|_{0,2}^2$ scale under T_ρ as $\rho^{(2p+4)/(p-2)}$, while $\|\phi\|_{0,2}^2$ and $\|\phi''\|_{0,1}$ scale as $\rho^{2(6-p)/(p-2)}$ and $\rho^{2p/(p-2)}$, respectively. Finally, we define $\psi_\rho \in H^1$ by

$$\psi_\rho(x) = (T_\rho \tilde{\phi})(x), \qquad x \in [-L/2, L/2]$$

Thus, under the hypotheses of (b) we will have, for some positive constants E, η_1, A, B, and ρ_0, and for all $\rho > \rho_0$,

$$\|\psi_\rho\|_2^2 < N - \eta_1 \tag{4.12a}$$

$$\|\psi_\rho\|_p^p \leqslant \rho^{(2p+4)/(p-2)} B \tag{4.12b}$$

$$-H(\psi_\rho) \geqslant \rho^{(2p+4)/(p-2)} E \tag{4.12c}$$

$$V(\psi'_\rho) = \|\psi''_\rho\|_1 + O(\rho) < \rho^{2p/(p-2)} A \tag{4.12d}$$

We now apply Lemma 4.2, with $\phi_1 = \psi_\rho$ and $\eta = \min\{\eta_1, E/(2B)\}$, to produce an ε for which the conclusions of that lemma hold. Then, from Lemma 4.3, for $\rho > \rho_0$,

$$\begin{aligned}
&\int_C F(\phi)\, d\mu_\beta(\phi) \\
&\quad \geqslant \int_{B_\varepsilon(\psi_\rho)} \exp\left(\frac{\beta}{p}\|\phi\|_p^p\right) d\mu_\beta(\phi) \\
&\quad \geqslant \left\{\exp\left[\frac{\beta}{p}\left[(1-\eta)\|\psi_\rho\|_p^p - \eta\right]\right]\right\} \int_{B_\varepsilon(\psi_\rho)} d\mu_\beta(\phi) \\
&\quad \geqslant \exp\left\{\beta\left[-H(\psi_\rho) - \frac{\eta}{p}\|\psi_\rho\|_p^p - \varepsilon V(\psi'_\rho) - \frac{\eta}{p}\right]\right\}
\end{aligned}$$

Then, as $\rho \to \infty$, we see from (4.12b)–(4.12d) that $F(\phi) \notin L^1(C; \mu_\beta)$. ∎

Finally, we give the following result:

Proof of Proposition 3.1. We modify the Hamiltonian of (1.1) by introducing a complex coupling constant λ:

$$H_\lambda(\phi) = \frac{1}{2}\int_0^L |\phi'(x)|_2^2\, dx - \frac{\lambda}{p}|\phi(x)|^p\, dx$$

Theorem 2.2 holds for this modified Hamiltonian, with the exception that if $p = 6$, then $F \in L^1$ requires

$$N^2 \operatorname{Re} \lambda \leqslant N_0^2 \tag{4.13}$$

Moreover, the Lebesgue dominated convergence theorem shows that the corresponding partition function

$$\tilde{Z}(\beta, N, \lambda) = \int_C \left[\exp\left(\frac{\lambda\beta}{p}\|\phi\|_p^p\right)\right] \chi_{\{\phi \mid \|\phi\|_2^2 \leqslant N\}}\, d\mu_\beta(\phi) \tag{4.14}$$

is analytic in λ for fixed β, N, for all $\lambda \in \mathbb{C}$ if $p < 6$, and for all λ satisfying (4.13) if $p = 6$.

Now a change of variables $\phi = \alpha^{1/2}\psi$ $(\alpha > 0)$ in (4.14) shows that

$$\tilde{Z}(\beta, N, \lambda) = \tilde{Z}(\alpha\beta, \alpha^{-1}N, \alpha^{(p-2)/2}\lambda)$$

In particular, taking $\lambda = \alpha^{-(p-2)/2}$ and $\beta = 1$ yields

$$\tilde{Z}(1, N, \alpha^{(2-p)/2}) = \tilde{Z}(\alpha, \alpha^{-1}N, 1) = Z(\alpha, \alpha^{-1}N)$$

The result is immediate. ∎

5. OPEN QUESTIONS

Many questions remain regarding the behavior of this and related models. Here are a few of them.

1. *Existence of equilibrium dynamics.* The Gibbs measure constructed in Section 2.2 lives on Wiener paths, which, although continuous, are typically not differentiable. On the other hand, the dynamical theory sketched in Section 2.1 works for differentiable (specifically, H^1) functions. It would be desirable to define a dynamics on the Wiener paths for which the measure would be invariant, but we have been unable to do so. Note that, at least for the definition of a local (in time) dynamics, the source of difficulty is the nonlinearity itself and not the "wrong" sign of the potential energy.

We remark that to apply a fixed-point method to (2.1) in a straightforward way (as in Section 2.1) one should replace the space $H^1(I)$ by a linear space X of functions on I such that: (i) the nonlinear map $\phi \mapsto |\phi|^{p-2}\phi$ is continuous on X; (ii) the free time evolution $U(t)$ is continuous on X; (iii) almost every Wiener path lies in X, i.e., $\mu_\beta(C\setminus X)=0$. The obvious candidates fail. $H^\alpha(I)$ satisfies (ii) for all α, but (i) only for $\alpha \geqslant 1/2$ and (iii) only for $\alpha < 1/2$; $L^\infty(I)$ satisfies (i) and (iii), but fails (ii) (as does C itself), as shown by the following argument, for which we thank E. Stein. Choose $s\in\mathbb{R}$ so that $2\pi s/L^2$ is irrational, and let

$$\Phi_N = \sum_{|n|\leqslant N} \exp[i(2\pi n/L)^2 s]\exp(2\pi inx/L)$$

so that

$$U(s)\,\Phi_N = \sum_{|n|\leqslant N} \exp(2\pi inx/L)$$

Then $\|U(s)\,\Phi_N\|_\infty = 2N+1$, but Hardy and Littlewood[24] show that $\|\Phi_N\|_\infty = o(N)$. Thus, $U(s)$ is unbounded on L^∞. [In fact, an application, given in Ref. 25, of the closed graph theorem shows that $U(s)$ cannot map L^∞ to itself.] Incidentally, $U(t)$ is bounded on L^∞ (or C) for $2\pi t/L^2$ rational, because it acts in Fourier space as multiplication by the Fourier coefficients of a complex measure of finite total variation.[25,26]

2. *Extension to higher dimensions.* The NLSE occurs naturally in various physical problems, including nonlinear optics as well as the plasma physics emphasized here, in which the position coordinate is two or three dimensional. The theory of smooth solutions extends readily to this case, at least on $\mathbb{R}^n$.[8] It would be interesting to extend the construction of the Gibbs measure to these cases for the appropriate values of p and N; the

standard free field measures could play the role that Wiener measure has played here. We would of course have to deal with ultraviolet problems. Mathematically, these would lead to normal ordering of the fields in $d=2$, as is usual in quantum field theory,[27] and to more complicated renormalizations in $d=3$, but we do not have a physical interpretation of such manipulations in, for example, the plasma physics case. Nor do we know whether the standard methods of constructive quantum field theory would suffice for Hamiltonians unbounded below.

3. *Properties of the measure* $\nu_{\beta,N}$. Various additional questions remain open for the measures constructed in this paper. Are physical quantities in fact analytic in β and N? What are the properties of the correlation functions? Is the measure normalizable in the critical case $p=6$, $N=N_0$? Can an ensemble be defined via (2.4), that is, by restricting the fields to lie on the surface of a sphere in L^2, and do the resulting measures differ significantly from those when the restriction is to the ball?

4. *Parameter ranges where collapse can occur.* The ensembles as constructed here do not exist for $p>6$ or for $p=6$ and $N>N_0$. Can a modified approach throw any light on the behavior of physical systems with parameters in this range? One approach would be the construction of a set X of nonzero Wiener measure, invariant under the dynamics, such that

$$\int_X \exp\left(\frac{\beta}{p}\int |\phi(x)|^p\,dx\right) d\mu_\beta < \infty$$

X should in some sense correspond to the set of field configurations giving rise to smooth solutions globally in time. (In the case $p=6$ one would require that the set $X\setminus\{\phi \mid \|\phi\|_2^2 \leqslant N_0\}$ have nonzero Wiener measure.) The Gibbs ensemble could then be defined on X. One could also study modified versions of the interaction energy, introducing a cutoff at high field amplitudes to prevent collapse. For example, the replacement of $\int |\phi|^4\,d^2x$ by

$$\int \frac{|\phi|^4}{1+\lambda|\phi|^2}\,d^2x$$

has been used in the study of laser propagation in a nonlinear medium[28] and a similar modification could be studied in one dimension.

5. *Stochastic dynamics.* It is possible to introduce a stochastic dynamics for the Zakharov model which preserves the L^2 norm of the u field and has the Gibbs state (3.5) as an invariant measure. The simplest such model, which must also include a dissipative term, is achieved by modifying only (3.3a), retaining (3.1) and (3.3b) in their original form. If

$\psi(k)$ denotes the kth Fourier coefficient of the field ψ, the modified (3.3a) is

$$V(k)_t = (-n_x + |u|_x^2)(k) - 2\rho(k)\, V(k) + \left(\frac{4\rho(k)}{\beta c}\right)^{1/2} \xi(k)$$

for $k = \pm 1, \pm 2,\ldots$. Here $\rho(k) > 0$ is the coefficient of dissipation in the kth mode and $\{\xi(k)\}$ are independently distributed white noise random variables with covariance

$$E[\xi(k)(t)\, \xi(k')(t')] = \delta_{k,k'}\, \delta(t-t')$$

In this model it seems possible that the Gibbs ensemble is the unique invariant measure. As already mentioned, however, this introduction of steady Gaussian "microscopic noise" into the ZE is not physically complete, since in a real plasma there are other, intermittent, dissipative processes associated with collapse that do not satisfy a fluctuation-dissipation relation.

ACKNOWLEDGMENTS

We thank Michael Aizenman, Josef Fritz, Sheldon Goldstein, Eli Stein, and Michael Weinstein for very helpful suggestions, and Joseph O'Connell for his investigation of the ensemble discussed in Remark 2.2. One of us (H. A. R.) acknowledges the support of the INLS at UCSD, where part of this research was performed. This work was supported in part by AFSOR grant 87-0010, by the U.S. DOE, and in part by ONR contract DP-F85-28.

REFERENCES

1. J.-P. Eckmann and D. Ruelle, *Rev. Mod. Phys.* **57**:617–656 (1985).
2. V. E. Zakharov, *Sov. Phys. JETP* **35**:908–914 (1972).
3. M. V. Goldman, *Rev. Mod. Phys.* **56**:709–735 (1984).
4. H. A. Rose, D. F. DuBois, and B. Bezzerides, *Phys. Rev. Lett.* **58**:2547–2550 (1985).
5. D. W. McLaughlin, G. C. Papanicolaou, C. Sulem, and P. L. Sulem, *Phys. Rev. A* **34**:1200–1210 (1986).
6. D. Ruelle, *Statistical Mechanics* (Benjamin, Reading, Massachusetts, 1974).
7. J. Glimm and A. Jaffe, *Quantum Physics* (Springer-Verlag, New York, 1981).
8. J. Ginibre and G. Velo, *Ann. Inst. Henri Poincaré* **28**:287–316 (1978).
9. V. E. Zakharov and A. B. Shabat, *Sov. Phys. JETP* **34**:62–69 (1972).
10. D. Russell, D. F. DuBois, and H. A. Rose, *Phys. Rev. Lett.* **56**:838–841 (1986).
11. M. Weinstein, *Commun. Math. Phys.* **87**:567–576 (1983).
12. S. H. Schochet and M. I. Weinstein, *Commun. Math. Phys.* **106**:569–580 (1986).
13. H. Tasso. *Phys. Lett. A* **120**:464–465 (1987).

14. G. Pelletier, *J. Plasma Phys.* **24**:287–297 (1980).
15. G. Pelletier, *J. Plasma Phys.* **24**:421–443 (1980).
16. G. Z. Sun, D. R. Nicholson, and H. A. Rose, *Phys. Fluids* **28**:2395–2405 (1985).
17. R. H. Kraichnan and D. Montgomery, *Rep. Prog. Phys.* **43**:547–619 (1979).
18. L. Nirenberg, *Commun. Pure Appl. Math.* **8**:648–674 (1955).
19. R. T. Glassey, *J. Math. Phys.* **18**:1794–1797 (1977).
20. O. Kavian, A Remark on the blowing-up of solutions to the Cauchy problem for the nonlinear Schrödinger equation, preprint, Lab. Ana. Nu., Universite P. & M. Curie.
21. B. Simon, *Functional Integration and Quantum Physics* (Academic Press, New York, 1979).
22. W. Feller, *An Introduction to Probability Theory and Its Applications*, Vol. II (Wiley, New York, 1971).
23. D. Dürr and A. Bach, *Commun. Math. Phys.* **60**:153–170 (1978).
24. G. H. Hardy and J. E. Littlewood, *Acta Mathematica* **37**:193–238 (1914).
25. E. Hille and R. S. Phillips, *Functional Analysis and Semi-Groups* (American Mathematical Society, Providence, Rhode Island, 1957).
26. A. Zygmund, *Trigonometrical Series* (Warsaw, 1935).
27. B. Simon, *The P(ϕ) Euclidean (Quantum) Field Theory* (Princeton University Press, Princeton, New Jersey, 1974).
28. G. Papanicolaou, private communication.

Incompressible Navier-Stokes and Euler Limits of the Boltzmann Equation

A. DE MASI
Universita dell'Aguila

R. ESPOSITO
Universita di Roma Tor Vergata

AND

J. L. LEBOWITZ
Rutgers University

Abstract

We consider solutions of the Boltzmann equation, in a d-dimensional torus, $d = 2,3$,

$$(*) \qquad \partial_\tau f(r, v, \tau) + v \cdot \nabla_r f = Q(f, f),$$

for macroscopic times $\tau = t/\varepsilon^N$, $\varepsilon \ll 1$, $t \geqq 0$, when the space variations are on a macroscopic scale $x = \varepsilon^{N-1} r$, $N \geqq 2$, x in the unit torus. Let $u(x, t)$ be, for $t \leqq t_0$, a smooth solution of the incompressible Navier Stokes equations (INS) for $N = 2$ and of the Incompressible Euler equation (IE) for $N > 2$.

We prove that $(*)$ has solutions for $t \leqq t_0$ which are close, to $O(\varepsilon^2)$ in a suitable norm, to the local Maxwellian $[\bar{\rho}/(2\pi\bar{T})^{d/2}]\exp\{-[v - \varepsilon u(x, t)]^2/2\bar{T}\}$ with constant density $\bar{\rho}$ and temperature $\bar{T}$. This is a particular case, defined by the choice of initial values of the macroscopic variables, of a class of such solutions in which the macroscopic variables satisfy more general hydrodynamical equations. For $N \geqq 3$ these equations correspond to variable density IE while for $N = 2$ they involve higher-order derivatives of the density.

1. Introduction

The Boltzmann Equation (BE) describes the time evolution of the positional and velocity density $f(r, v, \tau)$ of a gas. An old question is the relation of the Boltzmann equation to the Euler and Navier-Stokes equations of fluid dynamics. (For an historical background see Chapman and Cowling [1].) In particular, under what conditions does the BE lead to closed hydrodynamical type equations for the locally conserved quantities (collision invariants) which depend only on r and τ. These are the five moments of f corresponding to mass, momentum and

Communications on Pure and Applied Mathematics, Vol. XLII 1189–1214 (1989)

CCC 0010-3640/89/081189-26$04.00

energy densities,

$$\rho(r,\tau) = \int_{\mathbf{R}^d} f(r,v,\tau)\,dv, \qquad \rho u(r,\tau) = \int_{\mathbf{R}^d} v f(r,v,\tau)\,dv,$$
$$\rho T(r,\tau) = \frac{1}{d}\int_{\mathbf{R}^d} (v - u(r,\tau))^2 f(r,v,\tau)\,dv. \tag{1.1}$$

The possibility of such a "contracted description" is due to the big disparity (in many situations) between the microscopic scale of the BE, represented by the mean free path and mean free time and the macroscopic scales on which the (conserved) hydrodynamical variables change. This can be expected to lead to a state close to local equilibrium, i.e., $f(r,v,\tau)$ should be close to a local Maxwellian,

$$M(\rho, u, T; v) = \rho/(2\pi T)^{d/2} \exp\{-(v-u)^2/2T\}, \tag{1.2}$$

with the hydrodynamical variables changing "slowly" in space and time. M is singled out by the fact that it is the only function for which $Q(M,M) = 0$ (see [1]–[3]).

The disparity between microscopic and macroscopic scales is the central ingredient in the derivation of a variety of macroscopic laws from microscopic model systems (see [4]). In all these cases one follows the path initiated by Hilbert [5] and Enskog [6] who introduced the ratio of microscopic to macroscopic length scales, the Knudsen number ε, as an explicit parameter in the BE changing (*) to

$$\partial_\tau f_\varepsilon(r,\tau;v) + v\cdot\nabla_r f_\varepsilon = \varepsilon^{-1} Q(f_\varepsilon, f_\varepsilon), \tag{1.3}$$

where we have changed the order of the variables in f_ε to emphasize that v is to be thought of as a parameter.

It is now possible to make various formal expansions of f_ε in powers of ε about some local or global Maxwellian. In particular the well-known Hilbert-Chapman-Enskog method [1] leads to a sequence of equations for the hydrodynamical variables corresponding, respectively, to the (non-dissipative) Euler equations, the Navier-Stokes equations, etc. From now on we always consider a collision operator Q corresponding, a la Grad [2], to cut-off hard potentials.

An important step in the mathematical development of the theory was taken by Caflisch [7] (and by Caflisch and Papanicolau [8] for a discrete velocities model) who proved the following: let ρ, u, T be smooth solutions of Euler equations in a torus for $t \in [0, t_0]$. Then there exists a solution $f_\varepsilon(r,\tau;v)$ of the

BE (1.3) such that, for $\tau < t_0$, we have, in a suitable norm,

$$\|f_\varepsilon - M(\rho, u, T; v)\| < C\varepsilon. \tag{1.4}$$

This result has been extended by Lachowicz [9] to more general initial data $f_\varepsilon(r, 0; v)$ and more general domains, see also Bardos [10] and Bardos and Golse [11].

We now observe that equation (1.3) is a special case of scaled equations which can be obtained directly from (*) by looking explicitly at solutions which vary on different macroscopic space-time scales. More precisely we let ε^{-1} be the macroscopic scale, i.e., the size of the domain (torus) in which the system is confined and we let $r = \varepsilon^{-1}x$, $\tau = \varepsilon^{-\alpha}t$. The time evolution of $F^\varepsilon_{(\alpha)}(x, t; v) = f(\varepsilon^{-1}x, v, \varepsilon^{-\alpha}t)$ will satisfy the $(\mathrm{BE})_\alpha$ equation

$$\frac{\partial F^\varepsilon_{(\alpha)}}{\partial t} + \varepsilon^{(1-\alpha)} v \cdot \nabla_x F^\varepsilon_{(\alpha)} = \varepsilon^{-\alpha} Q\left(F^\varepsilon_{(\alpha)}, F^\varepsilon_{(\alpha)}\right), \tag{1.5}$$

with some initial condition $F^\varepsilon_{(\alpha)}(x, 0; v)$. Equation (1.5) is nothing more than a rewriting of (*) in terms of the new variables x and t. Any new results valid for "ε small" must therefore follow form (*) and the assumptions, if any, on the initial state $F^\varepsilon_{(\alpha)}(x, 0; v)$. This will be discussed further in Section 5; see also [12].

Equation (1.3) thus corresponds to $\alpha = 1$, i.e., space and time are scaled in the same way. That this is the right choice for the Euler equations is clearly related to the fact that these equations are invariant under this scaling. The reduction to Euler type equations under uniform space-time scaling can also be proven for some simple microscopic models (see [13]) and shown to occur, under the assumption of local equilibrium, for real particle systems evolving according to Hamiltonain equations of motion (see [4]).

A natural question to ask then is whether there is any different micro-macro scaling of space-time which produces the Navier-Stokes equations. An inspection of those equations shows that they are not invariant under any rescaling of space-time. This implies that there are no space-time scalings which produce the full Navier-Stokes equations as a well-defined limit. They have to be thought of instead as a correction to the Euler equations. On the other hand, if one also scales the macroscopic velocity u, i.e., one considers u's very small compared to microscopic velocities (or the Mach number), then there is a scaling which leaves a "subset" of the Navier Stokes equations, and in particular the INS equations, invariant; c.f. [14]. Similarly there are scalings of u which make the dissipative term in the INS vanish but leave a "subset" of the Euler equations, including the IE, invariant; see [15]. This suggests a derivation of these equations directly from the BE with proper scaling of the initial u's. We carry out this program for the INS and IE in Sections 2, 3, 4. We leave for Section 5 the discussion of the larger class of hydrodynamical evolutions, obtained by imposing some constraint on the $d + 2$ variables occurring in the compressible hydrodynamic equations, which can be obtained from the BE by choosing suitable initial conditions.

2. Statement of Results

Let $u(x, t)$ be the solution of the INS or IE in a unit d-dimensional torus Π, i.e., u satisfies the equations

$$\bar{\rho}\partial_t u + \bar{\rho} u \cdot \nabla_x u = -\nabla_x p + \eta_N \Delta u, \tag{2.1a}$$

$$\operatorname{div} u = 0, \tag{2.1b}$$

$$u(x,0) = u_0(x), \tag{2.1c}$$

where $\bar{\rho}$ is the constant mass density and η_N is the viscosity coefficient for $N = 2$ (INS), while $\eta_N = 0$ if $N > 2$ (IE). The vector field $\nabla_x p$ which corresponds physically to the pressure gradient is not specified a priori but is to be considered formally as an unknown function also determined from (2.1). The parameter N will be defined later. It is known (see for example [16]) that if u_0 is sufficiently smooth there is a t_0 depending on u_0 such that equation (2.1) has a unique smooth solution $(u(x, t), p(x, t))$ for $t \leq t_0$. In two dimensions, t_0 can be taken arbitrarily large.

In order to define precisely the smoothness we require, we introduce the Sobolev space $H_s(\Pi)$ of order s with scalar product

$$(h, g)_s = \sum_{j=0}^{s} \int_\Pi dx \, \nabla^j h(x) \cdot \nabla^j g(x). \tag{2.2}$$

Furthermore, we denote by B_{js} the space of the measurable functions on $\Pi \times \mathbb{R}^d$, such that

$$\|h\|_{js} = \sup_{v \in \mathbb{R}^d} (1 + v^2)^{j/2} |h(\cdot, v)|_s, \qquad |h|_s^2 = (h, h)_s, \tag{2.3}$$

is finite.

We wish to prove now that any smooth velocity and pressure field satisfying (2.1) can in fact be obtained from the solution to the $(\mathrm{BE})_\alpha$ equation (1.5) for suitable initial distribution and choices of the scaling α. The case $\alpha = 1$ is covered by the Caflisch theorem, equation (1.4). We shall now consider the case $1 < \alpha \leq 2$, and observe that by letting $\varepsilon^{\alpha-1} \to \varepsilon$ we can rewrite equation (1.5) as follows:

$$\frac{\partial F_N^\varepsilon}{\partial t} + \varepsilon^{-1} v \cdot \nabla_x F_N^\varepsilon = \varepsilon^{-N} Q(F_N^\varepsilon, F_N^\varepsilon), \tag{2.4}$$

with $N = \alpha/(\alpha - 1)$, so that $N \geq 2$. We assume that N is an integer: our proofs presumably work also for any rational number N, but we did not treat that case.

We prove the following

THEOREM 1. *Given a divergence-free field* $u_0 \in H_s(\Pi)$, $s \geqq 4$, $\bar{\rho} > 0$, $\bar{T} > 0$, $N \geqq 2$, *there are fixed* $t_0 > 0$ *and* $\varepsilon_0 > 0$ *such that equation* (2.4) *has a solution* $F_N^\varepsilon(x, t; v) \in B_{j2}$ *for* $t \leqq t_0$, $j > 4$, $\varepsilon < \varepsilon_0$, *satisfying*

$$\| F_N^\varepsilon(x, t; v) - M(\bar{\rho}, \varepsilon u(x, t), \bar{T}; v) \|_{j2} < C\varepsilon^2, \tag{2.5}$$

whenever $u(x, t)$ *is a solution to equation* (2.1) *in* $H_s(\Pi)$ *for* $t \leqq t_0$. *In the case* $N = 2$, *the viscosity coefficient* $\eta = \eta(\bar{\rho}, \bar{T})$ *is given by the equation* (3.32) *below.*

Remarks. (i) The bounds in the above statement are not optimal, but they are sufficient for $d = 2, 3$.

(ii) Note that u in equation (2.5) is not the same as the u in equation (1.1), $u \to \varepsilon u$.

(iii) We actually prove that there is a solution $F_N^\varepsilon(x, t; v)$ to equation (2.4) of the following form: we consider for concreteness the case $N = 2$ and write $F_2^\varepsilon = F^\varepsilon$,

$$F^\varepsilon(x, t; v) = M(\rho_\varepsilon, u_\varepsilon, T_\varepsilon) + \varepsilon^2 g(x, v, t) + \varepsilon^3 h_\varepsilon(x, t, v), \tag{2.6}$$

where

$$\rho_\varepsilon = \bar{\rho} + \varepsilon^2 \rho^{(2)}(x, t), \tag{2.7a}$$

$$u_\varepsilon = \varepsilon u(x, t) + \varepsilon^2 u^{(2)}(x, t), \tag{2.7b}$$

$$T_\varepsilon = \bar{T} + \varepsilon^2 T^{(2)}(x, t), \tag{2.7c}$$

$g(x, v, t)$ is a function orthogonal to the space generated by the collision invariants (see (3.25)) and all the functions ρ_ε, u_ε, T_ε, $g(x, v, t)$ and $h_\varepsilon(x, t, v)$ are uniformly (in ε) bounded in B_{j2}.

Finally, $u(x, t)$ in (2.7b) is the solution to the INS and the term proportional to ε^2 in $\rho_\varepsilon T_\varepsilon$, i.e., $\bar{T}\rho^{(2)}(x, t) + \bar{\rho} T^{(2)}(x, t)$ is "the pressure" $p(x, t)$ associated to the solution $u(x, t)$ of equation (2.1) with $N = 2$.

(iv) Our results involve choosing an initial distribution $F_N^\varepsilon(x, 0; v)$ which is of the "right form". This may in fact not be necessary. As was shown by Lachowitz [9] for the case $\alpha = 1$ considered by Caflisch (and by Ellis [17] and Ellis and Pinsky [12] in some simple linear case), choosing a "slightly wrong" distribution at $t = 0$ has the effect of producing an "initial layer" which then has to be "matched" to the hydrodynamic solution. Presumably this is also the case for $\alpha > 1$, at least within a certain class of distributions. We discuss this and other questions, including extensions of Theorem 1 to the case in which the gas is subject to a suitably scaled "smooth" external force, in Section 5. We also discuss there choices of initial conditions which lead to variable density IE and to some "higher-order" Navier-Stokes equations in which the pressure is constant up to order ε^2.

The idea of the proof of (2.5) is similar to that of Caflisch [7]; perform a truncated (Hilbert) series expansion in ε and control the remainder equation. All the estimates of the terms of the expansion and of the remainder are due to the good control of the appropriate hydrodynamic equations in the interval $[0, t_0]$. In the next section we develop the expansion, while Section 4 is devoted to the control of the remainder.

3. The Expansion

We first consider the

Case N = 2. In this case equation (2.1) becomes

$$\partial_t F^\varepsilon + \varepsilon^{-1} v \cdot \nabla F^\varepsilon = \varepsilon^{-2} Q(F^\varepsilon, F^\varepsilon). \tag{3.1}$$

We look for a solution of (3.1) of the form

$$F^\varepsilon = m_\varepsilon + \sum_{n=2}^{8} \varepsilon^n f_n + \varepsilon^4 f_R, \tag{3.2a}$$

where

$$m_\varepsilon(x, t, v) = M(\rho_0(x, t), \varepsilon u(x, t), T_0(x, t); v), \tag{3.2b}$$

for some functions u, ρ_0, T_0 to be determined. The functions f_n will be chosen to coincide with those obtained from the Hilbert expansion while the remainder term f_R will satisfy the equation below. We note that m_ε is such that $Q(m_\varepsilon, m_\varepsilon) = 0$ and that the first correction to m_ε is $O(\varepsilon^2)$.

We write

$$m_\varepsilon = m_0 + \sum_{n=1}^{k} \varepsilon^n \varphi_n + \varepsilon^{k+1} r_{k+1}, \tag{3.3}$$

where

$$\varphi_n = \frac{1}{n!} \frac{\partial^n}{\partial \varepsilon^n} m_\varepsilon \bigg|_{\varepsilon=0}, \qquad n \geq 1, \tag{3.4}$$

and r_{k+1} is a remainder of order $k + 1$,

$$\varphi_1 = m_0 \frac{u \cdot v}{T_0}, \qquad \varphi_2 = \frac{1}{2} m_0 \frac{(v_i v_j - T_0 \delta_{ij})}{T_0^2} u_i u_j, \quad i, j = 1, \cdots, d. \tag{3.5}$$

In (3.5) and below we sum over all repeated indices.

Substituting (3.2a) and (3.3) in equation (3.1) and equating terms of the same order in ε not containing f_R we get

$$\varepsilon^{-1}: \quad v \cdot \nabla m_0 = 0, \tag{3.6a}$$

$$\varepsilon^{0}: \quad \partial_t m_0 + v \cdot \nabla \varphi_1 = Lf_2, \tag{3.6b}$$

$$\varepsilon^{1}: \quad \partial_t \varphi_1 + v \cdot \nabla (f_2 + \varphi_2) = Lf_3 + L^{(1)} f_2 \tag{3.6c}$$

and, for $2 \leq k \leq 6$,

$$\begin{aligned} \varepsilon^{k}: \quad & \partial_t (f_k + \varphi_k) + v \cdot \nabla (f_{k+1} + \varphi_{k+1}) \\ & = Lf_{k+2} + L^{(1)} f_{k+1} + \sum_{\substack{n,\, m \geq 2 \\ n+m=k+2}} Q(f_n, f_m). \end{aligned} \tag{3.7}$$

Moreover we have the remainder equation

$$\partial_t f_R + \varepsilon^{-1} v \cdot \nabla f_R = \varepsilon^{-2} L_\varepsilon f_R + \widetilde{L_\varepsilon} f_R + \varepsilon^2 Q(f_R, f_R) + \varepsilon^3 A. \tag{3.8}$$

In equations (3.6), (3.7) and (3.8) we have used the following definitions:

$$L_\varepsilon f = 2Q(m_\varepsilon, f), \tag{3.9}$$

$$Lf = 2Q(m_0, f), \tag{3.10}$$

$$L^{(1)} f = 2Q(r_1, f), \tag{3.11}$$

$$\widetilde{L_\varepsilon} f = 2Q\left(\sum_{n=2}^{\infty} \varepsilon^{n-2} f_n, f \right), \tag{3.12}$$

$$\begin{aligned} A = & -\partial_t [(f_7 + \varphi_7) + \varepsilon (f_8 + r_8)] - v \cdot \nabla (f_8 + r_8) + L^{(1)} f_8 \\ & + \sum_{k \geq 9} \varepsilon^{k-9} \sum_{\substack{n,\, m \geq 2 \\ n+m=k}} Q(f_n, f_m). \end{aligned} \tag{3.13}$$

Now we analyze the consequences of equations (3.6) and put $d = 3$: equation (3.6a) implies,

$$\nabla \rho_0 = 0, \qquad \nabla T_0 = 0. \tag{3.14}$$

Equation (3.6b) implies

$$P(\partial_t m_0 + v \cdot \nabla \varphi_1) = 0, \tag{3.15}$$

$$(1 - P)(\partial_t m_0 + v \cdot \nabla \varphi_1) = Lf_2, \tag{3.16}$$

where

(3.17)

$$Pf(v) = \int_{\mathbf{R}^3} f(v')\, dv' + v \int_{\mathbf{R}^3} v' f(v')\, dv' + (v^2 - 3T_0) \int_{\mathbf{R}^3} (v'^2 - 3T_0) f(v')\, dv'.$$

Equation (3.15) thus becomes

$$\int_{\mathbf{R}^3} (\partial_t m_0 + v \cdot \nabla \varphi_1)\, dv = 0, \tag{3.18a}$$

$$\int_{\mathbf{R}^3} v(\partial_t m_0 + v \cdot \nabla \varphi_1)\, dv = 0, \tag{3.18b}$$

$$\int_{\mathbf{R}^3} (v^2 - 3T_0)(\partial_t m_0 + v \cdot \nabla \varphi_1)\, dv = 0. \tag{3.18c}$$

Using (3.14) and the relations

$$\int_{\mathbf{R}^3} m_0\, dv = \rho_0, \quad \int_{\mathbf{R}^3} v_i v_j m_0\, dv = \rho_0 T_0 \delta_{ij}, \quad \int_{\mathbf{R}^3} v_i \varphi_1\, dv = \rho_0 u_i, \tag{3.19}$$

(3.18a) and (3.18c) give

$$\partial_t \rho_0 + \operatorname{div} \rho_0 u = 0, \tag{3.20}$$

$$3\rho_0 \partial_t T_0 + 2 \operatorname{div} T_0 \rho_0 u = 0. \tag{3.21}$$

Integrating equations (3.20) and (3.21) on the volume Π it follows by (3.14) that $\partial_t \rho_0 = 0$ and $\partial_t T_0 = 0$. This, together with (3.14) implies that

$$\rho_0 = \bar{\rho}, \qquad T_0 = \bar{T}, \tag{3.22}$$

i.e., m_0 is a global Maxwellian,

$$m_0(x, t; v) = \bar{\rho} g_0 \equiv M(\bar{\rho}, 0, \bar{T}; v). \tag{3.23}$$

Moreover, by (3.22) and (3.20) we must have, for the expansion to be consistent,

$$\operatorname{div} u = 0 \qquad \text{(incompressibility)}. \tag{3.24}$$

Finally (3.18b) is satisfied by the properties of the Gaussian integrals.

Equation (3.16) yields, for all $t \leqq t_0$, including the initial distribution at $t = 0$,

$$\begin{aligned} f_2 = {} & L^{-1}\left(\bar{T}^{-1} g_0\left(v_i v_j - \tfrac{1}{3} v^2 \delta_{ij}\right)\right)\bar{\rho}\partial_j u_i \\ & + g_0\left[\rho^{(2)} + \rho^{(2)} u^{(2)} \cdot v + e^{(2)}\left(v^2 - 3\bar{T}\right)\right], \end{aligned} \tag{3.25}$$

where $\rho^{(k)}$, $u^{(k)}$, $e^{(k)}$ will denote here and below the components of f_k along the collision invariants. Since they are in the null space of L, they are not fixed by equation (3.25) and will be fixed later. We find (see for example [3])

$$L^{-1}\left(g_0\left(v_i v_j - \tfrac{1}{3} v^2 \delta_{ij}\right)\right) = -\Psi(v^2)\left(v_i v_j - \tfrac{1}{3} v^2 \delta_{ij}\right) g_0, \tag{3.26}$$

with Ψ a known positive function. Therefore,

$$\begin{aligned} f_2 = {} & -\frac{\bar{\rho}}{\bar{T}}\Psi(v^2)\left(v_i v_j - \tfrac{1}{3} v^2 \delta_{ij}\right) g_0 \partial_j u_i \\ & + g_0\left[\rho^{(2)} + \rho^{(2)} u^{(2)} \cdot v + e^{(2)}\left(v^2 - 3\bar{T}\right)\right]. \end{aligned} \tag{3.27}$$

The stress tensor and the heat current vector associated to f_k are defined by

$$\pi_{ij}^{(k)} = \int_{\mathbf{R}^3} v_i v_j f_k \, dv, \qquad q_i^{(k)} = \int_{\mathbf{R}^3} v_i \left(v^2 - 3\bar{T}\right) f_k \, dv. \tag{3.28}$$

To compute the ones associated to f_2 we note that

$$\begin{aligned} & \int_{\mathbf{R}^3} dv\, \Psi(v^2)\left(v_i v_j - \tfrac{1}{3} v^2 \delta_{ij}\right) v_k v_l g_0(v) \\ & \quad = \mu(\bar{T})\left(\delta_{ik}\delta_{jl} + \delta_{jk}\delta_{il}\right)\left(1 - \delta_{ij}\right) + \sigma(\bar{T})\delta_{ij}\delta_{kl}, \end{aligned} \tag{3.29}$$

with positive $\mu(\bar{T})$ and $\sigma(\bar{T})$. Moreover,

$$\begin{aligned} & \int_{\mathbf{R}^3} \rho^{(2)} u_i^{(2)} v_i v_k v_l g_0(v)\, dv = 0, \qquad \int_{\mathbf{R}^3} \rho^{(2)} v_k v_l g_0(v)\, dv = \rho^{(2)} \bar{T} \delta_{kl}, \\ & \int_{\mathbf{R}^3} e^{(2)} v_k v_l \left(v^2 - 3\bar{T}\right) g_0(v)\, dv = 2 e^{(2)} \bar{T}^2 \delta_{kl} \equiv \bar{\rho} T^{(2)} \delta_{kl}. \end{aligned} \tag{3.30}$$

Hence, using (3.24),

$$\pi_{ij}^{(2)} = -\tfrac{1}{2}\eta(\partial_i u_j + \partial_j u_i) + p\delta_{ij}, \tag{3.31}$$

where the viscosity coefficient and the correction to the pressure are given by

$$\frac{1}{2}\eta = \frac{\mu(\bar{T})\bar{\rho}}{\bar{T}}, \qquad p = (\rho^{(2)} + 2\bar{T}e^{(2)})\bar{T} = \rho^{(2)}\bar{T} + \bar{\rho}T^{(2)}. \tag{3.32}$$

On the other hand, since

(3.33)

$$\int_{\mathbf{R}^3} dv\, \Psi(v^2)\left(v_i v_j - \tfrac{1}{3}v^2\delta_{ij}\right)v_k v^2 g_0(v) = 0, \qquad \int_{\mathbf{R}^3} dv\, v_k v^2 g_0(v) = 0,$$

and

$$\int_{\mathbf{R}^3} (v^2 - 3\bar{T})v_i v_j g_0(v)\, dv = 2\bar{T}^2\delta_{ij}, \tag{3.34}$$

we have

$$q_i^{(2)} = 2\bar{T}^2\rho^{(2)}u_i^{(2)}. \tag{3.35}$$

Using the relations

$$\int_{\mathbf{R}^3} \varphi_2 v\, dv = 0, \qquad \int_{\mathbf{R}^3} \varphi_2 v(v^2 - 3\bar{T})\, dv = 0, \tag{3.36}$$

and

$$\int_{\mathbf{R}^3} \varphi_2 v_i v_j\, dv = \bar{\rho}u_i u_j, \tag{3.37}$$

we have by (3.6), to order ε,

$$\operatorname{div} \rho^{(2)}u^{(2)} = 0, \tag{3.38}$$

$$\partial_t \bar{\rho}u_1 + \partial_j \pi_{ij}^{(2)} + \partial_j \bar{\rho}u_i u_j = 0, \tag{3.39}$$

$$\begin{aligned} f_3 = {} & L^{-1}\left[(1 - P)(\partial_t \varphi_1 + v \cdot \nabla f_2) - L^{(1)}f_2\right] \\ & + g_0\left[\rho^{(3)} + \rho^{(3)}u^{(3)} \cdot v + e^{(3)}(v^2 - 3\bar{T})\right], \end{aligned} \tag{3.40}$$

with $\rho^{(3)}$, $u^{(3)}$, $e^{(3)}$ to be fixed later. Using the expression (3.31), equation (3.39) becomes the incompressible Navier-Stokes equation.

By (3.7), we have, for $2 \leq k \leq 6$,

$$P(\partial_t(f_k + \varphi_k) + v \cdot \nabla(f_{k+1} + \varphi_{k+1})) = 0, \tag{3.41}$$

$$\begin{aligned} f_{k+2} = L^{-1}\Big\{&(1 - P)[\partial_t(f_k + \varphi_k) + v \cdot \nabla(f_{k+1} + \varphi_{k+1})] \\ &- L^{(1)} f_{k+1} - \sum_{\substack{n, m \geq 2 \\ n+m=k+2}} Q(f_n, f_m)\Big\} \\ &+ g_0\big[\rho^{(k+2)} + \rho^{(k+2)} u^{(k+2)} \cdot v + e^{(k+2)}(v^2 - 3\overline{T})\big]. \end{aligned} \tag{3.42}$$

After equation (3.41) is satisfied, equation (3.42) determines f_{k+2} up to the projection on the space of the collision invariants. On the other hand, we can use the arbitrary functions $\rho^{(k)}$, $u^{(k)}$, $e^{(k)}$ to satisfy equation (3.41). The analysis of the conditions for $\rho^{(k)}$, $u^{(k)}$, $e^{(k)}$ is simple and we omit it: the functions f_k can be determined and will be considered as known functions of x, v, t below. It is a classical result of Grad [2] that they have the same spatial regularity as u and satisfy the bounds

$$|f_k(x, t, v)| \leq \mathscr{P}_k(x, t; v) M_0(v), \tag{3.43}$$

where the $\mathscr{P}_k(x, t; v)$ are some polynomials in v with coefficients depending on x, through $u(x, t)$.

In consequence of the expansion, if u is a solution of the incompressible Navier-Stokes equation and the f_k are chosen as above, equation (3.1) becomes equivalent to equation (3.8). This equation is much simpler than the original one, since the nonlinear term is multiplied by a power of ε. This fact allows then for control of the remainder term.

Case $N > 2$. We use the same procedure as before. We look for a solution of (2.1) of the form

$$F_N^\varepsilon = m_\varepsilon + \sum_{n=0}^{2N+4} \varepsilon^{n+N} f_{n+N} + \varepsilon^{N+2} f_R, \tag{3.44}$$

where

$$m_\varepsilon(x, v, t) = M(\rho_\varepsilon(x, t), u_\varepsilon(x, t), T_\varepsilon(x, t); v). \tag{3.45}$$

We assume that $\rho_\varepsilon(x, t)$ and $T_\varepsilon(x, t)$ have the following expansions:

$$\rho_\varepsilon(x,t) = \bar{\rho} + \sum_{n=2}^{N-1} \rho_n(x,t)\varepsilon^n, \tag{3.46}$$

$$T_\varepsilon(x,t) = \bar{T} + \sum_{n=2}^{N-1} T_n(x,t)\varepsilon^n,$$

and $u_\varepsilon = \varepsilon u$ for $N = 3$ having the expansion

$$u_\varepsilon(x,t) = \varepsilon u + \sum_{n=3}^{N-1} u_n(x,t)\varepsilon^n \tag{3.47}$$

for $N > 3$.

We could have chosen m_ε as in (3.45) also in the previous case: this would have only changed the definitions of the f_k in (3.2) for $k \geq 2$.

We have (see equations (3.4) and (3.5))

$$m_\varepsilon = m_0 + \varepsilon\varphi_1 + \sum_{n=2}^{N+5} \varepsilon^n(\varphi_n + G_n) + \varepsilon^{N+6} r_{N+6}, \tag{3.48}$$

where we denote by $\varphi_n + G_n$ the coefficients of the power series of m_ε, while the φ_n still denote the quantities (3.5). We now proceed as in the case $N = 2$ and consider the equations coming from the orders ε^{-1}, ε^0 and ε^1. The ε^{-1}-order is exactly the same, and therefore from it we get the same informations as before. The ε^0-order gives

$$\partial_t m_0 + v \cdot \nabla_{\varphi_1} = Lf_N. \tag{3.49}$$

Equation (3.49) is equal to (3.6b) and so as before it gives the incompressibility condition and the expression for f_N as in (3.25). The ε^1-order is

$$\partial_t \varphi_1 + v \cdot \nabla(G_2 + \varphi_2) = Lf_{N+1} + L^{(1)} f_N. \tag{3.50}$$

If we multiply by v_j both sides of (3.50) and take the v-integral, we get (see equation (3.37)),

$$\bar{\rho}\partial_t u + \bar{\rho} u \cdot \nabla_x u + \nabla_x p = 0, \qquad p = \bar{T}\rho_2 + \bar{\rho} T_2. \tag{3.51}$$

Equation (3.51) is the incompressible Euler equation. The functions f_n are determined as before and the functions ρ_n, T_n, u_n, $n = 3, \cdots, N-1$, are determined in such a way that the solvability conditions for f_{N+n-1} are satisfied,

$$P[\partial_t(\varphi_{n-1} + G_{n-1}) + v \cdot \nabla(\varphi_n + G_n)] = 0, \quad n = 3, \cdots, N-1. \tag{3.52}$$

Once all the functions appearing in the expansion are determined, we are left with the following equation for the remainder f_R.

Let

$$\widetilde{L_\varepsilon} f = 2Q\left(\sum_{n=2}^{2N+4} \varepsilon^{n-2} f_n, f\right) \tag{3.53}$$

and

$$\begin{aligned} A = & -\partial_t\left[\varphi_{N+5} + G_{N+5} + \varepsilon r_{N+6} + \sum_{n=N+5}^{2N+4} \varepsilon^{n-N-5} f_n\right] \\ & - v\cdot\nabla\left[r_{N+6^+} \sum_{n=N+6}^{2N+4} \varepsilon^{n-N-6} f_n\right] \\ & + L^{(1)} f_{2N+4} + \sum_{k\geqq 2N+5} \varepsilon^{k-2N-5} \sum_{\substack{n,m\geqq N \\ n+m=k}} Q(f_n, f_m). \end{aligned} \tag{3.54}$$

Then f_R must satisfy

$$\frac{\partial f_R}{\partial t} + \varepsilon^{-1} v\cdot\nabla_x f_R = \varepsilon^{-N} L_\varepsilon f_R + \widetilde{L_\varepsilon} f_R + \varepsilon^2 Q(f_R, f_R) + \varepsilon^3 A, \tag{3.55}$$

which is completely analogous to equation (3.8).

4. The Remainder Equation

The study of the remainder equation follows closely the works [7], [9]. Therefore we only sketch the proofs in the case $N = 2$. The main difficulty in the study of equation (3.8) comes from the negative powers of ε. The ε^{-1} in front of $v\cdot\nabla f_R$ may be easily managed, because the free stream operator generates an isometry in all the spaces we consider below, independently of ε. A way to control the ε^{-2} in front of L_ε is to use the nonpositivity of L_ε on $L_2[m_\varepsilon^{-1}(x, t; v)\, dv]$. Following Grad [2], one would consider the equation for $\widetilde{f_R} = (m_\varepsilon)^{-1/2} f_R$. Since m_ε depends on x and t, the equation for $\widetilde{f_R}$ would contain polynomials in velocity which are unbounded for large velocities. This difficulty has been solved by Caflisch [7], by means of a decomposition of f_R in low and high velocity parts. The low velocity part is then controlled by the usual Grad technique, while, to control high velocities, one introduces a global Maxwellian $M^*(v)$ with a temperature T^* larger than $\overline{T}$, in order to gain from it a convergence factor in the velocity space.

More precisely, following Caflisch [7], we look for the solution of equation (3.8) in the form

$$f_R = \sqrt{m_\varepsilon}\, g + \sqrt{M^*}\, h, \tag{4.1}$$

where

$$M^*(v) = M(\bar{\rho}, 0, T^*; v), \qquad T^* > \bar{T}, \tag{4.2}$$

and g and h (the low and high velocity parts, respectively) are defined as the solutions of the following coupled system of equations:

$$(\partial_t + \varepsilon^{-1} v \cdot \nabla) g = \varepsilon^{-2} \mathscr{L}_\varepsilon g + \varepsilon^2 \chi \sigma^{-1} \mathscr{K}^* h, \tag{4.3}$$

$$\begin{aligned}(\partial_t + \varepsilon^{-1} v \cdot \nabla) h = &- \mu \sigma g - \varepsilon^{-2} (\nu_\varepsilon - \bar{\chi} \mathscr{K}^*) h + \widetilde{L}_\varepsilon (\sigma g + h) \\ &+ \varepsilon^2 \nu^* \Gamma^* (\sigma g + h, \sigma g + h) + \varepsilon^3 a,\end{aligned} \tag{4.4}$$

$$g(x, v, 0) = h(x, v, 0) = 0, \tag{4.5}$$

where, given $\gamma > 0$,

$$\chi(v) = \begin{cases} 1, & |v| \leqq \gamma, \\ 0 & \text{otherwise}. \end{cases} \qquad \bar{\chi} = 1 - \chi, \tag{4.6}$$

$$\mu = \tfrac{1}{2} m_\varepsilon^{-1} (\partial_t + \varepsilon^{-1} v \cdot \nabla) m_\varepsilon, \qquad \sigma = \frac{m_\varepsilon}{M^*}, \tag{4.7}$$

$$\nu_\varepsilon = \int_{\mathbf{R}^3} dv_1 \int_{(v_1 - v)\cdot n \geqq 0} dn\, (v_1 - v) \cdot n\, m_\varepsilon(x, t, v_1), \tag{4.8}$$

$$\nu^* = \int_{\mathbf{R}^3} dv_1 \int_{(v_1 - v)\cdot n \geqq 0} dn\, (v_1 - v) \cdot n\, M^*(v_1), \tag{4.9}$$

$$\mathscr{L}_\varepsilon f = \left(\sqrt{m_\varepsilon}\right)^{-1} L_\varepsilon\left[\sqrt{m_\varepsilon} f\right] = -\nu_\varepsilon f + \mathscr{K}_\varepsilon f, \tag{4.10}$$

$$\mathscr{L}^* f = \left(\sqrt{M^*}\right)^{-1} L_\varepsilon\left[\sqrt{M^*} f\right] = -\nu^* f + \mathscr{K}^* f, \tag{4.11}$$

$$\widetilde{\mathscr{L}} f = \left(\sqrt{M^*}\right)^{-1} \widetilde{L}_\varepsilon\left[\sqrt{M^*} f\right], \tag{4.12}$$

$$\nu^* \Gamma^*(f, g) = \left(\sqrt{M^*}\right)^{-1} Q\left[\sqrt{M^*} f, \sqrt{M^*} g\right], \tag{4.13}$$

$$a = \frac{A}{\sqrt{M^*}}. \tag{4.14}$$

The term μ in equation (4.4) is unbounded in the velocity, but now it is controlled by the exponentially decaying factor σ.

To study this problem we consider a linear version of the equations (4.3)-(4.5), i.e., we replace $\nu^*\Gamma^*(f, g) + \varepsilon a$ by a given function b. This yields

$$(4.15) \qquad (\partial_t + \varepsilon^{-1} v \cdot \nabla) g = \varepsilon^2 \mathscr{L}_\varepsilon g + \varepsilon^{-2} \chi \sigma^{-1} \mathscr{K}^* h,$$

$$(4.16) \qquad \begin{aligned} &(\partial_t + \varepsilon^{-1} v \cdot \nabla) h \\ &\quad = -\mu \sigma g - \varepsilon^{-2}(\nu_\varepsilon - \bar{\chi} \mathscr{K}^*) h + \tilde{\mathscr{L}}(\sigma g + h) + \varepsilon^2 b, \end{aligned}$$

$$(4.17) \qquad g(x, v, 0) = h(x, v, 0) = 0,$$

Now we have a system quite similar to (6.1)–(6.3) of [7]. The main difference is that the factor ε^{-1} in the right-hand sides of (6.1) and (6.3) of [7] is replaced by ε^{-2} and that the left-hand sides of (4.15) and (4.16) contain the factor ε^{-1}. Moreover, the x variable is three-dimensional instead of one-dimensional. This requires the control of higher derivatives in x, because eventually the nonlinear term has to be bounded using the algebraic properties of $\mathscr{H}_s$ for $s > \frac{1}{2}d$. The ε^{-2} factor in front of $\mathscr{L}_\varepsilon$ is not relevant because the proof is based on the nonpositivity of $\mathscr{L}_\varepsilon$. Finally the factor ε^{-1} in front of $v \cdot \nabla$ does not make any problem as already mentioned (see Appendix). We therefore prove the following

LEMMA. *There is an* $\varepsilon_0 > 0$ *such that any solution to equations* (4.15), (4.16) *and* (4.17) *satisfies the bounds*

$$(4.18) \qquad \|g\|_{js} \leqq \varepsilon^{1/4} \|b/\nu_\varepsilon\|_{j+2s},$$

$$(4.19) \qquad \begin{aligned} &\|h\|_{js} \leqq \varepsilon^{5/4} \|b/\nu_\varepsilon\|_{js}, \\ &\text{for} \quad j > 3 \quad \text{and} \quad s \leqq 2 \quad \text{if} \quad \varepsilon < \varepsilon_0 \end{aligned}$$

The proof of the lemma is almost a repetition of the proof in [7], [9]; it is given in the appendix for the sake of completeness.

Now we assume

$$(4.20) \qquad b = v^*\Gamma^*(\sigma g + h, \sigma g + h) + \varepsilon a.$$

Since in three dimensions the space $\mathscr{H}_s$ is an algebra for $s > \frac{3}{2}$, using the Grad estimate for Γ^*,

$$(4.21) \qquad \|\Gamma^*(f, g)\|_{js} \leqq c \|f\|_{js} \|g\|_{js} \quad \text{for} \quad j > \tfrac{3}{2}, s > \tfrac{3}{2},$$

and the bound (3.43) to estimate a, we get, for $s > \frac{3}{2}$, $j > \frac{3}{2}$,

$$|b/\nu_\varepsilon|_{js} \leqq \left(\|g\|_{0s}^2 + \|h\|_{js}^2 + c\varepsilon\right), \tag{4.22}$$

where we have taken into account the exponentially decaying factor σ. Therefore, by the lemma,

$$\|h\|_{j+2s} \leqq \varepsilon^{5/4}\left(\|g\|_{0s}^2 + \|h\|_{j+2s}^2 + c\varepsilon\right) \tag{4.23}$$

which implies

$$\|h\|_{j+2s} \leqq \|g\|_{0s} + c\varepsilon, \tag{4.24}$$

for ε small enough. Then,

$$\|g\|_{js} \leqq \varepsilon^{1/4}\left(\|g\|_{0s}^2 + \|h\|_{j+2s}^2 + c\varepsilon\right) \leqq 3\varepsilon^{1/4}\left(\|g\|_{js}^2 + c\varepsilon\right) \tag{4.25}$$

and hence the estimate for g. Using this and (4.24) we get the estimate for h,

$$\|h\|_{js} \leqq C\varepsilon, \qquad \|g\|_{js} \leqq C\varepsilon, \tag{4.26}$$

for some constant C.

A standard iteration procedure gives the existence of the solution once the a priori estimate (4.26) has been established. This proves the theorem.

5. Generalizations and Discussion

Our main result, Theorem 1, can be restated, see comment (iii) in Section 2, as follows: Let $F_N^\varepsilon(x, t; v)$ be the solution of the BE (2.4) for the initial distribution

$$F_N^\varepsilon(x, 0; v) = M(\bar{\rho}, \varepsilon u_0(x), \bar{T}; v) + \sum_{n=N}^{2N+4} \varepsilon^n h_n(x, v), \tag{5.1}$$

$$\operatorname{div} u_0 = 0, \tag{5.2}$$

where the $h_n(x, v)$, $\rho_\varepsilon(x, 0)$, $u_\varepsilon(x, 0)$, $T_\varepsilon(x, 0)$ are determined from the initial values of the hydrodynamical variables satisfying (2.1). (E.g., for $N = 2$, $h_2(x, v)$ is given by the right side of (3.27) for $t = 0$ and ρ_ε, u_ε, T_ε are given by initial values of (2.7) for $N = 2$ and by (3.46), (3.47) for $N > 2$.) Then $F_N^\varepsilon(x, t; v)$ satisfies (2.5) for $t \leqq t_0$. General theory assures (or ought to) that this solution is unique.

A natural question then is, what happens if we replace (5.1) by

$$F_N^\varepsilon(x,0;v) = M(\rho_0(x), \varepsilon u_0(x), T_0(x); v) + \varepsilon h_1(x,v) + \sum_{n=2}^{2N+4} \varepsilon^n h_n(x,v), \tag{5.3}$$

and drop (5.2b)? The answer turns out to be rather complicated and more dependent on N in (2.4) than before. Let us first state the result for the case $N \geqq 4$. We assume, instead of (3.44), a solution of the form

$$F_N^\varepsilon = m_\varepsilon + \sum_{n=1}^{2N+4} \varepsilon^n f_n + \varepsilon^{N+2} f_R. \tag{5.4}$$

In view of (3.46) and (3.47) there is no loss of generality in assuming that the f_n are orthogonal to the collision invariants for $n = 1, \cdots, N-1$. The method of Section 3 then allows us to prove that F_N^ε is a solution to the (2.4) if $f_n = 0$ for $n = 1, \cdots, N-2$, and f_{N-1} satisfies the equation

$$v \cdot \nabla m_0 = L f_{N-1}. \tag{5.5}$$

The compatibility condition for (5.5) implies that ρ_0 and T_0 may depend on x, but the pressure $p_0 = \rho_0 T_0$ has to be spatially constant. Therefore one can prove that Theorem 1 is still true, but (2.5) is replaced by

$$\|F_N^\varepsilon - M(\rho_0(x,t), \varepsilon u(x,t), T_0(x,t); v)\| \leqq C\varepsilon^2, \tag{5.6}$$

with ρ_0 and u satisfying

$$\begin{aligned} &\partial_t \rho_0 + u \cdot \nabla \rho_0 = 0, \\ &\operatorname{div} u = 0, \\ &\rho_0(\partial_t u + u \cdot \nabla u) = -\nabla p, \end{aligned} \tag{5.7}$$

which is the IE with nonconstant density. Moreover, T_0 is given by $\rho_0 T_0 = p_0$ for some $p_0 > 0$.

It is clear from (5.5) that f_{N-1} depends on $\nabla \rho_0$. In particular, for $N = 3$, f_2 does not vanish and (3.5) has to be modified accordingly. However, it turns out that f_2 is an odd function of v and the stress tensor $\pi^{(2)}$ vanishes. Therefore, the hydrodynamical equations are the IE with nonconstant density, also in this case. Considering now the case $N = 2$, we find that f_1 does not vanish and (3.21), (3.27) and (3.39) have to be modified. The hydrodynamical equations which can be obtained along the same lines as in Section 3 become more complicated and involve additional terms, containing higher derivatives of ρ_0, which depend on extra transport coefficients. We do not go into details here but simply write them down and then make some remarks:

$$\partial_t \rho_0 + \operatorname{div} \rho_0 u = 0, \tag{5.8a}$$

$$\operatorname{div}\left(\frac{d+2}{d} \rho_0 T_0 u\right) = \partial_i\left(\kappa_{ij}(T_0)\partial_j T_0\right), \tag{5.8b}$$

$$\begin{aligned} &\partial_t \rho_0 u_i + \partial_j\left[p_2 \delta_{ij} + \rho_0 u_i u_j\right] \\ &= \partial_j\left(\mu(T_0)(\partial_i u_j + \partial_j u_i) + \nu(T_0)\delta_{ij} \operatorname{div} u\right) \\ &\quad + \partial_j\left(\alpha_{ijkl}(T_0)\partial_k \partial_l T_0 + \beta_{ijkl}(T_0) u_k \partial_l T_0 + \gamma_{ijkl}\partial_k T_0 \partial_l T_0\right), \end{aligned} \tag{5.8c}$$

$$T_0 = \frac{p_0}{\rho_0}, \tag{5.8d}$$

with $\kappa, \nu, \mu, \alpha, \beta, \gamma$ functions of T_0 which depend on the intermolecular interactions, while p_0 does not depend on x or t.

Remarks. (a) If the density and temperature of the initial distribution are constant up to terms of order ε, then we get the situation discussed in Section 3. The same is true if we consider isothermal or isoentropic flows.

(b) The scaling considered in this paper, for $N = 2$, leaves the new equations invariant. This means that the extra terms cannot be removed by scaling. On the other hand, if one assumes that κ vanishes, then (5.8b) implies

$$\operatorname{div} u = 0.$$

If we assume that ν does not depend on T_0 and that α, β, and γ vanish, we have

$$\operatorname{div} u = 0,$$

$$\partial_t \rho_0 + u \cdot \nabla \rho_0 = 0,$$

$$\rho_0 \partial_t u + \rho_0 u \cdot \nabla u = -\nabla p_2 + \nu \Delta u,$$

which represent the INS with nonconstant density.

There are various directions in which the present paper can or ought to be extended. We discuss some of them here.

5.1. Initial and asymptotic distributions. It would be desirable to prove a theorem of the following type. Given a "smooth" initial distribution $f(r, v, 0)$, the BE (*) will bring f, in times short compared to ε^{-1}, to a local equilibrium state with parameters ρ, u, T satisfying the compressible Euler equations for times $\varepsilon^{-1} < \tau < \varepsilon^{-\alpha}, \alpha > 1$, the IE equations for times $\tau \sim \varepsilon^{-\alpha}$, $1 < \alpha < 2$, and finally the INS for times $\tau \sim \varepsilon^{-2}$ leading eventually as $\tau \to \infty$ to the global Maxwellian distribution in which ρ, T, and u are constants determined by the initial state (u can be taken to be zero by a Galileian transformation).

As already mentioned, a first step in that direction was taken by Lachowitz [9] who extended the Caflisch theorem to the case when the initial distribution is not "too far" from a local Maxwellian. His result for the BE (1.5) with $\alpha = 1$ can be stated as follows:

THEOREM 2 (Caflisch-Lachowitz). *Let*

$$\|F^{\varepsilon}_{(1)}(x,0;v) - M(\rho_0, u_0, T_0; v)\| < \delta,$$

for some δ sufficiently small. Then there is a $\varepsilon_0 > 0$ s.t. for any $\delta' > 0$

$$\|F^{\varepsilon}_{(1)}(x,t;v) - M(\rho, u, T; v)\| \leqq C\varepsilon \quad \textit{for all} \quad \varepsilon < \varepsilon_0,$$

for all $t \in (\delta', t_0]$, where $\rho(x,t)$, $u(x,t)$ and $T(x,t)$ are the smooth solutions of the compressible Euler equations up to time t_0 with initial values $\rho_0(x)$, $u_0(x)$, $T_0(x)$.

We hope that the work in progress by Bardos, Golse and Levermore [18] which utilizes the recent work of di Perna and Lions [19] proving the global existence of weak solutions of the BE will in fact lead to results along the lines described above. For the moment however, it should be emphasized, we have no uniform bound for arbitrary times t even for the restricted class of initial distributions considered here. We expect however and, in cases where the deviation from a global Maxwellian is small, can prove that as $t \to \infty$ the solution of the INS (2.1) and of the BE (2.4) are $u(x,t) \to 0$ and $F^{\varepsilon}(x,t;v) \to M(\bar{\rho}, 0, \tilde{T}; v)$ as $t \to \infty$. Here

$$\tilde{T} = \bar{T} + \frac{1}{d\bar{\rho}}\left\{\int_{\Pi} dx\left[\int dv\, v^2\left[F^{\varepsilon}(x,0;v) - M(\bar{\rho}, \varepsilon u_0(x), \bar{T}; v)\right] + \varepsilon^2\bar{\rho}u_0^2(x)\right]\right\}$$

is determined by the conservation of energy, $\bar{\rho}$ and $\bar{T}$ being the initial density and temperature.

5.2. Fluid subject to external forces. The systems considered so far, approach, as $t \to \infty$, the state of global equilibrium with uniform ρ, u and T. We wish now to consider the more interesting situation in which the hydrodynamical flow is subject to an external force, i.e., when (2.1a) is replaced by

$$\bar{\rho}\partial_t u + \bar{\rho} u \cdot \nabla u = -\nabla p + \eta_N \Delta u + \bar{\rho} b(x), \tag{5.9a}$$

where $b(x)$ is smooth and periodic. (We consider for simplicity only the case $N = 2$, corresponding to the INS.) This corresponds to changing (2.4) to

$$\frac{\partial F^{\varepsilon}}{\partial t} + \varepsilon^{-1} v \cdot \nabla_x F^{\varepsilon} + \varepsilon b \cdot \nabla_v F^{\varepsilon} = \varepsilon^{-2} Q(F^{\varepsilon}, F^{\varepsilon}). \tag{5.9b}$$

Note that in the original Boltzmann equation (*) the force term would be $\varepsilon^3 b(\varepsilon r) \cdot \nabla_v f(r,v,t)$, i.e., the force varies on a macroscopic scale and is very small compared to the interparticle forces in the gas which give rise to the collision term. It is now straightforward to extend Theorem 1 to this case. Equations (3.2a) to (3.2b) are unchanged, while (3.6c), (3.1) and (3.8) are modified in obvious ways. The only difference is in the proof of the lemma (see Section 4) given in the appendix. We need to replace the free motion $(x, v) \to (x + vt, v)$ by the evolution in the force field

$$(x, v) \to \left(x + vt + \int_0^t ds \int_0^s ds' b(x(s')), v + \int_0^t ds\, b(x(s)\right),$$

which of course changes also the various constants in the bounds.

There are two simple special cases which are worth considering:

(a) *The conservative case.* We take $b = -\nabla_x U(x)$. In this case the energy of a gas particle $\frac{1}{2}v^2 + \varepsilon^2 U(x)$ is conserved and the equilibrium state of the gas is $M(\rho_\varepsilon(x), 0, \bar{T}; v)$ with the density $\rho_\varepsilon(x)$ given by

$$\rho_\varepsilon(x) = \bar{\rho} \exp\{-\varepsilon^2 U(x)/2\bar{T}\}. \tag{5.10}$$

We could therefore replace $\bar{\rho}$ by $\rho_\varepsilon(x)$ in (2.5) and state the theorem in terms of this local Maxwellian.

(b) *The uniform case.* We take $b = \bar{b}$ independent of x and consider an initial distribution $F^\varepsilon(x, 0; v) = M(\bar{\rho}, \varepsilon\bar{u}, \bar{T}; v)$, where $\bar{u}$ is a constant (which as already noted can be set equal to zero by a Galileian transformation). The solutions of the INS and of the BE (5.8) and (5.9) are then

$$u(x, t) = \bar{u} + \bar{b}t, \tag{5.11a}$$

$$F^\varepsilon(x, t; v) = M(\bar{\rho}, \varepsilon[\bar{u} + \bar{b}t], \bar{T}; v), \tag{5.11b}$$

which are seen to satisfy (2.5) with $C = 0$.

The interesting case is of course when $b \neq -\nabla U$ but $\int_\Pi b\, dx = 0$. We expect then that (5.8) will have a stationary solution $u_s(x)$ which will be approached asymptotically as $t \to \infty$ starting with any u_0. The gas distribution $F^\varepsilon(x, t; v)$ on the other hand does not approach a stationary state in this case. The reason for this is that the external force continues to do work on the system even in the stationary state which, since the system is finite for $\varepsilon \neq 0$, leads to an ever increasing temperature. This however does not appear at the level of the INS (5.8a) which only describes the organized flow field.

To see this explicitly we consider the case where $u_0(x) = u_s(x)$. Multiplying (5.8a) by $u_s(x)$, integrating over the torus Π using (5.8b), we get

$$\bar{\rho} \int_\Pi (u_s \cdot b)\, dx = \eta \int_\Pi \sum_{\alpha=1}^{d} \left(\frac{\partial u_{s\alpha}}{\partial x_\alpha}\right)^2 dx. \tag{5.12}$$

On the other hand, if we multiply (5.9) by $\frac{1}{2}v^2$ and integrate over both v and x we find

(5.13)

$$\frac{d}{dt}\int dx \int dv\, \tfrac{1}{2}v^2 F^\varepsilon(x,t;v) = \frac{d}{dt}\left[\rho\langle v\rangle^2 + \tfrac{1}{2}d\rho T\right] = \varepsilon\int_{\Pi}\rho\langle v\rangle \cdot b\,dx,$$

so that for $\langle v\rangle = \varepsilon u_s$ the rate of heat production in the system is proportional to ε^2 times the right-hand side of (5.12).

It appears therefore that in order to get a stationary distribution for the positions and velocities of the gas particle in the presence of a stationary macroscopic flow one needs either (i) provide contact with external heat reservoirs, e.g. through contact with walls at fixed temperatures, which will absorb the energy or (ii) consider an infinite system in which any finite heat generation can "dissipate to infinity". An example of case (i) is the study by Bardos, Caflisch and Nicolaenko [20] of the stationary distribution of a gas between two plates kept at different fixed temperatures. An example of the case (ii) is the study by Asano and Ukai [21] of the flow past an obstacle. They prove the existence of stationary distributions which are actually attracting for a certain class of initial distributions.

5.3 Linear Boltzmann and incompressible Stokes equations. Many of the questions raised in the last sections become easier to answer if instead of starting with (*) one starts with the linearized BE for f,

$$\partial_t f + v\cdot\nabla_x f = Lf, \tag{5.14}$$

where

$$Lf = 2Q[m_0, f], \tag{5.15}$$

$m_0 = M(\bar\rho, 0, \bar T; v)$ being the global Maxwellian with density $\bar\rho$ and temperature $\bar T$. Scaling variables like in Section 2 and assuming $N = 2$ we get, instead of (2.4), the equation

$$\frac{\partial F^\varepsilon}{\partial t} + \varepsilon^{-1}v\cdot\nabla_x F^\varepsilon = \varepsilon^{-2}LF^\varepsilon. \tag{5.16}$$

In this case, using the arguments of Section 3, it is quite easy to prove the following

THEOREM 3. *Given a divergence-free field* $u_0 \in H_s(\Pi)$, $s \geqq 2$, $\bar\rho > 0$, $\bar T > 0$, *equation* (5.16) *has a unique solution* $F^\varepsilon(x,t;v) \in L_2[m_0^{-1}(v)\,dv\,dx]$ *satisfying, for any finite* t,

$$\|F^\varepsilon(x,t;v) - M(\bar\rho, \varepsilon u(x,t), \bar T; v)\| < C\varepsilon^2, \tag{5.17}$$

whenever $u(x, t)$ *is a solution in* $\mathscr{H}_s(\Pi)$ *of the Stokes equation*:

$$\bar{\rho}\partial_t u = -\nabla p + \eta\Delta u, \tag{5.18a}$$

$$\operatorname{div} u = 0, \tag{5.18b}$$

$$u(x,0) = u_0(x), \tag{5.18c}$$

and η *is determined as in Theorem* 1.

To prove Theorem 3 it is enough to realize that following the lines of Section 2, equations (3.6) are modified as follows:

$$\varepsilon^{-1}: \quad v \cdot \nabla m_0 = 0, \tag{5.19a}$$

$$\varepsilon^{0}: \quad \partial_t m_0 + v \cdot \nabla\varphi_1 = L(f_2 + \varphi_2), \tag{5.19b}$$

$$\varepsilon^{1}: \quad \partial_t \varphi_1 + v \cdot \nabla(f_2 + \varphi_2) = L(f_3 + \varphi_3). \tag{5.19c}$$

From (5.19b) we get that $f_2 = \bar{f}_2 - \varphi_2$, where $\bar{f}_2$ is equal to the right-hand side of (3.27). Therefore, (5.19c) becomes

$$\partial_t \varphi_1 + v \cdot \nabla\bar{f}_2 = L(f_3 + \varphi_3). \tag{5.20}$$

From (5.20), equations (5.18) easily follow. The equation for the remainder is linear and this allows us to control it by the nonpositivity of the operator L on $L_2[m_0^{-1}(v)\, dv\, dx]$.

Appendix

We use the following estimates proved in [2], [7] and [9]:

$$\|\mathscr{K}_\varepsilon f\|_{js} \leqq c\|f\|_{j-1s}, \quad \|\mathscr{K}^* f\|_{js} \leqq c\|f\|_{j-1s} \quad \text{for} \quad j \geqq 1, s \geqq 0, \tag{A.1}$$

$$\|\mathscr{K}_\varepsilon f\|_{00} \leqq c|||f|||, \quad \|\mathscr{K}^* f\|_{00} \leqq c|||f|||, \tag{A.2}$$

$$0 < \nu_0 \leqq \nu_\varepsilon(x, v, t) \leqq \nu_1(I + |v|)^\beta, 0 < \nu_0 \leqq \nu^*(v) \leqq \nu_1(1 + |v|)^\beta, \quad 0 < \beta \leqq 1, \tag{A.3}$$

$$\left\| \frac{1}{\nu_\varepsilon} \widetilde{\mathscr{L}}_\varepsilon f \right\|_{js} \leqq c\|f\|_{js}, \tag{A.4}$$

$$|\nabla^s \nu_\varepsilon| \leqq c\nu_\varepsilon, \quad \left\| \frac{1}{\nu_\varepsilon}(\nabla^s \mathscr{K}_\varepsilon) f \right\|_{j0} \leqq \varepsilon^s \|f\|_{j+s}. \tag{A.5}$$

In (A.2) the $|||\cdot|||$ denotes L_2-norm in velocities and positions. Estimate (A.4) follows from the bounds (3.43) and (A.5) follow from a direct computation using the fact that the temperature is constant.

We denote by $x + vt$ the translation on the torus Π with velocity v in time t. Of course the Lebesgue measure is invariant under this translation. We write (4.15)–(4.17) in integral form:

$$h(x,v,t) = \int_0^t ds \left\{ \exp\left\{ -\int_s^t ds' \nu_\varepsilon(s') \varepsilon^{-2} \right\} \cdot \left[-\mu\sigma g + \varepsilon^2 \bar{\chi} \mathcal{K}^* h + \widetilde{\mathcal{L}}(\sigma g + h) + \varepsilon^2 b \right] \right\}^{\#}, \tag{A.6}$$

$$g(x,v,t) = \int_0^t ds \left\{ \exp\left\{ -\int_s^t ds' \nu_\varepsilon(s') \varepsilon^{-2} \right\} \cdot \left[\varepsilon^{-2} \mathcal{K}_\varepsilon g + \varepsilon^{-2} \chi \sigma^{-1} \mathcal{K}^* h \right] \right\}^{\#}, \tag{A.7}$$

where the symbol $\{\ \}^{\#}$ means that the functions in braces have to be computed at $(x - (t-s)v\varepsilon^{-1}, v, s)$. Since we compute integral norms in x, the presence of ε^{-1} makes no difference for the invariance of the Lebesgue measure on the torus. Perform the time integral which gives a factor $\varepsilon^2/\nu_\varepsilon$. Using (A.1) (A.4) and choosing γ large enough we get, by (A.6),

$$\|h\|_{j0} \leq \varepsilon^2 \|g\|_{00} + \varepsilon^4 \left\| \frac{1}{\nu_\varepsilon} b \right\|_{j0}. \tag{A.8}$$

Differentiating (4.16) and writing it in integral form we have

$$\nabla h(x,v,t) = \int_0^t ds \left\{ \exp\left\{ -\int_s^t ds' \nu_\varepsilon(s') \varepsilon^{-2} \right\} \cdot \left[-\nabla(\mu\sigma g) + \varepsilon^{-2} (\nabla \nu_\varepsilon) h + \varepsilon^{-2} \bar{\chi} \nabla (\mathcal{K}^* h) + \nabla\left(\widetilde{\mathcal{L}}_\varepsilon(\sigma g + h)\right) + \varepsilon^2 \nabla b \right] \right\}^{\#}. \tag{A.9}$$

Using (A.8) and (A.1), (A.2), (A.4), (A.5) we get, for γ large enough,

$$\|\nabla h\|_{j0} \leq \varepsilon^2 \|g\|_{01} + \varepsilon^4 \left\| \frac{1}{\nu_\varepsilon} b \right\|_{j1}. \tag{A.10}$$

By the same argument higher derivatives may be estimated and we get

$$\|h\|_{js} \leqq \varepsilon^2\|g\|_{0s} + \varepsilon^4\left\|\frac{1}{\nu_\varepsilon}b\right\|_{js}. \tag{A.11}$$

Now we estimate g. Following [2] we take advantage of the bound (A.1) to reduce the estimate of $\|g\|_{js}$ to an L_2 estimate which follows by the nonpositivity of the operator $\mathscr{L}_\varepsilon$. Namely, denoting by $C(\gamma)$ a constant diverging with γ, we have

$$\begin{aligned} \|g\|_{j0} &\leqq \|\mathscr{K}_\varepsilon g\|_{j0} + \|\chi\sigma^{-1}\mathscr{K}^*h\|_{j0} \leqq \|\mathscr{K}_\varepsilon g\|_{j0} + C(\gamma)\|\mathscr{K}^*h\|_{j0} \\ &\leqq \|g\|_{j-10} + C(\gamma)\|h\|_{j-10} \leqq \|g\|_{j-10} + C(\gamma)\left(\varepsilon^2\|g\|_{00} + \varepsilon^4\left\|\frac{1}{\nu_\varepsilon}b\right\|_{j-10}\right) \\ &\leqq \|g\|_{j-10} + \varepsilon^{7/2}\left\|\frac{1}{\nu_\varepsilon}b\right\|_{j-10}, \end{aligned} \tag{A.12}$$

for ε small enough. In the same way, using (A.2) we get

$$\|g\|_{00} \leqq |||g||| + \varepsilon^{7/2}\left\|\frac{1}{\nu_\varepsilon}b\right\|_{00}. \tag{A.13}$$

A recursive argument (Grad hierarchy) then gives

$$\|g\|_{j0} \leqq |||g||| + \varepsilon^{7/2}\left\|\frac{1}{\nu_\varepsilon}b\right\|_{j-10}. \tag{A.14}$$

The L_2 estimate is obtained using integration by part to cancel the streaming operator $\varepsilon^{-1}v\cdot\nabla$, and the nonpositivity of $\mathscr{L}_\varepsilon$ to get

$$\begin{aligned} \partial_t|||g||| &\leqq \varepsilon^{-2}|||\chi\sigma^{-1}\mathscr{K}^*h||| \leqq C(\gamma)\varepsilon^{-2}|||\mathscr{K}^*h||| \leqq C(\gamma)\varepsilon^{-2}|||h||| \leqq C(\gamma)\varepsilon^{-2}\|h\|_{20} \\ &\leqq C(\gamma)\left[\|g\|_{00} + \varepsilon^2\left\|\frac{1}{\nu_\varepsilon}b\right\|_{20}\right] \leqq C(\gamma)\left[|||g||| + \varepsilon^2\left\|\frac{1}{\nu_\varepsilon}b\right\|_{20}\right]. \end{aligned} \tag{A.15}$$

Therefore, by the Gronwall inequality we have

$$|||g||| \leqq \exp\{C(\gamma)t_0\}\varepsilon^2\left\|\frac{1}{\nu_\varepsilon}b\right\|_{20} \leqq \varepsilon^{3/2}\left\|\frac{1}{\nu_\varepsilon}b\right\|_{20}, \tag{A.16}$$

if ε is small enough. Therefore,

$$\|g\|_{j0} \leqq \varepsilon^{3/2}\left\|\frac{1}{\nu_\varepsilon}b\right\|_{j-10}, \qquad j \geqq 3. \tag{A.17}$$

The equation for the derivatives of g can be studied in the same way. Higher norms of b appear in consequence of (A.5).

Acknowledgments. We thank Uriel Frisch whose ideas about scaling the macroscopic velocities in addition to that of space-time scaling was the inspiration of our work. During the final stages of this work we learned that Claude Bardos, Francois Golse and David Levermore were independently engaged in a similar but more ambitious project [18]. We benefited from several discussions with them for which we thank them. We also thank Russel E. Caflisch, Errico Presutti and Mario Pulvirenti for many helpful discussions and suggestions.

ADM and RE thank the Math Dept. at Rutgers University, where this work was started, for its very kind hospitality.

Part of the work for this paper was done at the Institut des Hautes Etude Scientifiques, Bures-sur-Yvette, France. It was supported in part by AFOSR Grant No. 86-0010, CNR-GNFM, CNR-PSAITM and by Applied Mathematical Sciences subprogram of the Office of Energy Research, U.S.D.O.E. under Contract No. DE-FG02-88ER25053.

Bibliography

[1] Chapman, S., and Cowling, T. G., *The Mathematical Theory of Non-Uniform Gases*, 3d ed., Cambridge University Press, 1970.

[2] Grad, A., *Asymptotic theory of the Boltzmann equation*, in *Rarified Gas Dynamics* II, Paris, 1962, pp. 22–59.

[3] Cercignani, C., *The Boltzmann Equation and Its Applications*, Springer-Verlag, New York, 1988.

[4] DeMasi, A., Ianiro, N., Pellegrinotti, A., and Presutti, E., *A survey of the hydrodynamical behaviour of interacting particle systems*, Studies in Statistical Mechanics, Vol. 11, J. L. Lebowitz and E. W. Montrol, eds., North-Holland, 1984.

[5] Hilbert, D., *Begründung der kinetischen Gastheorie* Math. Ann. 22, 1912, p. 561–577.

[6] Enskorg, D., *The Kinetic Theory of Phenomena in Fairly Rare Gas*, Dissertation Upsala, 1917.

[7] Caflisch, R., *The fluid dynamic limit of the nonlinear Boltzmann equation*, Comm. Pure Appl. Math. 33, 1980, pp. 651–666.

[8] Caflisch, R., and Papanicolaou, G., *The fluid-dynamical limit of a nonlinear model Boltzmann equation*, Comm. Pure Appl. Math. 32, 1979, pp. 589–616.

[9] Lachowitz, M., *On the initial layer and the existence theorem for the nonlinear Boltzmann equation*, Math. Appl. Sci. 9, 1987, pp. 342–366.

[10] Bardos, C., *Une interpretation des relations entre les équations de Boltzmann, de Navier-Stokes et d'Euler à l'aide d'entropie*, Mat. Aplic. Comp. 6, 1987, pp. 97–117.

[11] Bardos, C., and Golse, F., C. R. Acad. Sci. Paris, t299, Serie I, n. 7, 1984, pp. 225–228.

[12] Ellis, R., and Pinsky, M., J., *Limit theorems for model Boltzmann equations with several conserved quantities*, Math. Pure Appl. 54, 1975, pp. 125–156.

[13] Lebowitz, J. L., Presutti, E., and Spohn, H., *Microscopic models of hydrodynamic behavior*, J. Stat. Phys. 51, 1988, pp. 841–862.

[14] Frish, U., *Lectures on Turbulence and Lattice Gas Hydrodynamics*, Summer School on Turbulence, NCAR, Boulder, Colorado, 1987.

[15] Majda, A., *Compressible Fluid Flow and Systems of Conservation Laws in Several Space Variables*, Springer-Verlag, New York, 1984.

[16] Kato, T., J. *Nonstationary flows of viscous and ideal fluids in R^3*, Func. Anal. 9, 1972, p. 296.

[17] Ellis, R., *Chapman-Enskog-Hilbert expansion for a Markovian model of the Boltzmann equations*, Comm. Pure Appl. Math. 26, 1973, pp. 327–359.

[18] Bardos, C., Golse, F., and Levermore, D., *Macroscopic limits of the kinetic equations*, Compte. Rendu., to appear.

[19] DiPerna, R. J., and Lions, P. L., *On the Cauchy problem for the Boltzmann equation: Global existence and weak stability results*, Annals of Math., to appear.
DiPerna, R. J., and Lions, P. L., *On the Cauchy problem for the Boltzmann equation. Global existence and weak stability*, preprint.

[20] Bardos, C., Caflisch, R., and Nicolaenko, B., *The Milne and Kramers problems for the Boltzmann equation of a hard sphere gas*, Comm. Pure Appl. Math. 39, 1986, pp. 323–352.

[21] Ukai, S., and Asano, K., *Steady solutions of the Boltzmann equation for a gas flow past an obstacle, I. Existence*, Arch. Rat. Mech. Anal. 84, 1983, pp. 248–291.
Ukai, S., *Solutions to the Boltzmann equation*, in *Pattern and Waves—Qualitative Analysis of Nonlinear Differential Equations*, North Holland, Amsterdam, 1986, pp. 37–96.

Received December, 1988.

Existence and Positivity of Solutions of a Fourth-Order Nonlinear PDE Describing Interface Fluctuations

PAVEL M. BLEHER
Institute for Advanced Study

JOEL L. LEBOWITZ
Rutgers University

AND

EUGENE R. SPEER
Institute for Advanced Study

Abstract

We study the partial differential equation

$$w_t = -w_{xxxx} + \left(\frac{w_x^2}{w}\right)_{xx},$$

which arose originally as a scaling limit in the study of interface fluctuations in a certain spin system. In that application x lies in $\mathbb{R}$, but here we study primarily the periodic case $x \in S^1$. We establish existence, uniqueness, and regularity of solutions, locally in time, for positive initial data in $H^1(S^1)$, and prove the existence of several families of Lyapunov functions for the evolution. From the latter we establish a sharp connection between existence globally in time and positivity preservation: if $[0, T^*)$ is a maximal half open interval of existence for a positive solution of the equation, with $T^* < \infty$, then $\lim_{t\to T^*} w(t,\cdot)$ exists in $C^1(S^1)$ but vanishes at some point. We show further that if $T^* > (1+\sqrt{3})/16\pi^2\sqrt{3}$ then $T^* = \infty$ and $\lim_{t\to\infty} w(t,\cdot)$ exists and is constant. We discuss also some explicit solutions and propose a generalization to higher dimensions. © 1994 John Wiley & Sons, Inc.

1. Introduction

In this paper we discuss the evolution of a function $w(t,x)$ under the partial differential equation

$$w_t = -w_{xxxx} + \left(\frac{w_x^2}{w}\right)_{xx}, \tag{1.1}$$

for $t > 0$, subject to an initial condition

$$w(0,x) = g(x). \tag{1.2}$$

Communications on Pure and Applied Mathematics, Vol. XLVII, 923–942 (1994)
 CCC 0010–3640/94/070923–20

We take $x \in X$, where X is either the real line $\mathbb{R}$ or the circle S^1 parameterized by a variable x satisfying $0 \leqq x \leqq 1$. Equation (1.1) may also be written as

$$w_t = -\left(w(\log w)_{xx}\right)_{xx} . \tag{1.3}$$

We will occasionally find it convenient to introduce a new dependent variable by setting $w = y^2$; $y = y(t,x)$ then satisfies the equation

$$y_t = -y_{xxxx} + \frac{y_{xx}^2}{y} . \tag{1.4}$$

Equation (1.1) arose (see [2], [3]) in the study, via a "collective variable approximation," of a family of random variables $\{M_n \mid n = 1,2,\ldots\}$ associated with a certain spin system. Each M_n takes integer values and has probability mass function W_n, that is, $W_n(m) = \Pr\{M_n = m\}$, $m \in \mathbb{Z}$. The functions W_n are determined by a recursion relation (together with the initial condition $W_n(m) = \delta_{mn}$) which we give for completeness in the Appendix but whose details will not concern us here. If for large n we approximate W by a scaled continuous function, $W_n(m) \simeq w(\varepsilon^4 n, \varepsilon m)$, then up to a constant time rescaling the partial differential equation (1.1) is the formal $\varepsilon \to 0$ limit of the recursion relation; see [2].

The recursion for the functions W_n manifestly preserves two characteristic properties of probability mass functions: positivity (either strict positivity or nonnegativity) and normalization ($\sum_{m \in \mathbb{Z}} W_n(m) = 1$). Clearly (1.1) also preserves normalization: $\int_X w\,dx$ is constant (assuming, if $X = \mathbb{R}$, that suitable boundary conditions are imposed). We have not been able to show that (1.1) is positivity preserving, but believe on several grounds that this is probably so. First, it is the scaling limit of a positivity preserving discrete recursion. Second, the form (1.4) of the equation is suggestive: at a small quadratic minimum of y the nonlinear term in (1.1) will be large and positive. Finally, we have solved (1.1) numerically, with $X = S^1$, for several positive initial conditions, and verified positivity preservation in each case. We observe from the numerical solutions, however, that the evolution does not preserve order ($g_1 > g_2$ does not imply $w_1(t,x) > w_2(t,x)$ for all t,x) and does not yield monotonic behavior of the minimum: a quartic minimum initially decreases, while a higher order minimum bifurcates into two smaller minima separated by a relative maximum. In Figure 1 we plot some solutions illustrating this behavior.

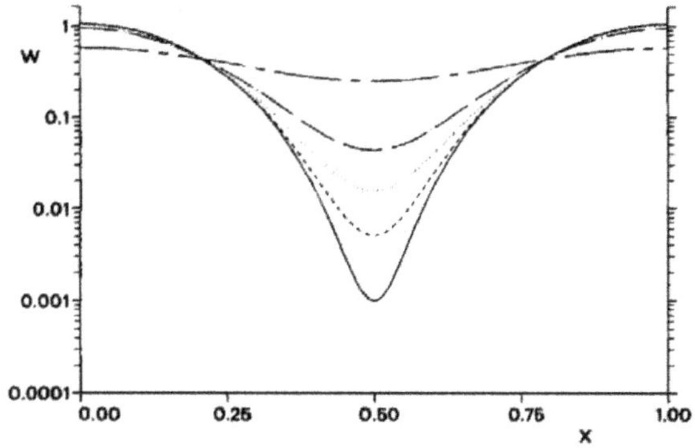

(a)

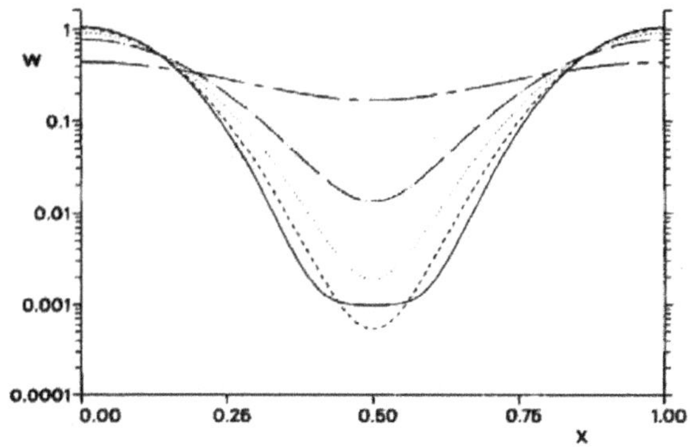

(b)

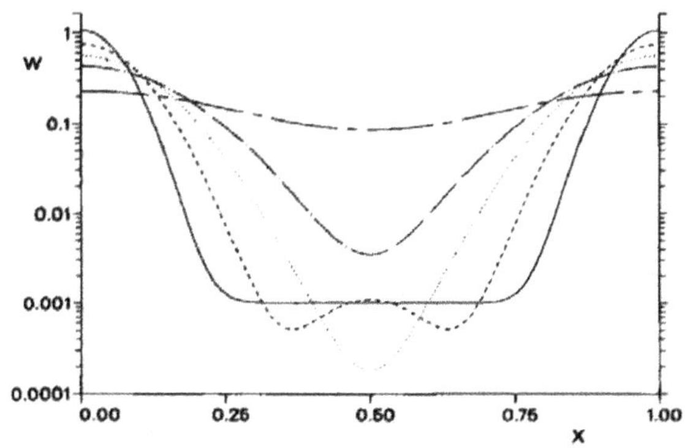

(c)

Figure 1. Solutions of equation (1.1) for three different initial conditions. In each case, $w(t)$ is plotted for $t = 0.0$ (solid), $t = 8.0 \times 10^{-6}$ (dashed), $t = 3.2 \times 10^{-5}$ (dotted), $t = 1.0 \times 10^{-4}$ (dash-dot), and $t = 7.2 \times 10^{-4}$ (dash-dash). Here $w(0, x) = \left(\varepsilon^{1/2} + [(1 + \cos 2\pi x)/2]^m\right)^2$, with $\varepsilon = 0.001$ and $m = 1$, 2, and 8 in Figures 1 (a), (b), and (c), respectively.

Although the scaling limit discussed above leads to (1.1) with $X = \mathbb{R}$, we have established rigorous control of the behavior of solutions only for $X = S^1$. In this case we prove, in Section 4, existence (locally in time) and uniqueness for solutions of (1.1) in the space $H^1(S^1)$, with positive initial conditions; we establish also the regularity of solutions and determine the asymptotic behavior as $t \to \infty$ when the solution exists for all time. Two questions remain unanswered: whether in general solutions must exist globally in time, and whether the evolution is positivity-preserving. We show in Section 5 that these questions are related by proving that the vanishing of the solution is the only possible impediment to global existence: if $[0, T^*)$ is the maximal interval of existence of some positive solution w of the initial value problem, and $T^* < \infty$, then $w(T^*) \equiv \lim_{t \nearrow T^*} w(t)$ exists and is in fact continuously differentiable, but vanishes at some point of S^1.

In Section 2 we discuss briefly a few explicit solutions of (1.1). Control of $\lim_{t \nearrow T^*} w(t)$ is obtained from certain Lyapunov functions for the evolution, which we discuss in Section 3. Finally, Section 6 describes some extensions of our results to a higher dimensional version of (1.1).

Our problem bears some resemblance to one arising in a recent study (see [1]) of droplet breakup in a Hele-Shaw cell. There the thickness h of a neck between two masses of fluid is described by the fourth-order equation $h_t = -(hh_{xxx})_x$, and the questions of global existence and positivity preservation are again closely related.

2. Explicit Solutions

We can readily identify some explicit solutions of (1.1). It should be noted that additional solutions can be derived from these through the invariance of the equation under the transformations

$$\begin{aligned}
&w \to Aw\,, \qquad A \neq 0\,;\\
&x \to x - x_0\,;\\
&t \to t - t_0\,;\\
&x \to ax\,, \quad t \to a^4 t\,, \quad a \neq 0\,.
\end{aligned}$$

The Gaussian solution

$$w(t,x) = t^{-1/4} \exp(-x^2/4t^{1/2})\,, \qquad x \in \mathbb{R}\,,$$

is the relevant solution for the original application to the spin system: the random variables M_n in that case are approximately Gaussian, with variance growing as $n^{1/2}$. A related solution is

$$w(t,x) = t^{-1/4}(x^2/t^{1/2}) \exp(-x^2/4t^{1/2})\,.$$

It was pointed out to us by M. Kruskal, [9], that general scaling solutions of the form $w(t,x) = t^{\alpha}\Psi(xt^{-1/4})$ may be found by requiring $\Psi(\xi)$ to satisfy the ordinary differential equation

$$\Psi^{(4)} - \left(\frac{\Psi'^2}{\Psi}\right)'' - \frac{\xi}{4}\Psi' + \alpha\Psi = 0\,.$$

Stationary solutions include 1, $\sin^2 x$, $\sinh^2 x$, $\cosh^2 x$, e^x, and x^2. A. Rokhlenko, [11], has noted that there are traveling solutions of the form $w(t,x) = u(x+t)$, where u is the square of an Airy function.

3. Lyapunov Functionals

In this section we discuss the existence of certain functionals $\Phi(f)$, where f is a function defined on X, such that when w is a solution of (1.1), $\Phi(w(t,\cdot))$ varies monotonically in time or is constant. We will refer to such functionals as *Lyapunov functionals* and *conserved quantities*, respectively. It will be clear from our proofs that this monotonicity can be verified whenever (1.1) holds in a sufficiently strong sense to justify various formal manipulations and, in the case $X = \mathbb{R}$, w and its derivatives vanish sufficiently rapidly at infinity to justify neglect of boundary terms in integration by parts. To make a precise statement, however, we will restrict ourselves to the case $X = S^1$. We begin by establishing some notation to be used here and in the next section.

For $1 \leqq p \leqq \infty$ and r a non-negative integer we let $W^{r,p} \equiv W^{r,p}(S^1)$ denote the space of functions on S^1 with finite norm

$$\|f\|_{W^{r,p}} = \begin{cases} \left[\int_0^1 \sum_{j=0}^{r} |f^{(j)}(x)|^p\,dx\right]^{1/p}, & \text{if } p < \infty\,, \\ \sup_{x\in S^1, 0\leqq j\leqq r} |f^{(j)}(x)|\,, & \text{if } p = \infty\,. \end{cases}$$

In particular, we write $L^p \equiv W^{0,p}$ and $H^r \equiv W^{r,2}$, and let C^r denote the subspace of $W^{r,\infty}$ consisting of r-times continuously differentiable functions. Let Y^r denote either C^r or $W^{r,p}$ for some fixed p. For I an interval of real numbers we let $C^m(I;Y^r)$ denote the set of functions $u : I \to Y^r$ with continuous (strong) derivatives of order m; if I is closed then $C^m(I;Y^r)$ is a Banach space with norm

$$\|u\|_{C^m(I;Y^r)} = \sup_{t\in I, 0\leqq k\leqq m} \left\|u^{(k)}(t)\right\|_{Y^r}\,.$$

We may omit the superscript m when $m = 0$. We let Y^r_+ denote the set of elements of Y^r which are (pointwise) strictly positive and $C^m_+(I;Y^r)$ the elements of $C^m(I;Y^r)$ with images contained in Y^r_+. Finally, if I is an open interval of real numbers, a *classical solution* of (1.1) in Y^r is an element of $w \in C^1_+(I;Y^r) \cap C(I;Y^{r+4})$ satisfying (1.1).

PROPOSITION 3.1. *If Y^r is as above and w is a (positive) classical solution of* (1.1) *in Y^0, then*

(a) $\int_{S^1} w(t,x)^\beta\, dx$ *is nondecreasing in time if* $0 < \beta < 1$, *is constant if* $\beta = 1$, *and is nonincreasing if* $1 < \beta < 3/2$; *moreover,*

(b) $\int_{S^1} \log(w(t,x))\, dx$ *and* $-\int_{S^1} w \log(w(t,x))\, dx$ *are nondecreasing in time.*

If $w(t,x)$ *is also a classical solution of* (1.1) *in* Y^1, *then*

(c) $\int_{S^1} \left[w_x(t,x)^2/w(t,x)\right]^\alpha\, dx$ *is nonincreasing in time if* $1 \leqq \alpha \leqq 3/2$.

Proof: Each functional $\Phi(w(t))$ referred to in Proposition 3.1 is of the form $\Phi(w(t)) = \int_{S^1} \phi(w(t,x), w_x(t,x))\, dx$, and the strong differentiability condition, together with one integration by parts in case (c), guarantees that $\Phi(w(t))$ is differentiable and satisfies

$$\frac{d}{dt}\Phi(w(t)) = \int_{S^1} \frac{\delta\phi}{\delta w}(w, w_x) w_t\, dx\,, \tag{3.1}$$

where $\delta\phi/\delta w = \partial\phi/\partial w - (\partial\phi/\partial w_x)_x$. Applying this to $\phi(w) = w^\beta$, using (1.1), and integrating twice by parts yields

$$\begin{aligned}\frac{d}{dt}\int_{S^1} w^\beta\, dx &= -\beta \int_{S^1} (w^{\beta-1})_{xx} \left(w_{xx} - \frac{w_x^2}{w}\right) dx \\ &= -\beta(\beta-1)\int_{S^1} w^{\beta-2}\left(w_{xx}^2 + (\beta-3)\frac{w_{xx}w_x^2}{w} - (\beta-2)\frac{w_x^4}{w^2}\right) dx\,.\end{aligned} \tag{3.2}$$

If to (3.2) we add $\beta(\beta-1)$ times the equation

$$0 = \frac{\beta-3}{9}\int_{S^1} \left(w_x^3 w^{\beta-3}\right)_x dx = \frac{\beta-3}{9}\int_{S^1} w^{\beta-2}\left(3\frac{w_{xx}w_x^2}{w} + (\beta-3)\frac{w_x^4}{w^2}\right) dx\,,$$

we arrive at

$$\begin{aligned}&\frac{d}{dt}\int_{S^1} w^\beta\, dx \\ &\quad = -\beta(\beta-1)\int_{S^1} w^{\beta-2}\left(w_{xx}^2 + \frac{2}{3}(\beta-3)\frac{w_{xx}w_x^2}{w} - \frac{1}{9}(\beta^2+3\beta-9)\frac{w_x^4}{w^2}\right) dx\,;\end{aligned}$$

the integrand is a quadratic form in the variables w_{xx} and w_x^2/w which is positive semi-definite for $0 \leqq \beta \leqq 3/2$. Part (a) of Proposition 3.1 follows at once.

The monotonicity of $\int_{S^1} \log w\, dx$ follows from the remark that the time derivative of this quantity is obtained from (3.2) by omitting the factor β and then setting $\beta = 0$; from this point on, the argument is a special case of the one above. Monotonicity of $\int_{S^1} w \log w\, dx$ is obtained easily by using the form (1.3) of w_t in (3.2).

To verify (c) it is most convenient to work in terms of the variable $y = w^{1/2}$; since w is strictly positive, y will be a strong solution of (1.4) in Y^1. Note that $w_x^2/w = y_x^2$, so that from (1.4) and the analogue of (3.1),

$$\begin{aligned}
\frac{d}{dt}\int_{S^1}(y_x^2)^\alpha\,dx
&= -2\alpha\int_{S^1}((y_x^2)^{\alpha-1}y_x)_x\left(-y_{xxxx}+\frac{y_{xx}^2}{y}\right)dx\\
&= -2\alpha\int_{S^1}\left[((y_x^2)^{\alpha-1}y_x)_{xx}\,y_{xxx}-(y_x^2)^{\alpha-1}y_x\left(\frac{y_{xx}^2}{y}\right)_x\right]dx\\
&= -2\alpha\int_{S^1}(y_x^2)^{\alpha-1}\left[(2\alpha-1)y_{xxx}^2-2y_{xxx}\frac{y_{xx}y_x}{y}+\left(\frac{y_{xx}y_x}{y}\right)^2\right]dx\\
&\quad -4\alpha(2\alpha-1)(\alpha-1)\int_{S^1}(y_x^2)^{\alpha-2}y_xy_{xx}^2y_{xxx}\,dx\,.
\end{aligned}$$

The integrand of the first term is, for $\alpha \geqq 1$, a positive semi-definite quadratic form in the variables y_{xxx} and $y_{xx}y_x/y$; the second term, after one more integration by parts, becomes

$$\frac{4}{3}\alpha(2\alpha-1)(\alpha-1)(2\alpha-3)\int_{S^1}(y_x^2)^{\alpha-2}y_{xx}^4\,dx\,,$$

which is non-positive for $1 \leqq \alpha \leqq 3/2$.

Remark 3.2. The origin of the of the evolution equation (1.1), discussed in the Introduction, and the exact conservation of $\int_X w\,dx$, make it natural to view w as a probability density function. In this case two of the Lyapunov functionals of Proposition 3.1 are familiar: $-\int_X w\log w\,dx$ is the *entropy* and $\int_X(w_x^2/w)\,dx$ the *Fisher information* of the probability distribution.

4. The Evolution on S^1

We now turn to the existence theory for the initial value problem (1.1)–(1.2) posed in $X = S^1$. As discussed in Remark 4.1 below, we will assume throughout this section that the initial value g in (1.2) lies in H_+^1. Equations (1.1)–(1.2) are formally equivalent to the integral equation

$$w(t) = e^{-At}g + \int_0^t e^{-A(t-s)}F(w(s))_{xx}\,ds\,, \tag{4.1}$$

where $A = \partial_x^4$ and $F(f) = f_x^2/f$, which may be written as

$$w(t) = e^{-At}g + \partial_x^2\int_0^t e^{-A(t-s)}F(w(s))\,ds\,. \tag{4.2}$$

By a *mild* solution of the initial value problem (1.1)–(1.2) on an interval $[0,T]$ we shall mean an element of $C_+([0,T];H^1)$ which satisfies (4.2); we write $T^* = T^*(g)$ when $[0,T^*)$ is the maximal half-open interval such that a mild solution exists on every closed subinterval $[0,T] \subset [0,T^*)$.

Remark 4.1. e^{-At} is an analytic semigroup on L^2 and (1.1) may be written as

$$\frac{dw}{dt} + Aw = \tilde{F}(w) ,$$

where $\tilde{F}$ maps H^3_+ to H^0. Since H^3 is the domain of $A^{3/4}$, standard theory (see [5] and [10]) assures us that for initial data $g \in H^3_+$ the integral equation (4.1) has a unique solution on some interval $[0,T]$, which is in fact a classical solution in H^0. In our discussion in Section 5 of the relation between global existence and strict positivity of the solution, however, we will need to know the existence of solutions for initial data with only one derivative — specifically, for data in C^1. In this section we take $g \in H^1$ since this slight additional generality does not complicate the proofs. It has been pointed out to us by T. Kato, [8], that solutions in fact exist for $g \in W^{1,1}$, but we will not write down this generalization. Our method is closely related to that of [5] and [10] and also to that of [6] and [4]; see also [7].

The next theorem and its corollary summarize our results.

THEOREM 4.2. *Suppose that $g \in H^1_+$. Then the following hold.*

(a) LOCAL EXISTENCE AND UNIQUENESS OF A MILD SOLUTION: *For some $T > 0$ there exists a unique $w \in C_+([0,T];H^1)$ satisfying* (4.2).

(b) REGULARITY OF MILD SOLUTIONS: *If $w \in C_+([0,T];H^1)$ satisfies* (4.2) *then $w \in C((0,T];H^r)$ for all r; moreover, w is a classical solution of* (1.1) *in H^r for any r.*

COROLLARY 4.3. LYAPUNOV FUNCTIONALS: *If $g \in H^1_+$ and $w \in C_+([0,T];H^1)$ is a mild solution of the initial value problem* (1.1)–(1.2), *then the* (*monotonicity*) *conclusions of Proposition* 3.1 *hold on the interval* $(0,T^*(g))$.

We begin our discussion with a series of lemmas, closely related to standard tools in the theory of analytic semigroups. In stating them we introduce the usual fractional powers A^ρ of the operator A. (For a general definition of A^ρ when A is the generator of an analytic semigroup; see, e.g., [10], here A^ρ is given by multiplication of the k-th Fourier coefficient by $(2\pi|k|)^{4\rho}$.)

LEMMA 4.4. *Suppose that $1 \leqq p \leqq q \leqq \infty$, that j is a non-negative integer, and that $\rho \geqq 0$. Then for $t > 0$,*

$$\|A^\rho \partial_x^j e^{-At} f\|_{L^q} \leqq C t^{-(4\rho+j+1/p-1/q)/4} \|f\|_{L^p} . \tag{4.3}$$

Moreover, if $4\rho + j + 1/p - 1/q < 4$ *and* $u \in C([0,T];L^p)$, *then*

$$\|A^\rho \partial_x^j \int_0^t e^{-A(t-s)t} u(s)\,ds\|_{L^q} \leqq C\|u\|_{C([0,T];L^p)} t^{1-(4\rho+j+1/p-1/q)/4} \,. \tag{4.4}$$

Here C *is a constant depending on* ρ, j, p, *and* q.

Proof: Equation (4.4) is an immediate consequence of (4.3). To prove the latter we first suppose that $\rho = 0$. The operator e^{-At} has kernel $G(x-y;t)$ with

$$G(\xi;t) = \sum_{k\in\mathbb{Z}} t^{-1/4} H\left(\frac{\xi + k}{t^{1/4}}\right),$$

where

$$H(\xi) = \frac{1}{2\pi}\int_{\mathbb{R}} e^{-p^4} e^{ip\xi}\,dp\,.$$

Thus $\|\partial_\xi^j G(\xi;t)\|_{L^r} = t^{-(j+1-1/r)/4}\|H\|_{L^r(\mathbb{R})}$, and (4.3) follows from Young's inequality by taking $1/r + 1/p = 1 + 1/q$. The case $p = q$, $j = 0$ of (4.3) is a consequence (see [10]) of the fact that e^{-At} is an analytic semigroup on L^p, and the general result then follows by writing $A^\rho \partial_x^j e^{-At} = A^\rho e^{-At/2}\partial_x^j e^{-At/2}$.

In stating the next lemmas we will use the following notation. For Y any Banach space of functions on S^1 and $0 < \rho < 1$ we let $C^\rho(I;Y)$ denote the elements of $C(I;Y)$ which are Hölder continuous of order ρ on I, and $C^{m+\rho}(I;Y)$ the elements u of $C^m(I;Y)$ such that $u^{(m)} \in C^\rho(I;Y)$.

LEMMA 4.5. *Suppose that* $u \in C([0,T];L^p)$ *for some* p, $1 \leqq p \leqq \infty$, *and that* $z(t) = \partial_x^2 \int_0^t e^{-A(t-s)} u(s)\,ds$. *Then* $z \in C^\rho([0,T];W^{1,q})$, *with Hölder constant depending only on* ρ, p, q, *and* $\|u\|_{C([0,T];L^p)}$, *whenever* ρ *and* q *satisfy* $p \leqq q \leqq \infty$ *and* $4\rho + 1/p - 1/q < 1$.

Proof: This is a standard argument from analytic semigroup theory; we recall the estimate (e.g., [10])

$$\|(e^{-At} - I)f\|_{L^q} \leqq Ct^\rho \|A^\rho f\|_{L^q}\,, \tag{4.5}$$

valid when $0 < \rho \leqq 1$ and $f \in L^q$ is in the domain of A^ρ. Now for $t, \tau > 0$ and $t + \tau \leqq T$ we write

$$\begin{aligned} z(t+\tau) - z(t) &= (e^{-A\tau} - 1)\partial_x^2 \int_0^t e^{-A(t-s)} u(s)\,ds \\ &\quad + \partial_x^2 \int_t^{t+\tau} e^{-A(t+\tau-s)} u(s)\,ds \\ &\equiv z_1 + z_2\,. \end{aligned}$$

From (4.5) and (4.4),

$$\|z_{1,x}(t)\|_{L^q} \leqq C_T\rho \left\| A^\rho \partial_x^3 \int_0^t e^{-A(t-s)} u(s)\, ds \right\|_{L^q}$$
$$\leqq C_T \rho T^{1-(4\rho+1/p-1/q)/4} \|u\|_{C([0,T];L^p)} ,$$

and from (4.4),

$$\|z_{2,x}(t)\|_{L^q} \leqq C_T^{(1-1/p+1/q)/4} \|u\|_{C([0,T];L^p)} .$$

LEMMA 4.6. *Suppose that $u \in C^\rho([0,T];L^p)$ for some ρ, p satisfying $0 < \rho < 1$ and $1 \leqq p \leqq \infty$, and that $z(t) = \int_0^t e^{-A(t-s)} u(s)\, ds$. Then $z \in C^{1+\rho}((0,T];L^p)$, $\partial_x^4 z \in C^\rho((0,T];L^p)$, and $z' = -\partial_x^4 z + u$.*

Proof: This is a slight generalization of Lemma 2.14 of [4]; the proof in this form is easily obtained from, for example, Theorems 4.2.4, 4.3.4, and 4.3.5 of [10].

In the next lemma we record some one-dimensional Sobolev estimates; constants are given explicitly since they will be used in Section 5 to derive sufficient conditions for global existence. Here, and throughout the remainder of the paper, we adopt the notation $\bar{f} = \int_{S^1} f\, dx$ for any $f \in L^1$.

LEMMA 4.7. *If $f \in W^{1,1}$ then $f \in C^0$; moreover, if $\bar{f} = 0$ then*

$$\|f\|_{C^0} \leqq \frac{1}{2} \|f_x\|_{L^1} , \tag{4.6}$$

$$\|f\|_{C^0} \leqq \frac{1}{2\sqrt{3}} \|f_x\|_{L^2} , \tag{4.7}$$

$$\|f\|_{L^2} \leqq \frac{1}{2\pi} \|f_x\|_{L^2} . \tag{4.8}$$

Proof: The first statement is well known. Suppose then that $\bar{f} = 0$; we may assume that $f(0) = 0$, so that

$$|f(x)| = \frac{1}{2} \left| \int_0^x f_x\, dx + \int_x^1 f_x\, dx \right| \leqq \frac{1}{2} \|f_x\|_{L^1} .$$

Moreover, $\bar{f} = 0$ and $\|f_x\|_{L^2} < \infty$ imply that $f' \in E^\perp$, where $E \subset L^2$ is the subspace generated by 1 and x, so that

$$|f(x)| = \left| \int_0^x f'\, dx \right| \leqq \|P\chi_x\|_{L^2} \|f_x\|_{L^2} ,$$

with P the orthogonal projection onto $E^\perp$ and χ_x the characteristic function of the interval $[0,x]$. The estimate (4.7) follows from $\sup_{x\in S^1}\|P\chi_x\| = 1/2\sqrt{3}$. Finally, (4.8) is obtained immediately from expansion in Fourier series. We remark that the constants in (4.6)–(4.8) are easily seen to be the best possible.

Finally, we give two lemmas on the behavior of the non-linearity in (1.1).

LEMMA 4.8. *Let* $S \equiv \{f \in W^{r,p} \mid \bar{f} = a, f(x) \geqq \nu, \|f\|_{W^{r,p}} \leqq M\}$ *for some* $a, \nu, M > 0$. *Then*

(a) *If* $r = 1$ *and* $p \geqq 2$ *then* F *is a bounded, Lipschitz mapping from* S *to* $L^{p/2}$;

(b) *If* $r \geqq 2$ *then* F *is a bounded, Lipschitz mapping from* S *to* $W^{r-1,p}$.

Proof: (a) For $f, h \in S$,

$$\|F(f)\|_{L^{p/2}} = \left\|\frac{f_x^2}{f}\right\|_{L^{p/2}} \leqq \frac{M^2}{\nu} \tag{4.9}$$

and

$$\begin{aligned}\|F(f) - F(h)\|_{L^{p/2}} &= \left\|\frac{(f_x + h_x)(f_x - h_x)}{f} - \frac{(h_x)^2(f-h)}{fh}\right\|_{L^{p/2}} \\ &\leqq \left(\frac{2M}{\nu} + \frac{M^2}{2\sqrt{3}\nu^2}\right)\|f_x - h_x\|_{L^p}\,,\end{aligned} \tag{4.10}$$

where we have used estimate (4.7).

(b) It suffices to give bounds for $\|\partial_x^{r-1}F(f)\|_{L^p}$ and $\|\partial_x^{r-1}(F(f)-F(h))\|_{L^p}$, for $f, h \in S$. Now $\partial_x^{r-1}F(f)$ is a sum of terms of the form

$$\frac{\partial_x^{k_1}f\cdots\partial_x^{k_m}f}{f^{m-1}}\,, \tag{4.11}$$

where $2 \leqq m \leqq r+1$, $1 \leqq k_j \leqq r$, $\sum_j k_j = r+1$. The estimates proceed as in (a), using $\|\partial_x^k f\|_{C^0} \leqq M$ and, from (4.6), $\|\partial_x^k(f-h)\|_{C^0} \leqq (1/2)\|\partial_x^{k+1}(f-h)\|_{L^{p/2}}$, for $k < r$.

LEMMA 4.9. *Suppose that* $w \in C_+([0,T];H^1)$. *If* $w_x \in C^\rho((0,T];L^4)$ *then* $F(w) \in C^\rho((0,T];L^2)$; *if* $\partial_x^r w \in C^\rho((0,T];L^2)$ *for some* $r \geqq 2$, *then* $\partial_x^{r-1}F(w) \in C^\rho((0,T];L^2)$.

Proof: This is an immediate consequence of Lemma 4.8.

We can now give the proofs of our main results.

Proof of Theorem 4.2 (a): Local existence and uniqueness. Define $v(t) = e^{-At}g$ and let $\mu = \inf_{x \in S^1} g(x)$; because e^{-At} is continuous on H^1 we may by (4.6) choose T_1 so that $v(t) \geqq \mu/2$ for $t \in [0, T_1]$. Now for positive $T \leqq T_1$ let R be the set of functions in $C([0,T]; H^1)$ satisfying $\overline{u(t)} \equiv \overline{g}$ and $\|u_x - v_x\|_{C([0,T];L^2)} \leqq \alpha$, where $\alpha = \sqrt{3}/2\mu$; from (4.7), $u \in R$ implies that $u(t) \geqq \mu/4$, $t \in [0,T]$. Thus for $u \in R$ we may define $B(u)$ by

$$B(u)(t) = \partial_x^2 \int_0^t e^{-A(t-s)} F(u(s))\, ds\ ,$$

and set $K(u) \equiv v + B(u)$. Fixed points of K yield mild solutions of the initial value problem.

We will show that, when T is suitably restricted, K maps R into R and is a contraction in the metric $\rho(u,w) = \|u_x - w_x\|_{L^2}$ on R; thus, K has a unique fixed point on R. Clearly $\overline{K(u)(t)} = \overline{v(t)} = \overline{g}$. We estimate $\|K(u)_x - v_x\|_{L^2}$ and $\|K(u)_x - K(w)_x\|_{L^2}$ from (4.4); applying (4.9) and (4.10) with $p = 2$, $\nu = \mu/4$, and $M = (\|g_x\|_{L^2} + \alpha)$ we see that it suffices to require that

$$\frac{4M^2CT^{1/8}}{\mu} \leqq \alpha \tag{4.12}$$

and

$$\frac{8CM(1 + M/\sqrt{3}\mu)T^{1/8}}{\mu} \leqq 1\ , \tag{4.13}$$

with C a constant from (4.4). Equations (4.12) and (4.13) will be satisfied if T is chosen sufficiently small.

Proof of Theorem 4.2 (b): Regularity of mild solutions. We first prove that $w_x \in C^\rho((0,T]; L^q)$ whenever $1 \leqq q < \infty$ and $0 < \rho < 1/4q$. Write

$$w(t) = v(t) + B(w)(t) \equiv e^{-At}g + \partial_x^2 \int_0^t e^{-A(t-s)} F(w(s))\, ds \tag{4.14}$$

as usual. Certainly $v \in C^k((0,T]; W^{r,q})$ for any k, r and q. On the other hand, since $w \in C_+([0,T]; H^1)$, Lemma 4.8 (a) implies that $F(w) \in C([0,T]; L^1)$, and Lemma 4.5 that $B(w) \in C^\rho([0,T]; W^{1,q})$.

We next show that $w \in C^\rho((0,T]; H^2)$; again it suffices to verify this for $B(w)$. By the result above and Lemma 4.9, $F(w) \in C^\rho((0,T]; L^2)$ and hence, since

$$\partial_x^2 B(w)(t) = \partial_x^4 \int_0^t e^{-A(t-s)} F(w(s))\, ds\ ,$$

Lemma 4.6 implies that $B(w) \in C^\rho((0,T];H^2)$.

We now verify inductively that $w \in C^\rho((0,T];H^r)$ for $r \geqq 3$. Fix $\varepsilon > 0$ and for $t > \varepsilon$ write

$$\partial_x^{r+1} w = \partial_x^{r+1} e^{-A(t-\varepsilon)} w(\varepsilon) + \partial_x^4 \int_\varepsilon^t e^{-A(t-s)} \partial_x^{r-1} F(w(s))\, ds\,.$$

The induction hypothesis and Lemma 4.9 imply that $\partial_x^{r-1}F(w) \in C^\rho((0,T];L^2)$, so that again Lemma 4.6 implies that $w \in C^\rho((0,T];H^r)$.

Finally, for any $r \geqq 0$ we may write

$$\partial_x^r w = \partial_x^r e^{-A(t-\varepsilon)} w(\varepsilon) + \int_\varepsilon^t e^{-A(t-s)} \partial_x^{r+2} F(w(s))\, ds\,.$$

The last statement of Lemma 4.6 now implies that $\partial_x^r w$ is differentiable in L^2 and satisfies $(\partial_x^r w)' = \partial_x^{r+4} w + \partial_x^{r+2} F(w)$, so that w is a classical solution of (1.1) in H^r.

Proof of Corollary 4.3: Lyapunov functionals. This is an immediate consequence of Proposition 3.1 and part (b) of Theorem 4.2. We remark also that all of the functionals referred to in Proposition 3.1, with the exception of $\int_{S^1} (w_x^2/w)^\alpha\, dx$ for $\alpha > 1$, are defined also on the half-closed interval $[0, T^*)$ and, by continuity, vary monotonically there.

5. Relation Between Positivity Preservation and Global Existence

As indicated in the Introduction, equation (1.1) arose from a positivity preserving discrete scheme. While we have not been able to prove either positivity preservation or global existence in time for the partial differential equation, we can give a sharp connection between the two.

THEOREM 5.1. *Suppose that* $g \in H^1_+$ *and that* $w \in C_+([0, T^*(g)); H^1)$ *is a mild solution of the initial value problem* (1.1)–(1.2) *defined on a maximal half-open interval. If* $T^*(g) < \infty$, *then* $\lim_{t \nearrow T^*} w(t)$ *exists in* C^1, *but the limiting function vanishes at at least one point of* S^1.

Proof: Suppose that $T^* \equiv T^*(g) < \infty$. Fix ε with $0 < \varepsilon < T^*$; we claim that $w(t)$ is uniformly Hölder continuous in C^1 on $[\varepsilon, T^*)$. Then for any sequence of times $t_n \nearrow T^*$, $\{w(t_n)\}$ is Cauchy in C^1; this implies the existence of $w(T^*) \equiv \lim_{t \nearrow T^*} w(t)$ in C^1. If $w(T^*)$ were strictly positive then the solution could be extended by Theorem 4.2 (a), so $w(T^*)$ must vanish at at least one point of S^1.

To verify the claim we write

$$w(t) = e^{-A(t-\varepsilon)} w(\varepsilon) + \partial_x^2 \int_\varepsilon^t e^{-A(t-s)} F(w(s))\, ds\,. \tag{5.1}$$

The first term is as regular as we wish. On the other hand, we know by Corollary 4.3 and Proposition 3.1 (c) that for $1 \leqq p \leqq 3/2$, $F(w(t)) \in L^p$ on $[\varepsilon, T^*)$, with $\|F(w(t))\|_{L^p}$ non-increasing. Thus for $\varepsilon < T < T^*$, $\|F(w)\|_{C([\varepsilon,T];L^p)}$ is finite and independent of T. We take $p > 1$ and apply Lemma 4.5 to conclude that the second term is uniformly Hölder continuous in $W^{1,\infty}$ on $[\varepsilon, T^*)$. Since Theorem 4.2 (b) implies that $w \in C([\varepsilon, T^*); C^1)$, w must also be uniformly Hölder continuous in C^1.

The fact that positivity of solutions suffices for global existence enables us to give several sufficient conditions for such existence.

THEOREM 5.2. *Suppose that $g \in H^1_+$. Then $T^*(g) = \infty$ and $\lim_{t\to\infty} \|w(t) - \bar{g}\|_{H^1} = 0$ if any of the following three conditions holds:*

$$\text{(i)} \quad \int_{S^1} \frac{g_x^2}{g}\,dx < 16\bar{g}\,, \quad \text{(ii)} \quad \|g_x\|_{L^2} < 4\bar{g}/\sqrt{3}\,, \quad \text{(iii)} \quad T^*(g) > \frac{(1+\sqrt{3})}{16\pi^2\sqrt{3}}\,.$$

Proof: Let w be a solution of the initial value problem (4.2) on $[0, T^*)$ as above. We introduce the variable $y(t,x) = \sqrt{w(t,x)}$ and define the quantity $\psi(t)$ by

$$\psi^2(t) = \frac{1}{\bar{g}} \int_{S^1} y_x^2(t)\,dx\,.$$

By Proposition 3.1 (c) and the remark in the proof of Corollary 4.3,

$$\psi(t) \leqq \psi(0) = \frac{1}{2\bar{g}^{1/2}} \left[\int_{S^1} \frac{g_x^2}{g}\,dx\right]^{1/2}, \tag{5.2}$$

for all $t < T^*$. Let $y_{\max}(t)$ and $y_{\min}(t)$ denote respectively the maximum and minimum values of y. From the argument leading to (4.6), and the Schwarz inequality,

$$y_{\max}(t) - y_{\min}(t) \leqq \frac{1}{2}\|y_x(t)\|_{L^1} < \frac{\bar{g}^{1/2}}{2}\psi(t) \leqq \frac{\bar{g}^{1/2}}{2}\psi(0)\,; \tag{5.3}$$

since certainly $y_{\max} \geqq \bar{g}^{1/2}$, (5.3) yields $y_{\min}(t) > \bar{g}^{1/2}(1 - \psi(0)/2)$. Now condition (i) of the theorem implies that this lower bound is strictly positive and hence, by Theorem 5.1, that $T^*(g) = \infty$. On the other hand, if condition (ii) holds, then $g > \bar{g} - \|g - \bar{g}\|_{C^0} \geqq \bar{g}/3$ from (4.7) and hence $\int g_x^2/g\,dx \leqq (3/\bar{g})\|g_x\|_{L^2}^2 < 16\bar{g}$, so that global existence follows from condition (i).

We next introduce the quantity

$$\phi(t) \equiv \frac{1}{\bar{g}^{1/2}} \int_{S^1} \frac{y_{xx}^2(t)}{y(t)}\, dx\,.$$

which by Theorem 4.2 (b) is finite for $t > 0$. From (4.7) we have

$$y(t) \leqq \bar{y}(t) + \frac{1}{2\sqrt{3}} \|y_x(t)\|_{L^2} \leqq \bar{g}^{1/2} \left[1 + \frac{\psi(t)}{2\sqrt{3}} \right].$$

and hence, from (4.8),

$$\psi(t)^2 = \frac{1}{\bar{g}} \int_{S^1} y_x^2(t)\, dx \leqq \frac{1}{4\pi^2 \bar{g}} \int_{S^1} y_{xx}^2(t)\, dx \leqq \frac{1}{4\pi^2} \phi(t) \left[1 + \frac{\psi(t)}{2\sqrt{3}} \right]. \tag{5.4}$$

Equation (5.4) implies that $\psi(t) \leqq G(\phi(t))$, where the function G is continuous and increasing on $[0, \infty)$ and satisfies $G(0) = 0$, $G(\phi_0) = 2$ for $\phi_0 = 16\pi^2\sqrt{3}/(1 + \sqrt{3})$. On the other hand, from (1.4) and an argument as in the proof of Proposition 3.1, or directly from the proof of that proposition, we find that

$$\frac{d}{dt} \int_{S^1} y(t)\, dx = \bar{g}^{1/2} \phi(t)\,. \tag{5.5}$$

Since $\int_{S^1} y(t)\, dx \leqq \|y(t)\|_{L^2} = \bar{g}^{1/2}$, (5.5) implies that for any $T < T^*$, $\int_0^T \phi(s)\, ds \leqq 1$: in particular, for some $t_0 \in [0, T]$, $\phi(t_0) \leqq 1/T$ and hence

$$\psi(t_0) \leqq G(1/T)\,. \tag{5.6}$$

Now suppose that condition (iii) holds. Then we may choose T so that $\phi_0^{-1} < T < T^*$, and (5.6) implies that $\psi(t_0) \leqq 2$, so that $T^* = \infty$ by condition (i) applied on the interval $[t_0, \infty)$. Finally, if $T^* = \infty$ we may take T arbitrarily large in (5.6), so that, from the monotonicity of $\psi(t)$ and the condition $G(0) = 0$, $\lim_{t\to\infty} \psi(t) = 0$, which implies that $w(t) \to \bar{g}$ in H^1.

6. Higher Dimensions

Among possible generalizations of (1.1) to higher dimensions the most interesting appears to be

$$w_t = -\Delta^2 w + \partial_i \partial_j \left(\frac{\partial_i w \partial_j w}{w} \right). \tag{6.1}$$

or equivalently

$$w_t = -\partial_i\partial_j(w\partial_i\partial_j \log w) \ . \tag{6.2}$$

Here $x \in X$, with X taken to be Euclidean space $\mathbb{R}^d$ or the d-torus $(S^1)^d$, and we have adopted the convention of summing over repeated indices and written $\partial_i \equiv \partial/\partial x_i$, $\Delta \equiv \partial_i\partial_i$. We choose (6.2) as our generalization of (1.3), rather than the apparently more natural equation $w_t = -\Delta(w\Delta \log w)$, for two reasons. First, the resulting modification of the equation (1.4) for the variable $y = \sqrt{w}$,

$$y_t = -\Delta^2 y + \frac{(\Delta y)^2}{y} \ , \tag{6.3}$$

has a simple form which again suggests the possible presence of a mechanism to preserve positivity of solutions. Second, as we will see shortly, (6.2) implies that certain of the functionals of Proposition 3.1 — specifically, the integrals of w and y, the entropy, and the Fisher information — are again Lyapunov functionals.

In this section we discuss briefly the extent to which our one-dimensional results generalize to the equation (6.1). As indicated above, some of the monotonicity results of Proposition 3.1 remain valid; moreover, we can give a result on existence, uniqueness, and regularity which is quite similar to Theorem 4.2. It is of course still necessary to control the denominator of the non-linear term in (6.1) by controlling the uniform norm of w; for $d > 1$, this (and other considerations in the proof) seem to necessitate working in a more restrictive space than H^1. On the other hand, the monotonic behavior of the Fisher information gives control only on the H^1 norm of the solution. As a result, we are not able to establish any connection between global existence and positivity preservation which would generalize Theorem 5.1.

We take $X = (S^1)^d$ throughout the remainder of this section and adapt to this case, without further elaboration, the notations for function spaces and the definitions of mild and classical solutions introduced earlier. In particular, a mild solution of (1.1) with initial condition $w(0) = g$ is a solution of the integral equation

$$w(t) = e^{-At}g + \partial_i\partial_j \int_0^t e^{-A(t-s)}F_{ij}(w(s))\,ds \ , \tag{6.4}$$

with $A = \Delta^2$ and $F_{ij}(f) = \partial_i f \partial_j f/f$. We begin with the generalization of Proposition 3.1; it is natural in view of the regularity result to be proved below to state it for spaces of continuous functions.

PROPOSITION 6.1. *If w is a (positive) classical solution of* (1.1) *in $C^0(X)$, then*

(a) *$\int_X w(t,x)\,dx$ is constant in time and $\int_X w(t,x)^{1/2}\,dx$ is nondecreasing; moreover,*

(b) *The entropy* $-\int_X w\log(w(t,x))\,dx$ *is nondecreasing in time.*

If $w(t,x)$ *is also a classical solution of* (1.1) *in* C^1, *then*

(c) *The Fisher information* $\int_X \partial_i w_x(t,x)\partial_i w_x(t,x)/w(t,x)\,dx$ *is nonincreasing in time.*

Proof sketch: As in the proof of Proposition 3.1, these results follow by simple manipulations and appropriate integrations by parts. It is convenient to work in terms of the variable y satisfying (6.3) when studying both $\int_X w(t,x)^{1/2}\,dx$ and the Fisher information.

To obtain existence, uniqueness, and regularity we may, according to the Sobolev inequality, control the denominator of $F(w(t)) = \partial_i w \partial_j w/w$ by taking $w(t)$ to lie in any space $W^{1,p}$ with $p > d$.

THEOREM 6.2. *Suppose that* $g \in W_+^{1,p}$ *for some* p *satisfying* $d < p \leqq \infty$. *Then the following hold.*

(a) LOCAL EXISTENCE AND UNIQUENESS OF A MILD SOLUTION: *For some* $T > 0$ *there exists a unique* $w \in C_+([0,T]; W^{1,p})$ *satisfying* (6.4).

(b) REGULARITY OF MILD SOLUTIONS: *If* $w \in C([0,T]; W^{1,p})$ *is a mild solution of* (6.1) *on* $[0,T]$ *for some* $p > d$ *then, for any integer* r *satisfying* $1 \leqq r \leqq p$, $w \in C((0,T]; W^{r,p/r})$. *In particular, if* $g \in W^{1,\infty}$ *then* $w \in C((0,T]; C^r)$, *and* w *is a classical solution of* (1.1) *in* C^r, *for all* r.

Proof sketch: The proof follows closely the lines of the proof of Theorem 4.2; we begin by noting the necessary changes in the auxiliary lemmas. The estimate (4.3) is replaced by

$$\|A^\rho \partial_x^j e^{-At} f\|_{L^q} \leqq C t^{-(4\rho+j+d/p-d/q)/4} \|f\|_{L^p}\,, \tag{6.5}$$

derived in the same way, and the estimates (4.6)–(4.8) by the Sobolev inequalities

$$\|f\|_{C^0} \leqq C\|f\|_{W^{r,p/r}}\,, \tag{6.6}$$

$$\|f\|_{W^{k,p/k}} \leqq C\|f\|_{W^{r,p/r}}\,, \qquad 1 \leqq k < r\,, \tag{6.7}$$

valid for $p > d$ and $p \geqq r \geqq 1$. It follows from the generalization of (4.11) to the higher dimensional case, under these same restrictions on p and r, that F_{ij} is a bounded, Lipschitz mapping from $\{ f \in W^{r,p} \mid f \geqq \nu > 0, \|f\|_{W^{r,p}} \leqq M < \infty \}$ to $W^{r-1,p/(r+1)}$, and thus that if $w \in C_+^\rho((0,T]; W^{r,p/r})$ then $F_{ij}(w) \in C^\rho((0,T]; W^{r-1,p/(r+1)})$. Finally, Lemma 4.5 is modified only in the condition that ρ must satisfy, which is now $4\rho + d/p - d/q < 1$, and Lemma 4.6 is unchanged.

With this machinery in place the proof of (a) is a straightforward application of the fixed point theorem on a ball in $C([0,T]; W^{1,p})$ with center ν, where again

$v(t) = e^{-At}g$. To verify regularity we first observe that if w is a mild solution on $[0,T]$ in $W^{1,p}$ then $F_{ij}(w) \in C([0,T];L^{p/2})$ and hence $w \in C^{\rho}((0,T];W^{1,p})$ whenever ρ satisfies $4\rho < (1 - d/p)$, and then prove by induction on r that $w \in C^{\rho}((0,T];W^{r,p/r})$ for all $r \leqq p$. If $p = \infty$ then we conclude immediately from the Sobolev inequalities that $w \in C^{\rho}((0,T];C^r)$ for all r, and Lemma 4.6 then implies that w is a classical solution.

Finally, we show that the results of Proposition 6.1 apply to the solutions we have constructed.

COROLLARY 6.3. LYAPUNOV FUNCTIONALS: *If* $g \in W_+^{1,p}$ *for some* $p > d$ *and* $w \in C_+([0,T];W^{1,p})$ *is a mild solution of* (6.1) *with initial value* g *then the conclusions of Proposition* 6.1 *hold on* $[0,T]$.

Proof of Corollary 6.3: If $p = \infty$ then this is an immediate consequence of Proposition 6.1 and part (b) of Theorem 6.2 (in fact, as long as $p \geqq 4$ for cases (a) and (b), and $p \geqq 5$ for case (c), then enough derivatives of w exist to justify directly the manipulations used in verifying Proposition 6.1). For general p we may verify that the solutions are continuous functions of the initial data and then approximate the initial value g by elements of $W^{1,\infty}$; we omit details.

Acknowledgments. We are pleased to thank H. Brezis, T. Kato, M. Kruskal, P. Lax, and A. Rokhlenko for helpful conversations and correspondence, and M. Shelley for providing the computer code used for numerical study of (1.1). The first and second authors were supported in part by National Science Foundation Grant DMR 89–18903. J. L. L. also thanks Peter Lax for hospitality at Courant Institute, where part of this work was done with support from the Applied Mathematical Sciences Program of the United States Department of Energy under Contract DE–FG–02–88ER25053. E. R. S. thanks Tom Spencer for hospitality at the Institute for Advanced Study.

Appendix

In this appendix we record the recursion relations which lead to (1.1) in a scaling limit. Further details are given in [2].

Recall that M_n denotes a random variable taking integer values and that $W_n(m)$ denotes the probability that $M_n = m$. In the recursion for W_n it is helpful to regard M_n as the position, at time n, of a particle undergoing a random walk with a local bias determined by a certain function $H_n(m)$ satisfying $-1 \leqq H_n(m) \leqq +1$:

$$W_{n+1}(m) = \frac{1}{2}[1 + H_n(m-1)]W_n(m-1) + \frac{1}{2}[1 - H_n(m+1)]W_n(m+1)\,. \tag{A.1}$$

The scheme involves also an auxiliary, non-negative, dynamically independent function $U_n(m)$, which satisfies the recursion

$$U_{n+1}(m) = \frac{(W_n(m) + U_n(m+1))(W_n(m) + U_n(m-1))}{U_n(m+1) + 2W_n(m) + U_n(m-1)}, \tag{A.2}$$

with the supplementary provision that $U_{n+1}(m) = 0$ if the denominator in (A.2) vanishes. Finally, the bias H is itself determined, non-recursively, by W and U:

$$H_n(m) = \frac{U_n(m+1) - U_n(m-1)}{U_n(m+1) + 2W_n(m) + U_n(m-1)}, \tag{A.3}$$

where again $H_n(m) = 0$ if the denominator vanishes. Note that if we define $V_n^{\pm}(m) \equiv W_n(m) + U_n(m \pm 1)$ then (A.3) may be rewritten as $H_n(m) = (V_n^+(m) - V_n^-(m))/(V_n^+(m) + V_n^-(m))$, so that certainly $-1 \leqq H_n(m) \leqq +1$, with strict inequalities as long as $W_n(m)$ or the product $U_n(m-1)U_n(m+1)$ is nonvanishing. It is then clear that the recursive scheme is well defined, preserves the normalization condition $\sum_{m \in \mathbb{Z}} W_n(m) = 1$, and preserves both non-negativity and strict positivity of W.

As indicated in the Introduction, (1.1) is obtained from this recursive scheme when we introduce smooth functions $w(t,x)$, $u(t,x)$, and $h(t,x)$, write $W_n(m) \simeq w(\varepsilon^4 n, \varepsilon m)$, $U_n(m) \simeq u(\varepsilon^4 n, \varepsilon m)$, and $H_n(m) \simeq h(\varepsilon^4 n, \varepsilon m)$, and take a formal $\varepsilon \to 0$ limit. (More precisely, the limiting equation thus obtained differs from (1.1) by a constant rescaling of time: $w_t = -(3/8)(w(\log w)_{xx})_{xx}$.) In the scaling limit $u = w + O(\varepsilon^2)$. At first glance (A.1) might suggest a diffusive scaling $t = \varepsilon^2 n$, $x = \varepsilon m$, leading to a second-order PDE, but in fact the bias H as defined by (A.3) introduces additional cancellations which lead to the fourth-order equation (1.1).

Bibliography

[1] Constantin, P., Dupont, T. F., Goldstein, R., Kadanoff, L. P., Shelley, M., and Zhou, S.-M., *Droplet breakup in a model of the Hele-Shaw cell*, Phys. Rev. E 47, 1993, pp. 4169–4181.
[2] Derrida, B., Lebowitz, J. L., Speer, E., and Spohn, H., *Dynamics of an anchored Toom interface*, J. Phys. A 24, 1991, pp. 4805–4834.
[3] Derrida, B., Lebowitz, J. L., Speer, E., and Spohn, H., *Fluctuations of a stationary nonequilibrium interface*, Phys. Rev. Lett. 67, 1991, pp. 165–168.
[4] Fujita, H., and Kato, Y., *On the Navier-Stokes initial value problem*, Arch. Rational Mech. Anal. 16, 1964, pp. 269–315.
[5] Henry, D., *Geometric Theory of Semilinear Parabolic Equations*, Lecture Notes in Mathematics No. 840, Springer-Verlag, New York, 1981.
[6] Kato, Y., and Fujita, H., *On the nonstationary Navier-Stokes system*, Rend. Sem. Mat. Univ. Padova 32, 1964, pp. 243–260.
[7] Kato, T., *Strong L^p-solution of the Navier-Stokes equation in $\mathbb{R}^m$, with applications to weak solutions*, Math. Z. 187, 1984, pp. 471–480.
[8] Kato, T., private communication, 1991.
[9] Kruskal, M., private communication, 1991.
[10] Pazy, A., *Semigroups of Linear Operators and Applications to Linear Equations*, Applied Mathematical Sciences Vol. 44, Springer-Verlag, New York, 1983.

[11] Rokhlenko, A., private communication, 1990.

PAVEL M. BLEHER
Department of Mathematics
Rutgers University
New Brunswick, New Jersey 08903

JOEL L. LEBOWITZ
Departments of Mathematics and Physics
Rutgers University
New Brunswick, New Jersey 08903

EUGENE R. SPEER
Department of Mathematics
Rutgers University
New Brunswick, New Jersey 08903

Received April 1992.

VOLUME 70, NUMBER 15 PHYSICAL REVIEW LETTERS 12 APRIL 1993

Derivation of Ohm's Law in a Deterministic Mechanical Model

N. I. Chernov,[1,2] G. L. Eyink,[3] J. L. Lebowitz,[3] and Ya. G. Sinai[4]

[1] *Center for Dynamical Systems and Nonlinear Studies, Georgia Institute of Technology, Atlanta, Georgia 30332*
[2] *Joint Institute of Nuclear Research, Dubna, P.O. Box 79, Head Post Office, Moscow, Russia*
[3] *Departments of Mathematics and Physics, Rutgers University, New Brunswick, New Jersey 08903*
[4] *L. D. Landau Institute for Theoretical Physics, Russian Academy of Sciences, 117940 GSP-1 Moscow V-334, Russia*

(Received 15 September 1992)

We study the Lorentz gas in small external electric and magnetic fields with the particle kinetic energy held fixed by a Gaussian "thermostat." Starting from any smooth initial density, a unique stationary, ergodic measure is approached for times $t \to \infty$. The steady-state electric current $\mathbf{J}(\mathbf{B},\mathbf{E})$ is given by a Kawasaki formula and the entropy production $\mathbf{J}\cdot\mathbf{E}/T$, with T the "temperature," is equal to both the asymptotic decay rate of the Gibbs entropy and minus the sum of the Lyapunov exponents. The Einstein and Kubo formulas hold, i.e., $\mathbf{J}(\mathbf{B},\mathbf{E}) = \boldsymbol{\sigma}(\mathbf{B})\cdot\mathbf{E}$ + higher order terms, with the diffusion matrix $\mathbf{D}(\mathbf{B})$ at $\mathbf{E}=0$ given by $k_B T$ times the symmetric part $\tilde{\sigma}(\mathbf{B})$ of the conductivity matrix.

PACS numbers: 05.60.+w, 05.45.+b

We consider the Lorentz model of a metallic conductor given by a classical point particle or a noninteracting gas of such particles ("electrons") moving in a periodic array of fixed, hard, convex scatterers ("ions"). Here we study the validity of *Ohm's law* in the presence of applied fields. At present, no general statistical mechanical theory can predict which microscopic dynamics will yield such transport laws, and their derivation has been referred to by Peierls as one of the outstanding unsolved problems of modern physics [1]. In addition to this fundamental interest, our work may also be of relevance to some beautiful experimental systems constructed recently by microfabrication—the so-called "antidots"—which are lattice structures with periodicities of order 10^2 nm [2]. The experimental observation of magnetoresistance in these systems, with Fermi wavelength smaller than the superlattice spacing, has been interpreted with reasonable success in terms of classical dynamics [2].

The Lorentz model, also known as the dispersed billiard, possesses very strong hyperbolic properties which allow the diffusive Brownian motion of (undriven) test particles to be rigorously established in a long length- and time-scale limit [3]. This requires that the scatterer configuration have *finite horizon*, so that any particle travels only a uniformly bounded time freely between collisions. (When there is infinite horizon—as is typically the case for the antidots [1]—the diffusion is expected to be anomalous with the mean-square displacement growing like $t \ln t$ [4].) However, under the influence of an applied electric field, the particles in this model accelerate without any limiting velocity, because there is no dissipative mechanism to absorb the energy put in as work by the electric field. The actual drift appears to be sublinear, with the net displacement in the direction of the field apparently growing like $t^{2/3}$. This was noted by Moran and Hoover [5], who proposed and studied numerically a modification of the Lorentz gas which incorporates a frictional term designed to model the interaction of the test particle with a "heat bath." The friction coefficient is chosen according to Gauss' "principle of least constraint" [6], so that the kinetic energy of the particle is held at a fixed constant value. This is a simple example of a new method in nonequilibrium molecular dynamics (NEMD) which has been developed in the past decade by Evans, Hoover, Morriss, Nosé, and others [7].

To state precisely the model, the motion between collisions is governed by the set of first-order equations

$$\dot{\mathbf{q}} = \mathbf{p}/m\,, \tag{1}$$

$$\dot{\mathbf{p}} = \mathbf{E} + \mathbf{p}\times\mathbf{B} - \zeta\mathbf{p}\,. \tag{2}$$

$\mathbf{q} = (q_1, \ldots, q_d)$ are the Cartesian coordinates of the particle, $\mathbf{p} = (p_1, \ldots, p_d)$ are the corresponding momenta, and we have added to the Moran-Hoover model a Lorentz force produced by an external magnetic field. The "friction coefficient" ζ is chosen as

$$\zeta = \mathbf{E}\cdot(\mathbf{p}/m)/(p^2/m)\,, \tag{3}$$

so that kinetic energy is fixed. Although our work can be generalized to $d>2$, we only discuss here $d=2$. The reduced phase space, at each fixed value of the particle speed v, has coordinates $X = (q_1, q_2, \theta)$, where θ is the angle of the particle velocity vector with respect to the 1 direction. It is easy to verify that the divergence of the dynamical velocity vector $\dot{X}$ is $-\zeta$, so that the Liouville measure $d\mu_0 = d\mathbf{q}\,d\theta$ is not preserved when $E \neq 0$. On the other hand, observe that Eqs. (1)–(3) define a flow on the phase space, running backward as well as forward, and reversible, i.e., starting from an initial phase point $X(0)$ we get $X(t)$ for all t and if one reverses velocities and magnetic fields at $t=\tau$ then $X(t+2\tau)$ is just $X(0)$ with velocities reversed.

It should be emphasized that for a correct application of the general NEMD method one should properly consider an N-particle system with N large. In that situation, holding the total kinetic energy $K = \frac{1}{2}\sum_{i=1}^{N} m_i \dot{\mathbf{q}}_i^2$ fixed should arguably be equivalent to holding the temperature fixed, with the identification $K = N d k_B T/2$.

Equations (1) and (2) would then hold for each $(\mathbf{q}_i,\mathbf{p}_i)$ with $\zeta=\mathbf{E}\cdot\sum_i\mathbf{v}_i/2K$. It is easy to check, with X now representing a point in $(2dN-1)$-dimensional phase space, that

$$\nabla_X\cdot\dot{X}=-\mathbf{E}\cdot\mathbf{J}(X)/k_BT \tag{4}$$

follows, with $\mathbf{J}(X)=\sum_{i=1}^N\mathbf{v}_i\sim N$ and with a correction of relative order unity neglected. Remarkably, the expression $\mathbf{E}\cdot\mathbf{J}(X)$ has the form of "force" times (microscopic) "flux," which is identical to the form of entropy production in phenomenological thermodynamics. It is this crucial relation that, in our opinion, is the important feature of Gaussian dynamics which accounts for its successful representation of the steady state. In our simple example with $N=1$ the identification of "temperature"—according to the above prescription—as the nonfluctuating kinetic energy of a single particle lacks physical justification and, also, fails to satisfy the crucial Eq. (4). Therefore, we here define "temperature" as twice that value, i.e., $k_BT=1/\beta=p^2/m$, just so that the equation $\nabla_X\cdot\dot{X}=-\mathbf{E}\cdot\mathbf{v}/k_BT$ analogous to Eq. (4) still holds.

We now present our results for the case $N=1$, using, however, a slightly different form than that in which they are rigorously obtained [8]. For the mathematical proofs we must consider a discrete-time description of the system in which time is counted by a number of successive collisions of the particle with scatterers, a so-called *special representation* of the flow. It is only in the discrete-time representation that we can presently obtain the good decay of correlations for Eq. (6) or (8) below, for example.

First, we show that starting from any probability distribution $f_0(X)dX$ on the three-dimensional phase space, a unique ergodic measure μ^+ is obtained by evolving forward in time. This result is obtained at least when E and B are sufficiently small and the condition of finite horizon holds for zero field. Furthermore, there is a very simple integral formula for the average $\langle\phi\rangle^+$ of any continuous function $\phi(X)$ in this stationary state. For a formal derivation, note that the average of ϕ at time t—starting from an initial constant ensemble density $f_0(X)=C$—is given by a simple calculation to be

$$\langle\phi\rangle_t=\langle\phi\rangle_0+\beta\mathbf{E}\cdot\int_0^t\langle\mathbf{v}\phi(X_{\mathbf{B},\mathbf{E}}(s))\rangle_0ds\,, \tag{5}$$

where $\langle\phi\rangle_t$ is the average with respect to the ensemble density at time t. For clarity the $\mathbf{B},\mathbf{E}$ dependence of the phase trajectory $X(t)$ under the dynamics is explicitly indicated. Under the stated assumptions, we show that $\langle\phi\rangle_t$ converges to $\langle\phi\rangle^+$ as $t\to+\infty$, and, furthermore, the integrand on the right-hand side of (5) has sufficiently good decay so that the limit $t\to+\infty$ may be taken there. We obtain finally

$$\langle\phi\rangle^+=\langle\phi\rangle_0+\beta\mathbf{E}\cdot\int_0^\infty\langle\mathbf{v}\phi(X_{\mathbf{B},\mathbf{E}}(t))\rangle_0dt\,. \tag{6}$$

While such expressions for steady-state measures have been known for a long time [9], they are generally purely formal since the convergence of the integral is not known.

From Eq. (6) with $\phi=\mathbf{v}$, one obtains a formula for the exact current response as a nonlinear function of field:

$$\mathbf{J}(\mathbf{B},\mathbf{E})\equiv\langle\mathbf{v}\rangle^+=\beta\mathbf{E}\cdot\int_0^\infty\langle\mathbf{v}\mathbf{v}_{\mathbf{B},\mathbf{E}}(t)\rangle_0dt\,. \tag{7}$$

This is an example of a so-called *Kawasaki formula* for the nonlinear response [10] and it describes a variety of phenomena, including the usual *Hall effect* of transverse electric currents in a magnetic field. The response to linear order in the field is then given by

$$\langle\phi\rangle^+=\langle\phi\rangle_0+\beta\mathbf{E}\cdot\int_0^\infty\langle\mathbf{v}\phi(X_{\mathbf{B},\mathbf{0}}(t))\rangle_0dt + \text{higher order terms in } E\,. \tag{8}$$

To prove this rigorously correct we establish a bound on the decay in t of $\langle\mathbf{v}\phi(X_{\mathbf{B},\mathbf{E}}(t))\rangle_0$ uniform in E, which allows standard convergence theorems to be applied to show that the remainder term is really of higher order in E. This is not just a fine point of rigor but is exactly where dynamical properties enter in the derivation of the transport law. From Eq. (8) one obtains directly *Ohm's law* $\mathbf{J}(\mathbf{B},\mathbf{E})=\sigma(\mathbf{B})\cdot\mathbf{E}+$higher order, with the conductivity tensor $\sigma(B)$ given by

$$\sigma(\mathbf{B})=\beta\int_0^\infty\langle\mathbf{v}\mathbf{v}_{\mathbf{B},\mathbf{0}}(t)\rangle_0dt\,, \tag{9}$$

a usual *Kubo formula* for the conductivity [11]. The formal linear response calculations given above can be made quite generally, and suggest that the integrable decay of the Green-Kubo formula (uniformly in the mechanical or thermodynamical force) is the necessary and sufficient condition for validity of the standard linear transport law [12]. Since the diffusion matrix $\mathbf{D}(\mathbf{B})$ describing the mean-square deviation of test particles from their expected motion in zero electric field can be shown to be given also by a Green-Kubo formula

$$\mathbf{D}(\mathbf{B})=\frac{1}{2}\int_{-\infty}^{+\infty}\langle\mathbf{v}\mathbf{v}_{\mathbf{B},\mathbf{0}}(t)\rangle_0dt$$

(generalizing that for $\mathbf{B}=\mathbf{0}$ in [3]), we obtain immediately also a (generalized) *Einstein relation*

$$\tilde{\sigma}(\mathbf{B})=\beta\cdot\mathbf{D}(\mathbf{B})\,, \tag{10}$$

where $\tilde{\sigma}$ denotes the symmetric part of $\boldsymbol{\sigma}$. Note that the β appearing here comes directly from the formal definition of T in (4): That relation is the key.

To prove the above results we show that the dynamical system may be very well approximated by a certain stochastic Markov chain. This approximation is obtained by means of a so-called *Markov sieve* [13]. The method of proof can be viewed as a realization at a sophisticated mathematical level of the "stochastization" of a dynamical particle system by a "coarse graining" of the phase space. Rather than follow the detailed trajectory in the phase space, one may consider just the transition of the phase point from cell to cell of the coarse-graining partition under successive time steps.

In the context of the Gaussian method there is a very remarkable relation, noted in [5], which arises between the time derivative of the Gibbs entropy, the Lyapunov exponents of the dynamics, and physical entropy production; see also [7]. We can rigorously establish this relation [using the definition of temperature which comes from Eq. (4)], as we briefly indicate here. First, with the usual definition of Gibbs entropy, $S(t) = -k_B \int f_t(X) \times \ln f_t(X) dX$, where f_t is the ensemble density at time t. It is a simple calculation that $\dot{S}(t) = k_B \int f_t(X)(\nabla_X \cdot \dot{X}) \times dX = -k_B \langle \zeta \rangle_t$. Using Eq. (4) we can therefore infer that

$$-\dot{S}(t) \to k_B \langle \zeta \rangle^+ = \mathbf{J} \cdot \mathbf{E}/T \tag{11}$$

as $t \to +\infty$, where the right-hand side is just the entropy production due to Ohmic dissipation. This can be understood if one imagines that the Gaussian dynamics model the effect of reservoir elements on the particle system, for which the *total* system, reservoir+particles, obeys the Liouville theorem. Hence, the decrease of particle entropy corresponds to the increase of reservoir entropy, and the latter represents the physical entropy production. This result gives some support, in fact, to the interpretation of Gaussian dynamics as an adequate model for the effects of a heat bath (at least, as discussed above, for large N). It does not constitute a proof that this is so, which would require a derivation of the dynamics in some sense from a microscopic model of a thermal reservoir. However, other more traditional methods, e.g., representation of reservoirs by stochastic boundary conditions, have never been derived either from a microscopic model and all the methods rest upon some intuitive ideas that they should be "adequate" to represent the interaction with a heat bath. It is certainly interesting to compare the different methods with each other, and, as much as possible, with real steady-state flows, to verify that they are physically equivalent and lead to the same results.

On the other hand, the divergence of the dynamical vector field can also be related to the rate of change of Liouville volume in the phase space. It is heuristically obvious that the steady-state expected rate of volume contraction should be equal to the sum of the stable and unstable Lyapunov exponents $\lambda^s_{\mathbf{B},\mathbf{E}} < 0$ and $\lambda^u_{\mathbf{B},\mathbf{E}} > 0$ of the flow, defined and almost surely constant with respect to the ergodic measure μ^+. In our case this can be proved, and we obtain the relation

$$-\dot{S}(t) \to k_B(\lambda^s_{\mathbf{B},\mathbf{E}} + \lambda^u_{\mathbf{B},\mathbf{E}}) \tag{12}$$

for $t \to +\infty$. Putting together the two expressions for the asymptotic decay of Gibbs entropy we obtain finally the relation

$$\mathbf{E} \cdot \mathbf{J}/T = -k_B(\lambda^s_{\mathbf{B},\mathbf{E}} + \lambda^u_{\mathbf{B},\mathbf{E}}) , \tag{13}$$

which directly relates entropy-production and Lyapunov exponents. In the limit $E \to 0$, this yields also a relation between diffusion coefficient and Lyapunov exponents of the Gaussian dynamics with respect to μ^+: $\hat{\mathbf{e}} \cdot \mathbf{D}(\mathbf{B}) \cdot \hat{\mathbf{e}} = \lim_{E \to 0} -k_B T^2(\lambda^s_{\mathbf{B},\mathbf{E}} + \lambda^u_{\mathbf{B},\mathbf{E}})/E^2$ (where $\hat{\mathbf{e}}$ is the unit vector in the direction of $\mathbf{E}$). Such relations are believed to also extend to other more complicated many-particle situations [14].

Since it is known that $\mathbf{D} > 0$, it follows from the relation like Eq. (11) that the Gibbs entropy starting from an initially smooth distribution has an asymptotically constant rate of decrease. Therefore, one expects S in the stationary state μ^+ to be $-\infty$, and, in particular, that μ^+ will be singular with respect to Liouville measure μ_0, i.e., it will have a support with fractal dimension less than 3, which is the dimension of the energy surface [5,7]. In fact, we can show, in agreement with the numerical results [5,15], that this is indeed the case. More precisely, we can show that *Young's formula* for the Hausdorff dimension (D_H) of the measure is valid [16] (see also [17]). This states that

$$D_H(\mu^+) = 1 + h[(\lambda^u)^{-1} - (\lambda^s)^{-1}]$$

in which $h = \lambda^u$ is the dynamical Kolmogorov-Sinai entropy of the flow defined by Eqs. (1)-(3). (The equality $h = \lambda^u$ is called Pesin's formula [17].) Using the relationship between Lyapunov exponents and transport coefficients, it then follows that for small E

$$D_H(\mu^+_{\mathbf{B},\mathbf{E}}) = 3 - \frac{\mathbf{E} \cdot \sigma(\mathbf{B}) \cdot \mathbf{E}/T}{k_B h_{\mathbf{B},0}} + \text{higher order corrections} . \tag{14}$$

In particular, it follows that $D_H(\mu^+) < 3$ when E is small but finite.

Although no rigorous or reliable numerical results are available for many-particle systems, we do *not* expect that this "dimensional reduction" will persist in an extensive sense at least in those systems where the particles are in a state of *local equilibrium*. That is, in the appropriate "hydrodynamic limit" for such systems in which particle number and volume both go to infinity, we expect the ratio of the Hausdorff dimension and phase-space dimension to go to unity, just because the system is then composed of small "local equilibrium regions" in which the distribution is (nearly) given by an absolutely continuous Gibbs measure. It should be noted anyway that the phenomenon of "dimensional reduction" is a property just of the representation of thermal reservoirs by Gaussian dynamics, and can be proved *not* to occur, for example, with suitable stochastic representation of heat baths [18]. On the other hand, we would not be surprised if the dimensional reduction per degree of freedom persists in the infinite volume limit—for the stationary measures of Gaussian-type dynamics—in those situations where local equilibrium breaks sown, e.g., where the velocity distribution is not close to a Maxwellian. This would be the case for current-carrying electrical conductors. For some further discussion, see [19]. The stationary measures μ^+ we construct here, although singular with respect to Liouville measure, have the property that they are smooth in the unstable directions of the flow. The property is clearly evidenced in the computer simulations of the density found from the time sampling of phase points, which shows a concentration along the unstable manifolds; see [5]. This property implies, in particular, the smoothness of the reduced measures for the system, e.g., the positional density or the probability density of the x component of the velocity. In fact, the measure μ^+ is just a particular case of the so-called Sinai-Ruelle-Bowen measures for our hyperbolic system, constructed previously for smooth systems (e.g., see [17]).

All of the formal arguments given above extend to the general N-particle model and lead to the reasonably expected results, so that we expect our theorems hold also in the general case. Unfortunately, because of the lack of strict hyperbolicity for the many-particle case, we cannot rigorously study by our method the small-field perturbation in those models. On the other hand, we have for this model a *rigorous* proof of the validity of linear response theory—despite the fact that the dynamics of individual trajectories are sensitively dependent upon the presence or absence of the field. Therefore, our work can be regarded as a counterexample to some of the objections raised by van Kampen against linear response theory [20]. For a fuller discussion of this point we refer to our longer work [8], in which the proofs of the main propositions announced here are also given.

The authors would like to thank E. G. D. Cohen, D. Evans, W. Hoover, G. Gallavotti, and L.-S. Young for many helpful conversations. N.C., G.E., and J.L. acknowledge Grants No. AF-0010E-91 and No. NSF-DMR-89-18903 for financial support.

[1] R. E. Peierls, in *Theoretical Physics in the Twentieth Century*, edited by M. Fiera and V. F. Weisskopf (Wiley, New York, 1961).
[2] D. Weiss *et al.*, Phys. Rev. Lett. **66**, 2790 (1991); R. Fleischmann *et al.*, Phys. Rev. Lett. **68**, 1367 (1992).
[3] L. A. Bunimovich and Ya. G. Sinai, Commun. Math. Phys. **73**, 247 (1980); **78**, 479 (1981).
[4] P. M. Bleher, J. Stat. Phys. **66**, 315 (1992).
[5] B. Moran and W. Hoover, J. Stat. Phys. **48**, 709 (1987).
[6] K. F. Gauss, J. Reine Angew. Math. **IV**, 232 (1829).
[7] W. G. Hoover, *Computational Statistical Mechanics* (Elsevier, Amsterdam, 1991); D. J. Evans and G. P. Morriss, *Statistical Mechanics of Nonequilibrium Liquids* (Academic, San Diego, 1990).
[8] N. I. Chernov, G. L. Eyink, J. L. Lebowitz, and Y. G. Sinai (to be published).
[9] J. L. Lebowitz, Phys. Rev. **114**, 1192 (1959); J. A. McLennan, Phys. Rev. **115**, 1405 (1959); D. N. Zubarev, Dokl. Akad. Nauk SSSR **140**, 92 (1962) [Sov. Phys. Dokl. **6**, 776 (1962)].
[10] T. Yamada and K. Kawasaki, Prog. Theor. Phys. **38**, 1031 (1967).
[11] M. Toda, R. Kubo, and N. Hashitume, *Non-equilibrium Statistical Mechanics* (Springer, Berlin, 1985).
[12] G. L. Eyink, J. L. Lebowitz, and H. Spohn, in *Chaos/Xoac. Soviet-American Perspectives on Nonlinear Science*, edited by D. K. Campbell (AIP, New York, 1990).
[13] L. A. Bunimovich, Ya. G. Sinai, and N. I. Chernov, Russ. Math. Surv. **46** (1991).
[14] D. J. Evans, E. G. D. Cohen, and G. P. Morriss, Phys. Rev. A **42**, 5990 (1990).
[15] W. G. Hoover, H. A. Posch, and C. G. Hoover, Chaos **2**, 245 (1992).
[16] L.-S. Young, Ergod. Theor. Dyn. Syst. **2**, 109 (1982).
[17] J.-P. Eckman and D. Ruelle, Rev. Mod. Phys. **57**, 617 (1985).
[18] S. Goldstein, C. Kipnis, and N. Ianiro, J. Stat. Phys. **41**, 915 (1985).
[19] G. L. Eyink and J. L. Lebowitz, in *Microscopic Simulations of Complex Hydrodynamic Phenomena* (Plenum, New York, 1992).
[20] N. van Kampen, Phys. Norv. **5**, 279 (1971).

Gaussian Fluctuation in Random Matrices

Ovidiu Costin and Joel L. Lebowitz
Department of Mathematics and Physics, Rutgers University,, New Brunswick, New Jersey 08903
(Received 18 November 1994)

Let $N(L)$ be the number of eigenvalues, in an interval of length L, of a matrix chosen at random from the Gaussian orthogonal, unitary, or symplectic ensembles of $\mathcal{N}$ by $\mathcal{N}$ matrices, in the limit $\mathcal{N} \to \infty$. We prove that $[N(L) - \langle N(L)\rangle]/\sqrt{\ln L}$ has a Gaussian distribution when $L \to \infty$. This theorem, which requires control of all the higher moments of the distribution, elucidates numerical and exact results on chaotic quantum systems and on the statistics of zeros of the Riemann zeta function.

PACS numbers: 05.45.+b, 03.65.-w

Ensembles of $\mathcal{N}$-dimensional Gaussian random matrices (GRM's) with invariances under the orthogonal, unitary, or symplectic groups correspond to the GOE, GUE, and GSE were introduced by Wigner and developed by Porter, Dyson, Mehta, and others [1,2]. Wigner's inspired surmise that the statistics of eigenvalues of these GRM can be used to model the statistical properties of the observed spectra of complex nuclei turned out to be exactly right. There is indeed good agreement between the observers high energy level spacings, pair correlations, and variance or Δ statistics and those calculated analytically from the GRM's in the limit $\mathcal{N} \to \infty$. Moreover, the GRM's have been found to be the very robust "renormalization group fixed points" of a large class of RM's [3] which play an important role in many areas of physics and mathematics [1–7].

In the present work we focus on the large L (long wavelength) behavior of the random variable $N(L)$ giving the number of eigenvalues of a GRM, chosen from any of the Gaussian ensembles, in an interval $(y, y + L)$: We always consider the $\mathcal{N} \to \infty$ limit when the distribution is translation invariant and use units in which the mean spacing is unity. It is well known that the variance of $N(L)$ grows like $\ln L$ as $L \to \infty$. We prove that all the moments of $\xi(L) \equiv [N(L) - L]/\sqrt{\ln L}$ approach for large L of a Gaussian distribution which implies (weak) convergence of $\xi(L)$ to a Gaussian random variable. We shall discuss later the connection of our result with the statistics of energy levels of quantum systems with generic chaotic classical Hamiltonians and of the zeros of the Riemann zeta function [4,5,7].

It is a remarkable fact that the distribution of energy levels of the G(O,U,S)E are given by the Gibbs canonical distribution of the positions of charged point particles on the line interacting via the (two dimensional) repulsive logarithmic Coulomb potential $v(r) = -\ln r$ at reciprocal temperatures $\beta = 1, 2, 4$, respectively [1,2]. The particles with positions x_i, $i = 1, \ldots, \mathcal{N}$, on the real line are confined by a harmonic potential. The total potential energy of the system is

$$V_G(x_1, \ldots, x_{\mathcal{N}}) = \frac{\pi^2}{4\mathcal{N}} \sum_{i=1}^{\mathcal{N}} x_i^2 - \frac{1}{2} \sum_{i \neq j}^{\mathcal{N}} \ln|x_i - x_j|. \quad (1)$$

In the corresponding circular ensembles of Dyson the x_i lie on a circle of length $\mathcal{N}$, and the energy is given by the second term on the right-hand (r.h.s.) of (1) with distance measured in the plane. The canonical Gibbs measures corresponding to Gaussian and circular ensembles become equivalent in the thermodynamic limit $\mathcal{N} \to \infty$, yielding the same k point, $k = 1, 2, 3, \ldots$, correlation functions for all $\beta > 0$ [1,2,8].

These infinite volume correlation functions are known explicitly for the "solvable" cases $\beta = 1, 2, 4$ corresponding to the GRM. Defining as usual $n_j(x_1, \ldots, x_j)$ as the joint density for j-tuples, the corresponding Ursell functions $U_k(x_1, \ldots, x_k)$ [9] are given by

$$U_1(x_1) \equiv n_1(x_1) \equiv 1, \quad (2)$$
$$U_2(x_1, x_2) \equiv n_2(x_1, x_2) - n_1(x_1)n_1(x_2)$$
$$\equiv n_2(x_1 - x_2) - 1,$$
$$U_k(x_1, \ldots, x_k) \equiv \sum_G (-1)^{m-1}(m-1)! \prod_{j=1}^m n_{G_j}(\{x_{G_j}\}),$$

where G is a partition of the indices $(1, 2, \ldots, k)$ into m subgroups $G_1, \ldots, G_m$ and $\{x_{G_j}\}$ are the x_i with indices in G_j. [In the GRM literature, $Y_k = (-1)^{k-1}U_k$ is usually called the kth cluster function.] The integrals $\bar{U}_k$ of U_k over a k-dimensional cube having sides of length L are directly related to the cumulants $C_j(L), j = 1, \ldots, k$ of the random variable $N(L)$, the number of points (or eigenvalues) in an interval of length L, which we shall take for definiteness to be the interval $[-t, t]$, $L \equiv 2t$. Thus

$$\bar{U}_1 = \int_{-t}^{t} U_1(x_1)\,dx_1 = 2t = \langle N(L)\rangle = C_1(L), \quad (3)$$
$$\bar{U}_2 = \int_{-t}^{t}\int_{-t}^{t} dx_1\,dx_2\,U_2(x_1, x_2) \equiv \langle N(N-1)\rangle - \langle N\rangle^2$$
$$= \langle(N - \langle N\rangle^2)\rangle - \langle N\rangle = C_2(L) - C_1(L),$$
$$\bar{U}_3 = C_3(L) - 3C_2(L) + 2C_1(L), \text{ etc.}$$

Using the generating function [9]

$$F(\mu) = \sum_{n=0}^{\infty} E_n(L)e^{n\mu}, \quad (4)$$

with $E_n(L)$ the probability of having exactly n particles in the interval L, we have

$$\ln F(\mu) = \sum_{n=1}^{\infty} \frac{C_n}{n!}\mu^n = \sum_{n=1}^{\infty} \frac{1}{n!}\bar{U}_n(e^{\mu} - 1)^n . \quad (5)$$

This gives

$$C_k = \left(\sum_{j=2}^{k-1} b_{k,j}C_j\right) + (-1)^k(k-1)!C_1 + \bar{U}_k , \quad (6)$$

where

$$b_{n,j} = b_{n-1,j-1} - (n-1)b_{n-1,j}, \quad 2 \le j \le n-1,$$
$$b_{n,n} = -1, \quad n \ge 2.$$

The U_k take on a particularly simple form for the GUE ($\beta = 2$) [1,2],

$$U_k(x_1,\ldots,x_k) = (-1)^{k-1} \sum_{\text{Perm}} \prod_{i=1}^{k} S(x_{i+1} - x_i), \quad (7)$$

where $S(x) = (\pi x)^{-1}\sin \pi x$ and $x_{k+1} = x_1$, so the indices are to be thought of as being on a circle. Using (6) one readily obtains [1,2]

$$C_2(L) = (\ln L)/\pi^2 + O(1). \quad (8)$$

Note that for a system with short range interactions $C_2(L)$ would grow like L but the logarithmic interactions between the (charged) particles induce a sort of local crystalline order reducing the variance to $\log L$. It is this strong correlation which produces level repulsion between the eigenvalues and makes the large scale behavior of the fluctuations far from obvious. Defining now the normalized random variable

$$\eta(L) = (N(L) - L)/\sqrt{(1/\pi^2)\ln L}, \quad (9)$$

the kth cumulant of $\eta(L)$ will be $c_k = C_k/[\log L/\pi^2]^{k/2}$. As is well known (cf [10]), $\eta(L)$ will approach a Gaussian random variable (with mean zero and unit variance) as $L \to \infty$, if and only if all c_k, $k \ge 3$, go to zero, i.e., if $C_k = o([\ln L/\pi^2]^{k/2})$. Using an induction argument based on the recurrence relation (6) and the equality (7), this corresponds to proving that

$$s_k(t) \equiv \int_{-t}^{t}\cdots\int_{-t}^{t} dx_1\cdots dx_k\, S(x_2 - x_1)S(x_3 - x_2)\cdots S(x_1 - x_k) = 2t + o[(\ln t)^{k/2}], \quad k \ge 3. \quad (10)$$

We shall actually prove that $s_k(t) = 2t + O(\ln t)$ which implies that for $k \ge 3$ $C_k(L) = O(\ln L)$; in fact, we believe that for $k \ge 3$ $C_k(L)$ stays bounded as $L \to \infty$, as suggested by the explicit asymptotic evaluation of the integrals,

$$s_3 = 2t - \frac{3}{2\pi^2}\ln t + O(1),$$
$$s_4 = 2t - \frac{11}{6\pi^2}\ln t + O(1), \quad (11)$$

which gives, using (6), that C_3 and C_4 are of $O(1)$.

To prove (10) we make use of the fact [1] that

$$s_k(t) = \sum_{i=1} \lambda_i^k(t) = \mathrm{Tr}\mathbf{S}^k(t), \quad (12)$$

where the $\lambda_i(t)$ are the eigenvalues of the integral operator $\mathbf{S}(t)$, $(\mathbf{S}f)(x) = \int_{-t}^{t} dy\, S(x-y)f(y)$. It is known [1] (and can be easily proven) that the spectrum of $\mathbf{S}$ lies in [0,1]. We can now use an induction argument to prove that $s_k(t) = 2t + O(\ln t)$. This is so for $k = 1, 2$ (also for 3,4), and for $k \ge 2$ we have

$$\mathrm{Tr}\mathbf{S}^{k+1} = \mathrm{Tr}\mathbf{S}^k - \mathrm{Tr}[\mathbf{S}^{k-1}(\mathbf{S} - \mathbf{S}^2)]. \quad (13)$$

The first term is of the desired form by the indication assumption while the terms in the parenthesis are positive operators, and $\| \mathbf{S}^{k-1} \| \le 1$ so it can be taken out of the product yielding $\mathrm{Tr}\mathbf{S}^{k+1} = \mathrm{Tr}\mathbf{S}^k + O(\ln t)$ and the proof is complete.

Our results extend to show that, if we divide up the real line into a union of intervals of length L, let $N_j(L)$ be the particle number in $[jL,(j+1)L$, and set $\eta_j(L) = [N_j(L) - L]/\sqrt{C_2(L)}$, $j \in Z$, then the $\{\eta_j(L)\}$ approach, as $L \to \infty$, jointly Gaussian random variables with mean zero and covariances $\langle \eta_j\eta_k \rangle = \delta_{j,k} - \frac{1}{2}\delta_{j\pm 1,k}$.

To prove the results for the GOE, $\beta = 1$, we use an identity conjectured by Dyson and proved by Gunson [11] (we thank Dyson for pointing this out to us). According to this identity, superimposing two noninteracting Coulomb gases, say, blue and red, in the circular ensemble at reciprocal temperature $\beta = 1$ and then looking only at alternate particles, e.g., at all the odd (or even) ones, yields the distribution at $\beta = 2$. Considering now the number of particles in an interval of length L gives

$$N_{\text{total}}(L) = N^{(1)}_{\text{blue}}(L) + N^{(1)}_{\text{red}}(L) = N^{(2)}_{\text{odd}}(L) + N^{(2)}_{\text{even}}(L)$$
$$= 2N^{(2)}(L) + (0, \pm 1), \quad (14)$$

where the superscripts (1,2) stand for the random variables obtained from the ensembles with $\beta = 1, 2$, and $N_{\text{blue}}, N_{\text{red}}$ are independent. This shows immediately that in the infinite $\mathcal{N}$ limit the variables $N^{(2)}(L)$ and $N^{(1)}(L)$ normalized by the square root of their variances have the same asymptotic behavior. Taking $\langle N^{(1)}_{\text{blue}}(L)\rangle = \langle N^{(1)}_{\text{red}}(L)\rangle = \langle N^{(1)}(L)\rangle = L$, we have

$$\tfrac{1}{2}\langle\{[N^{(1)}_{\text{blue}}(L) - L] + [N^{(1)}_{\text{red}}(L) - L]\}^2\rangle = \langle[N^{(1)}(L) - L]^2\rangle = 2\langle[N^{(1)}(L) - L]^2\rangle \sim \frac{2\ln L}{\pi^2}, \quad (15)$$

giving the well known variance of the GOE [1].

For the GSE $\beta = 4$ we use the equality between statistics of its eigenvalues and the odd eigenvalues of the GOE [1]. This again leads to Gaussian asymptotics with a variance given by $(1/2\pi^2)\log L + O(1)$. It seems very reasonable to expect and one can give strong heuristic arguments, based on the "long wavelength response" of "Coulomb" systems, that the Gaussian nature of the fluctuations, with variances $(2/\pi^2\beta)\log L$, holds for all β. (We thank Jancovici for pointing this out to us; see also [8].)

Using more detailed information on the spectrum of $\mathbf{S}$ (see [12]), it follows that $s_k(t) = 2t - \pi^{-2}\sum_{j=1}^{k-1} j^{-1}\ln t + o(\ln t)$ which using (6) implies that $C_k = o(\ln t)$ for all $k \geq 3$. (We are indebted to Widom for this information.) Widom also noted that our proof does not make any use of the specific form of $\mathbf{S}$. It only uses the property $\mathrm{spec}(S) \in [0, 1]$ and the fact that $\mathrm{Tr}(\mathbf{S} - \mathbf{S}^2) \to \infty$ (as $t \to \infty$). The conclusion therefore holds for a larger class of matrix models [3].

The local statistics of the eigenvalues ϵ_j, $j = 1, 2, \ldots, \infty$, $0 \leq \epsilon_1 \leq \epsilon_2 \leq \cdots$ of a classically chaotic quantum Hamiltonian (CQH), such as the geodesic flow on a (nonarithmetic) surface of constant negative curvature or the Sinai billiard, appear to coincide at high energies with those obtained from the GRM [4]. More precisely, if we consider the energy levels of a generic CQH, suitably scaled so that the mean distance between levels is unity in an interval $(y, y + L)$ then their statistics obtained by letting y vary uniformly in some interval (T_1, T_2), will coincide, for $T_2 \to \infty$, with that obtained from one of the standard GRM ensembles when the matrix size tends to infinity. Our result then predicts a Gaussian distribution of the fluctuations in the number of levels $N(L)$ when $L \to \infty$. A numerical check of this for some CQH will require using energy levels in a (scaled) energy interval L with $1 \ll L \ll (T_2 - T_1)$, T_1 large.

It is also interesting, as emphasized by Berry and co-workers [4], to consider in addition to the local statistics of quantum levels also their global statistics. These correspond, in our context, to fluctuations in the number of levels in an interval (y, x) whose length $L(x) = x - y(x)$ is not fixed but grows with x as x varies in the interval (T_1, T_2) with $T_2 \to \infty$. This includes, in particular, the case $y = 0$, $L(x) = x$ corresponding to the fluctuations in the number of eigenvalues less than x. This quantity, normalized by the square root of its variance, was conjectured in [5] [where it is denoted by $N_{fl}(x)$] to have a Gaussian distribution as $x \to \infty$ for all CQH. If true this would be a general characterization of CQH and distinguish them from integrable systems where it was found rigorously that the global distribution is non-Gaussian [13]. Quite generally, it was shown by Berry [5] that when $L(x) > L_{\max}(x)$, the variance of $N(L(x))$ saturates for $L > L_{\max}$. Berry also found that $N_{fl}^2(x)$ (averaged over some interval containing many eigenvalues but very small compared to x) grows for billiard systems like $\log x$.

As already noted, the distribution of eigenvalues in the GRM is translation invariant, when the matrix size $\mathcal{N}$ goes to infinity, so there is no analog of $L_{\max}$ in our considerations. One can, however, consider fluctuations in $N(L)$ for an interval $L(\mathcal{N})$ which contains a number of eigenvalues small compared to $\mathcal{N}$ but goes to infinity when $\mathcal{N} \to \infty$, e.g., in the circular ensemble we could have $L(N) \sim N^\gamma$, $\gamma < 1$ or even $\alpha\mathcal{N}$, $\alpha \ll 1$. For the Coulomb system with neutralizing background it is also possible to consider semi-infinite systems with various boundary conditions and/or nonuniform background. Some such systems have been considered in [8], and we believe that our results about Gaussian behavior would extend also to these systems which might model some of the saturation features of CQH.

We turn finally to the (nontrivial) zeros of the Riemann zeta function $\zeta(z) = \sum_{n=1}^{\infty} n^{-z}$ which are, according to the Riemann hypothesis, for the form $z_n = \frac{1}{2} + i\gamma_n$. As pointed out by Berry and co-workers [4], there are reasons to expect similarities between the statistics of the γ_n and of the energy levels of CQH. In fact Montgomery [14] proved that the pair correlation function of the γ_n agrees with that of the GUE, Eq. (7). Numerical calculation by Odlyzko [7] give striking evidence that the nearest neighbor level spacing distribution of the γ_n is, for large n, indeed the same as that obtained from the GUE. In a very interesting recent paper Rudnick and Sarnak [7] greatly extended the results of Montgomery by showing that the n-point correlation functions of these zeros converge on a large class of test functions to those of the GUE. Moreover, the normalized global fluctuations in these zeros corresponding to $N_{fl}(x)$ was shown by Selberg [15] to have a Gaussian distribution. The same arguments imply that the local fluctuation in their number in an interval $(y, y + L)$ averaged over (T_1, T_2) and properly scaled become Gaussian when $T_2 \to \infty$, followed by $L \to \infty$ (we are indebted to Sarnak [7] for explaining this to us). The results proved here thus fit completely with the picture of the statistics of Riemann zeros being in the same "universality class" as that of the GUE.

We have benefited much from discussion with F. Dyson, B. Jancovici, M. L. Mehta, and P. Sarnak. We also thank M. V. Berry, E. Brezin, and H. Widom for very useful comments and J. Bolte and F. Steiner for calling our attention to their work prior to publication.

[1] For a collection of original articles and an overview, see *Statistical Theories of Spectra: Fluctuation*, edited by C. E. Porter (Academic, New York, 1965); M. L. Mehta, *Random Matrices* (Academic Press, New York, 1990), 2nd ed.

[2] F. J. Dyson, J. Math. Phys. **3**, 166 (1962); **3**, 1191 (1962); **3**, 1199 (1962); Commun. Math. Phys. **19**, 235 (1970); **47**, 171 (1976).

[3] E. Brezin and A. Zee, Nucl. Phys. **B402**, [FS] 613 (1993); L. Pastur, Lett. Math. Phys. **25**, 259 (1992); (to be published). See also H. Leff, J. Math. Phys. **5**, 756 (1964); **5**, 763 (1964); D. Fox and P. B. Kahn, Phys. Rev. **134**, B1151 (1964).

[4] M. V. Berry, Proc. R. Soc. London A **400**, 229 (1985); *Nonlinearity* **1**, 399 (1988); see also articles by Berry and others in *Proceedings of the Summer School Les Houches Session LII*, edited by M. J. Giannoni, A. Voros, and J. Zinn-Justin (North Holland, Amsterdam, 1991); O. Bohigas, M. J. Giannoni, and C. S. Schmit in *Quantum Chaos and Statistical Nuclear Physics*, edited by T. H. Zelegman and H. Nishioka, Lecture Notes in Physics (Springer-Verlag, Berlin, 1986), Vol. 263; M. C. Gutzwiller, *Chaos in Classical and Quantum Mechanics*, (Springer, New York, 1990); H. Spohn, in *IMA Publication No. 9*, edited by G. Papanicolau (Springer, Berlin, 1987), p. 151.

[5] R. Aurich, J. Bolte, and F. Steiner, Phys. Rev. Lett. **73**, 1356 (1994).

[6] M. Moshe, H. Neuberger, and B. Shapiro, Phys. Rev. Lett. **73**, 1497 (1994); E. Basor, C. A. Tracy, and H. Widom, Phys. Rev. Lett. **69**, 5 (1992); G. Mahoux and M. L. Mehta, J. Phys. I (France) **3**, 697 (1993).

[7] A. M. Odlyzko, AT&T Report, 1989 (unpublished); Z. Rudnick and P. Sarnak, Princeton Report, 1994 (unpublished).

[8] P. J. Forester, Nucl. Phys. **B402**, 709 (1994); **416**, 377 (1994); B. Jancovici and P. J. Forester (to be published).

[9] J. L. Lebowitz and J. K. Percus, J. Math Phys. **4**, 248 (1963).

[10] Compare L. Breiman, *Probability* (Addison-Wesley, Reading, 1968).

[11] J. Gunson, J. Math. Phys **1**, 4 (1962); **1**, 752 (1962).

[12] H. J. Landau and H. Widom, J. Math. Anal. Appl. **77**, 469–481 (1980).

[13] P. Bleher, F. Dyson, and J. L. Lebowitz, Phys. Rev. Lett. **71**, 3047 (1993).

[14] H. L. Montgomery, in *Proceedings of the Symposia in Pure Mathematics*, edited by Harold G. Diamond (AMS, Providence, 1975), Vol. 24, p. 181.

[15] A. Selberg, Arch. Math. Naturvid. **B48**, 89 (1946); *Collected Papers* (Springer, New York, 1991), Vol. II, p. 47.

VOLUME 71, NUMBER 19 PHYSICAL REVIEW LETTERS 8 NOVEMBER 1993

Non-Gaussian Energy Level Statistics for Some Integrable Systems

Pavel M. Bleher,[1,2] Freeman J. Dyson,[1] and Joel L. Lebowitz[2]

[1]*School of Natural Science, Institute for Advanced Study, Princeton, New Jersey 08540*

[2]*Departments of Mathematics and Physics, Rutgers University, New Brunswick, New Jersey 08903*

(Received 30 August 1993)

The number of levels with energy less than E of an integrable quantum system with two degrees of freedom is equal to $\lambda E + sE^{1/4}$, where λ is a constant and s a fluctuating quantity with a non-Gaussian distribution. The probability distribution of s decreases roughly like $\exp(-s^4)$ when s is large. The number of levels between E and $E+z\sqrt{E}$ is equal to $\lambda z\sqrt{E}+rE^{1/4}$ where r is another fluctuating quantity. The distribution of r tends to a Gaussian distribution as $z \to 0$ and oscillates around some limiting non-Gaussian distribution as $z \to \infty$.

PACS numbers: 03.65.Sq, 02.50.−r

The nature of the distribution of quantum energy levels for systems with integrable or chaotic classical Hamiltonians is an active field of study using both numerical and analytic methods [1–7]. These studies strongly indicate the universality of the local statistics of eigenvalues of generic quantum Hamiltonians: For integrable systems the local statistics is Poissonian, while for chaotic systems it is the Wigner statistics of the ensemble of Gaussian matrices. In this Letter we summarize new rigorous results about the statistics of levels of simple integrable Hamiltonians [8–12]. These are related to the distribution of integer lattice points inside a "random" region of the plane, a problem of independent interest in number theory [13,14].

To see the connection with the lattice problem, consider a free particle on a torus. The eigenvalues are, in suitable units, $E_{\mathbf{n}} = n_1^2 + n_2^2$, with $\mathbf{n} = (n_1, n_2) \in \mathbb{Z}^2$. More generally we may consider integrable systems with eigenvalues

$$E_{\mathbf{n}} = I(n_1 - \alpha_1, n_2 - \alpha_2), \tag{1}$$

where $I(x_1, x_2)$ (maybe after some renormalization) is a smooth homogeneous function of second degree and $\boldsymbol{\alpha} = (\alpha_1, \alpha_2)$ is a point in the unit square [1]. The number of levels

$$N(E) = \{\mathbf{n} \mid E_{\mathbf{n}} \le E\}$$

is clearly the same as the number of lattice points inside the oval curve defined by $I(x_1 - \alpha_1, x_2 - \alpha_2) = E$. We are interested in the behavior of $N(E)$ and $N(E,S) = N(E+S) - N(E)$, the number of levels between E and $E+S$, for generic Hamiltonians in this class, e.g., for "typical" $\boldsymbol{\alpha}$ in (1) and also for individual Hamiltonians with given α. Thus if E is considered a random variable uniformly or otherwise smoothly distributed on the interval $[c_1 T, c_2 T]$, $c_2 > c_1 > 0$, we can ask for the variance and distribution of $N(E)$ and $N(E,S)$ when the energy or T is very large compared to the average spacing between levels, which is of order 1; cf. Fig. 1 for the behavior of $N(E) - \pi E$ as a function of $R = \sqrt{E}$ when $I(x_1, x_2) = x_1^2 + x_2^2$ and $\boldsymbol{\alpha} = (0.34367, 0.43037)$. The local statistics of energy levels in a fixed interval S, which does not grow with energy, has been investigated numerically, and there is strong evidence that the statistics are Poissonian [1,4,15]. When S grows with E in such a way that $S \gtrsim CE^{1/2}$, the levels are no longer "truly random" on this scale, and we can expect deviations from Poisson statistics [1,4].

We first describe the results informally in the language of energy levels and then give a brief indication of the proof in terms of lattice problems. Our statements are to be understood as referring to the limit $T \to \infty$ when E is uniformly (or otherwise smoothly) distributed in the interval $c_1 T < E \le c_2 T$, $c_2 > c_1 > 0$. Let λ be the area of the interior of the oval curve γ defined by $I(x_1, x_2) = 1$. We assume that γ is (at least 7 times) differentiable and has strictly positive radius of curvature everywhere.

The average number of levels per unit energy, $N(E)/E$,

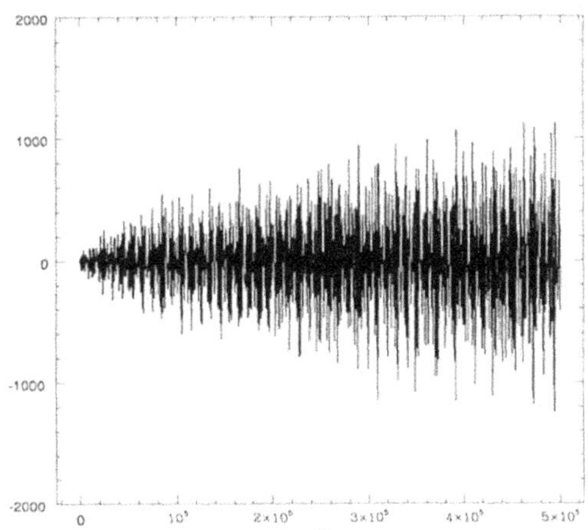

FIG. 1. The graph of the function $N_{\boldsymbol{\alpha}}(R^2) - \pi R^2 =$ number of $\{\mathbf{n} \in \mathbb{Z}^2 \mid |\mathbf{n} - \boldsymbol{\alpha}| \le R\} - \pi R^2$ for $\alpha = (0.34367, 0.43037)$.

0031-9007/93/71(19)/3047(4)$06.00

VOLUME 71, NUMBER 19 PHYSICAL REVIEW LETTERS 8 NOVEMBER 1993

is given asymptotically by $\lambda+O(E^{-1/2})$. The difference $N(E)-\lambda E$ has mean square fluctuation behaving like $\bar{V}E^{1/2}$ with a constant $\bar{V}$ as $E\to\infty$. This behavior can be understood from the fact that $N(E)$ counts the number of lattice points inside the curve $\sqrt{E}\gamma$, whose perimeter, around which the fluctuations occur, grows like $\sqrt{E}$. Furthermore the probability that the random variable $F(E)=[N(E)-\lambda E]/E^{1/4}$ takes on values in the interval between s and $s+ds$ has a limit $p(s)ds$ as E or $T\to\infty$. This density $p(s)$ is a real analytic function of s with mean zero and variance $\bar{V}$. The decay of $p(s)$ for large s is bounded above and below (roughly) by $\exp[-cs^4]$ where c and $\bar{V}$ depend on the shape of γ as well as on the shift $\boldsymbol{\alpha}$ in (1). This shows in particular that $p(s)$ cannot be Gaussian; cf. Fig. 2 where $p(s)$ is presented for

$$I(x_1,x_2)=(x_1-\alpha_1)^2/a_1^2+(x_2-\alpha_2)^2/a_2^2 \qquad (2)$$

with $a_1/a_2=\pi/10$, $\alpha_1=\alpha_2=0$ in graph (a) and $a_1/a_2=\pi/10$, $\alpha_1=0.34367$, $\alpha_2=0.43037$ in graph (b). As these examples illustrate, the distribution $p(s)$ is usually skew and sometimes bimodal (the parameters in the second example were specially chosen to demonstrate a case of bimodal distribution).

Let us consider now the fluctuations in $N(E,S)$. These depend on the rate of growth of S when $E\to\infty$. When $S/E\to 0$ but $S/E^{1/2}\to\infty$, then $N(E+S)$ and $N(E)$ are asymptotically independent, so their variances add and $\langle[N(E,S)-\lambda S]^2\rangle\sim 2\bar{V}E^{1/2}$. The distribution of $[N(E,S)-\lambda S]/E^{1/4}$ then converges to the distribution of a difference of two independent identically distributed

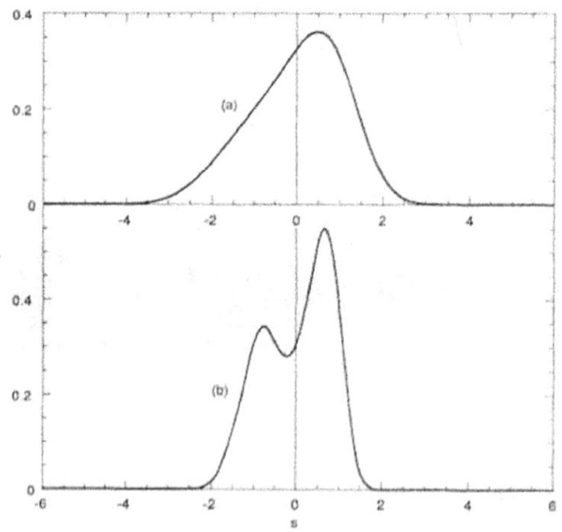

FIG. 2. The density $p(s)$ of the limit distribution of $F(E)=[N(E)-\lambda E]/E^{1/4}$ for the energy function $I(x_1,x_2)$ given in (2) with (a) $a_1/a_2=\pi/10$, $\alpha_1=\alpha_2=0$ and (b) $a_1/a_2=\pi/10$, $\alpha_1=0.34367$, $\alpha_2=0.43037$.

random variables, with distribution $p(s)$, the limit distribution of $F(E)$.

When $S/E^{1/2}\to z>0$, the variance has a scaling behavior,

$$\langle[N(E,S)-\lambda S]^2\rangle\sim E^{1/2}V(E^{-1/2}S).$$

The scaling function $V(z)$ is an almost periodic function of z, so it is oscillating and has no limit at infinity; its average $(1/L)\int_0^L V(z)dz$ approaches, as $L\to\infty$, its value in the previous case, when $S/E^{1/2}\to\infty$, i.e., $2\bar{V}$. In the limit $z\to 0$ we show that in typical cases, corresponding to no systematic degeneracy of the E_n,

$$V(z)\sim\lambda z. \qquad (3)$$

This implies that when $z\to 0$, the variance of $N(E,S)$ becomes equal to its average value λS, which is consistent with (but does not imply) a Poisson distribution. The relation (3) is violated in degenerate cases. For instance, for the circle with center at the origin [or at any rational point $\boldsymbol{\alpha}=(\alpha_1,\alpha_2)$] the behavior of $V(z)$ is given by

$$V(z)\sim Cz|\ln z|, \quad z\to 0. \qquad (4)$$

This anomalous behavior of $V(z)$ is related to an arithmetic degeneracy of the circle problem: For some integers k there are many representations of k as a sum of two squares, and on the average, their number grows as $\ln k$, which shows up in the log-correction to linear asymptotics of $V(z)$ as $z\to\infty$. For a circle with center at a very irrational (Diophantine) point (α_1,α_2), the behavior is normal, satisfying (3). For an ellipse centered on the origin with a transcendental ratio of the axes there is a fourfold degeneracy and $V(z)\sim 4\lambda z$.

We prove also the existence of a limit distribution of

$$F(E;S)=[N(E,S)-\langle N(E,S)\rangle]/\sqrt{\mathrm{Var}N(E,S)}$$

in the regime $S/E^{1/2}\to z$. The limit distribution is not Gaussian and in a generic case its density decays at infinity roughly as $\exp[-c(z)x^4]$. However, when $z\to 0$ this limit distribution converges to a standard Gaussian distribution, which can be taken as signaling an approach to a random regime.

When S stays fixed we expect to get a Poisson distribution with mean λS as $E\to\infty$. There is strong numerical and analytic evidence that this, and the corresponding exponential distribution of distances between neighboring levels, is indeed the case, for typical values of $\boldsymbol{\alpha}$ [15]. The only rigorous result in that direction, however, is due to Sinai and Major [16] for the number of lattice points in a narrow strip surrounding a typical "very random curve," so random in fact that it is not even twice differentiable, so that its relevance to real systems is questionable. For the type of smooth oval curves γ considered here all that can be shown at present is that the first and second moments of $N(E,S)$ are, after averaging over $\boldsymbol{\alpha}$, indeed given by λS, as they would be for a Poisson distribution [12].

To give a brief sketch of the proof we shall for simplicity consider the case of a circle centered at $\boldsymbol{\alpha}$. We then have $E=R^2$ and $N_{\boldsymbol{\alpha}}(R)$ is the number of lattice points in the shifted circle, $|\mathbf{n}-\boldsymbol{\alpha}|\le R$. The classical circle problem which goes back to Gauss is to prove a uniform bound for the fluctuations in $N_0(R)$. The best result in this direction at present [17], $|N_0(R)-\pi R^2|\le C_\epsilon \times R^{46/73+\epsilon}$, is quite far from the expected bound $R^{1/2+\epsilon}$ or even $R^{1/2}(\ln R)^\delta$ for some δ. (Numerical studies suggest $\delta<1$.)

Our results show that the situation is much better when we abandon the attempt to prove uniform bounds and consider the statistical behavior of the fluctuations. Using the Poisson summation formula $\sum_{\mathbf{n}} f(\mathbf{n})=\sum_{\mathbf{n}}\tilde{f}(2\pi\mathbf{n})$, where $\tilde{f}$ is the two-dimensional Fourier transform of f and the summation goes over $\mathbf{n}\in\mathbb{Z}^2$, yields

$$\begin{aligned} F_{\boldsymbol{\alpha}}(R) &= [N_{\boldsymbol{\alpha}}(R)-\pi R^2]/\sqrt{R} = \sum_{\mathbf{n}\neq 0} |\mathbf{n}|^{-1} e(\mathbf{n}\cdot\boldsymbol{\alpha}) J_1(2\pi|\mathbf{n}|R)\sqrt{R} \\ &= \pi^{-1}\sum_{\mathbf{n}\neq 0}|\mathbf{n}|^{-3/2} e(\mathbf{n}\cdot\boldsymbol{\alpha})\cos[2\pi|\mathbf{n}|R-3\pi/4]+O(R^{-1}), \end{aligned} \tag{5}$$

where $J_1(t)$ is the Bessel function and $e(t)=e^{2\pi i t}$.

The basic idea, due originally to Heath-Brown [14], is to rewrite (5) as a sum of terms representing a flow on an infinite-dimensional torus with incommensurate frequencies, R playing the role of time. The problem of finding a limiting distribution is then a problem in ergodic theory [18]. To do this we first group terms in (5) with commensurate frequencies, i.e., all those $\mathbf{n}\in\mathbb{Z}^2$ for which $|\mathbf{n}|=k\sqrt{m}, k=1,2,\ldots$, where $m>0$ is a fixed square-free natural number, i.e., $m\neq k^2 l$ with $k>1$. This gives

$$F_{\boldsymbol{\alpha}}(R)=\sum_{\text{square-free } m} f_m(\sqrt{m}R;\boldsymbol{\alpha})+O(R^{-1}), \tag{6}$$

where

$$f_m(t;\boldsymbol{\alpha})=\pi^{-1}m^{-3/4}\sum_{k=1}^{\infty}k^{-3/2}\cos(2\pi k t-3\pi/4)\sum_{\mathbf{n}:|\mathbf{n}|=k\sqrt{m}} e(\mathbf{n}\cdot\boldsymbol{\alpha}) \tag{7}$$

is periodic in t with period 1.

Since the $\sqrt{m}$'s with square-free m are linearly independent over the field of rational numbers we are led to the study of limit distributions of almost periodic functions of the form

$$F(t)=\sum_{n=1}^{\infty}a_n(\gamma_n t) \tag{8}$$

with rationally independent γ_n. If the sum (8) were finite then ergodic theory would indeed imply the existence of a limiting distribution of $F(t)$ corresponding to the distribution of the finite series $\sum_{n=1}^{N}a_n(\theta_n)$, where the θ_n are independent random variables uniformly distributed on [0,1] (see, e.g., [19]). This means, roughly speaking, that the $\gamma_n t \bmod 1$, $n=1,\ldots,N$, behave like independent random variables uniformly distributed on [0,1], as t varies over a sufficiently large range. Our problem is that the series in (6) is infinite, and in fact only conditionally convergent, since $f_m(t;\boldsymbol{\alpha})\sim m^{-3/4}$. To prove the existence of a limit distribution of such a series we generalize ideas of Heath-Brown to a large class of almost periodic functions (the Besicovitch space B^1). This permits us to find the distribution of lattice points in regions bounded by smooth convex curves.

The nature of the limiting distribution $p(s)$ for large $|s|$ is determined by the large m behavior of the f_m in (7). The proof is based on the following result [8,9,11]. Let θ_m, $m=1,2,\ldots$, be independent uniformly distributed random variables in [0,1] and consider the distribution of $\sum_m a_m(\theta_m)$ where the $a_m(\theta)$ are continuous functions of period one with

$$\int_0^1 a_m(\theta)d\theta=0,\quad \sum_{m=1}^{\infty}\int_0^1|a_m(\theta)|^2 d\theta<\infty \tag{9}$$

and

$$\begin{aligned} c_1 m^{-2\nu} &> \sup_\theta |a_m(\theta)|^2 \\ &\ge \int_0^1 |a_m(\theta)|^2 d\theta > c_2 m^{-2\nu}, \end{aligned} \tag{10}$$

with $c_1,c_2>0$ and $0<\nu<1$. Then

$$\begin{aligned} C'\exp[-c's^{1/(1-\nu)}] &> \Pr\left\{\left|\sum_m a_m(\theta_m)\right|>s\right\} \\ &> C''\exp[-c''s^{1/(1-\nu)}] \end{aligned} \tag{11}$$

for some $C',C'',c',c''>0$. For the f_m in (7), $\nu=3/4$, and so we obtain the bound $\exp(-\lambda s^4)$ on $p(s;\boldsymbol{\alpha})$, for every $\boldsymbol{\alpha}$.

While the limiting distribution $p(s;\boldsymbol{\alpha})$ of $F(R;\boldsymbol{\alpha})$ is independent of the details of the distribution of R in the interval $[c_1T,c_2T]$, it has a very nonsmooth dependence on $\boldsymbol{\alpha}$. In particular, the variance $\bar{V}_{\boldsymbol{\alpha}}$ is nondifferentiable at every rational point $\boldsymbol{\beta}=(p_1/q_1,p_2/q_2)$, behaving like $\bar{V}_{\boldsymbol{\alpha}}\sim\bar{V}_{\boldsymbol{\beta}}+C|\boldsymbol{\alpha}-\boldsymbol{\beta}|\ln|\boldsymbol{\alpha}-\boldsymbol{\beta}|$ as $\boldsymbol{\alpha}\to\boldsymbol{\beta}$ (see [10]). The variance $\bar{V}_{\boldsymbol{\alpha}}$ can be obtained either from the second moment of the limiting distribution $p(s)$ or directly from the limit

$$\langle F_{\boldsymbol{\alpha}}^2(R)\rangle=\lim_{T\to\infty}(1/T)\int_0^T F_{\boldsymbol{\alpha}}^2(R)dR.$$

The validity of this interchange of limits is an open problem for higher moments of $F_{\alpha}(R)$. In fact while the decay of $p(s;\alpha)$ guarantees the existence of all moments $\int s^k p(s;\alpha)ds$, we do not even know whether $\langle |F_{\alpha}(R)|^k\rangle$ exists for $k>9$ (cf. [14]).

The analysis of the distribution of $N(E,S)$ proceeds in a similar way. Considering the variance $v_{\alpha}(\delta)$ of $[N_{\alpha}(R+cR^{\delta})-N_{\alpha}(R)]/\sqrt{R}$ we find that $v_{\alpha}(\delta)\rightarrow 2\bar{V}$ for $0<\delta<1$, and $v_{\alpha}(\delta)\rightarrow V_{\alpha}(z)$ for $c=z/2\pi$ and $\delta=0$.

Our results depend strongly on the assumption that the curvature of γ does not vanish anywhere. The situation can be different in the presence of inflection points [20]. In the extreme case when γ is a rectangle of unit area with a very irrational orientation with respect to the x axis, e.g., $\tan\theta=\sqrt{2}$, Beck [21] has shown that the fluctuations in $N_{\alpha}(R)-R^2$ have a variance which grows like $\ln R$. After normalization by $(\ln R)^{1/2}$ the distribution of fluctuations converges to a Gaussian distribution.

If we keep γ fixed and rescale the lattice $\mathbb{Z}^2$ by R^{-1}, we may think of $R^{-2}[N_{\alpha}(R)-\lambda R^2]$ as the difference between the Riemann sum and the integral for the function $\psi(\mathbf{r})=1$ inside γ. Our results generalize to the case where $\psi(\mathbf{r})$ is a general smooth function of $\mathbf{r}$ which does not vanish on γ. The main contribution to the deviation of the Riemann sum from the integral then comes from the boundary of γ and will be of order $R^{-3/2}$ when the lattice spacing R^{-1} tends to zero.

We want to thank our collaborators Zheming Cheng and Peter Major for all theor help. The work was supported in part by the Ambrose Monell Foundation and NSF Grant No. DMR 92-13424.

[1] M. V. Berry and M. Tabor, Proc. R. Soc. London A **356**, 375 (1977); M. V. Berry, *ibid.* **400**, 229 (1985).

[2] M. V. Berry, in *Chaotic Behavior of Deterministic Systems*, Proceedings of the Les Houches Lectures Vol. 36, edited by G. Iooss, R. H. G. Helleman, and R. Stora (North-Holland, Amsterdam, 1983).

[3] O. Bohigas, in *Chaos and Quantum Physics*, Proceedings of the Les Houches Lectures Vol. 52, edited by M.-J. Giannoni, A. Voros, and J. Zinn-Justin (Elsevier Science, B.V., New York, 1991).

[4] G. Casati, I. Guarneri, and F. Valz-Gris, Phys. Rev. A **30**, 1586 (1984); G. Casati, B. V. Chirikov, and I. Guarneri, Phys. Rev. Lett. **54**, 1350 (1985).

[5] Ya. G. Sinai, in *Geometric Aspects of Functional Analysis*, edited by J. Lindenstrauss and V. C. Milman, Lecture Notes in Mathematics Vol. 1469 (Springer-Verlag, Berlin, 1991).

[6] E. B. Bogomolny, B. Georgeot, M.-J. Giannoni, and C. Schmit, Phys. Rev. Lett. **69**, 1477 (1992).

[7] H.-D. Gräf *et al.*, Phys. Rev. Lett. **69**, 1296 (1992).

[8] P. M. Bleher, Zh. Cheng, F. J. Dyson, and J. L. Lebowitz, Commun. Math. Phys. **154**, 433 (1993).

[9] P. M. Bleher, Duke Math. J. **67**, 461 (1992); **70**, 655 (1993).

[10] P. M. Bleher and F. J. Dyson, Institute for Advanced Study, Princeton, Report No. IASSNS-HEP-92/83, 1992 (to be published).

[11] P. M. Bleher and J. L. Lebowitz, Institute for Advanced Study, Princeton, Report No. IASSNS-HEP-93/14, 1993 (to be published).

[12] Z. Cheng, J. L. Lebowitz, and P. Major, Rutgers University report, 1993 (to be published).

[13] D. G. Kendall, Quart. J. Math. (Oxford) **19**, 1 (1948).

[14] D. R. Heath-Brown, Oxford University report (to be published).

[15] Z. Cheng and J. L. Lebowitz, Phys. Rev. A **44**, R3399 (1991).

[16] Ya. G. Sinai, in *Advances in Soviet Mathematics*, edited by Ya. G. Sinai (American Mathematical Society, Providence, 1991), Vol. 3; P. Major, Probab. Theory Relat. Fields **92**, 423 (1992).

[17] M. N. Huxley, University of Wales College of Cardiff report, 1992 (to be published).

[18] I. P. Cornfeld, S. V. Fomin, and Ya. G. Sinai, *Ergodic Theory* (Springer-Verlag, New York, 1982).

[19] Y. Colin de Verdière, Math. Z. **171**, 51 (1980).

[20] P. M. Bleher, Institute for Advanced Study, Princeton, Report No. IASSNS-HEP-93/35, 1993 (to be published).

[21] J. Beck, Rutgers University report, 1992 (to be published).

VOLUME 76, NUMBER 7 PHYSICAL REVIEW LETTERS 12 FEBRUARY 1996

Exact Macroscopic Description of Phase Segregation in Model Alloys with Long Range Interactions

Giambattista Giacomin*
Institut für Angewandte Mathematik, Universität Zürich-Irchel, Winterthurer Strasse 190, CH-8057 Zürich, Switzerland

Joel L. Lebowitz†
Department of Mathematics, Hill Center, Rutgers University, New Brunswick, New Jersey 08903
(Received 25 August 1995)

We derive an exact nonlinear nonlocal macroscopic equation for the time evolution of the conserved order parameter $\rho(\mathbf{r}, t)$ of a microscopic model binary alloy undergoing phase segregation: a d-dimensional lattice gas evolving via Kawasaki exchange dynamics, satisfying detailed balance for a Hamiltonian with a long range pair potential $\gamma^d J(\gamma|x|)$. The macroscopic evolution is on the spatial scale γ^{-1} and time scale γ^{-2}, in the limit $\gamma \to 0$. The domain coarsening, described by interface motion, is similar to that obtained from the Cahn-Hilliard equation.

PACS numbers: 64.60.Cn, 02.30.Jr, 64.70.Kb, 64.75.+g

The process of phase segregation in alloys, following a quench (sudden cooling) from a high temperature into the miscibility gap, is a problem of great practical importance and theoretical interest. The time evolution of the local macroscopic order parameter, e.g., the concentration of A atoms in a binary A-B alloy, is commonly described by the nonlinear Cahn-Hilliard equation (CHE) [1]. While the derivation of this equation is based on arguments embodying deep physical insight, there does not exist (to our knowledge) any microscopic model system, for which the CHE, or any modifications of it proposed so far [1], gives the exact macroscopic description.

In this Letter we rigorously derive a macroscopic equation, Eq. (7), describing phase segregation in systems with long range interactions, the same models whose equilibrium properties are described by the van der Waals equation of state (with the Maxwell construction) [2]. This equation, while structurally different from the CHE, yields similar behavior for the late stages of the coarsening process; this regime appears to be quite universal; see [1].

We begin by giving a simple heuristic derivation of a general CHE as an adaptation of the nonlinear diffusion equation

$$\begin{aligned} \frac{\partial \rho(\mathbf{r}, t)}{\partial t} &= \nabla \cdot [D(\rho) \nabla \rho] \\ &= \nabla \cdot [\sigma(\rho) \nabla \mu_{\text{eq}}(\rho)] \\ &= \nabla \cdot \left[\sigma(\rho) \nabla \left(\frac{\delta F_{\text{eq}}}{\delta \rho}\right)\right] \end{aligned} \tag{1}$$

describing the evolution of a nonuniform macroscopic concentration profile $\rho(\mathbf{r}, t)$ toward the uniform equilibrium state. In (1) $F_{\text{eq}} = \int f_{\text{eq}}(\rho)\, d\mathbf{r}$, D is the diffusion coefficient, $\mu_{\text{eq}}(\rho) = f'_{\text{eq}}(\rho)$ is the chemical potential, given by the derivative of the equilibrium Helmholtz free energy per unit volume $f_{\text{eq}}(\rho)$ in a system with uniform density ρ, and σ is the Onsager mobility $\sigma = \chi D$, where $\chi = (\partial \mu_{\text{eq}}/\partial \rho)^{-1}$ is the compressibility (or susceptibility).

Equation (1) can be derived rigorously for a variety of microscopic model systems when $\mathbf{r}$ and t are in units scaled by ϵ^{-1} and ϵ^{-2}, respectively, compared to the microscopic units in the limit $\epsilon \to 0$ [3]. The dynamics in these models are generally stochastic (although some deterministic examples are also available), conserve (only) particle number and have the equilibrium Gibbs distribution as the only stationary state. This includes the Ising model with Kawasaki dynamics for $T > T_c$ in $d = 2$ [4], and the continuous spin Ginzburg-Landau model with conservative Langevin dynamics, for all T and all dimensions [5]. In the coexistence region the mobility $\sigma(\rho)$, which is always given by a Green-Kubo formula, remains finite while $\mu_{\text{eq}}(\rho)$ is constant, so $\chi^{-1} = 0$ and $D(\rho) = \sigma(\rho)/\chi(\rho) = 0$ there [6]. Hence, when the range of densities in a nonuniform system lies within the miscibility gap, the right-hand side of (1) vanishes. This may seem at first paradoxical but one should remember that we are considering in (1) density variations on the macroscopic spatial scale ϵ^{-1} which is very large compared to the interaction range. This means that regions of size ϵ^{-1} can contain essentially equilibrated domains of both phases. Variations of the density on the spatial scale ϵ^{-1} can therefore be compensated for by changes in the volume fraction of the two phases. Of course the system will still try to aggregate further to reduce interfacial area but this evolution will occur on a time scale larger than ϵ^{-2}, so it is not seen in (1).

The CHE with some mobility $\tilde{\sigma}(\rho)$ can be obtained formally [1] by replacing F_{eq} [and thus $\mu_{\text{eq}}(\rho)$] on the right side of (1) by a free energy functional $\tilde{F}(\{\rho(\mathbf{r}, t)\})$

$$\begin{aligned} \tilde{F} &= \int d\mathbf{r} [\tilde{f}(\rho) + \tfrac{1}{2} \zeta (\nabla \rho)^2], \\ \mu &= \tilde{f}'(\rho) - \zeta \nabla^2 \rho\,, \end{aligned} \tag{2}$$

 0031-9007/96/76(7)/1094(4)$06.00

where $\tilde{f}(\rho)$ is a *constrained equilibrium* free energy density which has a single minimum above T_c and a double well shape at $T < T_c$, and ζ is a constant related to the surface tension between the phases. This gives the general CHE

$$\frac{\partial \rho}{\partial t} = \nabla \cdot \{\tilde{\sigma}(\rho)\nabla[\tilde{f}'(\rho) - \zeta\nabla^2\rho]\}. \quad (3)$$

$\tilde{f}(\rho)$ is often taken to be a quadratic polynomial in ρ and $\tilde{\sigma}(\rho)$ set equal to a constant: One then obtains the *standard* CHE [1,7].

As noted, however, the model systems yielding (1) give $\partial\rho/\partial t = 0$ in the miscibility gap, where $f''_{\rm eq}(\rho) = 0$, so (3) does not seem to have any microscopic derivation. To overcome this problem and find a macroscopic equation which describes exactly the phase segregation of some microscopic model, we consider the dynamics of a system in which the interaction responsible for phase segregation has a range γ^{-1}, which is large compared to the interparticle spacings. The equilibrium properties of such a system are well known [2]: Here we consider the evolution of the order parameter on the spatial scale γ^{-1} (or a hundred times as large).

The particles live in a cube Λ_γ of side $l\gamma^{-1}$, on the d-dimensional lattice $\mathbb{Z}^d$. The time evolution of the configuration η_t at time t, specified by $\eta_t(x) = 0$ or 1 for all $x \in \Lambda_\gamma$, proceeds via a Kawasaki type exchange dynamics, specified by giving the rate for the exchange of the occupancy of nearest neighbor sites x and y (using periodic boundary conditions) in the configurations η, i.e., for the transition $\eta \to \eta^{x,y}$. We take this rate to be

$$c_\gamma(x,y;\eta) = \exp\{-\tfrac{1}{2}\beta[H_\gamma(\eta^{x,y}) - H_\gamma(\eta)]\}, \quad (4)$$

where $\beta \geq 0$ is the reciprocal temperature, and

$$H_\gamma(\eta) = -\frac{1}{2}\sum_{x,y\in\Lambda_\gamma} \gamma^d J(\gamma(x-y))\eta(x)\eta(y) \quad (5)$$

and $J(\mathbf{r}) \geq 0$ is a smooth function with $\int J(\mathbf{r})\,d\mathbf{r} = \alpha$. We choose l to be much larger than the range of $J(\mathbf{r})$, so we do not have to worry about periodicity, and for simplicity let $J(\mathbf{r})$ depend only on $|\mathbf{r}|$; the anisotropic case can be treated too [8]. Assume furthermore that J is decreasing in $|\mathbf{r}|$. The dynamics (4) satisfies detailed balance with respect to the Gibbs measure $Z_N^{-1}\exp[-\beta H_\gamma(\eta)]$, with any fixed total number of particles $N = \sum_{x\in\Lambda_\gamma}\eta(\mathbf{x})$, which is the unique stationary measure (for N fixed).

We are interested in initial conditions described by a probability distribution (Prob_γ) on the configuration space (Prob_γ can be concentrated on one configuration) such that when $\gamma \to 0$, the typical microscopic configuration resembles more and more a smooth profile ρ_0 stretched by γ^{-1}, where ρ_0 is a smooth function from T^d, the d-dimensional torus of length l, to [0,1]. More precisely, we shall require that for any continuous function $\phi(\mathbf{r})$ on T^d and every $\delta > 0$

$$\lim_{\gamma\to 0}\text{Prob}_\gamma\left(\left|\gamma^d\sum_{x\in\Lambda_\gamma}\phi(\gamma x)\eta(x) - \int_{T^d}\phi(\mathbf{r})\rho_0(\mathbf{r})\,d\mathbf{r}\right| > \delta\right) = 0. \quad (6)$$

Condition (6) is clearly satisfied if Prob_γ is such that $\langle\eta(x)\rangle = \rho_0(\gamma x)$ for all x in Λ_γ and the occupation numbers of the sites are independent but this is not required. We could, for example, also have initial conditions concentrated on single periodic configurations, etc. Our theorem now says that (6) holds also in the limit $\gamma \to 0$ for $\eta_{t\gamma^{-2}}$, if we replace ρ_0 with $\rho(\mathbf{r},t)$, the solution of the integro-differential equation:

$$\partial_t\rho(\mathbf{r},t) = \nabla\cdot\left[\nabla\rho(\mathbf{r},t) - \beta\rho(\mathbf{r},t)(1-\rho(\mathbf{r},t)) \times \int_{T^d}\nabla J(\mathbf{r}-\mathbf{r}')\rho(\mathbf{r}')\,d\mathbf{r}'\right] \quad (7)$$

with initial value $\rho(\mathbf{r},0) = \rho_0(\mathbf{r})$. Equation (7) is valid for all β and $t \in [0,\infty)$, including the miscibility gap. Existence and uniqueness of a suitable weak solution of (7) globally in time are part of the proof of this theorem. Moreover, it can be shown that $\rho(\mathbf{r},t)$ is bounded between 0 and 1 and that if $\rho_0(\mathbf{r})$ is smooth, then $\rho(\mathbf{r},t)$ is smooth. Further results on the behavior of this particle system on a macroscopic scale can be found in [9].

The result stated here is proven in [8] using the technique introduced in [10]; see also [9]. It is essentially a consequence of the fact that, in a nearest neighbor exchange, due to the long range interactions, the energy difference is of order γ. Hence, the dynamics is a superposition of a Laplacian and a weak $[O(\gamma)]$ asymmetric perturbation which generates, on the macroscopic scale, a *force* term given by the gradient of the (long range) energy density at $\mathbf{r}$, $\int J(\mathbf{r}-\mathbf{r}')\rho(\mathbf{r}')\,d\mathbf{r}'$. This force term is multiplied by a conductivity $\beta\rho(1-\rho)$: β measures the bias introduced in the exchange rates c_γ, given in (4), by energy differences, while $\rho(1-\rho)$ gives the rate at which exchanges actually take place when a particular site is picked at random and the system is *locally described* by a product measure with average density ρ. We note that the product measure is the correct equilibrium measure for our *reference* system, i.e., the system without the weak long range interaction J.

The analysis and (7) remain unchanged if we replace (4) with a general rate $c_\gamma(x,y;\eta) = \Phi(\beta[H_\gamma(\eta^{x,y}) - H_\gamma(\eta)])$ which satisfies the detailed balance condition [3] $\Phi(E) = \Phi(-E)\exp(-E)$ as long as we rescale our time by $\Phi(0)$ and Φ is smooth.

It is now an observation, which at first sight appears surprising, that (7) can be rewritten in the form

$$\frac{\partial \rho(\mathbf{r},t)}{\partial t} = \nabla \cdot \left[\sigma^0(\rho)\nabla\left(\frac{\delta \mathcal{F}^0}{\delta \rho}\right)\right], \quad (8)$$

where $\sigma^0(\rho) = \beta\rho(1-\rho)$ and

$$\mathcal{F}^0(\{\rho\}) = \int f_c^0(\rho(\mathbf{r}))\,d\mathbf{r} + \frac{1}{4}\int\int J(\mathbf{r}-\mathbf{r}')[\rho(\mathbf{r})-\rho(\mathbf{r}')]^2\,d\mathbf{r}\,d\mathbf{r}' \quad (9)$$

with $f_c^0(\rho) = f_{\rm eq}^0(\rho) - \frac{1}{2}\alpha\rho^2$ and $f_{\rm eq}^0(\rho) = -\beta^{-1}s(\rho)$; $s(\rho) = -\rho\ln\rho - (1-\rho)\ln(1-\rho)$. $f_{\rm eq}^0(\rho)$ is clearly just the free energy density of the reference system while $\mathcal{F}^0(\{\rho\})$, which is a nonlocal functional of the density, is just the total constrained free energy of this system in the limit $\gamma \to 0$; see [2,11]. For ρ constant the second term in (9) vanishes, and $f_c^0(\rho)$ is in fact the correct equilibrium free energy density as long as $\beta \leq \beta_c = 1/T_c = 4\alpha^{-1}$. If $\beta > \beta_c$, $f_c^0(\rho)$ has a double minimum at $\rho = \rho_\beta^\pm$, the two nontrivial solutions of $\ln[\rho/(1-\rho)] = \beta\alpha(\rho - 1/2)$, and the correct free energy is then obtained by the double tangent construction [2,11].

The mobility $\sigma^0(\rho)$, as in (1), is just χD, with $\chi = \chi^0 = \beta\rho(1-\rho)$ and $D = D^0 = 1$ for the reference system [3]. It is this, at first sight fortuitous, coincidence of χ^0 with the *conductivity* appearing in front of the long range force term on the right side of (7) which makes it possible to transform (7) into the physically more revealing form (8). Further thought shows, however, that the form (8) is really forced by the requirement that the Gibbs measure $\exp\{-\beta H_\gamma\}/Z_N$ be stationary for the dynamics. This is most easily seen by adding to the reference system not H_γ but a one body external potential $U_\gamma(\eta) = \sum_x U(\gamma x)\eta(x)$, in which case the integral in (7) is replaced by $\nabla U(\mathbf{r})$. The coincidence is now simply the Einstein relation (or linear response theory) between the conductivity for the current induced by the weak perturbation U_γ and the diffusion coefficient D^0 of the reference system; cf. Refs. [2], p. 233–234 and [12].

The superscript zero in (8) and (9) is to remind us that our reference system is one in which there are no short range interactions (except for the hard core exclusion preventing multiple occupancy at a site). We consider now the macroscopic dynamics of a system in which there is also a short range interaction $H_s(\eta)$, making the total energy $H = H_s + H_\gamma$. The system evolves microscopically according to a Kawasaki exchange rate $c_\gamma(x,y;\eta)$ which satisfies detailed balance with respect to the Gibbs measure, $\exp\{-\beta[H_s + H_\gamma]\}/Z_N$ (Z_N is the corresponding canonical partition function). The dynamics of this system, at a temperature β^{-1} above the critical temperature of the reference system, will again be a superposition of diffusive dynamics for the reference system and a current linear in the force coming from the weak long range interaction. Using the same reasoning as before, the conductivity coefficient in front of this force, given by the integral in (7), must, by the stationarity of equilibrium measure (Einstein relation), be equal to the mobility σ of the reference system. It is known [2,11] that the constrained free energy of such a system is obtained from those with $H_s = 0$ by replacing in the integrand of (9) $f_{\rm eq}^0(\rho)$ by $f_{\rm eq}(\rho)$, the equilibrium free energy density of the reference system with short range Hamiltonian $H_s(\eta)$. Hence we expect that the macroscopic evolution of such a system will again be given by an equation of the form (8) and (9) without the superscript zero, i.e., $f_c^0 \to f_{\rm eq} - (1/2)\alpha\rho^2$ (which may have two minima, so that there will be phase segregation) and $\sigma^0 \to \sigma = \chi D$, where D is the diffusion coefficient (generally ρ dependent) and χ the susceptibility of the reference system. This has already been proven explicitly in some cases [13] and it is surely quite general.

We shall now compare solutions of the evolution (7) with those of the CHE [(2) and (3)] in the late stage of the domain coarsening process [1]: At this stage there are well defined interfaces between domains of linear sizes L, large compared to that of the interface width, in which the densities are close to those of the pure phases $\rho_\beta^\pm$. In this *sharp* interface limit, the CHE has been studied analytically, mostly by means of formal matched asymptotic expansions [7,14,15] and this type of analysis is extended in [8] to Eq. (7). Since in this regime the mean curvature of the boundary of the domains is $O(1/L)$, the local structure of the interface is given in first approximation by the one dimensional *instantonic* solutions of (7). These are stationary solutions $U_\beta(z)$ of (7) which depend only on one coordinate of $\mathbf{r}$ (in this case $\mathbf{r}$ is in $\mathbb{R}^d$), say $z = r_1$, which connect in a monotone increasing way the low (ρ_β^-) and high (ρ_β^+) density phases, i.e., $U_\beta(z) \to \rho_\beta^\pm$ as $z \to \pm\infty$. Once we fix $U_\beta(0) = 1/2$, there is only one (smooth) *instantonic* profile [16], given by the solution of

$$m_\beta(z) = \tanh\left\{\frac{\beta}{2}\int \tilde{J}(z-z')m_\beta(z')\,dz'\right\}, \quad (10)$$

where $m_\beta(z) = 2U_\beta(z) - 1$ and $\tilde{J}(r_1) = \int_{\mathbb{R}^{d-1}} J(\mathbf{r})\,dr_2\cdots dr_d$.

Let us now consider the evolution Eq. (7) on a torus $L\Omega$, where Ω is a fixed d-dimensional torus. Setting $\mathbf{x} = L^{-1}\mathbf{r}$ and $\tau_q = L^{-q}t$, $\rho^L(\mathbf{x},\tau_q) = \rho(L\mathbf{x}, L^q\tau_q)$, the equation becomes

$$L^{2-q}\partial_{\tau_q}\rho^L(\mathbf{x},\tau_q) = \nabla[\sigma(\rho^L(\mathbf{x},\tau_q))\nabla\mu^L(\mathbf{x},\tau_q)] \quad (11)$$

in which $\mu^L(\rho^L)(\mathbf{x},\tau) = -s'(\rho^L(\mathbf{x},\tau)) - \beta\int J_L(\mathbf{x}-\mathbf{x}')\rho^L(\mathbf{x}',\tau)\,d\mathbf{x}'$ and $J_L(\mathbf{x}) = L^d J(L\mathbf{x})$.

We are interested in cases in which $\rho^L(\mathbf{x},0) = \rho_\beta^\pm + O(1/L)$ unless $\mathbf{x}$ is at a distance $O(1/L)$ from a (hyper)surface $\Gamma_0 \subset \Omega$, which is the interface. In the thin $[O(1/L)]$ layer around the interface, ρ^L will be approximated by a stationary instanton. The relevant time scale for such an initial condition is given by $q = 3$, as for the CHE [7,14]. A formal asymptotic expansion, based mainly on the assumption that the interface is stable, can be used (as in [7] for the CHE) to study the evolution of the interface Γ_{τ_3} for τ_3 positive. One obtains [8] that the normal velocity of a point x on Γ_{τ_3} is given by

$$V_3(\mathbf{x}) = \frac{\sigma_\beta}{2\rho_\beta^+ - 1}[\boldsymbol{\nu} \cdot \boldsymbol{\nabla}\mu_1] \pm (\mathbf{x}), \qquad (12)$$

where $\nu(\mathbf{x})$ is the unit normal at the point $\mathbf{x}$ of the interface, pointing in the direction of the denser phase; $\sigma_\beta = \beta\rho_\beta^+(1 - \rho_\beta^+) = \beta\rho_\beta^-(1 - \rho_\beta^-)$ is the mobility on the homogeneous phase. The expression between square brackets is the jump of the normal derivative of the first order correction to the chemical potential $[\mu^L = \mu_0 + L^{-1}\mu_1 + O(L^{-2})]$ across the interface, in the direction ν. Such a correction can be computed by solving the following static problem:

$$\begin{aligned} \Delta\mu_1(\mathbf{x}) &= 0 \qquad &&\text{for } \mathbf{x} \in \Omega_{\tau_3}^+ \cup \Omega_{\tau_3}^-, \\ \mu_1(\mathbf{x}) &= -S_\beta K(\mathbf{x})/(2\rho_\beta^+ - 1) \qquad &&\text{for } \mathbf{x} \in \Gamma_{\tau_3}, \end{aligned} \qquad (13)$$

in which we have used the natural decomposition of Ω into an interface and the domains $\Omega_{\tau_3}^\pm$ of the opposite phases ($\Omega = \Omega_{\tau_3}^+ \cup \Omega_{\tau_3}^- \cup \Gamma_{\tau_3}$). Moreover, in (13) μ_1 is a continuous function on Ω which is at least twice differentiable in $\Omega_{\tau_3}^+ \cup \Omega_{\tau_3}^-$, $K(\mathbf{x})$ is $d - 1$ times the mean curvature of the interface at $\mathbf{x}$ and S_β is the surface tension for the model (i.e., the free energy density associated with the instantonic solution U_β, see [17]) which can be written as

$$S_\beta = \frac{1}{2}\int_{\mathbb{R}\times\mathbb{R}} \partial_z U_\beta(z) U_\beta(z')(z' - z)\tilde{J}(z - z')\,dz\,dz'. \qquad (14)$$

The geometric motion generated by $V_3(\mathbf{x})\nu(\mathbf{x})$ is known as the *symmetric solidification model* or *Mullins-Sekerka* motion [18]. A trivial extension of [7], to include a mobility term, shows that in the same scaling the CHE gives rise to the same interface evolution (12) and (13). The only change is in the surface tension, which in the CHE case is given by $S_\beta = \zeta \int [\partial_z U_\beta^{\mathrm{CH}}(z)]^2\,dz$, where $U_\beta^{\mathrm{CH}}(z)$ is the instantonic solution for the CHE with β dependence. A way of introducing the temperature in the CHE is given in [15] and, for β finite, U_β and U_β^{CH} are qualitatively similar. However (see [8]), U_β approaches a step function as β approaches infinity, which is not the case for U_β^{CH}, for fixed ζ.

In [8] we argue that also on the diffusive time scale [$q = 2$ in (11)], that is the case in which the density in the bulk is not yet relaxed to the equilibrium one; the CHE equation and (7) show a very similar behavior. And the same is true in the limit $\beta \to \infty$, in which V_3 [given in (12)] vanishes and the motion (surface diffusion) has to be observed on the time scale $q = 4$: The analogous result for CHE is shown in [15].

We conclude by noting that we have derived rigorously a macroscopic equation describing phase segregation in alloys with long range interactions $\gamma^d J(\gamma x)$, whose only inputs are the equilibrium free energy density and diffusion constant of the reference system. The late stage coarsening described by this equation, in which the details of J enter only in determining the surface tension, is in the same universality class as the CHE [1].

We would like to thank A. Asselah, P. Buttà, R. Kohn, J. Percus, S. Puri, J. Quastel, H. Spohn, L. Xu, H. T. Yau, and particularly E. Presutti for fruitful discussions. This work was supported in part by NSF Grant No. 92-13424 4-20946, NASA Grant No. NAG3-1414, and the Swiss National Science Foundation.

*Electronic address: gbg@amath.unizh.ch
†Electronic address: Lebowitz@math.rutgers.edu

[1] J. W. Cahn and J. E. Hilliard, J. Chem. Phys. **28**, 258 (1958). For recent reviews, see J. S. Langer, in *Solids Far from Equilibrium,* edited by C. Godreche (Cambridge University Press, Cambridge, England, 1992); A. J. Bray, Adv. Phys. **43**, 357 (1994).
[2] J. L. Lebowitz and O. Penrose, J. Math. Phys. (N.Y.) **7**, 98 (1966).
[3] H. Spohn, *Large Scale Dynamics of Interacting Particles* (Springer-Verlag, Berlin, 1991).
[4] S. R. S. Varadhan and H. T. Yau, "Diffusive Limit of Lattice Gas with Mixing Conditions" (to be published).
[5] F. Rezakhanlou, Commun. Math. Phys. **129**, 445 (1990).
[6] H. Spohn and H. T. Yau, J. Stat. Phys. **79**, 231 (1995).
[7] R. L. Pego, Proc. R. Soc. London A **422**, 261 (1989).
[8] G. Giacomin and J. L. Lebowitz (to be published).
[9] J. L. Lebowitz, E. Orlandi, and E. Presutti, J. Stat. Phys. **63**, 933 (1991); G. Giacomin, J. Stat. Phys. **65**, 217 (1991); H. T. Yau, J. Stat. Phys. **74**, 705 (1994).
[10] C. Kipnis, S. Olla, and S. R. S. Varadhan, Commun. Pure Appl. Math. **42**, 115 (1989).
[11] O. Penrose and J. L. Lebowitz, J. Stat. Phys. **3**, 211 (1971).
[12] A. Asselah *et al.* (to be published).
[13] J. Quastel, "Large Deviations from a Hydrodynamic Scaling Limit for a Nongradient System" (to be published).
[14] G. Caginalp, IMA J. Appl. Math. **44**, 77 (1990).
[15] J. W. Cahn, C. M. Elliott, and A. Novick-Cohen (to be published).
[16] A. De Masi, E. Orlandi, L. Triolo, and E. Presutti, J. Stat. Phys. **73**, 543 (1993).
[17] P. Buttà, J. Stat. Phys. **72**, 1401 (1993).
[18] W. W. Mullins and R. F. Sekerka, J. Appl. Phys. **34**, 323 (1963).

Journal of Statistical Physics, Vol. 94, Nos. 5/6, 1999

Liquid–Vapor Phase Transitions for Systems with Finite-Range Interactions

J. L. Lebowitz,[1] A. Mazel,[1,2] and E. Presutti[3]

Received May 12, 1998; final November 19, 1998

We consider particles in $\mathbb{R}^d$, $d \geqslant 2$, interacting via attractive pair and repulsive four-body potentials of the Kac type. Perturbing about mean-field theory, valid when the interaction range becomes infinite, we prove rigorously the existence of a liquid–gas phase transition when the interaction range is finite but long compared to the interparticle spacing for a range of temperature.

KEY WORDS: Continuum particle system; liquid–gas phase transition; mean-field theory; Pirogov–Sinai theory; cluster expansion; Dobrushin uniqueness.

1. INTRODUCTION

An outstanding problem in equilibrium statistical mechanics is to derive rigorously the existence of a liquid–vapor phase transition in a continuum particle system. While such transitions are observed in all types of macroscopic systems there is at present no proof from first principles of their existence in particles interacting with any kind of reasonable potential, say Lennard–Jones or hard core plus attractive square well. Such potentials are known, by comparison of experiment with low density expansions, to accurately describe the observed behavior of gases. Furthermore the properties of these systems at higher densities, obtained via approximate integral equations, are in good agreement with those of liquids in the ranges of temperatures and pressures where boiling and condensation takes

[1] Departments of Mathematics and Physics, Rutgers University, New Brunswick, New Jersey 08903; e-mail: lebowitz@math.rutgers.edu.
[2] International Institute of Earthquake Prediction Theory and Theoretical Geophysics, 113556 Moscow, Russia; e-mail:mazel@math.rutgers.edu.
[3] Dipartimento di Matematica, Università di Roma Tor Vergata, 00133 Rome, Italy; e-mail: presutti@axp.mat.utovrm.it.

place. In fact, computer simulations using classical statistical mechanics of such systems, containing several hundred to several million particles, convincingly show, via extrapolations which take into account finite size effects, that systems with these type of interactions will have true liquid–vapor phase transitions in the thermodynamic (infinite volume) limit. In this paper we go some way toward a proof of such a transition, i.e., we prove for the first time the existence of a liquid–vapor transition in a continuum particle system with finite range interactions and no special symmetry.

Historically the first proof of liquid–vapor type phase transitions was given for lattice systems (which are isomorphic to Ising spins). These systems can be thought of as idealizations of real fluids in which however the natural continuous spatial translation invariance symmetry is replaced by that of the lattice $\mathbb{Z}^d$, $d \geqslant 2$. It was Peierls [P] who first gave a convincing argument (later made fully rigorous by Dobrushin [D1] and Griffiths [G]) of the coexistence of different phases in such systems. The power of Gibbsian statistical mechanics to produce such rigorous results was brought home to the general science community by the dramatic work of Onsager [O] who exhibited the detailed structure of the critical point associated with this transition by explicitly solving the two dimensional Ising model (or lattice gas), with nearest neighbor interactions. Since that time there have been found many other exactly solvable two dimensional lattice systems [Bax]. At the same time the development of various types of inequalities as well as the powerful Pirogov–Sinai formalism [PS] have resulted in a comprehensive rigorous theory of phase transitions in lattice systems, at sufficiently low temperatures.

By contrast, much less is known rigorously about phase transitions in continuum systems. There is basically only the case of the two component Widom–Rowlinson [WR] model, in which the interaction is a hard core repulsion between particles of different species, where Ruelle [R1] was able to generalize the Peierls argument to prove phase coexistence in this system. Ruelle's proof strongly exploits the symmetry between the two components present in the WR model. The same is true, at least to some extent, of various extensions of this model [LL], [GH], [CCK]: see, however, [BKL] where some special multicomponent models without symmetry were treated by an extension of the Pirogov–Sinai (hereafter denoted by P–S) formalism. For general continuum systems without some special symmetry the only proofs of phase transitions so far are for systems with interactions which decay very slowly or not at all. Such one dimensional models, with many particle interactions, were analyzed and proven to exhibit phase transitions by Fisher and Felderhof [FF]. More recently Johansson [J] has considered interactions in one dimension which decay as $r^{-\alpha}$, $\alpha \in (1, 2)$, proving that at low temperatures there is phase transition in the sense that the pressure is not differentiable.

In a genuine continuum particle system with the type of pair interactions discussed above one expects several phase transitions. At high temperatures and small densities we have the gaseous phase, then, lowering the temperature, we first find the liquid and then the solid, crystalline phases, at the appropriate values of the density. We will concentrate here on the vapor–liquid phase transition, which is simpler to investigate than the formation of the periodic structures of crystals. Of course we also want to take advantage of the powerful techniques developed for proving phase transition in lattice systems, but these are generally applicable only at temperatures much smaller than those at which the vapor–liquid transition occurs. We will overcome this problem by performing a coarse graining which may also be considered a one step renormalization group transformation. To do this we divide space into large cubes of side ℓ and introduce variables ϱ_x, which are the particle densities in the cubes labeled by x. After integrating out all the other variables, we will be left with a new system described by the variables ϱ_x, their distribution still Gibbsian with a new Hamiltonian and temperature. The essential point is that the ratio between the new and old temperatures scales as ℓ^{-d}.

In general the verification of such statements is a very hard if not impossible task, but it can be accomplished in a relatively simple fashion if we consider interactions of Kac type in which the range γ^{-1} of the interaction is small. By suitably choosing the side of the cubes we will then enter into the low temperature regime where the Peierls and the P–S methods apply. In such a new perturbative scheme, the unperturbed state is described by mean field (formally $\gamma = 0$) and the small parameter of the expansion is the inverse interaction range γ, instead of the temperature in the traditional approach. By choosing a suitable range of values of chemical potential and temperature we will then be able to put ourselves at the vapor–liquid phase coexistence.

The mean field or van der Waals type of phase transition was first derived rigorously by Kac, Uhlenbeck and Hemmer [KUH] for a class of one dimensional models, hard rods of radius one with an added pair potential

$$\phi_\gamma(q_i, q_j) = -\alpha \tfrac{1}{2} \gamma \exp[-\gamma |q_i - q_j|], \qquad \gamma, \alpha > 0 \tag{1.1}$$

in the limit $\Gamma \to 0$, see also van Kampen [vK]. This was later generalized by Lebowitz and Penrose [LP] to d-dimensional systems with suitable short range interactions and Kac potentials of the form

$$\phi_\gamma(q_i, q_j) = -\alpha\gamma^d J(\gamma |q_i - q_j|) \tag{1.2}$$

with

$$\int_{\mathbb{R}^d} J(r)\,dr = 1, \qquad J(r) > 0 \tag{1.3}$$

In the thermodynamic limit followed by the limit $\gamma \to 0$ the Helmholtz free energy a takes the form, for a fixed temperature β^{-1},

$$\lim_{\gamma \to 0} a(\varrho, \gamma) = CE\{a_0(\varrho) - \tfrac{1}{2}\alpha\varrho^2\} \tag{1.4}$$

Here ϱ is the particle density, a_0 is the free energy density of the reference system, i.e., the system with $\alpha = 0$ in (1.2). a_0 is convex in ϱ (by general theorems) and $CE\{f(x)\}$ is the largest convex lower bound of f. For α large enough the term in the curly brackets in (1.4) has a double well shape and the *CE* corresponds to the Gibbs double tangent construction. This is equivalent to Maxwell's equal area rule applied to a van der Waals' type equation of state where it gives the coexistence of liquid and vapor phases [LP]; see also [vK].

As already discussed we prove the coexistence of liquid and gas phases for systems with finite range interactions as small perturbations, at finite $\gamma > 0$, from the mean field behavior at $\gamma = 0$. The same approach was recently applied to the lattice case for Ising models, [CP], [BZ], [BP], where the Peierls argument applies directly, because of the spin flip symmetry of the model. The absence of symmetries in our case requires instead the whole machinery of the Pirogov–Sinai theory. To insure stabilization against collapse, which would be induced by a purely attractive pair potential, the natural choice is to replace the point particles by hard spheres. Our approach however does not work in such a case, as we need a cluster expansion for the unperturbed reference system (i.e., without the Kac interaction) at values of the chemical potential or density for which it is not proved to hold. We avoid the problem by considering point particles and insuring stability by introducing a positive four body potential of the same range as the negative two body potential. In this way we avoid having to control strong short range interactions, something beyond our present day abilities for dense continuum systems. A preliminary announcement of our results is contained in [LMPI] and a fuller description of the proofs (with some variations) can be found in [Pr].

2. DEFINITIONS AND RESULTS

For ease of reference we collect below the main definitions in consecutive subsections.

Particle Configurations and Phase Space

We consider a system of identical point particles in $\mathbb{R}^d$, $d \geqslant 2$ and call *particle configurations* the locally finite subsets of $\mathbb{R}^d$. The collection of all particle configurations in Λ, $\Lambda \subset \mathbb{R}^d$, is the phase space $\mathcal{Q}^{(\Lambda)}$ and we simply write $\mathcal{Q}$ when $\Lambda = \mathbb{R}^d$. The particle configurations are denoted by q and sometimes by $q^{(\Lambda)}$ when we want to underline that they are in $\mathcal{Q}^{(\Lambda)}$. We write $q = (q_1,..., q_n)$ when the configuration consists of $|q| = n$ particles positioned at points $q_1,..., q_n \in \mathbb{R}^d$.

Free measure

For a bounded measurable subset Λ of $\mathbb{R}^d$ the free measure dq on $\mathcal{Q}^{(\Lambda)}$ (also called the Liouville measure) is

$$\int_{\mathcal{Q}^{(\Lambda)}} dq\, f(q) = \sum_{n=0}^{\infty} \frac{1}{n!} \int_{\Lambda} dq_1 \cdots \int_{\Lambda} dq_n\, f(q_1,..., q_n) \tag{2.1}$$

where f is any bounded measurable function on $\mathcal{Q}^{(\Lambda)}$.

Hamiltonian

The energy of the configuration $q = (q_i)$ is given by the formal Hamiltonian

$$\begin{aligned} H_{\gamma,\lambda}(q) = {} & -\lambda\, |q| - \frac{1}{2!} \sum_{i_1} \sum_{i_2 \neq i_1} J_\gamma^{(2)}(q_{i_1}, q_{i_2}) \\ & + \frac{1}{4!} \sum_{i_1} \sum_{i_2 \neq i_1} \sum_{i_3 \neq i_1, i_2} \sum_{i_4 \neq i_1, i_2, i_3} J_\gamma^{(4)}(q_{i_1}, q_{i_2}, q_{i_3}, q_{i_4}) \end{aligned} \tag{2.2}$$

Here λ is the chemical potential while $J_\gamma^{(2)}(\cdot, \cdot)$ and $J_\gamma^{(4)}(\cdot, \cdot, \cdot, \cdot)$ are respectively two and four-body potentials. For notational simplicity we choose $\gamma \in \{2^{-n}, n \in \mathbb{N}\}$ and γ is a scaling factor in the definition of $J_\gamma^{(2)}(\cdot, \cdot)$ and $J_\gamma^{(4)}(\cdot, \cdot, \cdot, \cdot)$. These are chosen of the form

$$J_\gamma^{(2)}(q_{i_1}, q_{i_2}) = \gamma^{2d} \int dr \prod_{j=1}^{2} \mathbb{1}_{|r - q_{i_j}| \leqslant \gamma^{-1} R_d} \tag{2.3}$$

$$J_\gamma^{(4)}(q_{i_1}, q_{i_2}, q_{i_3}, q_{i_4}) = \gamma^{4d} \int dr \prod_{j=1}^{4} \mathbb{1}_{|r - q_{i_j}| \leqslant \gamma^{-1} R_d} \tag{2.4}$$

where $\mathbb{1}_A$ is the characteristic function of the set A and R_d is the radius of the ball in $\mathbb{R}^d$ having a unit volume. Denoting by $B_\gamma(r)$ a ball of radius $\gamma^{-1}R_d$ centered at $r \in \mathbb{R}^d$ we have

$$J_\gamma^{(2)}(q_{i_1}, q_{i_2}) = \gamma^{2d} \, |B_\gamma(q_{i_1}) \cap B_\gamma(q_{i_2})| \tag{2.5}$$

and

$$J_\gamma^{(4)}(q_{i_1}, q_{i_2}, q_{i_3}, q_{i_4}) = \gamma^{4d} \left| \bigcap_{j=1}^{4} B_\gamma(q_{i_j}) \right| \tag{2.6}$$

Let $J^{(2)} = J_1^{(2)}$ and $J^{(4)} = J_1^{(4)}$. Then the scaling properties of $J_\gamma^{(2)}$ and $J_\gamma^{(4)}$ can be expressed by

$$J_\gamma^{(2)}(q_{i_1}, q_{i_2}) = \gamma^d J^{(2)}(\gamma q_{i_1}, \gamma q_{i_2}) \tag{2.7}$$

and

$$J_\gamma^{(4)}(q_{i_1}, q_{i_2}, q_{i_3}, q_{i_4}) = \gamma^{3d} J^{(4)}(\gamma q_{i_1}, ..., \gamma q_{i_4}) \tag{2.8}$$

which is similar to (1.2). It is also clear that

$$\int J_\gamma^{(2)}(0, r)\, dr = \int J_\gamma^{(4)}(0, r_1, r_2, r_3)\, dr_1\, dr_2\, dr_3 = 1 \tag{2.9}$$

as in (1.3).

As both $J_\gamma^{(2)}$ and $J_\gamma^{(4)}$ are positive we have in (2.2) a competition between an attractive pair and a repulsive four-body potential. When the scaling parameter γ is small (but finite), the model has a large but finite interaction radius, $2R_d\gamma^{-1}$, and a small interaction strength between any given two or four particles. Nevertheless, the total strength of the interaction between a given particle and all other particles in a configuration q of bounded nonvanishing density is of order 1. These are characteristic properties of the Kac potentials which, as noted earlier, usually reproduce the van der Waals theory [LP] in the scaling limit $\gamma \to 0$. The specific form (2.3)–(2.4) of the interaction $J^{(2)}$ and $J^{(4)}$ makes the analysis simpler; for more general potentials see (2.14) and the discussion in the beginning of Section 3.

Finite and Infinite Volume Gibbs Measures

For any two locally finite configurations $q = (q_i)$ and $\bar{q} = (\bar{q}_j)$ denote by $q \cup \bar{q} = (q_i, \bar{q}_j)$ the configuration including all particles from both q and $\bar{q}$. The *conditional energy* of $q^{(\Lambda)}$ given a configuration $\bar{q} \in \mathcal{Q}$ is

$$H_{\gamma,\lambda}(q^{(\Lambda)} \mid \bar{q}) = H_{\gamma,\lambda}(q^{(\Lambda)} \cup \bar{q}) - H_{\gamma,\lambda}(\bar{q}) \tag{2.10}$$

This conditional energy consist of two parts: the energy, $H_{\gamma,\lambda}(q^{(\Lambda)})$, of the configuration $q^{(\Lambda)}$ itself and the *interaction energy*

$$U_\gamma(q^{(\Lambda)} \mid \bar{q}) = H_{\gamma,\lambda}(q^{(\Lambda)} \mid \bar{q}) - H_{\gamma,\lambda}(q^{(\Lambda)}) \tag{2.11}$$

between configurations $q^{(\Lambda)}$ and $\bar{q}$. Both (2.10) and (2.11) are finite sums because of the finite range of the potential.

Let $\Lambda^c = \mathbb{R}^d \setminus \Lambda$ be the complement of Λ. The *Gibbs measure* $\mu^{(\Lambda)}_{\gamma,\beta,\lambda}(dq^{(\Lambda)} \mid \bar{q}^{(\Lambda^c)})$ in the bounded measurable set Λ with boundary condition $\bar{q}^{(\Lambda^c)}$ and inverse temperature $\beta > 0$, is the probability measure on $\mathcal{Q}^{(\Lambda)}$ given by

$$\mu^{(\Lambda)}_{\gamma,\beta,\lambda}(dq^{(\Lambda)} \mid \bar{q}^{(\Lambda^c)}) = \frac{e^{-\beta H_{\gamma,\lambda}(q^{(\Lambda)} \mid \bar{q}^{(\Lambda^c)})}}{\Xi_{\gamma,\beta,\lambda}(\Lambda \mid \bar{q}^{(\Lambda^c)})}\, dq^{(\Lambda)} \tag{2.12}$$

where $\Xi_{\gamma,\beta,\lambda}(\Lambda \mid \bar{q}^{(\Lambda^c)})$ is the *partition function*

$$\Xi_{\gamma,\beta,\lambda}(\Lambda \mid \bar{q}^{(\Lambda^c)}) = \int_{\mathcal{Q}^{(\Lambda)}} dq^{(\Lambda)}\, e^{-\beta H_{\gamma,\lambda}(q^{(\Lambda)} \mid \bar{q}^{(\Lambda^c)})} \tag{2.13}$$

The infinite volume Gibbs measures $\mu_{\gamma,\beta,\lambda}(dq)$ are probabilities on $\mathcal{Q}$ such that for any bounded measurable set Λ and $\mu_{\gamma,\beta,\lambda}$-almost any $\bar{q}^{(\Lambda^c)} \in \mathcal{Q}^{(\Lambda^c)}$ the conditional measure $\mu_{\gamma,\beta,\lambda}(dq^{(\Lambda)} \mid \bar{q}^{(\Lambda^c)})$ is equal to $\mu^{(\Lambda)}_{\gamma,\beta,\lambda}(dq^{(\Lambda)} \mid \bar{q}^{(\Lambda^c)})$ given by (2.12).

We say that a translation invariant Gibbs measure has a particle density $\varrho > 0$ if for any bounded set Λ the expectation of $|q \cap \Lambda|$ is equal to $\varrho\,|\Lambda|$.

The Main Result

Our main purpose is to investigate the phase diagram of the model (2.2) and to prove that a phase transition of the liquid–gas type takes place for some values of the parameters. Let $\beta_c = (\frac{3}{2})^{3/2}$ and $\beta_0 > \beta_c$ be a number defined via (3.14) below.

Theorem 2.1. For any $\beta \in (\beta_c, \beta_0)$ there exist functions $\gamma_0(\beta)$ and $\lambda(\gamma, \beta)$ such that for $0 < \gamma < \gamma_0(\beta)$ the model (2.2) has at least two distinct Gibbs measures $\mu^{\pm}_{\gamma,\beta,\lambda(\gamma,\beta)}(dq)$. These measures are translation invariant and ergodic, with an exponential decay of correlations. They have particle densities respectively equal to $\varrho^*_{\gamma,\beta,-} > 0$ and $\varrho^*_{\gamma,\beta,+} > \varrho^*_{\gamma,\beta,-}$.

We will prove Theorem 2.1 by using the P–S theory [PS]. The restriction $\beta \in (\beta_c, \beta_0)$ rather than $\beta > \beta_c$ allows for a simplified proof, we hope however to present in a forthcoming paper a proof for all $\beta > \beta_c$. While technically different and applied to different settings our methods are nevertheless conceptually close to those of [DZ].

The quantities $\lambda(\gamma, \beta)$ and $\varrho^*_{\gamma,\beta,\pm}$ in Theorem 2.1 have limits as $\gamma \to 0$, denoted by $\lambda(\beta)$ and $\varrho_{\beta,\pm}$, which, as already noted, are given exactly by mean field type formulas [LP]. Our proof of Theorem 2.1 is a perturbation theory constructed around the mean field picture.

We could extend Theorem 2.1 to slightly more general interactions where the range of the two and four body potentials are not necessarily the same. In particular we can make the range of the attractive two body forces greater than that of the repulsive four body ones, according to the physical intuition that the force is repulsive at short distances and attractive at longer distances. For brevity we do not give the details of the proofs and only introduce the new class of potentials. The four body interaction is unchanged while $J^{(2)}_\gamma(q_1, q_2)$ in (2.3) is replaced by

$$W^{(2)}_\gamma(q_1, q_2) = \gamma^{2d} \int dr \int dr' \, \mathbb{1}_{|r - q_2| \leqslant \gamma^{-1} R_d} \mathbb{1}_{|r' - q_2| \leqslant \gamma^{-1} R_d} \gamma^d w(\gamma\, |r - r'|) \qquad (2.14)$$

with w a smooth, non negative function having compact support and integral equal to 1. $\gamma^d w(\gamma\, |q_1 - q_2|)$ is the classical choice for a Kac pair potential (with the right scaling properties), what we do in (2.14) is simply to average such a function with the two arguments varied respectively on the balls $B_\gamma(q_1)$ and $B_\gamma(q_2)$. The choice $w = \delta(r - r')$ in (2.14) gives back (2.3).

Outline of the Remaining Sections

In Section 3 we give an outline of the proof formulating a number of statements which are proven in the next sections. In Section 4 we prove Peierls estimates on contours, and in Section 5 we use a cluster expansion to investigate the properties of the effective Hamiltonian obtained after the coarse graining transformation. In Section 6 we study the restricted ensembles proving the basic property of the P–S scheme, namely that it is possible to adjust the value of the chemical potential in such a way that the pressures in the two restricted ensembles are equal. We will conclude from this the proof of Theorem 2.1. Sections 7 and 8 contain more technical material.

3. SCHEME OF PROOF

The starting point of our approach, which as explained in the previous section is of crucial importance, consists of coarse graining or one step renormalization: we partition space into cubes $C_x^{(\ell_2)}$, x the centers of the cubes, of side $\ell_2 := \gamma^{-1+\alpha_2}$, α_2 a small positive number, (the subscript 2 foresees the use of other scales that will be introduced later in the proof). Given a particle configuration q, we call ϱ_x, the particle densities in each cube and we integrate out all the other variables, i.e., we consider the marginal over the variables ϱ_x. The new measure is still Gibbsian and its new effective Hamiltonian (a function of the ϱ_x) can be characterized with remarkable accuracy by cluster expansion techniques, see Section 5. As mentioned earlier, the main point of this procedure is that the new effective inverse temperature becomes $\beta\ell_2^d$. We can thus enter into the very low temperature regime by taking γ small enough and ℓ_2 correspondingly large (in Section 5 we will absorb the temperature into the Hamiltonian). we are then in the right setup for the P–S theory. An analysis à la [LP], which we here omit, would show that the new effective Hamiltonian converges formally, in the limit $\gamma \to 0$, to the mean field free energy functional $\mathscr{F}_{\beta,\lambda}(\varrho)$ given in (3.1) below. We will begin our analysis by characterizing the ground states of $\mathscr{F}_{\beta,\lambda}(\varrho)$ and thereafter use the P–S theory to investigate the perturbations at $\gamma > 0$.

Mean Field Free Energy Functional and Ground States

The mean field Gibbs free energy functional $\mathscr{F}_{\beta,\lambda}(\varrho)$, $\varrho = \varrho(r)$ denoting a non-negative bounded measurable function in $\mathbb{R}^d$, is

$$\mathscr{F}_{\beta,\lambda}(\varrho) = \int dr\, \frac{\varrho(r)}{\beta}(\log \varrho(r) - 1) - \int dr\, \lambda\varrho(r)$$

$$- \frac{1}{2!}\int dr_1\, dr_2\, J^{(2)}(r_1, r_2)\, \varrho(r_1)\, \varrho(r_2)$$

$$+ \frac{1}{4!}\int dr_1 \cdots dr_4\, J^{(4)}(r_1,..., r_4)\, \varrho(r_1) \cdots \varrho(r_4) \tag{3.1}$$

The first integral is the entropy contribution to the free energy (more precisely the product of the temperature times the entropy of the ideal gas with the sign changed). The other three terms arise from the corresponding interaction terms in (2.2). More details concerning the relation between (3.1) and (2.2) can be found in Section 4.

The *mean field ground states* are, by definition, the minimizers of $\mathscr{F}_{\beta,\lambda}(\varrho)$. To find them we set

$$\mathscr{R}(r,\varrho)=\int_{|r-r_1|\leqslant R_d} dr_1\,\varrho(r_1) \tag{3.2}$$

and

$$\mathscr{S}(r,\varrho)=\int_{|r-r_1|\leqslant R_d} dr_1\,\frac{\varrho(r_1)}{\beta}(\log\varrho(r_1)-1) \tag{3.3}$$

$\mathscr{R}(r,\varrho)$ and $\mathscr{S}(r,\varrho)$ are respectively the mean density and the mean entropy of ϱ in the ball $B(r)$. With this notation we can rewrite (3.1) as

$$\mathscr{F}_{\beta,\lambda}(\varrho)=\int dr\left(\mathscr{S}(r,\varrho)-\lambda\mathscr{R}(r,\varrho)-\frac{1}{2!}\mathscr{R}(r,\varrho)^2+\frac{1}{4!}\mathscr{R}(r,\varrho)^4\right) \tag{3.4}$$

which is true only because of the special form, (2.3) and (2.4), of $J^{(2)}$ and $J^{(4)}$. In fact (3.4) is the main reason for choosing $J^{(2)}$ and $J^{(4)}$ of this form. Observe however that if the potential $J^{(2)}$ is replaced by $W^{(2)}$ as given in (2.14) (with $\gamma=1$) then the r.h.s. of (3.4) simply becomes, after some easy manipulations,

$$\begin{aligned}&\int dr\left(\mathscr{S}(r,\varrho)-\lambda\mathscr{R}(r,\varrho)-\frac{1}{2!}\mathscr{R}(r,\varrho)^2+\frac{1}{4!}\mathscr{R}(r,\varrho)^4\right)\\&\quad+\frac{1}{4}\int dr\int dr'\,w(|r-r'|)[\mathscr{R}(r,\varrho)-\mathscr{R}(r',\varrho)]^2\end{aligned} \tag{3.5}$$

By convexity

$$\mathscr{S}(r,\varrho)\geqslant\frac{\mathscr{R}(r,\varrho)}{\beta}(\log\mathscr{R}(r,\varrho)-1) \tag{3.6}$$

and equality is achieved only if $\mathscr{R}(r,\varrho)=\varrho(r)$.

It follows then from (3.4) and (3.6) that if $s\geqslant 0$ is a minimizer of the function

$$F_{\beta,\lambda}(s)=\frac{s}{\beta}(\log s-1)-\lambda s-\frac{s^2}{2!}+\frac{s^4}{4!} \tag{3.7}$$

then $\varrho(r) = \mathscr{R}(r, \varrho) \equiv s$ is a minimizer of $\mathscr{F}_{\beta,\lambda}(\varrho)$, (and also of the functional in (3.5), corresponding to $\mathscr{F}$ with the modified potential (2.14)). The second derivative of $F_{\beta,\lambda}(s)$,

$$F''_{\beta,\lambda}(s) = \frac{1}{\beta s} - 1 + \frac{s^2}{2} \tag{3.8}$$

is positive if $\beta < \beta_c = (\frac{3}{2})^{3/2}$. Hence for $\beta < \beta_c$ and any λ the function $F_{\beta,\lambda}(s)$ is convex and has a unique minimizer which is the root of the equation

$$-s + \frac{1}{3!} s^3 + \frac{1}{\beta} \log s - \lambda = 0 \tag{3.9}$$

On the contrary, for $\beta > \beta_c$ this equation has three positive roots with two of them, $s = \varrho_{\beta,\lambda,-}$ and $s = \varrho_{\beta,\lambda,+}$, local minimizers of $F_{\beta,\lambda}(s)$. Furthermore, there exists a unique $\lambda = \lambda(\beta)$ for which both local minima are the global ones and the function $F_{\beta,\lambda}(s)$ has a "double well" shape. We set $\varrho_{\beta,\pm} = \varrho_{\beta,\lambda(\beta),\pm}$. Clearly for $\beta > \beta_c$ and $\lambda = \lambda(\beta)$ the densities $\varrho(r) \equiv \varrho_{\beta,-}$ and $\varrho(r) \equiv \varrho_{\beta,+}$ are distinct mean field gound states. For later purposes we remark that

$$-1 < \lambda(\beta) < 0 \tag{3.10}$$

which can be checked by direct calculation.

Mean Field Equations and Contraction Property

As the ground states are minimizers of $\mathscr{F}_{\beta,\lambda}(\varrho)$, they satisfy the mean field equation $\delta\mathscr{F}_{\beta,\lambda}(\varrho)/\delta\varrho(r) = 0$. By an explicit calculation we then find that they are fixed points of the transformation

$$\varrho(\cdot) \to \Phi(\varrho(\cdot), \cdot) \tag{3.11}$$

where

$$\Phi(\varrho(\cdot), r) := \exp\left\{\lambda + \int dr_1\, J^{(2)}(r, r_1)\, \varrho(r_1) - \frac{1}{3!} \int dr_1 \cdots \int dr_3\, J^{(4)}(r,..., r_3)\, \varrho(r_1)\, \varrho(r_2)\, \varrho(r_3)\right\} \tag{3.12}$$

Setting $\varrho^* = \varrho_{\beta,-}$ or $\varrho^* = \varrho_{\beta,+}$ the derivative $\psi(r) = \delta\Phi(\varrho(\cdot), r)/\delta\varrho(r)$ computed at $\varrho(r) \equiv \varrho^*$ is equal to

$$\psi(r) = \beta\varrho^* J^{(2)}(0, r)[1 - (\varrho^*)^2/2] \tag{3.13}$$

We define β_0 in Theorem 2.1 so that for all $\beta \in (\beta_c, \beta_0)$

$$\int dr\, |\psi(r)| = \beta\varrho^*\, |1 - \varrho^{*2}/2| < 1 \tag{3.14}$$

If (3.14) holds, the transformation Φ is a contraction (in a neighborhood of the ground states and using sup norms).

The existence of $\beta_0 > \beta_c$ follows by noticing that since $F''_{\beta,\lambda}(\varrho^*) > 0$ then by (3.8) $\beta\varrho < 1$ and the only condition to check for (3.14) to hold is $\beta\varrho^*[1 - \varrho^{*2}/2] > -1$. By (3.8) it can be rewritten as

$$F''_{\beta,\lambda}(\varrho^*) < \frac{2}{\beta\varrho^*} \tag{3.15}$$

and $F''_{\beta,\lambda}(\varrho^*) \to 0$ as $\beta \to \beta_c$ while both $\varrho_{\beta,\pm}$ remain bounded.

We will see in Section 5 that for $\beta \in (\beta_c, \beta_0)$ and $\gamma > 0$ small enough, the Dobrushin uniqueness condition is satisfied by the effective Hamiltonian in the system restricted to the ground state ensemble (defined later in Section 3). That property is the $\gamma > 0$ analogue of the contraction property (3.14). We will in the sequel restrict ourselves to $\beta \in (\beta_c, \beta_0)$ which ensures the following technical conditions.

For $\beta_c < \beta < \beta_0$ there exists a positive number $\zeta(\beta)$ such that

$$\max_{\sigma_1 = \pm 1, \sigma_2 = \pm 1} \beta(\varrho_{\beta,\sigma_1} + 2\zeta(\beta)) \left| 1 - \frac{(\varrho_{\beta,\sigma_1} + \sigma_2 2\zeta(\beta))^2}{2} \right| = a(\beta) < 1 \tag{3.16}$$

Consequently there exist a positive number $\delta(\beta)$ such that for any $\lambda \in (\lambda(\beta) - \delta(\beta), \lambda(\beta) + \delta(\beta))$ one has

$$\max_{\sigma_1 = \pm 1, \sigma_2 = \pm 1} \beta(\varrho_{\beta,\lambda,\sigma_1} + \zeta(\beta)) \left| 1 - \frac{(\varrho_{\beta,\lambda,\sigma_1} + \sigma_2 2\zeta(\beta))^2}{2} \right| \leqslant a(\beta) \tag{3.17}$$

Note $(\varrho_{\beta,\lambda,\sigma_1} + \zeta(\beta))$ in (3.17) instead of $(\varrho_{\beta,\sigma_1} + 2\zeta(\beta))$ in (3.16). This allows for the same $a(\beta)$ in (3.17) and (3.16) provided $\zeta(\beta)$ and $\delta(\beta)$ are small enough.

We expect that for γ small enough the particle configurations will have densities close to the mean field ground states. To investigate the issue we need a spatial scale for computing particle densities and a notion of closeness between densities. We will then have a picture of the particle configurations in terms of spatial regions where there is agreement or disagreement with the ground states.

The Partitions $\mathscr{D}^{(\ell)}$, the Cubes $C_x^{(\ell)}$ and the Densities $\varrho_x^{(\ell)}$

Let $\mathscr{D}^{(\ell)}$, $\ell \in \{2^n, n \in \mathbb{N}\}$, be decreasing partitions of $\mathbb{R}^d$ into cubes $C^{(\ell)}$ of side ℓ, i.e., $\mathscr{D}^{(\ell)}$ is coarser than $\mathscr{D}^{(\ell')}$ if $\ell > \ell'$. For any $r \in \mathbb{R}^d$ we denote by $C^{(\ell)}(r)$ the cube of $\mathscr{D}^{(\ell)}$ that contains r. We suppose that the centers of the cubes $C^{(1)}$ are in $\mathbb{Z}^d$. Consequently the centers of cubes $C^{(\ell)}$ are in $\mathbb{Z}_\ell^d = \ell\mathbb{Z}^d + ((\ell-1)/2, ..., (\ell-1)/2)$. For $x \in \mathbb{Z}_\ell^d$ we denote $C_x^{(\ell)}$ the cube of $D^{(\ell)}$ centered at x.

Given a region Λ, $[\Lambda]^{(\ell)}$ is the maximal $D^{(\ell)}$ measurable subset of Λ, i.e., the union of all $C^{(\ell)}$ contained in Λ. We also identify $[\Lambda]^{(\ell)}$ with $\{x \in \mathbb{Z}_\ell^d \mid x \in [\Lambda]^{(\ell)}\}$. The $\mathbb{R}^d$ volume of $[\Lambda]^{(\ell)}$ is denoted by $|[\Lambda]^{(\ell)}|$ while its $\mathbb{Z}_\ell^d$ volume, i.e., the number of lattice points, is denoted by $\|[\Lambda]^{(\ell)}\|$.

The density of a configuration q in a box $C_x^{(\ell)}$ is

$$\varrho_x^{(\ell)}(q) = \ell^{-d}\, |q \cap C_x^{(\ell)}| \tag{3.18}$$

Accordingly $\varrho^{(\ell)}(q) = (\varrho_x^{(\ell)}(q))$ is called a ($\mathscr{D}^{(\ell)}$ measurable) *density configuration* corresponding to a particle configuration q. We denote by $\varrho^{(\ell)}(\Lambda) = (\varrho_x^{(\ell)}(\Lambda))$ a density configuration in $[\Lambda]^{(\ell)}$ not related to any particle configuration.

The notation $[q]^{(\ell)}$ is used for the configuration obtained from $q = (q_i)$ by shifting all q_i to the centers of the corresponding boxes $C^{(\ell)}(q_i)$. For a configuration $q^{(\Lambda)}$ we set

$$H_{\gamma,\lambda}(\varrho^{(\ell)}(q^{(\Lambda)})) = H_{\gamma,\lambda}([q^{(\Lambda)}]^{(\ell)}) \tag{3.19}$$

and

$$H_{\gamma,\lambda}(\varrho^{(\ell)}(q^{(\Lambda)}) \mid \bar{q}^{(\Lambda^c)}) = H_{\gamma,\lambda}([q^{(\Lambda)}]^{(\ell)} \mid \bar{q}^{(\Lambda^c)}) \tag{3.20}$$

where Λ is a region and $\bar{q}^{(\Lambda^c)}$ is a boundary condition.

Spatial Scales, η Functions and Ground-State Configurations

Several scales are of special interest for us

$$\ell_1 = \ell_{1,\gamma} = \gamma^{-1+\alpha_1}, \qquad \ell_2 = \ell_{2,\gamma} = \gamma^{-1+\alpha_2}, \qquad \ell_3 = \ell_{3,\gamma} = \gamma^{-1-\alpha_3} \tag{3.21}$$

where the numbers $0 < \alpha_i < 1$ are rational, $\alpha_1 = 1/2$ and $\alpha_2 + \alpha_3 \ll (2d)^{-1}$.

The scale ℓ_2 is our coarse graining scale. Agreement or disagreement with the ground state is indicated by the η-functions, $\eta = \eta(q) = (\eta_x(q))$, $x \in \mathbb{Z}^d_{\ell_2}$:

$$\eta_x(q) = \begin{cases} -1, & \text{for } |\varrho_x^{(\ell_2)}(q) - \varrho_{\beta,-}| \leqslant \zeta \\ +1, & \text{for } |\varrho_x^{(\ell_2)}(q) - \varrho_{\beta,+}| \leqslant \zeta \\ 0, & \text{otherwise} \end{cases} \tag{3.22}$$

$\zeta = \zeta(\beta)$ being taken from (3.16). Particle configurations q or density configurations $\varrho^{(\ell)}(q)$ are called compatible with η if $\eta = \eta(\varrho^{(\ell)}(q)) = \eta(q)$.

We then say that a particle configuration q belongs to the *ground state ensemble* of the $\pm$ *phase* or equivalently *liquid/vapor phase* if $\eta(q) \equiv \pm 1$. For brevity we will sometimes simply say that q is a $\pm$ ground state configuration.

For a scale ℓ we define the corresponding *standard ground state configurations* $q_-^{(\ell)}$ and $q_+^{(\ell)}$ as those which have $\varrho^{(\ell)}(q_-^{(\ell)}) \equiv \varrho_{\beta,-}$ and $\varrho^{(\ell)}(q_+^{(\ell)}) \equiv \varrho_{\beta,+}$ with all the particles placed at the centers of the corresponding boxes $C^{(\ell)}$. Of course for the irrational $\varrho_{\beta,\pm}$ and even for some rational $\varrho_{\beta,\pm}$ it is impossible to achieve these densities exactly. With an abuse of the notation we set $\varrho^{(\ell)}(q_\pm^{(\ell)}) \equiv \varrho_{\beta,\pm}$ when the densities are the closest possible to $\varrho_{\beta,\pm}$. When $\ell = \ell_2$ we will drop the superscript (ℓ).

The other scales, ℓ_3 and ℓ_1, are used to construct contours (see below) and to do some approximate calculations (see Section 4) respectively.

Our proof of the phase transition or coexistence of phases is based on the following qualitative picture describing typical particle configurations in the two pure phases. In the gas phase a typical configuration q coincides in most of $\mathbb{R}^d$ with some typical configuration in the gas ground state ensemble. Inside this "sea" of the gas ground state there are rare "islands" occupied by the liquid ground state. These two types of ground states are separated from each other by regions in $\mathbb{R}^d$ which are called Peierls contours. The "excess" free energy of q with respect to the free energy of the gas ground state occupying all of $\mathbb{R}^d$ is concentrated in the vicinity of these contours and is proportional to their volume. A similar, inverse, picture describes typical configurations of the liquid phase; the liquid ground state now forms a "sea" with rare "islands" of the gas ground state. Thus the typical configurations of the gas and liquid phases are distinct with the density of particles being an order parameter distinguishing them. A rigorous verification of the above picture is fairly straightforward for simple lattice gases or Ising systems, in which the gas and liquid ground states correspond respectively to "all sites empty" and "all sites occupied". It requires however considerable work for continuum systems and we start with precise definitions.

Correct and Incorrect Sets, Contours

A family of cubes from the partition $\mathscr{D}^{(\ell)}$ is called *$*$-connected* if the closure of the union of the cubes in this family is a connected subset of $\mathbb{R}^d$; we consider from now on $d \geqslant 2$. We then say that r is a $(+)$-*correct* point of a configuration q if $\eta_x(q) = +1$ whenever $C_x^{(\ell_2)}(r) \subset \bar{C}^{(\ell_3)}(r)$ where the later set is the union of all cubes of scale ℓ_3 that are $*$-connected to $C^{(\ell_3)}(r)$. Similarly one defines $(-)$-*correct* points. Finally, if r is not a correct point of q it is an incorrect one. The connected components of the incorrect points of q form the supports of the contours. The pair consisting of the support of the contour and the restriction of $\eta(q)$ to this support is called a contour of q and is denoted $\Gamma(q)$. Observe that any q uniquely defines its contours $\Gamma_i(q)$.

Axiomatically a *contour* $\Gamma = (\text{Supp}(\Gamma), \eta^\Gamma)$ is defined as a pair which consists of a bounded $*$-connected $\mathscr{D}^{(\ell_3)}$ measurable set $\text{Supp}(\Gamma)$ called the spatial *support* of Γ and a spin valued function η^Γ on $\text{Supp}(\Gamma)$, with the condition that there exists at least one configuration q that gives rise to Γ.

To motivate the next definitions and to outline further constructions let us consider first an oversimplified problem.

The Case of a Single Contour

We restrict in this examples the system to have only one contour Γ of the type that we define with reference to Fig. 1 where its typical representative is drawn. To be more precise we consider the ensembles of contours inside the finite domain Λ containing exactly one contour of the special form shown in Fig. 1.

We choose boundary conditions $\bar{q}^{(\Lambda^c)}$ in the $+$ phase and we impose that in the region A of Fig. 1 the configuration is in the $+$ phase, in D it is in the $-$ phase while in $\text{Supp}(\Gamma)$ it agrees with η^Γ. We assume that in Fig. 1 $\text{Supp}(\Gamma)$ is the whole region extending between the bold lines.

The simplified contour ensemble gives rise to the simplified partition function Z (we are using here a notation that will be discontinued in the sequel)

$$Z := \sum_{\Gamma} \int_{\mathscr{Q}^{(\Lambda)}} dq^{(\Lambda)} \, \mathbb{1}_{\eta(q^{(A)}) \equiv 1} \, \mathbb{1}_{\eta(q^{(D)}) \equiv -1} \, \mathbb{1}_{\eta(q^{(\text{Supp}(\Gamma))}) \equiv \eta^\Gamma} \, e^{-\beta H_{\gamma,\lambda}(q^{(\Lambda)} \mid \bar{q}^{(\Lambda^c)})} \tag{3.23}$$

Here the external sum is taken over all contours Γ of the type described by Fig. 1 and the goal now is to rewrite Z as an integral over the $+$ ground

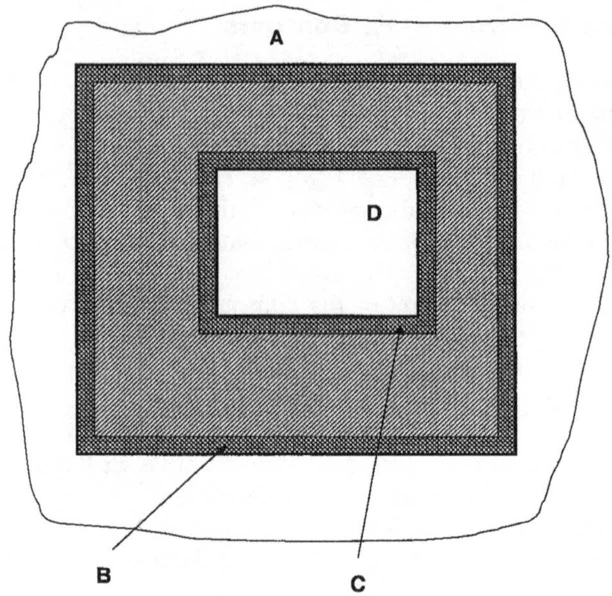

Fig. 1. A typical contour from he simplified contour ensemble.

state ensemble, i.e. as the integral over $q^{(A)}$ with $\{\eta(q^{(A)}) \equiv 1\}$. Let B and C be the strips of width γ^{-1} in Supp(Γ), as in Fig. 1, then we write

$$Z = \sum_{\Gamma} \int_{\mathscr{Q}^{(A)}} dq^{(A)} \, \mathbb{1}_{\eta(q^{(A)}) \equiv 1} \, e^{-\beta H_{\gamma,\lambda}(q^{(A)} \mid \bar{q}^{(A^c)})} W(\Gamma \mid q^{(B)}) \tag{3.24}$$

where the statistical weight $W(\Gamma \mid q^{(B)})$ is

$$W(\Gamma \mid q^{(B)}) = \frac{\int_{\mathscr{Q}(\mathrm{Supp}(\Gamma)\setminus B)} dq \, \mathbb{1}_{\eta(q) = \eta^\Gamma} \, e^{-\beta H_{\gamma,\lambda}(q \mid q^{(B)})} \int_{\mathscr{Q}^{(D)}} dq^{(D)} \, e^{-\beta H_{\gamma,\lambda}(q^{(D)} \mid q^{(C)})} \, \mathbb{1}_{\eta(q^{(D)}) \equiv -1}}{\int_{\mathscr{Q}(\mathrm{Supp}(\Gamma)\setminus B)} dq \, \mathbb{1}_{\eta(q) \equiv 1} \, e^{-\beta H_{\gamma,\lambda}(q \mid q^{(B)})} \int_{\mathscr{Q}^{(D)}} dq^{(D)} \, e^{-\beta H_{\gamma,\lambda}(q^{(D)} \mid q^{(C)})} \, \mathbb{1}_{\eta(q^{(D)}) \equiv 1}} \tag{3.25}$$

In (3.24) the configuration $q^{(A)}$ is not related to η^Γ and the contours Γ appear only through their statistical weights as extra variables in the partition functions. This is the contour model in the present over-simplified context. The advantage of the contour model is to work in a ground state ensemble, the price is the extra variables Γ. The game, in the general case with many contours, is to get good estimates on their statistical weights (3.25) in order to control their proliferation.

Let us examine this issue in the context of the oversimplified example. We should think of the integral in (3.25) over the region $\mathrm{Supp}(\Gamma)\backslash B$ as a "surface term" and of the integral over D as a "volume term". The second one looks therefore more likely to take large values and we consider it first, with reference to the numerator in (3.25) shorthanded by N. The integral is over all configurations in the region D that are in the ground state ensemble of the $-$ phase. The boundary conditions for this partition function are fixed in the region C and they are also configurations of the $-$ phase. In fact the $-$ phase extends over the whole ℓ_3 cubes $*$-connected to D, recall the definition of contours.

The intuition (which is made rigorous in Section 6) is that since C is well inside the $-$ phase, then the configurations in a neighborhood of C are actually not only in the $-$ ground state ensemble but they are very close to the standard ground state configuration. We thus write (recall that N denotes the numerator in (3.25))

$$N = T_1 T_2 T_3 \tag{3.26}$$

where T_1, T_2 and T_3 are defined as follows:

Splitting 1/2-1/2 the interaction between $\mathrm{Supp}(\Gamma)$ and D and recalling the intuition that the configurations involved in such interactions are very close to the $-$ ground state configuration, we write

$$T_1 := \int_{\mathscr{Q}(\mathrm{Supp}(\Gamma)\backslash B)} dq\, \mathbb{1}_{\eta(q) = \eta^\Gamma}\, e^{-\beta H_{\gamma,\lambda}(q \mid q^{(B)}, q_-^{(D)}) + \beta U_\gamma(q_-^{(C)} \mid q_-^{(D)})/2} \tag{3.27}$$

Denoting by $f_{\lambda,-}$ the thermodynamic limit of the pressure associated with the partition function restricted to the $-$ ground state ensemble we set

$$T_2 = e^{\beta f_{\lambda,-} |D|} \tag{3.28}$$

The third term T_3 is implicitly defined so that (3.26) holds. T_3 takes into account the errors made by replacing in (3.27) the interaction

$$\tfrac{1}{2} U_\gamma(q^{(C)} \mid q^{(D)}) \to U_\gamma(q^{(C)} \mid q_-^{(D)}) - \tfrac{1}{2} U_\gamma(q_-^{(C)} \mid q_-^{(D)}) \tag{3.29}$$

and the error which comes from T_2 where the partition function is replaced by its thermodynamic limit in the sense of (3.28).

The error term T_3 appears to be bounded by

$$|\log T_3| \leqslant c\gamma^{1/4} \,|\text{surface of } D| \tag{3.30}$$

where c is a constant. The above inequality is proven in Section 6 in the general case.

After an analogous decomposition in the denominator, we see that the dangerous volume terms simplify between numerator and denominator if we are able to choose the chemical potential in such a way that $f_{\lambda,-} = f_{\lambda,+}$. Once again, the possibility to solve this equation relies on a rather explicit representation of the pressures which one is able to derive.

The ratio of the terms T_1 (from numerator and denumerator) is bounded by

$$\exp(-c\ell_2^d \ell_3^{-d} \zeta^2 \,|\text{Supp}(\Gamma)|) \tag{3.31}$$

This is the famous Peierls estimate which is proven in the general case in Section 4. Its proof is relatively simpler than the proof of (3.30) and relies directly on properties of the mean free field energy functional. This step is similar to the analogous one in Ising models [BP].

Combining (3.30) and (3.31) we conclude that the whole statistical weight $W(\Gamma \mid q^{(B)})$ is bounded as in (3.31).

Let us now go from the oversimplified example to the general case.

More About Contours

Given a (large enough) region V and $\ell > 0$ we call

$$\begin{aligned} \partial^{(\ell)} V &= \{r \in V^c : \text{dist}(r, V) \leqslant \ell\}; \\ \delta^{(\ell)} V &= \{r \in V : \text{dist}(r, V^c) \leqslant \ell\} \end{aligned} \tag{3.32}$$

and we set $\partial V = \partial^{(\gamma^{-1})} V$ and $\delta V = \delta^{(\gamma^{-1})} V$.

Denoting by $\text{Ext}(\Gamma)$ and by $\text{Int}_m(\Gamma)$ respectively the unbounded and the bounded connected components of $(\text{Supp}(\Gamma))^c$ we observe that η^Γ takes the same values $\sigma(\Gamma)$ and $\sigma_m(\Gamma)$ on each of the sets $\partial^{(\ell_3)} \text{Ext}(\Gamma)$ and $\partial^{(\ell_3)} \text{Int}_m(\Gamma)$ respectively. The regions $\text{Ext}(\Gamma)$ and $\text{Int}(\Gamma) = \bigcup_m \text{Int}_m(\Gamma)$ are respectively called the *exterior* and the *interior* of Γ, in the previous example they are A and D.

For a contour Γ denote

$$\delta_m(\Gamma) = \partial \,\text{Int}_m(\Gamma) \tag{3.33}$$

(the set C in the example of Fig. 1)

$$\delta^{=}(\Gamma)=\partial\,\mathrm{Ext}(\Gamma)\cup\left(\bigcup_{m:\,\sigma_m(\Gamma)=\sigma(\Gamma)}\delta_m(\Gamma)\right) \tag{3.34}$$

(the set B in Fig. 1, the last term being absent in Fig. 1)

$$\delta^{\neq}(\Gamma)=\bigcup_{m:\,\sigma_m(\Gamma)\neq\sigma(\Gamma)}\delta_m(\Gamma) \tag{3.35}$$

(this is the set C in Fig. 1) and

$$\delta(\Gamma)=\delta^{=}(\Gamma)\cup\delta^{\neq}(\Gamma)=\delta\,\mathrm{Supp}(\Gamma) \tag{3.36}$$

Two contours are called *disjoint* if their supports are not $*$-connected. We say that a contour *belongs* to a region if its support is a subset of that region. In any collection of pairwise disjoint contours there is a uniquely defined subcollection of *external contours*, i.e., contours which do not belong to the interior of any other contour in the collection.

A contour Γ *surrounds a point* r if $r\in\mathrm{Supp}(\Gamma)\cup\mathrm{Int}(\Gamma)$. A contour $\Gamma_3(q)$ *separates* contours $\Gamma_1(q)$ and $\Gamma_2(q)$ if $\mathrm{Supp}(\Gamma_1(q))\in\mathrm{Int}(\Gamma_3(q))$ while $\mathrm{Supp}(\Gamma_2(q))\in\mathrm{Ext}(\Gamma_3(q))$.

Consider an arbitrary configuration q which differs from a ground state configuration of a given phase a only inside a bounded region Λ. Then all contours of q belong to Λ and for any external contour $\Gamma^{\mathrm{ext}}(q)$ one has $\sigma(\Gamma^{\mathrm{ext}})=\sigma$. Note that non-external contours $\Gamma(q)$ may have $\sigma(\Gamma(q))=-\sigma$. Moreover, the contours of q satisfy the following *matching condition.* If $\mathrm{Supp}(\Gamma_1(q))\in\mathrm{Int}_m(\Gamma_2(q))$ and there is no $\Gamma_3(q)$ separating $\Gamma_1(q)$ and $\Gamma_2(q)$ then $\sigma(\Gamma_1(q))=\sigma_m(\Gamma_2(q))$. This matching condition is highly non local and represents the main difficulty in understanding contour statistics.

The idea of the P–S theory is to remove matching condition and to construct an equivalent contour model with no matching rules present. Generally this can be done by modification of the statistical weights of contours and shifting the difficulty to the estimate of these modified statistical weights.

The study of the distribution of contours is important because to prove the existence of distinct translation invariant + and − phases it is enough to show that the probability of the event that an external contour surrounds the origin is sufficiently small. In a translation invariant measure this implies that only finitely many contours surround any given point r. Then one of the ground states occupies the infinite region $\bigcap_i \mathrm{Ext}(\Gamma_i^{\mathrm{ext}}(q))$ which is exactly the "sea" discussed earlier. Furthermore, the probability of the event that $\eta_0(q)=1$ can be taken as the order parameter distinguishing two phases. This probability is less than $\frac{1}{2}$ for the $-$ phase and it is larger than $\frac{1}{2}$ for the $+$ phase as follows from a bound on the probability of a contour of the form (3.31) (which will be shown to be true for all contours). As we will see even in the general case when there is more than just one single contour in the contour model the bound (3.31) implies that the contours are very rare and one can control them with an analysis not too distant from the oversimplified example.

Let us recall now that the bound (3.31) on the statistical weight of a contour was obtained after adjusting the chemical potential so that the pressures in the two ground states ensembles are equal. In the general case this is not a simple task because there are contours inside contours, i.e., inside the regions of type D of the example in Fig. 1 there are other contours and so on. One needs to treat contours recursively as in the one-contour example to obtain at the end of the recursion the contour model with the properly modified statistical weights of contours. Only after that one may try to equalize pressures in these contour models. The problem is that even if the bound (3.31) holds, the pressures in the contour models will depend on the contours and we are then in a sort of loop: to have the good bound (3.31) on the contours we need to make the pressures equal, but to control the pressures we need a good bound on the contours.

As explained by the P–S theory it is possible to overcome such an impasse; we will do it by following the Zahradnik approach [Z]. we will define a new cut-offed statistical weight which by its definition cannot exceed the value (3.31) (with a suitable constant c). In this context it will be easy to iterate the procedure of the example with one contour to derive a contour model representation of the true partition function. Such *auxiliary* partition functions (one for each phase) give rise to the corresponding pressures and we will adjust the chemical potential so that these two pressures (one for each phase) are equal. It will then turn out (see [Z]) that for this particular value of the chemical potential the contours satisfy the bound (3.31) without the need of the cutoff so that for this special value of the chemical potential the auxiliary and true partition functions are equal.

The next subsection contains exact definitions and statements which are necessary for application of the general P–S theorem (see [PS], [Z]).

Auxiliary Partition Functions

Given a phase σ and boundary condition $\bar{q}^{(\Lambda^c)}$ belonging to the ground state ensemble of this phase we define the *auxiliary partition functions* and the *truncated statistical weight* as

$$Z^A_{\gamma,\beta,\lambda}(\Lambda \mid \bar{q}^{(\Lambda^c)}) = \int_{\mathscr{Q}(\Lambda)} dq\, \mathbb{1}_{\eta(q) \equiv \sigma}\, e^{-\beta H_{\gamma,\lambda}(q^{(\Lambda)} \mid \bar{q}^{(\Lambda^c)})} \times \sum_{\{\Gamma_i\}^\sigma \in \Lambda} \prod_i W^T(\Gamma_i \mid q^{(\delta^-(\Gamma_i))}) \tag{3.37}$$

and

$$W^T(\Gamma \mid q^{(\delta^-(\Gamma))}) = \min(W^A(\Gamma \mid q^{(\delta^-(\Gamma))}), e^{-(c/3)\ell_2^d \ell_3^{-d} \zeta^2 |\mathrm{Supp}(\Gamma)|}) \tag{3.38}$$

where

$$W^A(\Gamma \mid \bar{q}^{(\delta^-(\Gamma))}) = \frac{\left(\int_{\mathscr{Q}(\mathrm{Supp}(\Gamma) \setminus \delta^-(\Gamma))} dq\, \mathbb{1}_{\eta(q) = \eta^\Gamma}\, e^{-\beta H_{\gamma,\lambda}(q \mid \bar{q}^{(\delta^-(\Gamma))})} \times \prod_{m:\, \sigma_m(\Gamma) \neq \sigma(\Gamma)} Z^A_{\gamma,\beta,\lambda}(\mathrm{Int}_m(\Gamma) \mid q^{(\partial\, \mathrm{Int}_m(\Gamma))}) \right)}{\left(\int_{\mathscr{Q}(\mathrm{Supp}(\Gamma) \setminus \delta^-(\Gamma))} dq\, \mathbb{1}_{\eta(q) = \sigma(\Gamma)}\, e^{-\beta H_{\gamma,\lambda}(q \mid \bar{q}^{(\delta^-(\Gamma))})} \times \prod_{m:\, \sigma_m(\Gamma) \neq \sigma(\Gamma)} Z^A_{\gamma,\beta,\lambda}(\mathrm{Int}_m(\Gamma) \mid q^{(\partial\, \mathrm{Int}_m(\Gamma))}) \right)} \tag{3.39}$$

To prove that the definition is well posed, we first say that a statement related to a $\mathscr{D}^{(\ell)}$ measurable region is proved by *induction in volume* if this statement is true for a $\mathscr{D}^{(\ell)}$ measurable region Λ as soon as it is true for all $\mathscr{D}^{(\ell)}$ measurable subsets of Λ. We then observe that (3.37)–(3.39) constitute a single inductive definition. At the initial step of this induction in volume one considers only contours with empty interior and uses (3.38)–(3.39) to define their truncated statistical weights. Then (3.37) allows one to calculate the auxiliary partition functions for sufficiently small regions admitting inside them only contours without interior. Then (3.38)–(3.39) are used again to define the truncated statistical weights of contours having sufficiently small but non empty interiors and so on.

We hope that a notational similarity of the numerator and the denominator of (3.39) does not hide from the reader the fact that these expressions are very different. For example, in the numerator of (3.39) configurations $q^{(\partial\, \mathrm{Int}_m(\Gamma))}$ belong to the ground state ensemble of the phase $\sigma = \sigma(\Gamma)$ while in the denominator of (3.39) these configurations belong to the ground state ensemble of the opposite phase, $-\sigma$.

Statement 3.2. If for any contour Γ the truncated statistical weight $W^T(\Gamma \mid q^{(\delta^-(\Gamma))})$ is smaller than the second term in the argument of the min in (3.88) then the auxiliary partition function coincides with the true one.

The proof of the statement is standard in the P–S theory and we omit it (see [Z]).

Another standard observation is that for any region Λ with the boundary condition $\bar{q}^{(\partial\Lambda)}$ belonging to the ground state ensemble of the phase σ the statistics of external contours in the contour model coincides with that of the true particle model. Note that the statistics of non-external contours are different in the contour and particle models. For example, the contours of the opposite phase, $-\sigma$, never appear in the contour model. As we explained before the absence of the matching rules makes the analysis of the contour model much easier than the analysis of the initial particle model. The price paid for this simplification is a rather involved expression for $W(\Gamma \mid \bar{q}^{(\delta^-(\Gamma))})$.

Statement 3.3. For all γ small enough and all chemical potentials $\lambda \in (\lambda(\beta) - \gamma^\alpha, \lambda(\beta) + \gamma^\alpha)$, $\alpha \geqslant \frac{1}{2}$ there are $f^A_{+,\lambda,\gamma}$, resp. $f^A_{-,\lambda,\gamma}$, such that for any sequence of cubes $\Lambda \to \mathbb{R}^d$ and any sequence of boundary conditions $\bar{q}^{(\Lambda^c)}$ belonging to the $+$ (resp. $-$) ground state ensemble

$$\lim_{\Lambda \to \mathbb{R}^d} \frac{1}{\beta\,|\Lambda|} \log Z^A_{\gamma,\beta,\lambda}(\Lambda \mid \bar{q}^{(\Lambda^c)}) = f^A_{\pm,\lambda,\gamma} \tag{3.40}$$

Moreover there exists $\lambda(\gamma, \beta)$ such that

$$\lim_{\gamma \to 0} \lambda(\gamma, \beta) = \lambda(\beta) \tag{3.41}$$

and

$$f^A_{+,\lambda(\gamma,\beta),\gamma} = f^A_{-,\lambda(\gamma,\beta),\gamma} \tag{3.42}$$

The statement is proved in Section 6 using cluster expansion techniques, together with a rather explicit representation of the auxiliary pressures $f^A_{\pm,\lambda,\gamma}$.

In analogue to the first term T_1 in the decomposition (3.26) of the statistical weight we introduce the function

$$w(\Gamma \mid \bar{q}^{(\delta^=(\Gamma))})$$

$$= \frac{\left(\begin{array}{l}\int_{\mathcal{Q}(\mathrm{Supp}(\Gamma)\backslash\delta(\Gamma))} dq\, \mathbb{1}_{\eta(q)=\eta^\Gamma}\, e^{-\beta H_{\gamma,\lambda}(q \mid \bar{q}^{\delta^=(\Gamma)}),\, q^{(\delta^{\neq}(\Gamma))}_{-\sigma(\Gamma)}} \\ \qquad \times \prod_{m:\, \sigma_m(\Gamma)\neq\sigma(\Gamma)} e^{(\beta/2) U_\gamma(q^{(\Gamma)}_{-\sigma(\Gamma)} \mid q^{(\delta^{\neq}(\Gamma))}_{-\sigma(\Gamma)})}\end{array}\right)}{\left(\begin{array}{l}\int_{\mathcal{Q}(\mathrm{Supp}(\Gamma)\backslash\delta(\Gamma))} dq\, \mathbb{1}_{\eta(q)=\sigma(\Gamma)}\, e^{-\beta H_{\gamma,\lambda}(q \mid \bar{q}^{\delta^=(\Gamma)}),\, q^{(\delta^{\neq}(\Gamma))}_{\sigma(\Gamma)}} \\ \qquad \times \prod_{m:\, \sigma_m(\Gamma)\neq\sigma(\Gamma)} e^{(\beta/2) U_\gamma(q^{(\Gamma)}_{\sigma(\Gamma)} \mid q^{(\delta^{\neq}(\Gamma))}_{\sigma(\Gamma)})}\end{array}\right)} \tag{3.43}$$

In Section 4 we will prove the Peierls bound (which is the analogue of (3.31)):

Statement 3.4. For any contour Γ there is $c>0$ such that for all γ and $\lambda \in (\lambda(\beta)-\gamma^\alpha, \lambda(\beta)+\gamma^\alpha)$, $\alpha \geqslant \frac{1}{2}$

$$w(\Gamma \mid \bar{q}^{(\delta^=(\Gamma))}) \leqslant \exp(-c\ell_2^d \ell_3^{-d} \zeta^2\, |\mathrm{Supp}(\Gamma)|) \tag{3.44}$$

Finally in Section 6 we will prove

Statement 3.5. For any contour Γ there is $c>0$ such that for all sufficiently small γ and $\lambda \in (\lambda(\beta)-\gamma^\alpha, \lambda(\beta)+\gamma^\alpha)$, $\alpha \geqslant \frac{1}{2}$

$$\Bigg|\log W^A(\Gamma \mid \bar{q}^{(\delta^=(\Gamma))}) - \log w(\Gamma \mid \bar{q}^{\delta^=(\Gamma)}) - \sum_{m:\, \sigma_m(\Gamma)\neq\sigma(\Gamma)} (\beta\, |\mathrm{Int}_m(\Gamma)| [f^A_{-\sigma(\Gamma),\lambda,\gamma} - f^A_{\sigma(\Gamma),\lambda,\gamma}])\Bigg| \leqslant c\, |\mathrm{Supp}(\Gamma)|\, \gamma^{1/4} \tag{3.45}$$

Note that the factor $\gamma^{1/4}$ is not optimal but it is enough for our purposes.

By combining the above statements and choosing $\lambda = \lambda(\beta, \gamma)$ we then obtain that the auxiliary and the true partition functions are equal and consequently that the probability of a contour in the original system is bounded as in (3.44). This proves the existence of two different phases and together with the exponential decay established in Sections 5 and 6 concludes the proof of Theorem 2.1.

4. PEIERLS ESTIMATE

In this section we prove Statement 3.4, i.e., we compare the partition functions in the numerator and denominator of (3.43).

The partition function in the denominator of (3.43) is taken over ground state configurations of the phase $\sigma = \sigma(\Gamma)$ placed in the region $\mathrm{Supp}(\Gamma)\backslash\delta(\Gamma)$ with the boundary condition $\bar{q}^{(\delta(\Gamma))}$ specified on $\delta(\Gamma)$. This boundary condition, $\bar{q}^{(\delta(\Gamma))}$, is different in $\delta^{=}(\Gamma)$ and in $\delta^{\neq}(\Gamma)$: it is a configuration of the ground state ensemble of the phase σ in the region $\delta^{=}(\Gamma)$ and it is the standard configuration of the phase σ in $\delta^{\neq}(\Gamma)$.

The partition function from the numerator of (3.43) is taken over contour configurations, i.e., ones compatible with η^{Γ}, in the same region $\mathrm{Supp}(\Gamma)\backslash\delta(\Gamma)$ with the same boundary condition $\bar{q}^{(\delta^{=}(\Gamma))}$ in $\delta^{=}(\Gamma)$ but with the different boundary condition given on $\delta^{\neq}(\Gamma)$. Contrary to the denominator this boundary condition is the standard configuration of the opposite phase, $-\sigma$.

Scheme of Estimate

The estimate of the ratio in (3.43) is performed in several steps. At each step we achieve a certain simplification at the price of a non essential error which we define as a positive factor not exceeding $\exp(\gamma^{\alpha}\ell_2^d\ell_3^{-d}\zeta^2\,|\mathrm{Supp}(\Gamma)|)$ with $\alpha > 0$. For γ small enough the product of the finite number of such factors is dominated by $\exp(-c\ell_2^d\ell_3^{-d}\zeta^2\,|\mathrm{Supp}(\Gamma)|)$. The last factor appears at the last step of the estimate and the preliminary simplifications make this main step of the Peierls estimate easier.

First we exploit a superstability of the potential and in Lemmas 4.1–4.4 we show that only particle configurations of bounded density need to be taken into account.

As soon as all considerations, are reduced to the case of bounded density it becomes clear that one needs to check (3.44) only for $\lambda = \lambda(\beta)$. Indeed, varying λ in the interval $\lambda \in (\lambda(\beta) - \gamma^{\alpha}, \lambda(\beta) + \gamma^{\alpha})$ with $\alpha \geqslant 1/2$ produces an error in $H_{\gamma,\lambda}(q \mid \bar{q}^{(\delta^{=}(\Gamma))}, q_{\pm\sigma(\Gamma)}^{(\delta^{\neq}(\Gamma))})$ which does not exceed $\gamma^{1/2}\,|\mathrm{Supp}(\Gamma)|$ in absolute value. Hence the total error in (3.43) is negligible with respect to the right hand side of (3.44).

Then in Lemma 4.5 we replace every configuration $q = (q_i)$ contributing to the numerator or denominator of (3.43) by $[q]^{(\ell_1)}$. After shifting particles into the centers of the corresponding boxes integrals over the particle configurations q become sums over density configurations $\varrho^{(\ell_1)}(q) = (\varrho_x^{(\ell_1)}(q))$. Moreover, the entropy factor coming from the summation over different density configurations $\varrho^{(\ell_1)}$ can be neglected and one can consider only a contribution of two density configurations providing the minimum of the energy in the numerator and denominator of (3.43) respectively. These arguments are presented in (4.19)–(4.23) below.

The next observation is that the energy of density configuration is very close to a discrete version, (4.26), of the mean field free energy functional (3.1) with function $\varrho(r)$ of a continuous argument $r \in \mathbb{R}^d$ replaced by its lattice version $\varrho_x^{(\ell_1)}$, $x \in \mathbb{Z}_{\ell_1}^d$. The corresponding errors are estimated in Lemma 4.6.

Thus, we arrive to the variational problem of finding minima of the discrete mean field functional under the constraints corresponding to the numerator and denominator of (3.43). The simplest of the constraints is the discreteness of our density variables. Lemma 4.7 shows that one can treat the density as a continuous variable for the price of non essential error. A more serious constraint is the presence of boundary conditions. (Our density configurations vary only inside the finite region, the support of the contour, being fixed outside this region.) Moreover, these boundary conditions are different for the numerator and denominator of (3.43). To handle this problem in Lemma 4.8 we show that the dependence of the minimizer on the boundary conditions decays exponentially with the distance from the boundary. In particular, at the distance of order $\gamma^{-1-\alpha_3}$ the boundary condition is practically not felt. Our proof of Lemma 4.8 is based on estimate (3.17).

Using Lemma 4.8 we perform the last of our simplifications. In (4.57)–(4.58) we show that modulo non essential errors one can replace the initial variational problem by a similar one in a smaller region but with *standard* boundary conditions only.

Finally we come to the key Lemma 4.9 expressing the essence of the Peierls estimate. It is obvious that for the denominator of (3.43) the minimizing density is the constant density configuration coinciding with the boundary condition $\varrho_{\beta,\sigma}$. The minimizing density configuration for the numerator of (3.43) is not a constant and may have a complicated structure. While not trying to calculate it exactly we estimate (from below) the corresponding value of the discrete free energy functional. This is a rather straightforward calculation which takes into account the fact that inside $\mathrm{Supp}(\Gamma)\backslash\delta(\Gamma)$ there are sufficiently many "wrong boxes" $C^{(\ell_2)}$ where the indicator function $\mathbb{1}_{\eta(q)=\eta^\Gamma}$ forces the minimizing configuration to be different from $\varrho_{\beta,\pm}$. Each "wrong box" produces an excess of energy of order ζ^2 and "wrong boxes" are distributed inside $\mathrm{Supp}(\Gamma)$ with the density $c\ell_2^d\ell_3^{-d}$. This finally gives us a factor $\exp(-c\ell_2^d\ell_3^{-d}\zeta^2\,|\mathrm{Supp}(\Gamma)|)$ dominating all error estimates discussed earlier.

The details are given in the rest of this section.

Stability Estimates

First we check that the accumulation of an infinite number of particles in a bounded region is impossible. This is a consequence of

Lemma 4.1. For any particle q_0 and any configuration $q=(q_1, q_2,...)$ the interaction energy

$$U_\gamma(q_0 \mid q) = -\sum_{q_{i_1} \in q} J_\gamma^{(2)}(q_0, q_{i_1}) + \sum_{q_{i_1}, q_{i_2}, q_{i_3} \in q} J_\gamma^{(4)}(q_0, q_{i_1}, q_{i_2}, q_{i_3}) \tag{4.1}$$

is bounded from below

$$U_\gamma(q_0 \mid q) \geqslant H_{\min} \tag{4.2}$$

by an absolute constant

$$H_{\min} = \min_{\mathcal{N}>0,\, 1>\gamma>0} \frac{1}{3!}(\mathcal{N}^3 - 3\gamma^d \mathcal{N}^2 + 2\gamma^{2d}\mathcal{N}) - \mathcal{N} \tag{4.3}$$

Proof. Given q let

$$\mathcal{N}(r, q) = \gamma^d \sum_{q_i \in q} \mathbb{1}_{|r-q_i| \leqslant \gamma^{-1} R_d} \tag{4.4}$$

be the mean number of particles of q situated inside $B_\gamma(r)$, $r \in \mathbb{R}^d$. Then

$$\begin{aligned} -\sum_{q_{i_1} \in q} J_\gamma^{(2)}(q_0, q_{i_1}) &= -\gamma^{2d} \sum_{q_i \in q} \int dr\, \mathbb{1}_{|r-q_0| \leqslant \gamma^{-1} R_d} \mathbb{1}_{|r-q_i| \leqslant \gamma^{-1} R_d} \\ &= -\gamma^d \int dr\, \mathbb{1}_{|r-q_0| \leqslant \gamma^{-1} R_d} \mathcal{N}(r, q) \end{aligned} \tag{4.5}$$

Similarly

$$\begin{aligned} &\sum_{q_{i_1}, q_{i_2}, q_{i_3} \in q} J_\gamma^{(4)}(q_0, q_{i_1}, q_{i_2}, q_{i_3}) \\ &\quad = \gamma^d \int dr\, \mathbb{1}_{|r-q_0| \leqslant \gamma^{-1} R_d} \frac{1}{3!}(\mathcal{N}^3(r, q) - 3\gamma^d \mathcal{N}^2(r, q) + 2\gamma^{2d}\mathcal{N}(r, q)) \end{aligned} \tag{4.6}$$

Hence

$$U_\gamma(q_0 \mid q) \geqslant \gamma^d \int dr\, \mathbb{1}_{|r-q_0| \leqslant \gamma^{-1} R_d} H_{\min} = H_{\min} \tag{4.7}$$

which proves the lemma. ∎

Lemma 4.1 provides the lower bound

$$H_{\gamma,\lambda}(q^{(\Lambda)} \mid \bar{q}^{(\Lambda^c)}) \geqslant H_{\min} |q^{(\Lambda)}| = H_{\min} |\Lambda|\, \varrho \tag{4.8}$$

where $\varrho = |q^{(\Lambda)}|/|\Lambda|$ is the corresponding density. To obtain an upper bound take some positive $\varrho_{\max} > \varrho_{\beta,+}$.

Lemma 4.2. Consider configurations $q^{(\Lambda)}$ and $\bar{q}^{(\Lambda^c)}$ such that $q^{(\Lambda)} \cup \bar{q}^{(\Lambda^c)}$ has at most $\varrho_{\max}(2\gamma^{-1})^d$ particles in any intersecting Λ cube with the side $2\gamma^{-1}$. Then

$$H_{\gamma,\lambda}(q^{(\Lambda)} \mid \bar{q}^{(\Lambda^c)}) \leqslant H_{\max}(\varrho_{\max})\, |\Lambda| \tag{4.9}$$

where

$$H_{\max}(\varrho_{\max}) = |\lambda(\beta)|\, \varrho_{\max} + 2^{3d} \varrho_{\max}^4 \tag{4.10}$$

Proof. It is clear that $|q^{(\Lambda)}| \leqslant |\Lambda|\, \varrho_{\max}$. The strength of the four-body interaction between any four particles is less than γ^{3d}. The number of interacting quadruples of particles such that one of the elements of the quadruple is a given particle is less than $(\varrho_{\max} 2^d \gamma^{-d})^3$. Hence the total four-body interaction contributing to $H_{\gamma,\lambda}(q^{(\Lambda)} \mid \bar{q}^{(\Lambda^c)})$ is less than $|\Lambda|\, \varrho_{\max}^4 2^{3d}$. The two-body interaction is negative and does not contribute to the estimate. ∎

Bad Boxes

In this subsection we treat boxes containing too many particles.

Lemma 4.3. Consider a box $C^{(\ell)}$, $\ell < \frac{1}{2}\gamma^{-1}$, a configuration $\bar{q} \in \mathcal{Q}$ and an integer

$$N \geqslant |C^{(\ell)}|\, e^{c\beta} = \ell^d e^{c\beta} \tag{4.11}$$

then

$$\int_{\mathcal{Q}(C^{(\ell)})} dq\, \mathbb{1}_{|q^{(C^{(\ell)})}| = N}\, e^{-\beta H_{\gamma,\lambda}(q^{(C^{(\ell)})} \mid \bar{q})} \leqslant e^{-N} \tag{4.12}$$

and

$$\int_{\mathscr{Q}(C^{(\ell)})} dq\, \mathbb{1}_{|q^{(C^{(\ell)})}| \geqslant |C^{(\ell)}|\, e^{c\beta}}\, e^{-\beta H_{\gamma,\lambda}(q^{(C^{(\ell)})} \mid \bar{q})} \leqslant 2e^{-|C^{(\ell)}|\, e^{c\beta}} \tag{4.13}$$

Proof. According to Lemma 4.1 the interaction $U_\gamma(q^{(C^{(\ell)})} \mid \bar{q})$ between particles of $q^{(C^{(\ell)})}$ and $\bar{q}$ satisfies the estimate

$$U_\gamma(q^{(C^{(\ell)})} \mid \bar{q}) \geqslant H_{\min} N \tag{4.14}$$

It is clear that

$$H_{\gamma,\lambda}(q^{(C^{(\ell)})}) \geqslant -\gamma^d \frac{N(N-1)}{2!} + \frac{1}{2d}\gamma^{3d} \frac{N(N-1)(N-2)(N-3)}{4!} - |\lambda|\, N \tag{4.15}$$

as the maximal volume of the intersection of two balls of radius $\gamma^{-1}R_d$ is γ^{-d} and the minimal volume of the intersection of four such balls with the centers in $C^{(\ell)}$ is larger than $(1/2^d)\,\gamma^{-d}$, Thus the logarithm of the left hand side of (4.12) does not exceed

$$-N\log N + N + N\log|C^{(\ell)}| - \beta H_{\min} N + \beta\,|\lambda|\, N$$
$$+\beta\gamma^d \frac{N(N-1)}{2!} - \beta\frac{1}{2^d}\gamma^{3d}\frac{N(N-1)(N-2)(N-3)}{4!} \tag{4.16}$$

It is not hard to check that for sufficiently small γ and sufficiently large absolute constant $c_{(4.11)}$ the last expression is smaller than $-N$ as soon as $N \geqslant |C^{(\ell)}|\, e^{c_{(4.11)}\beta}$. Note that the most dangerous term in (4.16) is $\beta\gamma^d(N(N-1)/2!)$. It is dominated by $\beta(1/2^d)\,\gamma^{3d}(N(N-1)(N-2)(N-3)/4!)$ for $N > c\gamma^{-d}$ and by $N\log N - N\log|C^{(\ell)}|$ for $|C^{(\ell)}|\, e^{c_{(4.11)}\beta} \leqslant N \leqslant c\gamma^{-d}$. This implies (4.12) and hence (4.13). ∎

Set $\varrho_{\max} = 2e^{c_{(4.11)}\beta}$. An easy consequence of Lemma 4.3 is

Lemma 4.4. For any contour Γ and any $\ell \leqslant \ell_2$ the partition function in the numerator of (3.43) does not exceed

$$e^{\ell^{-d}|\mathrm{Supp}(\Gamma)|} \int_{\mathscr{Q}(\mathrm{Supp}(\Gamma)\setminus\delta(\Gamma))} dq\, \mathbb{1}_{\eta(q)=\eta^\Gamma,\, \varrho^{(\ell)}(q)\leqslant \varrho_{\max}}\, e^{-\beta H_{\gamma,\lambda}(q \mid \bar{q}^{(\delta^{=}(\Gamma))},\, q_{-\sigma}^{(\delta^{\neq}(\Gamma))})}$$
$$\times \prod_{m:\,\sigma_m(\Gamma)\neq\sigma(\Gamma)} e^{(\beta/2)\, U_\gamma(q^{(\Gamma)}_{-\sigma(\Gamma)} \mid q^{(\delta^{\neq}(\Gamma))}_{-\sigma(\Gamma)})} \tag{4.17}$$

Proof. To obtain (4.17) we perform a partial integration over the configurations in the bad boxes. Suppose that q is fixed outside the cube $C^{(\ell)} \in \mathrm{Supp}(\Gamma)$ where q has more than $\varrho_{\max} |C^{(\ell)}|$ particles. Now applying Lemma 4.3 we integrate over all such q's leaving exactly $\frac{1}{2}\varrho_{\max} |C^{(\ell)}|$ particles in $C^{(\ell)}$. That means that we integrate over the particles in excess of $\frac{1}{2}\varrho_{\max} |C^{(\ell)}|$, assuming that the total number of particles in $C^{(\ell)}$ is larger than $\varrho_{\max} |C^{(\ell)}|$. This produces an extra factor $1 + 2e^{-|C^{(\ell)}| e^{c\beta}} \leqslant e$. We continue this procedure box by box and the total number of boxes in Supp(Γ) is $\ell^{-d} |\mathrm{Supp}(\Gamma)|$. ∎

Reduction to Variational Problem

It was shown in the previous subsection that we may restrict our considerations to configurations with bounded density. Now we estimate the error which is produced by shifting particles in such a configuration into the centers of the corresponding boxes $C^{(\ell)}$.

Lemma 4.5. Take $\ell < \gamma^{-1}$ and configurations $q^{(\Lambda)}$ and $\bar{q}^{(\Lambda^c)}$ with $\varrho^{(\ell)}(q^{(\Lambda)} \cup \bar{q}^{(\partial\Lambda)})) \leqslant \varrho_{\max}$. Then

$$|H_{\gamma,\lambda}(q^{(\Lambda)} \mid \bar{q}^{(\Lambda^c)}) - H_{\gamma,\lambda}(\varrho^{(\ell)}(q^{(\Lambda)}) \mid \bar{q}^{(\Lambda^c)})| \leqslant \ell\gamma\, |\Lambda|\, (2^d \varrho_{\max}^2 + 2^{3d} \varrho_{\max}^4) \tag{4.18}$$

Proof. It is clear that $|q^{(\Lambda)}| \leqslant |\Lambda|\, \varrho_{\max}$. Given two interacting particles the absolute value of the error in their interaction due to shifting these particles to the centers of the corresponding boxes is less than $\gamma^d \ell\gamma$. Given four interacting particles the absolute value of the error in their interaction due to shifting these particles to the centers of the corresponding boxes is less than $\gamma^{3d}\ell\gamma$. As in Lemma 4.2 the number of interacting quadruples of particles such that one of the elements of the quadruple is a given particle is less than $(\varrho_{\max} 2^d \gamma^{-d})^3$. Similarly the number of pairs of interacting particles such that one of the elements of the pair is a given particle is less than $(\varrho_{\max} 2^d \gamma^{-d})$. Hence the total error is less than $\ell\gamma\, |\Lambda|\, (2^d \varrho_{\max}^2 + 2^{3d} \varrho_{\max}^4)$. ∎

This lemma allows us to replace the integrals over dq in the numerator and denominator of (3.43) by the sums over density configurations $\varrho^{(\ell_1)}$. Namely, consider the partition function on the right hand side of (4.17). The integral over $q \in \mathcal{Q}^{(\mathrm{Supp}(\Gamma) \setminus \delta(\Gamma))}$ with $\eta(q) = \eta^\Gamma$ and $\varrho^{(\ell_1)}(q) \leqslant \varrho_{\max}$ can be calculated in two steps.

First one can fix a density configuration $\varrho^{(\ell_1)} = (\varrho_x^{(\ell_1)})$, $x \in [\mathrm{Supp}(\Gamma) \setminus (\Gamma)]^{(\ell_1)}$ and integrate over configurations $q^{(\mathrm{Supp}(\Gamma) \setminus \delta(\Gamma))}$ with $\varrho^{(\ell_1)}(q) = \varrho^{(\ell_1)}$.

Afterward one can sum over all $\varrho^{(\ell_1)} \leqslant \varrho_{\max}$ compatible with η^Γ. The obvious upper estimate for this sum is the total number of density configurations $\varrho_x^{(\ell_1)} \leqslant \varrho_{\max}$, $x \in [\mathrm{Supp}(\Gamma)\backslash\delta(\Gamma)]^{(\ell_1)}$ times the maximal contribution given by a single density configuration.

The number of density configurations $\varrho^{(\ell_1)} \leqslant \varrho_{\max}$ compatible with η^Γ is less than

$$(\varrho_{\max} |C^{(\ell_1)}|)^{|\mathrm{Supp}(\Gamma)\backslash\delta(\Gamma)|/|C^{(\ell_1)}|} \tag{4.19}$$

Given $\varrho^{(\ell_1)}$ it follows from Lemma 4.5 that the integral over configurations q with $\varrho^{(\ell_1)}(q) = \varrho^{(\ell_1)}$ does not exceed

$$\begin{aligned}\exp(&-\tilde{\mathscr{F}}_{\gamma,\beta,\lambda}(\varrho^{(\ell_1)} \mid \bar{q}^{(\delta^=(\Gamma))}, q_{-\sigma}^{(\delta^{\neq}(\Gamma))}, *) \\ &+ \beta\ell_1\gamma\, |\mathrm{Supp}(\Gamma)\backslash\delta(\Gamma)|\, (2^d\varrho_{\max}^2 + 2^{3d}\varrho_{\max}^4))\end{aligned} \tag{4.20}$$

where

$$\begin{aligned}&\tilde{\mathscr{F}}_{\gamma,\beta,\lambda}(\varrho^{(\ell_1)} \mid \bar{q}^{(\delta^=(\Gamma))}, q_{-\sigma}^{(\delta^{\neq}(\Gamma))}, *) \\ &\quad = \beta H_{\gamma,\lambda}(\varrho^{(\ell_1)} \mid \bar{q}^{(\delta^=(\Gamma))}, q_{-\sigma}^{(\delta^{\neq}(\Gamma))}) - \frac{\beta}{2} \sum_{m:\, \sigma_m(\Gamma) \neq \sigma(\Gamma)} U_\gamma(q_{-\sigma(\Gamma)}^{(\Gamma)} \mid q_{-\sigma(\Gamma)}^{(\delta^{\neq}(\Gamma))}) \\ &\quad\quad - |C^{(\ell_1)}| \log |C^{(\ell_1)}| \sum_{x \in [\mathrm{Supp}(\Gamma)\backslash\delta(\Gamma)]^{(\ell_1)}} \varrho_x^{(\ell_1)} \\ &\quad\quad + \sum_{x \in [\mathrm{Supp}(\Gamma)\backslash\delta(\Gamma)]^{(\ell_1)}} \log((\varrho_x^{(\ell_1)} |C^{(\ell_1)}|)!)\end{aligned} \tag{4.21}$$

The notations with $\sim$ and $*$ foresee forthcoming simplifications and variations. In particular $\tilde{\mathscr{F}}_{\gamma,\beta,\lambda}(\varrho^{(\ell_1)} \mid \bar{q}^{(\delta^=(\Gamma))}, q_\sigma^{(\delta^{\neq}(\Gamma))})$ is defined as $\tilde{\mathscr{F}}_{\gamma,\beta,\lambda}(\varrho^{(\ell_1)} \mid \bar{q}^{(\delta^=(\Gamma))}, q_\sigma^{(\delta^{\neq}(\Gamma))}, *)$ without the term $(\beta/2) \sum_{m:\, \sigma_m(\Gamma) \neq \sigma(\Gamma)} U_\gamma(q_{-\sigma(\Gamma)}^{(\Gamma)} \mid q_{-\sigma(\Gamma)}^{(\delta^{\neq}(\Gamma))})$ in (4.20).

Suppose that $\bar{\varrho}^{(\ell_1)}$ gives the minimum of $\tilde{\mathscr{F}}_{\gamma,\beta,\lambda}(\varrho^{(\ell_1)} \mid \bar{q}^{(\delta^=(\Gamma))}, q_{-\sigma}^{(\delta^{\neq}(\Gamma))}, *)$ among all $\varrho^{(\ell_1)} \leqslant \varrho_{\max}$ compatible with η^Γ. Then

$$\begin{aligned}&-\tilde{\mathscr{F}}_{\gamma,\beta,\lambda(\beta)}(\hat{\varrho}^{(\ell_1)} \mid \bar{q}^{(\delta^=(\Gamma))}, q_{-\sigma}^{(\delta^{\neq}(\Gamma))}, *) + |\mathrm{Supp}(\Gamma)\backslash\delta(\Gamma)| \\ &\quad \times (\gamma^{1/2} + \ell_1^{-d} + \ell_1^{-d} \log(\varrho_{\max} |C^{(\ell_1)}|) + \beta\ell_1\gamma(2^d\varrho_{\max}^2 + 2^{3d}\varrho_{\max}^4))\end{aligned} \tag{4.22}$$

is the upper bound for the log of the numerator of (3.43). Note that $\lambda \in (\lambda(\beta) - \gamma^\alpha, \lambda(\beta) + \gamma^\alpha)$, $\alpha \geqslant 1/2$ in (4.21) while $\lambda = \lambda(\beta)$ in (4.22). The corresponding error is covered by the $\gamma^{1/2}$ term in (4.22).

Similarly if $\hat{\varrho}^{(\ell_1)}$ gives the minimum of $\tilde{\mathscr{F}}_{\gamma,\beta,\lambda(\beta)}(\varrho^{(\ell_1)} \mid \bar{q}^{(\delta^=(\Gamma))}, q_\sigma^{(\delta^{\neq}(\Gamma))}, *)$ among all density configurations $\varrho^{(\ell_1)}$ from the ground state ensemble of the phase σ then

$$-\tilde{\mathscr{F}}_{\gamma,\beta,\lambda(\beta)}(\hat{\varrho}^{(\ell_1)} \mid \bar{q}^{(\delta^=(\Gamma))}, q_\sigma^{(\delta^{\neq}(\Gamma))}, *) \\ - |\mathrm{Supp}(\Gamma)\backslash\delta(\Gamma)|\,(\gamma^{1/2} + \beta\ell_1\gamma(2^d\varrho^2_{\max} + 2^{3d}\varrho^4_{\max})) \tag{4.23}$$

is the lower bound for the log of the denominator of (3.43).

Now denoting $b_{\gamma,\ell} = \|[B_\gamma(\cdot)]^{(\ell)}\|$ let

$$I_\gamma^{(2)}(x_1, x_2) = b_{\gamma,\ell}^{-2}\, \|[B_\gamma(x_1)]^{(\ell)} \cap [B_\gamma(x_2)]^{(\ell)}\| \tag{4.24}$$

and

$$I_\gamma^{(4)}(x_1,..., x_4) = b_{\gamma,\ell}^{-4} \left\| \bigcap_{j-1}^{4} [B_\gamma(x_j)]^{(\ell)} \right\| \tag{4.25}$$

be discrete versions of $J_\gamma^{(2)}(x_1, x_2)$ and $J_\gamma^{(4)}(x_1,..., x_4)$. For any region $[\Lambda]^{(\ell)}$ (we do note exclude the case $[\Lambda]^{(\ell)} = \mathbb{Z}_\ell^d$) and any density configuration $\varrho^{(\ell)}(\Lambda) = (\varrho_x^{(\ell)})$, $x \in [\Lambda]^{(\ell)}$ define a functional

$$\mathscr{F}_{\gamma,\beta,\lambda}(\varrho^{(\ell)}(\Lambda)) = \ell^d \Bigg(\sum_{x \in [\Lambda]^{(\ell)}} \frac{\varrho_x^{(\ell)}}{\beta} (\log \varrho_x^{(\ell)} - 1) - \sum_{x \in [\Lambda]^{(\ell)}} \lambda \varrho_x^{(\ell)} \\ - \frac{1}{2!} \sum_{x_2, x_1 \in [\Lambda]^{(\ell)}} I_\gamma^{(2)}(x_1, x_2)\, \varrho_{x_1}^{(\ell)} \varrho_{x_2}^{(\ell)} \\ + \frac{1}{4!} \sum_{x_1, x_2, x_3, x_4 \in [\Lambda]^{(\ell)}} I_\gamma^{(4)}(x_1,..., x_4)\, \varrho_{x_1}^{(\ell)} \cdots \varrho_{x_4}^{(\ell)} \Bigg) \tag{4.26}$$

which is a discrete analogue of the mean field free energy functional (3.1). We also define a conditional functional

$$\mathscr{F}_{\gamma,\beta,\lambda}(\varrho^{(\ell)}(\Lambda) \mid \bar{\varrho}^{(\ell)}(\Lambda^c)) = \mathscr{F}_{\gamma,\beta,\lambda}(\varrho^{(\ell)}(\Lambda) + \bar{\varrho}^{(\ell)}(\Lambda^c)) - \mathscr{F}_{\gamma,\beta,\lambda}(\bar{\varrho}^{(\ell)}(\Lambda^c)) \tag{4.27}$$

with the boundary condition $\bar{\varrho}^{(\ell)} = (\bar{\varrho}_x^{(\ell)})$, $x \in [\Lambda^c]^{(\ell)}$. Here the similarity with (2.10) is obvious and the meaning of $\varrho^{(\ell)} + \bar{\varrho}^{(\ell)}$ becomes straightforward if we set $\varrho^{(\ell)} \equiv 0$, $x \in [\Lambda^c]^{(\ell)}$ and $\bar{\varrho}^{(\ell)} \equiv 0$, $x \in [\Lambda]^{(\ell)}$. Setting

$$U_{\gamma,\beta}(\varrho^{(\ell)}(\Lambda) \mid \bar{\varrho}^{(\ell)}(\Lambda^c)) = \mathscr{F}_{\gamma,\beta,\lambda}(\varrho^{(\ell)}(\Lambda) \mid \bar{\varrho}^{(\ell)}(\Lambda^c)) - \mathscr{F}_{\gamma,\beta,\lambda}(\varrho^{(\ell)}(\Lambda)) \tag{4.28}$$

we introduce

$$\mathscr{F}_{\gamma,\beta,\lambda}(\varrho^{(\ell)}(\Lambda) \mid \bar{\varrho}^{(\ell)}(\Lambda^c), *)$$
$$= \mathscr{F}_{\gamma,\beta,\lambda}(\varrho^{(\ell)}(\Lambda) \mid \bar{\varrho}^{(\ell)}(\Lambda^c)) - \tfrac{1}{2} U_{\gamma,\beta}(\varrho^{(\ell)}_{\beta,\sigma}(\Lambda) \mid \varrho^{(\ell)}_{\beta,\sigma}(\Lambda^c)) \tag{4.29}$$

where $\varrho^{(\ell)}_{\beta,\sigma} \equiv \varrho_{\beta,\sigma}$ and $\sigma = +$ or $\sigma = -$. With some abuse of notation we use for $\mathscr{F}_{\gamma,\beta,\lambda}(\varrho^{(\ell)}(\Lambda) \mid \varrho^{(\ell)}(\bar{q}^{(\partial\Lambda)}))$ the alternative notations $\mathscr{F}_{\gamma,\beta,\lambda}(\varrho^{(\ell)}(\Lambda) \mid \bar{q}^{(\partial\Lambda)})$ or $\mathscr{F}_{\gamma,\beta,\lambda}(\varrho^{(\ell)}(\Lambda) \mid \bar{\varrho}^{(\ell)}(q^{(\partial\Lambda)}))$. The role which is played by the functional (4.27) is clarified in

Lemma 4.6. Consider a $\mathscr{D}^{(\ell)}$ measurable region Λ with the boundary condition $\bar{q}^{(\partial\Lambda)}$ which is a ground state configuration on every connected component of $\partial\Lambda$. Then for $\ell \leqslant \gamma^{-1}$ and any $\varrho^{(\ell)} = (\varrho^{(\ell)}_x)$, $x \in [\Lambda]^{(\ell)}$ such that $\varrho^{(\ell)}_x \leqslant \varrho_{\max}$ one has

$$|\tilde{\mathscr{F}}_{\gamma,\beta,\lambda}(\varrho^{(\ell)}(\Lambda) \mid \bar{q}^{(\partial\Lambda)}) - \mathscr{F}_{\gamma,\beta,\lambda}(\varrho^{(\ell)}(\Lambda) \mid \bar{\varrho}^{(\ell)}(q^{(\partial\Lambda)}))|$$
$$\leqslant \ell\gamma 5(2^d \varrho^2_{\max} + 2^{3d}\varrho^4_{\max} + \varrho^4_{\max})\,|\Lambda| \tag{4.30}$$
$$|\tilde{\mathscr{F}}_{\gamma,\beta,\lambda}(\varrho^{(\ell)}(\Lambda) \mid \bar{q}^{(\partial\Lambda)}, *) - \mathscr{F}_{\gamma,\beta,\lambda}(\varrho^{(\ell)}(\Lambda) \mid \bar{\varrho}^{(\ell)}(q^{(\partial\Lambda)}), *)|$$
$$\leqslant \ell\gamma 5(2^d \varrho^2_{\max} + 2^{3d}\varrho^4_{\max} + \varrho^4_{\max})\,|\Lambda| \tag{4.31}$$

Proof. Estimate (4.31) is an obvious consequence of (4.30) and we concentrate on (4.30). The difference between $\tilde{\mathscr{F}}_{\gamma,\beta,\lambda}$ and $\mathscr{F}_{\gamma,\beta,\lambda}$ has two sources. The first one is due to replacement of balls $B_\gamma(\cdot)$ in the definition of $J^{(\cdot)}_\gamma$ by their lattice versions $[B_\gamma(\cdot)]^{(\ell)}$ in the definition of $I^{(\cdot)}_\gamma$. Clearly the difference between $|B_\gamma(\cdot)| = \gamma^{-d}$ and $|[B_\gamma(\cdot)]^{(\ell)}|$ is less than $\gamma^{-d}\ell\gamma$. Hence the error produced by the discretization of $B_\gamma(\cdot)$ can be estimated exactly as in Lemma 4.5.

The second source is due to not properly counted contribution of pairs x_1, x_2 with $x_1 = x_2$ and quadruples x_1, x_2, x_3, x_4 with not all x_{i_j} being different. To estimate from above the absolute value of this error let us consider the following five contributions to the energy of $\varrho^{(\ell)}$.

(i) Self-interaction of $C^{(\ell)}_x$ due to the pair interaction of particles in $C^{(\ell)}_x$.

(ii) Self-interaction of $C^{(\ell)}_x$ due to the four-body interaction of particles in $C^{(\ell)}_x$.

(iii) Interaction between $C^{(\ell)}_{x_1}$ and $C^{(\ell)}_{x_2}$ due to the four-body interaction of two particles in $C^{(\ell)}_{x_1}$ with two particles in $C^{(\ell)}_{x_2}$.

(iv) Interaction between $C_{x_1}^{(\ell)}$ and $C_{x_2}^{(\ell)}$ due to the four-body interaction of three particles in $C_{x_1}^{(\ell)}$ with one particle in $C_{x_2}^{(\ell)}$.

(v) Interaction between $C_{x_1}^{(\ell)}$, $C_{x_2}^{(\ell)}$ and $C_{x_3}^{(\ell)}$ due to the four-body interaction of two particles in $C_{x_1}^{(\ell)}$ with one particle in $C_{x_2}^{(\ell)}$ and one particle in $C_{x_3}^{(\ell)}$.

All five contributions can be estimated by similar arguments. For that reason we present these arguments only for cases (i) and (v).

The strength of self-interaction of $C_x^{(\ell)}$ due to the pair interaction of particles in $C_x^{(\ell)}$ is less than $\gamma^d(\varrho_{\max}\ell^d)^2$. The number of boxes $C_x^{(\ell)}$ in the region Λ is $\ell^{-d}|\Lambda|$. Hence the total contribution is less than $(\ell\gamma)^d\varrho_{\max}^2|\Lambda|$.

The strength of interaction between $C_{x_1}^{(\ell)}$, $C_{x_2}^{(\ell)}$ and $C_{x_3}^{(\ell)}$ due to the four-body interaction of two particles in $C_{x_1}^{(\ell)}$ with one particle in $C_{x_2}^{(\ell)}$ and one particle in $C_{x_3}^{(\ell)}$ is less than $\gamma^{3d}(\varrho_{\max}\ell^d)^4$. The number of boxes $C_{x_2}^{(\ell)}$ and $C_{x_3}^{(\ell)}$ interacting with given box $C_{x_1}^{(\ell)}$ is less than $(\gamma\ell)^{-2d}$. The number of boxes $C_{x_1}^{(\ell)}$ is $\ell^{-d}|\Lambda|$. Hence the total contribution is less than $(\ell\gamma)^d\varrho_{\max}^4|\Lambda|$. ∎

Looking for the minima of $\mathscr{F}_{\gamma,\beta,\lambda}(\varrho^{(\ell)} \mid \bar{\varrho}^{(\ell)}(q^{(\partial\Lambda)}))$ it is simpler to understand $\varrho_x^{(\ell)}$ as continuous variables. Therefore if the minimum of $\mathscr{F}_{\gamma,\beta,\lambda}(\varrho^{(\ell)} \mid \bar{\varrho}^{(\ell)}(q^{(\partial\Lambda)}))$ is achieved on density configuration $\hat{\varrho}^{(\ell)}$ then it may happen that at least for some $x\in[\Lambda]^{(\ell)}$ the number $\ell^d\hat{\varrho}_x^{(\ell)}$ is not an integer. The solution to this problem is given by

Lemma 4.7. For the density configuration $\tilde{\varrho}_x^{(\ell)}=\ell^{-d}[\ell^d\hat{\varrho}_x^{(\ell)}]$ one has

$$|\mathscr{F}_{\gamma,\beta,\lambda}(\hat{\varrho}^{(\ell)}(\Lambda) \mid \bar{\varrho}^{(\ell)}(q^{(\partial\Lambda)}))-\mathscr{F}_{\gamma,\beta,\lambda}(\tilde{\varrho}^{(\ell)}(\Lambda) \mid \bar{\varrho}^{(\ell)}(q^{(\partial\Lambda)}))|\leqslant c\ell^{-d}\varrho_{\max}|\Lambda| \tag{4.32}$$

(Here $[\,\cdot\,]$ denotes the integer part of a number.)

Proof. We proceed as in the proof of Lemma 4.5. The total number of points $x\in[\Lambda]^{(\ell)}$ is $\ell^{-d}|\Lambda|$. Given point x the absolute value of the difference in the corresponding selfinteraction is less than $\beta^{-1}\log\varrho_{\max}+|\lambda|\varrho_{\max}$. The difference in the two-point interaction between points $x_1, x_2\in[\Lambda]^{(\ell)}$ does not exceed in absolute value $2\gamma^d\varrho_{\max}\ell^d$. The number of points x_2 interacting with given x_1 is less than $(\gamma\ell)^{-d}$. The difference in four-point interaction between points $x_1, x_2, x_4, x_4\in[\Lambda]^{(\ell)}$ does not exceed in absolute value $4\gamma^{3d}(\varrho_{\max}\ell^d)^3$. The number of points x_2, x_3, x_4 interacting with given x_1 is less than $(\gamma\ell)^{-3d}$. Combining these estimates one obtains the lemma. ∎

Variational Problem (Dependence on Boundary Condition)

The results of the previous subsection reduce the Peierls estimate to the following variational problem:

(i) Find the minimum of $\mathcal{F}_{\gamma,\beta,\lambda(\beta)}(\varrho^{(\ell_1)}(\mathrm{Supp}(\Gamma)\backslash\delta(\Gamma)) \mid \bar{q}^{(\delta^-(\Gamma))}, q_{\sigma}^{(\delta^{\neq}(\Gamma))}, *)$ over the density configurations $\varrho^{(\ell_1)} = (\varrho_x^{(\ell_1)})$, $x \in [\mathrm{Supp}(\Gamma)\backslash\delta(\Gamma)]^{(\ell_1)}$ such that $\varrho^{(\ell_1)}$ belongs to the ground state ensemble of the phase σ.

(ii) Find the minimum of $\mathcal{F}_{\gamma,\beta,\lambda(\beta)}(\varrho^{(\ell_1)}(\mathrm{Supp}(\Gamma)\backslash\delta(\Gamma)) \mid \bar{q}^{(\delta^-(\Gamma))}, q_{-\sigma}^{(\delta^{\neq}(\Gamma))}, *)$ over the density configurations $\varrho^{(\ell_1)} = (\varrho_x^{(\ell_1)})$, $x \in [\mathrm{Supp}(\Gamma)\backslash\delta(\Gamma)]^{(\ell_1)}$ such that $\varrho^{(\ell_1)}$ is compatible with η^{Γ}.

(iii) Estimate from below the difference between the minimal value of $\mathcal{F}_{\gamma,\beta,\lambda(\beta)}(\varrho^{(\ell_1)}(\mathrm{Supp}(\Gamma)\backslash\delta(\Gamma)) \mid \bar{q}^{(\delta^-(\Gamma))}, q_{-\sigma}^{(\delta^{\neq}(\Gamma))}, *)$ and the minimal value of $\mathcal{F}_{\gamma,\beta,\lambda(\beta)}(\varrho^{(\ell_1)}(\mathrm{Supp}(\Gamma)\backslash\delta(\Gamma)) \mid \bar{q}^{(\delta^-(\Gamma))}, q_{\sigma}^{(\delta^{\neq}(\Gamma))}, *)$.

The existence of the minima above is obvious as $0 < \varrho_x^{(\ell_1)} < \varrho_{\max}$. Following the approach of Section 3 define

$$\mathcal{R}_{\gamma,\ell}(x, \varrho^{(\ell)}) = b_{\gamma,\ell}^{-1} \sum_{x_1 \in [B_\gamma(x)]^{(\ell)}} \varrho_{x_1}^{(\ell)} \tag{4.33}$$

and

$$\mathcal{I}_{\gamma,\ell}(x, \varrho^{(\ell)}) = b_{\gamma,\ell}^{-1} \sum_{x_1 \in [B_\gamma(x)]^{(\ell)}} \frac{\varrho_{x_1}^{(\ell)}}{\beta} (\log \varrho_{x_1}^{(\ell)} - 1) \tag{4.34}$$

where we deliberately use a general scale (ℓ) instead of (ℓ_1) as the constructions below are of general origin. In complete similarity with (3.4)

$$\begin{aligned}\mathcal{F}_{\gamma,\beta,\lambda}(\varrho^{(\ell)}) = \ell^d \sum_{x \in \mathbb{Z}_\ell^d} \Big(&\mathcal{I}_{\gamma,\ell}(x, \varrho^{(\ell)}) - \lambda \mathcal{R}_{\gamma,\ell}(x, \varrho^{(\ell)}) \\ &- \frac{1}{2!} \mathcal{R}_{\gamma,\ell}(x, \varrho^{(\ell)})^2 + \frac{1}{4!} \mathcal{R}_{\gamma,\ell}(x, \varrho^{(\ell)})^4 \Big)\end{aligned} \tag{4.35}$$

implying that $\varrho^{(\ell)}(\mathbb{Z}_\ell^d) \equiv \varrho_{\beta,\lambda,-}$ and $\varrho^{(\ell)}(\mathbb{Z}_\ell^d) \equiv \varrho_{\beta,\lambda,+}$ are the global minimizers for $\mathcal{F}_{\gamma,\beta,\lambda}(\varrho^{(\ell)})$. The local minimizers of $\mathcal{F}_{\gamma,\beta,\lambda}(\cdot)$, i.e., the minima of $\mathcal{F}_{\gamma,\beta,\lambda}(\cdot \mid \bar{\varrho}^{(\ell)}(\Lambda^c))$ or $\mathcal{F}_{\gamma,\beta,\lambda}(\cdot \mid \bar{\varrho}^{(\ell)}(\Lambda^c), *)$, are studied in the lemma below. Note that in this lemma we consider not only $\lambda = \lambda(\beta)$ but all $\lambda \in (\lambda(\beta) - \delta(\beta), \lambda(\beta) + \delta(\beta))$. Though the lemma discusses $\mathcal{F}_{\gamma,\beta,\lambda}(\cdot \mid \bar{\varrho}^{(\ell)}(\Lambda^c))$ the same argument covers the case of $\mathcal{F}_{\gamma,\beta,\lambda}(\cdot \mid \bar{\varrho}^{(\ell)}(\Lambda^c), *)$.

Lemma 4.8. Consider a $\mathscr{D}^{(\ell)}$ measurable region Λ and take a boundary condition $\bar{\varrho}^{(\ell)}(\Lambda^c)$ with $\max_{x \in [\partial\Lambda]^{(\ell)}} |\bar{\varrho}_x^{(\ell)} - \varrho_{\beta,\lambda,\sigma}| \leqslant \zeta$, where σ is one of the phases, + or −. Then the unique minimum of $\mathscr{F}_{\gamma,\beta,\lambda}(\varrho^{(\ell)}(\Lambda) \mid \bar{\varrho}^{(\ell)}(\Lambda^c))$ is achieved on the density configuration $\hat{\varrho}^{(\ell)} = (\hat{\varrho}_x^{(\ell)})$, $x \in [\Lambda]^{(\ell)}$ such that

$$|\hat{\varrho}_x^{(\ell)} - \varrho_{\beta,\lambda,\sigma}| \leqslant \frac{\zeta a(\beta)^{[\gamma \operatorname{dist}(x,\Lambda^c)]}}{1 - a(\beta)} \tag{4.36}$$

where $[\,\cdot\,]$ denotes the integer part.

Proof. Calculating $(\partial/\partial\varrho_x^{(\ell)})\, \mathscr{F}_{\gamma,\beta,\lambda}(\varrho^{(\ell)}(\Lambda) \mid \bar{\varrho}^{(\ell)}(\Lambda^c))$, $x \in [\Lambda]^{(\ell)}$ one obtains the necessary condition for $\hat{\varrho}^{(\ell)}$

$$\begin{aligned} 0 = {} & \frac{1}{\beta} \log \hat{\varrho}_x^{(\ell)} - \lambda - \sum_{x_1} I_\gamma^{(2)}(x, x_1)\, \hat{\varrho}_{x_1}^{(\ell)} \\ & + \frac{1}{3!} \sum_{x_1, x_2, x_3} I_\gamma^{(4)}(x, x_1, x_2, x_3)\, \hat{\varrho}_{x_1}^{(\ell)} \hat{\varrho}_{x_2}^{(\ell)} \hat{\varrho}_{x_3}^{(\ell)} \end{aligned} \tag{4.37}$$

It is clear that for $\bar{\varrho}_x^{(\ell)} \equiv \varrho_{\beta,\lambda,\sigma}$, $x \in [\Lambda^c]^{(\ell)}$ one has $\hat{\varrho}_x^{(\ell)} \equiv \varrho_{\beta,\lambda,\sigma}$, $x \in [\Lambda]^{(\ell)}$. Introduce an auxiliary parameter $t \in [0, 1]$ and an *interpolated boundary condition*

$$\bar{\varrho}_x^{(\ell)}(t) = (1 - t)\, \varrho_{\beta,\lambda,\sigma} + t \bar{\varrho}_x^{(\ell)}, \qquad x \in [\Lambda^c]^{(\ell)} \tag{4.38}$$

Let $\hat{\varrho}^{(\ell)}(t)$ be the solution of (4.37) with the boundary condition $\bar{\varrho}^{(\ell)}(t)$. Then

$$\begin{aligned} \hat{\varrho}_x^{(\ell)}(t) = \exp\Bigg(& \beta\lambda + \beta \sum_{x_1 \in [B_\gamma(x)]^{(\ell)}} I_\gamma^{(2)}(x, x_1)\, \hat{\varrho}_{x_1}^{(\ell)}(t) \\ & - \frac{\beta}{3!} \sum_{x_1, x_2, x_3 \in [B_\gamma(x)]^{(\ell)}} I_\gamma^{(4)}(x, x_1, x_2, x_3)\, \hat{\varrho}_{x_1}^{(\ell)}(t)\, \hat{\varrho}_{x_2}^{(\ell)}(t)\, \hat{\varrho}_{x_3}^{(\ell)}(t) \Bigg) \end{aligned} \tag{4.39}$$

for all $x \in [\Lambda]^{(\ell)}$. Taking the derivative with respect to t one obtains

$$\begin{aligned} \frac{d}{dt} \hat{\varrho}_x^{(\ell)}(t) = \beta \hat{\varrho}_x^{(\ell)}(t) \Bigg(& \sum_{x_1 \in [B_\gamma(x)]^{(\ell)}} I_\gamma^{(2)}(x, x_1) \frac{d}{dt} \hat{\varrho}_{x_1}^{(\ell)}(t) \\ & - \frac{1}{2!} \sum_{x_1, x_2, x_3 \in [B_\gamma(x)]^{(\ell)}} I_\gamma^{(4)}(x, x_1, x_2, x_3) \frac{d}{dt} \hat{\varrho}_{x_1}^{(\ell)}(t)\, \hat{\varrho}_{x_2}^{(\ell)}(t)\, \hat{\varrho}_{x_3}^{(\ell)}(t) \Bigg) \end{aligned} \tag{4.40}$$

Set

$$I_\gamma^{(2)}(x_1, x_2 \mid \hat{\varrho}^{(\ell)}(t)) = \sum_{x_3, x_4} I_\gamma^{(4)}(x_1, x_2, x_3, x_4)\, \hat{\varrho}_{x_3}^{(\ell)}(t)\, \hat{\varrho}_{x_4}^{(\ell)}(t) \tag{4.41}$$

It is clear that

$$\begin{aligned} I_\gamma^{(2)}(x_1, x_2)(\varrho_{\beta,\lambda,\sigma} - \zeta)^2 &\leqslant I_\gamma^{(2)}(x_1, x_2 \mid \hat{\varrho}^{(\ell)}(t)) \\ &\leqslant I_\gamma^{(2)}(x_1, x_2)(\varrho_{\beta,\lambda,\sigma} + \zeta)^2 \end{aligned} \tag{4.42}$$

as soon as $\max_x |\hat{\varrho}_x^{(\ell)}(t) - \varrho_{\beta,\lambda,\sigma}| \leqslant \zeta$.

Introduce a symmetric matrix

$$A(x_1, x_2 \mid \hat{\varrho}^{(\ell)}(t)) = -I_\gamma^{(2)}(x_1, x_2) + \tfrac{1}{2} I_\gamma^{(2)}(x_1, x_2 \mid \hat{\varrho}^{(\ell)}(t)) \tag{4.43}$$

and a diagonal matrix

$$D(x_1, x_2 \mid \hat{\varrho}^{(\ell)}(t)) = (\beta \hat{\varrho}_{x_1}^{(\ell)}(t))^{-1}\, \mathbb{1}_{x_1 = x_2} \tag{4.44}$$

where $x_1, x_2 \in [\Lambda]^{(\ell)}$. It is not hard to see that the inverse matrix $B = (D - A)^{-1}$ exists if $\max_{x \in [\Lambda]^{(\ell)}} |\hat{\varrho}_x^{(\ell)}(t) - \varrho_{\beta,\lambda,\sigma}| \leqslant \zeta$. Indeed, for such $\hat{\varrho}_x^{(\ell)}(t)$ this matrix is given by a convergent series

$$B = \left(\sum_{n=0}^{\infty} (D^{-1}A)^k \right) D^{-1} \tag{4.45}$$

because

$$\begin{aligned} \|D^{-1}A\| &= \max_{x_1} \left(\beta \hat{\varrho}_{x_1}^{(\ell)}(t) \sum_{x_2} |A(x_1, x_2 \mid \hat{\varrho}^{(\ell)}(t))| \right) \\ &\leqslant a(\beta) < 1 \end{aligned} \tag{4.46}$$

where $a(\beta)$ is defined in (3.16). Moreover, the representation (4.45) and the fact that $A(x_1, x_2 \mid \hat{\varrho}^{(\ell)}(t)) = 0$ if $\mathrm{dist}(x_1, x_2) > \gamma^{-1}$ imply that

$$|B(x_1, x_2 \mid \hat{\varrho}^{(\ell)}(t))| \leqslant \frac{\beta(\varrho_{\beta,\lambda,\sigma} + \zeta)}{1 - a(\beta)}\, a(\beta)^{[\gamma\, \mathrm{dist}(x_1, x_2)]} \tag{4.47}$$

Iterating (4.40) and observing that $(d/dt)\, \bar{\varrho}_x^{(\ell)}(t) = -\varrho_{\beta,\lambda,\sigma} + \bar{\varrho}_x^{(\ell)}$ for $x \in [\Lambda^c]^{(\ell)}$ we rewrite (4.40) as

$$\frac{d}{dt} \hat{\varrho}_x^{(\ell)}(t) = \sum_{x_1 \in [\Lambda]^{(\ell)}} \sum_{x_2 \in [\Lambda^c]^{(\ell)}} B(x, x_1 \mid \hat{\varrho}^{(\ell)}(t))\, A(x_1, x_2 \mid \hat{\varrho}^{(\ell)}(t))(\bar{\varrho}_{x_2}^{(\ell)} - \varrho_{\beta,\lambda,\sigma}) \tag{4.48}$$

The right hand side of (4.48) is an absolutely convergent series in terms of $\hat{\varrho}_x^{(\ell)}(t)$, $x \in [\Lambda]^{(\ell)}$ and solving this differential equation one finds $\hat{\varrho}_x^{(\ell)}(t)$. The solution exists at least up to $t = t_1$ at which the condition

$$\max_{x \in [\Lambda]^{(\ell)}} \left| \frac{d}{dt} \hat{\varrho}_x^{(\ell)}(t) - \varrho_{\beta, \lambda, \sigma} \right| < \zeta \tag{4.49}$$

is violated. Suppose that $t_1 < 1$, i.e.,

$$\max_{x \in [\Lambda]^{(\ell)}} \left| \frac{d}{dt} \hat{\varrho}_x^{(\ell)}(t) - \varrho_{\beta, \lambda, \sigma} \right| < \zeta \tag{4.50}$$

for $t < t_1$ and

$$\left| \frac{d}{dt} \hat{\varrho}_{x_1}^{(\ell)}(t_1) - \varrho_{\beta, \lambda, \sigma} \right| = \zeta \tag{4.51}$$

for some $x_1 \in [\Lambda]^{(\ell)}$. Then the representation (4.48) is valid for $t \in [0, t_1]$ and

$$\max_{x \in [\Lambda]^{(\ell)}} |\hat{\varrho}_x^{(\ell)}(t) - \varrho_{\beta, \lambda, \sigma}| < \zeta \tag{4.52}$$

for $t \leqslant t_1$ as follows from the obvious identity

$$\hat{\varrho}_x^{(\ell)}(t) = \varrho_{\beta, \lambda, \sigma} + \int_0^t \frac{d}{ds} \hat{\varrho}_x^{(\ell)}(s) \, ds \tag{4.53}$$

Plugging (4.50)–(4.52) into (4.40) and using (3.17) one concludes that

$$\max_{x \in [\Lambda]^{(\ell)}} \left| \frac{d}{dt} \hat{\varrho}_x^{(\ell)}(t) - \varrho_{\beta, \lambda, \sigma} \right| < \zeta \tag{4.54}$$

for $t \leqslant t_1$ which contradicts (4.51). Hence

$$\max_{x \in [\Lambda]^{(\ell)}} \left| \frac{d}{dt} \hat{\varrho}_x^{(\ell)}(t) - \varrho_{\beta, \lambda, \sigma} \right| \leqslant \zeta \tag{4.55}$$

and

$$\max_{x \in [\Lambda]^{(\ell)}} |\hat{\varrho}_x^{(\ell)}(t) - \varrho_{\beta, \lambda, \sigma}| \leqslant \zeta \tag{4.56}$$

for all $t \in [0, 1]$. Thus representation (4.48) and estimate (4.47) are valid and being joined with (4.53) they give (4.36).

The density configuration $\hat{\varrho}(\Lambda)$ was defined as the solution of (4.37) and we just checked that such a solution exists and satisfies (4.36). On the other hand, it follows from (4.42) and (3.17) that the Hessian matrix $(\partial^2/\partial\varrho_x^{(\ell)}\,\partial\varrho_y^{(\ell)})\,\mathscr{F}_{\gamma,\beta,\lambda}(\varrho^{(\ell)}(\Lambda)\mid\bar{\varrho}^{(\ell)}(\Lambda^c))$, $x, y \in [\Lambda]^{(\ell)}$ is positive for any $\varrho^{(\ell)}(\Lambda) \in (\varrho_{\beta,\lambda,\sigma}-\zeta, \varrho_{\beta,\lambda,\sigma}+\zeta)^\Lambda$ with the mass bounded away from 0 independently on $\varrho^{(\ell)}(\Lambda)$ and $\bar{\varrho}^{(\ell)}(\Lambda^c)$. Thus the function $\mathscr{F}_{\gamma,\beta,\lambda}(\cdot\mid\bar{\varrho}^{(\ell)}(\Lambda^c))$ with convex domain $(\varrho_{\beta,\lambda,\sigma}-\zeta, \varrho_{\beta,\lambda,\sigma}+\zeta)^\Lambda$ is convex and therefore $\hat{\varrho}(\Lambda)$ is its unique minimum. ∎

Variational Problem (Comparison of Two Minima)

In this subsection we return back to the case $\lambda=\lambda(\beta)$. For any contour Γ the set

$$S=\partial^{(\ell_3)}\,\mathrm{Ext}(\Gamma)\cup\Big(\bigcup_{m:\,\sigma_m(\Gamma)=\sigma(\Gamma)}\partial^{(\ell_3)}\mathrm{Int}_m(\Gamma)\Big) \tag{4.57}$$

is occupied by the ground state of the phase $\sigma=\sigma(\Gamma)$. Consider a strip $\tilde{\delta}^=(\Gamma)$ of width γ^{-1} situated in the middle of S. According to Lemma 4.8 inside $\tilde{\delta}^=(\Gamma)$ the density configurations minimizing $\mathscr{F}_{\gamma,\beta,\lambda}(\varrho^{(\ell_1)}\mid\bar{q}^{(\delta^=(\Gamma))}, q_\sigma^{(\delta^{\neq}(\Gamma))}, *)$ and $\mathscr{F}_{\gamma,\beta,\lambda}(\varrho^{(\ell_1)}\mid\bar{q}^{(\delta^=(\Gamma))}, q_{-\sigma}^{(\delta^{\neq}(\Gamma))}, *)$ differ from $\varrho_{\beta,\sigma}$ by at most $(1-a(\beta))^{-1}\,\zeta a(\beta)^{[\gamma^{1/3}\gamma^{-1-\alpha_3}]}$. Therefore the density configuration which minimizes $\mathscr{F}_{\gamma,\beta,\lambda}(\varrho^{(\ell_1)}\mid\bar{q}^{(\delta^=(\Gamma))}, q_\sigma^{(\delta^{\neq}(\Gamma))}, *)$ under the additional condition that this configuration coincides with $\varrho_{\beta,\sigma}$ in $\tilde{\delta}^=(\Gamma)$ gives the value of the true minimum of $\mathscr{F}_{\gamma,\beta,\lambda}(\varrho^{(\ell_1)}\mid\bar{q}^{(\delta^=(\Gamma))}, q_\sigma^{(\delta^{\neq}(\Gamma))}, *)$ up to an nonessential error. The same is true for the density configuration minimizing $\mathscr{F}_{\gamma,\beta,\lambda}(\varrho^{(\ell_1)}\mid\bar{q}^{(\delta^=(\Gamma))}, q_{-\sigma}^{(\delta^{\neq}(\Gamma))}, *)$ with the same additional condition. Obviously these modified minimizing configurations coincide with each other not only in $\tilde{\delta}^=(\Gamma)$, where they both are equal to $\varrho_{\beta,\sigma}$, but in the whole part of S stretching from $\delta^=(\Gamma)$ to $\tilde{\delta}^=(\Gamma)$, i.e., in

$$S_1=\partial^{(\ell_3/2)}\,\mathrm{Ext}(\Gamma)\cup\Big(\bigcup_{m:\,\sigma_m(\Gamma)=\sigma(\Gamma)}\partial^{(\ell_3/2)}\,\mathrm{Int}_m(\Gamma)\Big)\cup\tilde{\delta}^=(\Gamma) \tag{4.58}$$

Estimating from below the difference between the minimal value of $\mathscr{F}_{\gamma,\beta,\lambda}(\varrho^{(\ell_1)}\mid\bar{q}^{(\delta^=(\Gamma))}, q_{-\sigma}^{(\delta^{\neq}(\Gamma))}, *)$ and the minimal value of $\mathscr{F}_{\gamma,\beta,\lambda}(\varrho^{(\ell_1)}\mid\bar{q}^{(\delta^=(\Gamma))}, q_\sigma^{(\delta^{\neq}(\Gamma))}, *)$ the contributions corresponding to S_1 cancel each other. Hence the initial variational problem in $\mathrm{Supp}(\Gamma)\backslash\delta(\Gamma)$ with a general boundary condition imposed on $\delta(\Gamma)$ is reduced to a similar problem in a smaller volume $\mathrm{Supp}(\Gamma)\backslash S_1$.

For the sake of notational simplicity we suppose from now on that the boundary condition is a standard one already for the initial problem, i.e., it is equal to $\varrho_{\beta,\sigma}$ in $\delta^{=}(\Gamma)$. Thus all we need to finish the Peierls estimate is the lower bound for

$$\Delta\mathscr{F}_{\gamma,\beta,\lambda}(\varrho^{(\ell_1)},\Gamma)=\ell_1^d\sum_{x\in[\mathrm{Supp}(\Gamma)\backslash\delta(\Gamma)]^{(\ell_1)}}\Big(\mathscr{I}_{\gamma,\ell_1}(x,\varrho^{(\ell_1)})-\lambda\mathscr{R}_{\gamma,\ell_1}(x,\varrho^{(\ell_1)})$$
$$-\frac{1}{2!}\mathscr{R}_{\gamma,\ell_1}(x,\varrho^{(\ell_1)})^2+\frac{1}{4!}\mathscr{R}_{\gamma,\ell_1}(x,\varrho^{(\ell_1)})^4-F_{\beta,\lambda}(\varrho_{\beta,\sigma})\Big) \tag{4.59}$$

We note that for x such that $\mathrm{dist}(x,\delta(\Gamma))\leqslant\gamma^{-1}$ the values of $\mathscr{I}_{\gamma,\ell}(x,\varrho^{(\ell)})$ and $\mathscr{R}_{\gamma,\ell}(x,\varrho^{(\ell)})$ depend on the boundary condition which is equal to $\varrho_{\beta,\sigma}$ in $\delta^{=}(\Gamma)$ and $\varrho_{\beta,-\sigma}$ in $\delta^{\neq}(\Gamma)$.

Lemma 4.9. For any contour $\Gamma=(\mathrm{Supp}(\Gamma),\eta^\Gamma)$

$$\min_{\varrho^{(\ell_1)}:\,\eta(\varrho^{(\ell_1)})=\eta^\Gamma}\Delta\mathscr{F}_{\gamma,\beta,\lambda}(\varrho^{(\ell_1)},\Gamma)\geqslant c\ell_2^d\ell_3^{-d}\zeta^2\,|\mathrm{Supp}(\Gamma)| \tag{4.60}$$

Proof. Rewrite $\Delta\mathscr{F}_{\gamma,\beta,\lambda}(\varrho^{(\ell_1)},\Gamma)$ in the form

$$\Delta\mathscr{F}_{\gamma,\beta,\lambda}(\varrho^{(\ell_1)},\Gamma)=\ell_1^d\sum_{x\in[\mathrm{Supp}(\Gamma)\backslash\delta(\Gamma)]^{(\ell_1)}}\Big(F_{\beta,\lambda}(\mathscr{R}_{\gamma,\ell_1}(x,\varrho^{(\ell_1)}))-F_{\beta,\lambda}(\varrho_{\beta,\sigma})$$
$$+b_{\gamma,\ell_1}^{-1}\sum_{x_1\in[B_\gamma(x)]^{(\ell_1)}}\frac{\varrho_{x_1}^{(\ell_1)}}{\beta}\log\varrho^{(\ell_1)}{}_{x_1}$$
$$-\frac{\mathscr{R}_{\gamma,\ell_1}(x,\varrho^{(\ell_1)})}{\beta}\log\mathscr{R}_{\gamma,\ell_1}(x,\varrho^{(\ell_1)})\Big) \tag{4.61}$$

and observe that

$$F_{\beta,\lambda}(\mathscr{R}_{\gamma,\ell_1}(x,\varrho^{(\ell_1)}))-F_{\beta,\lambda}(\varrho_{\beta,\sigma})\geqslant 0 \tag{4.62}$$

by definition of $\varrho_{\beta,\sigma}$ and

$$b_{\gamma,\ell_1}^{-1}\sum_{x_1\in[B_\gamma(x)]^{(\ell_1)}}\frac{\varrho_{x_1}^{(\ell_1)}}{\beta}\log\varrho_{x_1}^{(\ell_1)}\geqslant\frac{\mathscr{R}_{\gamma,\ell_1}(x,\varrho^{(\ell_1)})}{\beta}\log\mathscr{R}_{\gamma,\ell_1}(x,\varrho^{(\ell_1)}) \tag{4.63}$$

by convexity.

We call $C_x^{(\ell_2)} \in \mathrm{Supp}(\Gamma)$ a *wrong box* if either $\eta_x^\Gamma = 0$ or $\eta_x^\Gamma \eta_{x_1}^\Gamma = -1$ for at least one cube $C_{x_1}^{(\ell_2)}$ adjacent to $C_x^{(\ell_2)}$. According to the definition of the contour there exist at least $(3\ell_3)^{-d} |\mathrm{Supp}(\Gamma)|$ wrong boxes $C_x^{(\ell_2)} \in \mathrm{Supp}(\Gamma)$ such that they are at a distance greater than $5\gamma^{-1}$ from each other. In particular $C_{x_1}^{(2\gamma^{-1})} \cap C_{x_2}^{(2\gamma^{-1})} = \varnothing$ for any two such boxes $C_{x_1}^{(\ell_2)}$ and $C_{x_2}^{(\ell_2)}$.

Consider a wrong box $C_x^{(\ell_2)}$ for which $\eta_x^\Gamma = 0$. If inside $C_x^{(2\gamma^{-1})}$ there exist at least $\ell_1^{-d} \ell_2^d$ points x_1 with

$$\mathcal{R}_{\gamma, \ell_1}(x_1, \varrho^{(\ell_1)}) \notin \left(\varrho_{\beta, -} - \frac{\zeta}{2}, \varrho_{\beta, -} + \frac{\zeta}{2} \right) \cup \left(\varrho_{\beta, +} - \frac{\zeta}{2}, \varrho_{\beta, +} + \frac{\zeta}{2} \right) \tag{4.64}$$

then $C_x^{(2\gamma^{-1})}$ contributes to $\Delta \mathcal{F}_{\gamma, \beta, \lambda}(\varrho^{(\ell_1)}, \Gamma)$ by at least

$$c \ell_2^d \zeta^2 \min(F''_{\beta, \lambda}(\varrho_{\beta, -}), F''_{\beta, \lambda}(\varrho_{\beta, +})) \tag{4.65}$$

This contribution comes from the terms

$$\ell_1^d F_{\beta, \lambda}(\mathcal{R}_{\gamma, \ell_1}(x_1, \varrho^{(\ell_1)})) - \ell_1^d F_{\beta, \lambda}(\varrho_{\beta, \sigma}) \tag{4.66}$$

in (4.61).

In the opposite situation when

$$\mathcal{R}_{\gamma, \ell_1}(x_1, \varrho^{(\ell_1)}) \notin \left(\varrho_{\beta, -} - \frac{\zeta}{2}, \varrho_{\beta, -} + \frac{\zeta}{2} \right) \cup \left(\varrho_{\beta, +} - \frac{\zeta}{2}, \varrho_{\beta, +} + \frac{\zeta}{2} \right) \tag{4.67}$$

for not more than $\ell_1^{-d} \ell_2^d$ points $x_1 \in C_x^{(2\gamma^{-1})}$ we will extract a contribution similar to (4.65) from the terms

$$\ell_1^d b_{\gamma, \ell_1}^{-1} \sum_{x_2 \in [B_\gamma(x_1)]^{(\ell_1)}} \frac{\varrho_{x_2}^{(\ell_1)}}{\beta} \log \varrho_{x_2}^{(\ell_1)} - \ell_1^d \frac{\mathcal{R}_{\gamma, \ell_1}(x_1, \varrho^{(\ell_1)})}{\beta} \log \mathcal{R}_{\gamma, \ell_1}(x_1, \varrho^{(\ell_1)}) \tag{4.68}$$

in (4.61).

Indeed, consider $x_1 \in C_x^{(2\gamma^{-1})}$ such that $C_x^{(\ell_2)} \in [B_\gamma(x_1)]^{(\ell_1)}$ and observe that

$$\varrho_x^{(\ell_2)} = \| [C_x^{(\ell_2)}]^{(\ell_1)} \|^{-1} \sum_{x_2 \in [C_x^{(\ell_2)}]^{(\ell_1)}} \varrho_{x_2}^{(\ell_1)} \tag{4.69}$$

does not belong to $(\varrho_{\beta, -} - \zeta, \varrho_{\beta, -} + \zeta) \cup (\varrho_{\beta, +} - \zeta, \varrho_{\beta, +} + \zeta)$ as $\eta_x^\Gamma = 0$. Denote

$$\mathcal{R}_{x_1} = \| [B_\gamma(x_1)]^{(\ell_1)} \backslash [C_x^{(\ell_2)}]^{(\ell_1)} \|^{-1} \sum_{x_2 \in [B_\gamma(x_1)]^{(\ell_1)} \backslash [C_x^{(\ell_2)}]^{(\ell_1)}} \varrho_{x_2}^{(\ell_1)} \tag{4.70}$$

Then by convexity

$$\sum_{x_2 \in [B_\gamma(x_1)]^{(\ell_1)}} \varrho_{x_2}^{(\ell_1)} \log \varrho_{x_2}^{(\ell_1)} \geqslant \|[B_\gamma(x_1)]^{(\ell_1)} \backslash [C_x^{(\ell_2)}]^{(\ell_1)}\| \, \mathscr{R}_{x_1} \log \mathscr{R}_{x_1} + \|[C_x^{(\ell_2)}]^{(\ell_1)}\| \, \varrho_x^{(\ell_2)} \log \varrho_x^{(\ell_2)} \tag{4.71}$$

implying the lower bound

$$\frac{\|[B_\gamma(x_1)]^{(\ell_1)} \backslash [C_x^{(\ell_2)}]^{(\ell_1)}\|}{\beta b_{\gamma, \ell_1}} \mathscr{R}_{x_1} \log \mathscr{R}_{x_1} + \frac{\|[C_x^{(\ell_2)}]^{(\ell_1)}\|}{\beta b_{\gamma, \ell_1}} \varrho_x^{(\ell_2)} \log \varrho_x^{(\ell_2)} - \frac{\mathscr{R}_{\gamma, \ell_1}(x_1, \varrho^{(\ell_1)})}{\beta} \log \mathscr{R}_{\gamma, \ell_1}(x_1, \varrho^{(\ell_1)}) \tag{4.72}$$

for (4.68). Now we apply the inequality

$$(1-\alpha)\, a \log a - \alpha b \log b - ((1-\alpha)\, a + \alpha b) \log((1-\alpha)\, a + \alpha b) \geqslant \frac{\alpha}{2} \frac{(a-b)^2}{\max(a, b)} - \alpha^2 \frac{(a-b)^2}{a} \tag{4.73}$$

which is true for any $a, b > 0$ and $0 < \alpha < 1$. This leads to the lower bound

$$\frac{\|[C_x^{(\ell_2)}]^{(\ell_1)}\|}{2\beta b_{\gamma, \ell_1}} \frac{(\mathscr{R}_{x_1} - \varrho_x^{(\ell_2)})^2}{\max(\mathscr{R}_{x_1}, \varrho_x^{(\ell_2)})} - \left(\frac{\|[C_x^{(\ell_2)}]^{(\ell_1)}\|}{\beta b_{\gamma, \ell_1}}\right)^2 \frac{(\mathscr{R}_{x_1} - \varrho_x^{(\ell_2)})^2}{\mathscr{R}_{x_1}} \tag{4.74}$$

for (4.72). Observing that $(\beta b_{\gamma, \ell_1})^{-1} \|[C_x^{(\ell_2)}]^{(\ell_1)}\|$ is of order $\gamma^{\alpha_2 d}$ we conclude that (4.74) exceeds

$$c \frac{\|[C_x^{(\ell_2)}]^{(\ell_1)}\|}{\beta b_{\gamma, \ell_1} \varrho_{\beta, -}} (\mathscr{R}_{\gamma, \ell_1}(x_1, \varrho^{(\ell_1)}) - \varrho_x^{(\ell_2)})^2 \geqslant c \gamma^{\alpha_2 d} \frac{\zeta^2}{\beta \varrho_{\beta, -}} \tag{4.75}$$

for γ sufficiently small. The number of points $x_1 \in C_x^{(2\gamma^{-1})}$ for which (4.75) is true is not less than $b_{\gamma, \ell_1}/2$. Remind that for these points

$$\mathscr{R}_{\gamma, \ell_1}(x_1, \varrho^{(\ell_1)}) \in \left(\varrho_{\beta, -} - \frac{\zeta}{2}, \varrho_{\beta, -} + \frac{\zeta}{2}\right) \cup \left(\varrho_{\beta, +} - \frac{\zeta}{2}, \varrho_{\beta, +} + \frac{\zeta}{2}\right) \tag{4.76}$$

and $[B_\gamma(x_1)]^{(\ell_1)} \ni C_x^{(\ell_2)}$. Thus (4.75) again gives us a lower bound

$$c \ell_2^d \zeta^2 (\beta \varrho_{\beta, -})^{-1} \tag{4.77}$$

similar to (4.65).

Finally for the wrong box $C_x^{(\ell_2)}$ of the second type, i.e., when $\eta_x^\Gamma \eta_{x_1}^\Gamma = -1$ for an adjacent box $C_{x_1}^{(\ell_2)}$, we can consider $C_x^{(\ell_2)} \cup C_{x_1}^{(\ell_2)}$ instead of $C_x^{(\ell_2)}$ and repeat all the arguments above. They will work perfectly because

$$\|[C_x^{(\ell_2)}]^{(\ell_1)} \cup [C_{x_1}^{(\ell_2)}]^{(\ell_1)}\|^{-1} \sum_{x_2 \in [C_x^{(\ell_2)}]^{(\ell_1)} \cup [C_{x_1}^{(\ell_2)}]^{(\ell_1)}} \varrho_{x_2}^{(\ell_1)} \tag{4.78}$$

is again outside $(\varrho_{\beta,-} - \zeta, \varrho_{\beta,-} + \zeta) \cup (\varrho_{\beta,+} - \zeta, \varrho_{\beta,+} + \zeta)$. ∎

To finish the proof of the Peierls estimate we need to compare all the errors which are at most of the order $c\ell_1 \gamma\, |\mathrm{Supp}(\Gamma)| = c\gamma^{\alpha_1}\, |\mathrm{Supp}(\Gamma)|$ (see (4.22) and (4.23)) with the contribution coming from Lemma 4.9. The last is of order $c\ell_2^d \ell_3^{-d}\, |\mathrm{Supp}(\Gamma)| = c\gamma^{d\alpha_2 + d\alpha_3}\, |\mathrm{Supp}(\Gamma)|$ and dominates $c\gamma^{\alpha_1}\, |\mathrm{Supp}(\Gamma)|$ since $d\alpha_2 + d\alpha_3 < \alpha_1$.

5. AUXILIARY MODEL

In this section we study metastable models and we prove that the corresponding measures satisfy the Dobrushin uniqueness condition and hence exhibit an exponential decay of correlations. We note again that the configuration $q^{(\Lambda)}$ in a $\mathscr{D}^{(\ell_2)}$ measurable region Λ belongs to the ground state ensemble of the phase σ iff in every cube $C_x^{(\ell_2)} \in \Lambda$ the density of particles

$$\varrho_x^{(\ell_2)}(q^{(\Lambda)}) = \ell_2^{-d}\, |q^{(\Lambda)} \cap C_x^{(\ell_2)}| \tag{5.1}$$

belongs to the interval

$$(\varrho_{\beta,\sigma} - \zeta, \varrho_{\beta,\sigma} + \zeta) \tag{5.2}$$

The partition function of the auxiliary model is given by (3.37), i.e., it is the integral over ground state configurations of the corresponding contour partition function.

In the previous section we have shown that some partition functions initially defined in terms of particle configurations can be approximated by partition functions written in terms of density configurations related to some scale ℓ. Now we go further and show that the auxiliary model for each of the two phases can be *equivalently* rewritten in terms of density configurations. Such an equivalent model is defined on the lattice $\mathbb{Z}_{\ell_2}^d$ with the density variables $\varrho_x^{(\ell_2)}$ taking discrete values $n\ell_2^{-d}$, $n = 1, 2...$ from the bounded interval (5.2). The corresponding Hamiltonian is of infinite range but with sufficiently fast decaying interactions. Quite naturally this Hamiltonian is close to (4.27) (with $\ell = \ell_2$) and it can be understood as a small perturbation of a positive definite quadratic form. The treatment of the equivalent model is based on a specific approach [COPP] to the Dobrushin uniqueness theorem developed initially for unbounded lattice spin systems.

From now on we fix ℓ_2 as the scale at which we define density configurations. Thus all regions are assumed to be $\mathscr{D}^{(\ell_2)}$ measurable and we often drop the superscript (ℓ_2) from notations.

In the next subsection we construct the effective Hamiltonian, the exact form of which is stated in Lemma 5.2 at the very end of the next subsection. Then in the last two subsections we prove Lemma 5.3 saying that the effective Hamiltonian satisfies the Dobrushin uniqueness condition.

Reduction to Density Model

The technical part of the reduction is based on the cluster or polymer expansion technique. For the convenience of the reader Section 7 quotes a version of the general cluster expansion theorem which is suitable for our purposes.

Consider a region Λ with the boundary condition $\bar{q}^{(\Lambda^c)}$ belonging to the ground state ensemble of the phase σ. For the partition function $Z^{\Lambda}_{\gamma,\beta,\lambda}(\Lambda \mid \bar{q}^{(\Lambda^c)})$ we decompose the integral in (3.37) into a sum of integrals. In this decomposition the external sum is taken over all density configurations ϱ_x, $x \in [\Lambda]^{(\ell_2)}$ satisfying (5.2). Given such a density configuration $\varrho(\Lambda)$ an internal integral is taken over all particle configurations $q^{(\Lambda)}$ such that $\varrho_x^{(\ell_2)}(q^{(\Lambda)}) = \varrho_x^{(\ell_2)}(\Lambda)$ for all $x \in [\Lambda]^{(\ell_2)}$. The reduction we perform is nothing but the calculation of

$$\log \int_{\mathscr{Q}(\Lambda)} dq^{(\Lambda)}\, \mathbb{1}_{\varrho^{(\ell_2)}(q) = \varrho(\Lambda)}\, e^{-\beta H_{\gamma,\lambda}(q^{(\Lambda)} \mid \bar{q}^{(\Lambda^c)})} \sum_{\{\Gamma_i\}^\sigma \in \Lambda} \prod_i W^T(\eta^{\Gamma_i} \mid q^{(\delta^-(\Gamma_i))}) \tag{5.3}$$

as a function of $\varrho_x(\Lambda)$. Denote by $q_{x,i}$ the particles of $q^{(\Lambda)}$ situated inside $C_x^{(\ell_2)}$ and set $n_x = \ell_2^d \varrho_x = |q^{(\Lambda)} \cap C_x^{(\ell_2)}|$. Then $\int_{\mathscr{Q}(\Lambda)} dq$ is

$$\left(\prod_{x \in [\Lambda]^{(\ell_2)}} \frac{\ell_2^{dn_x}}{n_x!} \ell_2^{-dn_x} \right) \int \cdots \int \left(\prod_{x \in [\Lambda]^{(\ell_2)}} \prod_{i=1}^{n_x} \mathbb{1}_{q_{x,i} \in C_x^{(\ell_2)}}\, dq_{x,i} \right) e^{-\beta H_{\gamma,\lambda}(q^{(\Lambda)} \mid \bar{q}^{(\Lambda^c)})}$$

$$\times \sum_{\{\Gamma_i\}^\sigma \in \Lambda} \prod_i W^T(\eta^{\Gamma_i} \mid q^{(\delta^-(\Gamma_i))}) \tag{5.4}$$

where the integral can be understood as an expectation with respect to the system of independent particles $q_{x,i}$ uniformly distributed in the corresponding boxes $C_x^{(\ell_2)}$.

The Hamiltonian $H_{\gamma,\lambda}(q^{(\Lambda)} \mid \bar{q}^{(\Lambda^c)})$ can be decomposed into the sum of a ϱ-dependent Hamiltonian

$$H_{\gamma,\lambda}(\varrho^{(\ell_2)}(Q^{(\Lambda)}) \mid \bar{q}^{(\Lambda^c)}) = H_{\gamma,\lambda}([q^{(\Lambda)}]^{(\ell_2)} \mid \bar{q}^{(\Lambda^c)}) = H_{\gamma,\lambda}(\varrho(\Lambda) \mid \bar{q}^{(\Lambda^c)}) \tag{5.5}$$

and error terms

$$\Delta H_{\gamma,\lambda}(q^{(\Lambda)} \mid \bar{q}^{(\Lambda^c)}) = H_{\gamma,\lambda}(q^{(\Lambda)} \mid \bar{q}^{(\Lambda^c)}) - H_{\gamma,\lambda}(\varrho^{(\ell_2)}(q^{(\Lambda)}) \mid \bar{q}^{(\Lambda^c)}) \tag{5.6}$$

Clearly $\exp(-\beta H_{\gamma,\lambda}(\varrho \mid \bar{q}^{(\Lambda^c)}))$ can be taken outside the integral in (5.4) leaving us with the calculation of the log of the partition function

$$\left(\prod_{x\in[\Lambda]^{(\ell_2)}} \ell_2^{-dn_x}\right)\int\cdots\int\left(\prod_{x\in[\Lambda]^{(\ell_2)}}\prod_{i=1}^{n_x} \mathbb{1}_{q_{x,i}\in C_x^{(\ell_2)}}\, dq_{x,i}\right) e^{-\beta\,\Delta H_{\gamma,\lambda}(q^{(\Lambda)} \mid \bar{q}^{(\Lambda^c)})}$$
$$\times \sum_{\{\Gamma_i\}^\sigma\in\Lambda}\prod_i W^T(\eta^{\Gamma_i} \mid q^{(\delta^-(\Gamma_i))}) \tag{5.7}$$

where the first product is included for the convenience of treating this partition function as an expectation over a system of independent particles.

Observe that the error part of the Hamiltonian is given by

$$\Delta H_{\gamma,\lambda}(q^{(\Lambda)} \mid \bar{q}^{(\Lambda^c)})$$
$$= -\sum_{q_{i_1},q_{i_2}\in q} \Delta J_\gamma^{(2)}(q_{i_1},q_{i_2}) + \sum_{q_{i_1},q_{i_2},q_{i_3},q_{i_4}\in q} \Delta J_\gamma^{(4)}(q_{i_1},q_{i_2},q_{i_3},q_{i_4}) \tag{5.8}$$

where $\Delta J_\gamma^{(2)}(q_{i_1},q_{i_2})$ and $\Delta J_\gamma^{(4)}(q_{i_1},q_{i_2},q_{i_3},q_{i_4})$ are much smaller than 1. The estimates

$$0 < \Delta J_\gamma^{(2)}(q_{i_1},q_{i_2}) < c\gamma^{\alpha_2} J_\gamma^{(2)}(q_{i_1},q_{i_2}) \tag{5.9}$$

and

$$0 < \Delta J_\gamma^{(4)}(q_{i_1},q_{i_2},q_{i_3},q_{i_4}) < c\gamma^{\alpha_2} J_\gamma^{(4)}(q_{i_1},q_{i_2},q_{i_3},q_{i_4}) \tag{5.10}$$

are true unless some of the interacting particles are at the distance larger than $\gamma^{-1}-\gamma^{-1+\alpha_2}$ from each other. In the last case $\Delta J_\gamma^{(2)}(q_{i_1},q_{i_2})$ and $\Delta J_\gamma^{(4)}(q_{i_1},q_{i_2},q_{i_3},q_{i_4})$ are extremely small

$$0<\gamma^{-d}\,\Delta J_\gamma^{(2)}(q_{i_1},q_{i_2}), \qquad \gamma^{-3d}\,\Delta J_\gamma^{(4)}(q_{i_1},q_{i_2},q_{i_3},q_{i_4}) < c\gamma^{3d/2-\alpha_2/2} \tag{5.11}$$

Denoting

$$w_\gamma^{(2)}(q_{i_1}, q_{i_2}) = e^{\beta \Delta J_\gamma^{(2)}(q_{i_1}, q_{i_2})} - 1 \tag{5.12}$$

and

$$w_\gamma^{(4)}(q_{i_1}, q_{i_2}, q_{i_3}, q_{i_4}) = e^{\beta \Delta J_\gamma^{(4)}(q_{i_1}, q_{i_2}, q_{i_3}, q_{i_4})} - 1 \tag{5.13}$$

we have

$$\begin{aligned} e^{-\beta \Delta H_{\gamma, \lambda}(q^{(\Lambda)} \mid \bar{q}^{(\Lambda^c)})} = & \prod_{q_{i_1}, q_{i_2} \in q} (1 + w_\gamma^{(2)}(q_{i_1}, q_{i_2})) \\ & \times \prod_{q_{i_1}, q_{i_2}, q_{i_3}, q_{i_4} \in q} (1 + w_\gamma^{(4)}(q_{i_1}, q_{i_2}, q_{i_3}, q_{i_4})) \end{aligned} \tag{5.14}$$

First we obtain the polymer expansion for the log of the partition function

$$\left(\prod_{x \in [\Lambda]^{(\ell_2)}} \ell_2^{-dn_x} \right) \int \cdots \int \left(\prod_{x \in [\Lambda]^{(\ell_2)}} \prod_{i=1}^{n_x} \mathbb{1}_{q_{x,i} \in C_x^{(\ell_2)}} \, dq_{x,i} \right) e^{-\beta \Delta H_{\gamma, \lambda}(q^{(\Lambda)} \mid \bar{q}^{(\Lambda^c)})} \tag{5.15}$$

containing no contours. Opening all brackets in (5.14) we rearrange the expression under the integral in (5.15) in the following way.

Let a *2-link*, $L^{(2)} = (q_1, q_2)$, be a couple of particles q_1, q_2 such that $W(L^{(2)}) = w_\gamma^{(2)}(q_1, q_2) \neq 0$. Similarly a *4-link*, $L^{(4)} = (q_1, q_2, q_3, q_4)$, is a quadruple of particles q_1, q_2, q_3, q_4 such that $W(L^{(4)}) = w_\gamma^{(4)}(q_1, q_2, q_3, q_4) \neq 0$. The quantities $W(L^{(2)})$ and $W(L^{(4)})$ are called the statistical weights of the 2-link and 4-link respectively. Two links are *connected* if they have a common particle. We stress that if two links have no common particles but have a common space point occupied by one particle from the first link and another particle from the second link then these links are not connected. A *pre-diagram*, θ, is a connected set of links. Denote by $q(\theta) = (q_i(\theta))$ the particles of $q^{(\Lambda)}$ which influence θ, i.e., the endpoints of links of θ. The *statistical weight*, $w(\theta)$, of the pre-diagram is the product of the statistical weights of the contributing links. Two pre-diagrams are *compatible* if they are not connected. Finally, a *compatible collection of pre-diagrams* consists of mutually compatible pre-diagrams.

The definitions above justify the representation for (5.15) of the form

$$\left(\prod_{x \in [\Lambda]^{(\ell_2)}} \ell_2^{-dn_x} \right) \int \cdots \int \left(\prod_{x \in [\Lambda]^{(\ell_2)}} \prod_{i=1}^{n_x} \mathbb{1}_{q_{x,i} \in C_x^{(\ell_2)}} \, dq_{x,i} \right) \sum_{\{\theta_j\} \notin \Lambda^c} \prod_j w(\theta_j) \tag{5.16}$$

where the sum goes over compatible collections, $\{\theta_j\}$, of pre-diagrams and $\{\theta_j\} \notin \Lambda^c$ means that every θ_j has at least one particle inside Λ.

We call two pre-diagrams *equivalent* if they can be transformed one into another by shifting some particles such that every shifted particle $q_{x,i}$ remains in its initial box $C_x^{(\ell_2)}$. The corresponding equivalence classes, Θ, are called *diagrams*. To have a geometrical interpretation of the diagram Θ we identify it with the pre-diagram $\theta \in \Theta$ having all particles at the centers of the corresponding cubes $C_x^{(\ell_2)}$. We say that two diagrams Θ_1 and Θ_2 are compatible if any $\theta_1 \in \Theta_1$ and $\theta_2 \in \Theta_2$ are compatible. Setting

$$W(\Theta) = \ell_2^{-d\,|q(\Theta)|} \int \cdots \int w(\theta) \prod_{i=1}^{|q(\Theta)|} \mathbb{1}_{q_i(\theta) \in C^{(\ell_2)}_{q_i(\Theta)}}\, dq_i(\theta) \tag{5.17}$$

we rewrite the partition function (5.15) in the so called cluster form

$$\sum_{\{\Theta_j\} \notin \Lambda^c} \prod_j W(\Theta_j) \tag{5.18}$$

where the sum is extended to all compatible collections of diagrams. The transition from (5.16) to (5.18) relies on the fact that for compatible θ_{j_1} and θ_{j_2} the corresponding sets of particles $q(\theta_{j_1})$ and $q(\theta_{j_2})$ do not intersect each other.

Lemma 5.1. Let

$$a(\Theta) = |q(\Theta)| \tag{5.19}$$

Then

$$\sum_{\Theta' \not\sim \Theta} |W(\Theta')|\, e^{a(\Theta')} \leqslant a(\Theta) \tag{5.20}$$

where $\Theta' \not\sim \Theta$ denotes a diagram Θ' not compatible with a given diagram Θ.

Proof. We say that a diagram Θ is not compatible with a particle q and we denote it $\Theta \not\sim q$ if $q \in q(\Theta)$. From our definition of compatibility of diagrams it is clear that (5.20) follows from

$$\sum_{\Theta \not\sim q} |W(\Theta)|\, e^{a(\Theta)} \leqslant 1 \tag{5.21}$$

Note that for each box $C_x^{(\ell_2)}$, $x \in [\Lambda^{(\ell_2)}]$ the number of particles inside $C_x^{(\ell_2)}$ is fixed once the density ϱ of the configuration is fixed. Hence when ϱ

belongs to the interval (5.2) there are at most $c(\varrho_{\beta,\lambda,\sigma}+\zeta)\,\gamma^{-d}$ 2-links and at most $c(\varrho_{\beta,\lambda,\sigma}+\zeta)^3\,\gamma^{-3d}$ 4-links passing through any given particle. The statistical weights of links satisfy

$$|W(L^{(2)})| \leqslant c\gamma^{\alpha_2}\gamma^{d} \tag{5.22}$$

and

$$|W(L^{(4)})| \leqslant c\gamma^{\alpha_2}\gamma^{3d} \tag{5.23}$$

as follows from (5.9)–(5.11) and (2.3)–(2.4).

We provide a diagram with an abstract tree structure according to the following algorithm. The root of the tree is the particle q_1. Links which start from q_1 are called links of the first level. Links which start at endpoints of the links of the first level and are different from them are the links of the second level. Generally, links which starts at the endpoints of nth level links and are different from all links of levels $1, 2,..., n$ are called links of level $n+1$.

Denote by $n(\Theta)$ the maximal level of links in Θ. It is clear that for γ small enough

$$\begin{aligned}&\sum_{\Theta \not\sim q:\, n(\Theta)=1} |W(\Theta)|\, e^{a(\Theta)}\\ &\quad\leqslant \prod_{L^{(2)}\ni q} (1+e^2\,|W(L^{(2)})|) \prod_{L^{(4)}\ni q} (1+e^4\,|W(L^{(4)})|)-1 \leqslant 1\end{aligned} \tag{5.24}$$

By induction suppose that

$$\sum_{\Theta \not\sim q:\, n(\Theta)\leqslant N} |W(\Theta)|\, e^{a(\Theta)} \leqslant 1 \tag{5.25}$$

and consider Θ with $n(\Theta)\leqslant N+1$. Take a link of the first level in such Θ. From every non root endpoint of this link "grows" a subdiagram Θ_1 with $n(\theta_1)\leqslant N$. Hence

$$\begin{aligned}\sum_{\Theta \not\sim q:\, n(\Theta)\leqslant N+1} |W(\Theta)|\, e^{a(\Theta)} &\leqslant \prod_{L^{(2)}\ni q} (1+|W(L^{(2)})|\,(e+1)^2)\\ &\quad\times \prod_{L^{(4)}\ni q} (1+|W(L^{(4)})|\,(e+1)^4)-1\\ &\leqslant 1\end{aligned} \tag{5.26}$$

Here e correspond to the case when nothing is "growing" from a given endpoint of the first level link while 1 is the inductive estimate for the case when nonempty subdiagram Θ_1 with $n(\theta_1) \leqslant N$ is "growing" from this endpoint. ∎

From Lemma 5.1 applying Theorem 7.1 one obtains the polymer expansion

$$\sum_{\pi \notin \Lambda^c} W(\pi) \tag{5.27}$$

for the log of (5.18). The precise definition of the polymer $\pi = [\Theta_j^{e_j}]$ and its statistical weight $W(\pi)$ can be found in Section 7. Geometrically a polymer is again a diagram-like object probably with some links entering it more than once. We underline that constructing pre-diagrams, diagrams and polymers one considers all particles entering these object as distinct, say having unique indices or labels. For that reason we call polymers from (5.27) *labeled polymers*. To clarify the dependence of (5.27) on ϱ_x we perform another factorization and define unlabeled polymers.

Suppose that from the total $n_x = \varrho_x \ell_2^d$ particles situated inside a box $C_x^{(\ell_2)}$ exactly $k(\pi)$ particles contribute to the labeled polymer π. Replacing these $k(\pi)$ particles with another $k(\pi)$ particles from the same box $C_x^{(\ell_2)}$ one obtains different labeled polymer with the same statistical weight. Two labeled polymers which can be transformed one into another after several replacements, possibly taking place in different boxes, are called *equivalent*. The corresponding equivalence classes are called *unlabeled polymers* and are denoted by τ. In other words, an unlabeled polymer is obtained from a labeled one by dropping the labels of the particles.

Denote by $X(\tau) \subseteq [\Lambda]^{(\ell_2)}$ the set of the centers of all boxes $C_x^{(\ell_2)}$ containing particles from τ. For $x \in X(\tau)$ let $k_x(\tau)$ be the number of particles from $C_x^{(\ell_2)}$ contributing to τ. Then the total number of different labeled polymers $\pi \in \tau$ is a polynomial function of $\varrho_x \ell_2^d$, $x \in X(\tau)$

$$0 < P(\tau) \leqslant \prod_{x \in X(\tau)} (\varrho_x \ell_2^d)^{k_x(\tau)} \tag{5.28}$$

Setting $W(\tau) = W(\pi)$, where π is an arbitrary labeled polymer from τ, we obtain the expression for the log of the partition function (5.15)

$$\sum_{\tau \in \Lambda} W(\tau)\, P(\tau) \tag{5.29}$$

written in terms of ϱ_x. Despite its involved structure we need only a few simple estimates on this sum.

It follows from Corollary 7.2 in Section 7 that the sum of statistical weights of all labeled (and hence unlabeled) polymers passing through a given particle and containing not less than k links does not exceed $\gamma^{-k\alpha_2/2}$. Hence the sum of the statistical weights of all labeled (or unlabeled) polymers containing two given particles q_1 and q_2 at a distance $r > \gamma^{-1}$ from each other satisfies the bound

$$\sum_{\pi \ni q_1, q_2} |W(\pi)| = \sum_{\tau \ni q_1, q_2} |W(\tau)|\, P(\tau) \leqslant \gamma^{-[\gamma r]\,\alpha_2/2} \tag{5.30}$$

Similarly for any given particle q and sufficiently large absolute constant c

$$\sum_{\pi \ni q:\, L(\tau) \geqslant c} |W(\pi)| = \sum_{\tau \ni q:\, L(\tau) \geqslant c} |W(\tau)|\, P(\tau) \leqslant \gamma^{4d} \tag{5.31}$$

where $L(\tau)$ denotes the number of links contributing to π or τ.

Polymer sum (5.29) can be viewed as a Hamiltonian which we separate into two parts. The first one is

$$\Delta H^{(1)}_{\gamma,\lambda}(\varrho \mid \bar{q}^{(\Lambda^c)}) = \sum_{\tau \notin \Lambda^c:\, L(\tau) < c} W(\tau)\, P(\tau) \tag{5.32}$$

and the second is

$$\Delta H^{(2)}_{\gamma,\lambda}(\varrho \mid \bar{q}^{(\Lambda^c)}) = \sum_{\tau \notin \Lambda^c:\, L(\tau) \geqslant c} W(\tau)\, P(\tau) \tag{5.33}$$

It is not hard to see that $\Delta H^{(1)}_{\gamma,\lambda}(\varrho \mid \bar{q}^{(\Lambda^c)})$ is simply a finite radius Hamiltonian of the polynomial type

$$\Delta H^{(1)}_{\gamma,\lambda}(\varrho \mid \bar{q}^{(\Lambda^c)}) = \ell_2^d \sum_{D \notin \Lambda^c} W(D) \prod_{q \in D} \varrho_q \tag{5.34}$$

Here the sum is taken over connected sets of links (with no restriction for a given link to enter this set more than once) containing less than $c_{(5.31)}$ links and the product is over all endpoints of the links. The notation ϱ_q instead of ϱ_x is not ambiguous as all endpoints of the links are assumed to be at the centers of the corresponding boxes $C_x^{(\ell_2)}$. The statistical weights $W(D)$ are obtained by resummation from (5.32).

From (5.15) we pass to partition function (5.7) containing contours. For the log of the ratio between (5.7) and (5.15) we also obtain a polymer expansion exploiting the theory of contour models with interactions [DS], [BKL]. This expansion has the form

$$\sum_{\xi \notin \Lambda^c} W(\xi) \tag{5.35}$$

where ξ are other polymers constructed from labeled polymers π and contours Γ in the same way as polymers π are constructed from diagrams Θ. The statistical weights $W(\xi)$ are local functions of ϱ_x and we interpret the whole sum as the Hamiltonian

$$\Delta H^{(3)}_{\gamma,\lambda}(\varrho \mid \bar{q}^{(\Lambda^c)}) = \sum_{\xi \notin \Lambda^c} W(\xi) \tag{5.36}$$

The only property of this Hamiltonian used later is the estimate

$$\sum_{\xi \ni q} |W(\xi)| \leqslant \gamma^{4d} \tag{5.37}$$

Since all details can be found in [DS] and [BKL] we give only a sketch of the proofs pointing out few technically important points.

Denote by $\nu(\cdot \mid q^{(\Lambda^c)})$ the Gibbs distribution of $\sum_{x \in [\Lambda]^{(\ell_2)}} \ell_2^d \varrho_x$ particles given by the Hamiltonian $\Delta H_{\gamma,\lambda}(q^{(\Lambda)} \mid \bar{q}^{(\Lambda^c)})$ with the corresponding partition function (5.15). For every contour Γ consider region $R(\Gamma) = \partial^{(\ell_3/3)} \operatorname{Supp}(\Gamma)$ with empty boundary condition $\varnothing^{(R(\Gamma)^c)}$ and define a modified statistical weight

$$\tilde{W}^T(\eta^\Gamma \mid q^{(\delta^-(\Gamma))}) = \nu(q^{(\delta^-(\Gamma))} \mid \varnothing^{(R(\Gamma)^c)})\, W^T(\eta^\Gamma \mid q^{(\delta^-(\Gamma))}) \tag{5.38}$$

Then the ratio of partition functions (5.7) and (5.15) can be rewritten as

$$\sum_{\{\Gamma_i\}^\sigma \in \Lambda} \int \prod_i dq^{(\delta^-(\Gamma_i))} \exp\Bigg(\sum_{\substack{\pi \notin \Lambda^c:\, \exists i,\, \pi \cap \delta^-(\Gamma_i) \neq \varnothing \\ \pi \cap R(\Gamma_i)^c \neq \varnothing}} \big(W(\pi \mid q^{(\cup_i \delta^-(\Gamma_i))}) - W(\pi)\big)\Bigg)$$
$$\times \prod_i \tilde{W}^T(\eta^{\Gamma_i} \mid q^{\delta^-(\Gamma_i))}) \tag{5.39}$$

Here the statistical weight $W(\pi \mid q^{(\cup_i \delta^-(\Gamma_i))})$ is defined respecting an additional boundary condition $q^{(\cup_i \delta^-(\Gamma_i))}$ which is imposed in $\bigcup_i \delta^-(\Gamma_i)$. We note that only $\bar{q}^{(\Lambda^c)}$ affects $W(\pi)$ if $\pi \cap \Lambda^c \neq \varnothing$.

The polymer sum in (5.39) describes the interaction between contours Γ_i. It is important that every polymer contributing to (5.39) is sufficiently long and contains at least $[\gamma \ell_3/3]$ links. In view of (5.30) such a polymer has a very small statistical weight. Therefore, in complete similarity with (5.12)–(5.14) expanding

$$e^{W(\pi \mid q^{(\bigcup_i \delta^{=}(\Gamma_i))}) - W(\pi)} = 1 + (e^{W(\pi \mid q^{(\bigcup_i \delta^{=}(\Gamma_i))}) - W(\pi)} - 1) \tag{5.40}$$

and integrating over $\prod_i dq^{(\delta^{=}(\Gamma_i))}$ one can derive for (5.39) a representation analogous to (5.18). In this representation newly defined diagrams are constructed from contours Γ connected via polymers π. Taking logarithms and applying Theorem 7.1 one obtains (5.35) with correspondingly defined polymers ξ. The key fact ensuring the condition (7.3) of Theorem 7.1 is that for γ small enough the sum of absolute values of statistical weights of all polymers π containing a given particle and being longer than $\ell_3/3$ is much smaller than the quantity $c\ell_2^d\ell_3^{-d}\zeta$ entering the Peierls estimate (3.44).

The route to (5.35) looks rather involved and tedious but it is a standard one in the cluster expansion technique. On the other hand we need only minor knowledge about ξ, namely (5.37). This estimate is obtained by the methods of [DS] and [BKL] along the following way.

The sum over all polymers ξ containing a given particle q_1 is equal to

$$\sum_{\Gamma:\, \delta^{=}(\Gamma) \ni q_1} \ \sum_{\xi:\, \xi \ni \Gamma} |W(\xi)| + \sum_{\pi:\, \pi \ni q_1,\, L(\pi) \geqslant c_{(5.31)}} \ \sum_{\xi:\, \xi \ni \pi} |W(\xi)| \tag{5.41}$$

By (7.6) the first internal sum does not exceed $cW^T(\eta^\Gamma \mid q^{(\delta^{=}(\Gamma))})$ and the second internal sum does not exceed $c\,|W(\pi)|$. In turn

$$\sum_{\Gamma:\, \delta^{=}(\Gamma) \ni q_1} cW^T(\eta^\Gamma \mid q^{(\delta^{=}(\Gamma))}) + \sum_{\pi:\, \pi \ni q_1,\, L(\pi) \geqslant c_{(5.31)}} c\,|W(\pi)| \leqslant \gamma^{4d} \tag{5.42}$$

because of (3.38) and (5.31).

The final result of this subsection can be stated now as

Lemma 5.2. The expression (5.3) is equal to

$$\sum_{x \in [\Lambda]^{(\ell_2)}} \log\left(\frac{\ell_2^{dn_x}}{n_x!}\right) - \beta H_{\gamma,\lambda}(\varrho \mid \bar{q}^{(\Lambda^c)}) + \sum_{i=1}^{3} \Delta H^{(i)}_{\gamma,\lambda}(\varrho \mid \bar{q}^{(\Lambda^c)}) \tag{5.43}$$

Expression (5.43) without the last sum is nothing but $\beta\tilde{\mathscr{F}}_{\gamma,\beta,\lambda}(\varrho^{(\ell_2)} \mid \bar{q}^{(\Lambda^c)})$ defined in (4.21) with its main part $\beta\mathscr{F}_{\gamma,\beta,\lambda}(\varrho^{(\ell_2)} \mid \bar{q}^{(\Lambda^c)})$ given by (4.26)–(4.27).

Dobrushin Uniqueness (Basic Calculation)

Lemma 5.3. The effective Hamiltonian (5.43) satisfies the Dobrushin uniqueness condition.

Proof. For the sake of simplicity we first check that the Dobrushin uniqueness condition is true for the Hamiltonian $\mathscr{F}_{\gamma,\beta,\lambda}(\varrho^{(\ell_2)})$. Then in the next subsection we show that only minor modifications are necessary to treat (5.43).

Take any site $x \in \mathbb{Z}^d_{\ell_2}$ and consider two boundary conditions $\bar{\varrho}^{(0)}$ and $\bar{\varrho}^{(1)}$ on $\mathbb{Z}^d_{\ell_2} \backslash x$ such that they both belong to the interval (5.2) and differ only at a site $y \in \mathbb{Z}^d_{\ell_2}$. For definiteness we assume that $\bar{\varrho}^{(1)}_y > \bar{\varrho}^{(0)}_y$. Denote by $\nu^{(0)}(d\varrho_x)$ and $\nu^{(1)}(d\varrho_x)$ conditional Gibbs distributions defined by the Hamiltonians $\mathscr{F}_{\gamma,\beta,\lambda}(\varrho_x \mid \bar{\varrho}^{(0)})$ and $\mathscr{F}_{\gamma,\beta,\lambda}(\varrho_x \mid \bar{\varrho}^{(1)})$ respectively. The *Vasserstein distance* between $\nu^{(0)}(d\varrho_x)$ and $\nu^{(1)}(d\varrho_x)$ is

$$\begin{aligned} R(\nu^{(0)}, \nu^{(1)}) &= \int_{-\infty}^{\infty} dz \left| \int_{-\infty}^{z} (\nu^{(0)}(d\varrho_x) - \nu^{(1)}(d\varrho_x)) \right| \\ &= \int_0^{\infty} dz \left| \int_0^{z} (\nu^{(0)}(d\varrho_x) - \nu^{(1)}(d\varrho_x)) \right| \end{aligned} \tag{5.44}$$

where the last equality utilizes the positivity of ϱ_x.

The *Dobrushin uniqueness condition* is satisfied if one is able to find a function r_{xy} such that

$$R(\nu^{(0)}, \nu^{(1)}) \leqslant r_{xy} \, |\bar{\varrho}^{(0)}_y - \bar{\varrho}^{(1)}_y| \tag{5.45}$$

and

$$\sum_y r_{xy} < 1 \tag{5.46}$$

To check this condition we follow the strategy of [COPP] and define $\nu_t(d\varrho_x)$, $t \in [0, 1]$ as a Gibbs measure corresponding to the interpolated Hamiltonian

$$\mathscr{F}_{\gamma,\beta,\lambda}(\varrho_x \mid t) = \mathscr{F}_{\gamma,\beta,\lambda}(\varrho_x \mid \bar{\varrho}^{(0)}) + t(\mathscr{F}_{\gamma,\beta,\lambda}(\varrho_x \mid \bar{\varrho}^{(1)}) - \mathscr{F}_{\gamma,\beta,\lambda}(\varrho_x \mid \bar{\varrho}^{(0)})) \tag{5.47}$$

This measure has the density

$$p(\varrho_x \mid t) = \frac{\exp(-\beta \mathscr{F}_{\gamma,\beta,\lambda}(\varrho_x \mid t))}{\int_0^\infty \exp(-\beta \mathscr{F}_{\gamma,\beta,\lambda}(\varrho_x \mid t)) \, d\varrho_x} \tag{5.48}$$

which is differentiable in t and therefore from (5.44) we have

$$R(\nu^{(0)}, \nu^{(1)}) \leqslant \int_0^1 dt \int_0^\infty dz \left| \int_0^z \frac{\partial}{\partial t} p(\varrho_x \mid t) \, d\varrho_x \right| \tag{5.49}$$

Denoting

$$\Delta\mathscr{F}_{\gamma,\beta,\lambda}(\varrho_x \mid \bar\varrho^{(0)}, \bar\varrho^{(1)}) = (\mathscr{F}_{\gamma,\beta,\lambda}(\varrho_x \mid \bar\varrho^{(1)}) - \mathscr{F}_{\gamma,\beta,\lambda}(\varrho_x \mid \bar\varrho^{(0)})) \tag{5.50}$$

one has

$$\frac{\partial}{\partial t} p(\varrho_x \mid t) = \beta p(\varrho_x \mid t)(\Delta\mathscr{F}_{\gamma,\beta,\lambda}(\varrho_x \mid \bar\varrho^{(0)}, \bar\varrho^{(1)}) - \langle \Delta\mathscr{F}_{\gamma,\beta,\lambda}(\varrho_x \mid \bar\varrho^{(0)}, \bar\varrho^{(1)})\rangle_t) \tag{5.51}$$

where $\langle \cdot \rangle_t$ denotes the expectation with respect to $p(\varrho_x \mid t)\, d\varrho_x$. Observe that

$$\Delta\mathscr{F}_{\gamma,\beta,\lambda}(\varrho_x \mid \bar\varrho^{(0)}, \bar\varrho^{(1)}) = \ell_2^d \left(I_\gamma^{(2)}(x, y) - \frac{1}{2!} I_\gamma^{(2)}(x, y \mid \bar\varrho) \right) (\bar\varrho_y^{(1)} - \bar\varrho_y^{(0)})\, \varrho_x \tag{5.52}$$

where $I_\gamma^{(2)}(x, y \mid \bar\varrho)$ is defined by (4.41) (with $\bar\varrho(\Lambda \setminus x \setminus y) = \bar\varrho^{(0)}(\Lambda \setminus x \setminus y) = \bar\varrho^{(1)}(\Lambda \setminus x \setminus y)$ instead of $\hat\varrho$) and satisfies (4.42). by direct calculation

$$\int_0^\infty dz \left| \int_0^z (\varrho_x - \langle \varrho_x \rangle_t)\, p(\varrho_x \mid t)\, d\varrho_x \right| = \langle \varrho_x^2 \rangle_t - \langle \varrho_x \rangle_t^2 \tag{5.53}$$

For sufficiently small γ and therefore sufficiently large ℓ_2^d the last expression can be estimated by the Laplace method and it is equal to

$$\ell_2^{-d} \varrho_*(t) + O(\ell_2^{-2d}) \tag{5.54}$$

where $\ell_2^d \varrho_*(t)^{-1}$ is the value of $\beta(\partial^2/\partial\varrho_x^2)\, \mathscr{F}_{\gamma,\beta,\lambda}(\varrho_x \mid t)$ at the point $\varrho_*(t)$ of the minimum of $\mathscr{F}_{\gamma,\beta,\lambda}(\varrho_x \mid t)$. As we know from Lemma 4.8 such a minimum exists, is unique and lies strictly inside the interval (5.2). Hence $O(\ell_2^{-2d})$ in (5.54) is uniform in t (see [F]).

Combining (5.49)–(5.54) we obtain that

$$R(v^{(0)}, v^{(1)}) \leqslant \int_0^1 dt(\beta\varrho_* + O(\ell_2^{-d})) \left(I_\gamma^{(2)}(x, y) - \frac{1}{2!} I_\gamma^{(2)}(x, y \mid \bar\varrho) \right) (\bar\varrho_y^{(1)} - \bar\varrho_y^{(0)}) \tag{5.55}$$

Setting

$$r_{xy} = \max_{\bar\varrho} \left| I_\gamma^{(2)}(x, y) - \frac{1}{2!} I_\gamma^{(2)}(x, y \mid \bar\varrho) \right| \int_0^1 dt(\beta\varrho_*(t) + O(\ell_2^{-d})) \tag{5.56}$$

one concludes that (5.46) is true for sufficiently small γ because of (3.17), (4.42) and the fact that $\sum_y I_\gamma^{(2)}(x, y) = 1$.

Dobrushin Uniqueness (General Case)

We turn now to the complete effective Hamiltonian (5.43) and show that only nonessential modifications of the calculation of the previous subsection are required to cover this case.

The first correction is due to the difference between $\mathscr{F}_{\gamma,\beta,\lambda}(\varrho^{(\ell_2)} \mid \bar{q}^{(\Lambda^c)})$ and $\tilde{\mathscr{F}}_{\gamma,\beta,\lambda}(\varrho^{(\ell_2)} \mid \bar{q}^{(\Lambda^c)})$. This difference is discussed in detail in Lemma 4.6. The corresponding modifications of our previous arguments are the following.

The difference between volumes of $B_\gamma(\cdot)$ and $[B_\gamma(\cdot)]^{(\ell_2)}$ produces a factor $(1+O(\gamma^\alpha)), \alpha>0$, in front of $I_\gamma^{(2)}(\cdot, \cdot)$ and $I_\gamma^{(4)}(\cdot, \cdot, \cdot, \cdot)$ which clearly is not essential.

The self interactions like $I_\gamma^{(2)}(\varrho_x, \varrho_x)$ or $I_\gamma^{(4)}(\varrho_x, \varrho_x, \varrho_y, \varrho_z)$ produce nonlinear terms in $\tilde{\mathscr{F}}_{\gamma,\beta,\lambda}(\varrho_x \mid t)$ which is defined similarly to (5.47). Their contribution to $\tilde{\mathscr{F}}_{\gamma,\beta,\lambda}(\varrho_x \mid t)$ has the form

$$c_2\varrho_x^2 + c_3\varrho_x^3 + c_4\varrho_x^4 + t(c_5\varrho_x^2 + c_6\varrho_x^3)(\bar{\varrho}_y^{(1)} - \bar{\varrho}_y^{(0)}) \tag{5.57}$$

where c_i does not depend on t, ϱ_x and $\varrho_y^{(i)}$. Only the last, t-dependent, part from (5.57) contributes to $\Delta\tilde{\mathscr{F}}_{\gamma,\beta,\lambda}(\varrho_x \mid \bar{\varrho}^{(0)}, \bar{\varrho}^{(1)})$ which is defined similarly to (5.50). The corresponding effect on $\int_0^\infty dz\, |\int_0^z (\partial/\partial t)\, p(\varrho_x \mid t)\, d\varrho_x|$ result in terms of the form

$$\begin{aligned} &\beta c_5(\bar{\varrho}_y^{(1)} - \bar{\varrho}_y^{(0)}) \int_0^\infty dz \left| \int_0^z (\varrho_x^2 - \langle \varrho_x^2 \rangle_t)\, p(\varrho_x \mid t)\, d\varrho_x \right| \\ &\quad = \beta c_5(\bar{\varrho}_y^{(1)} - \bar{\varrho}_y^{(0)})(\langle \varrho_x^3 \rangle_t - \langle \varrho_x^2 \rangle_t \langle \varrho_x \rangle_t) \end{aligned} \tag{5.58}$$

and

$$\begin{aligned} &\beta c_6(\bar{\varrho}_y^{(1)} - \bar{\varrho}_y^{(0)}) \int_0^\infty dz \left| \int_0^z (\varrho_x^3 - \langle \varrho_x^3 \rangle_t)\, p(\varrho_x \mid t)\, d\varrho_x \right| \\ &\quad = \beta c_6(\bar{\varrho}_y^{(1)} - \bar{\varrho}_y^{(0)})(\langle \varrho_x^4 \rangle_t - \langle \varrho_x^3 \rangle_t \langle \varrho_x \rangle_t) \end{aligned} \tag{5.59}$$

in addition to $\ell_2^d \beta(I_\gamma^{(2)}(x, y) - (1/2!)\, I_\gamma^{(2)}(x, y \mid \bar{\varrho}))(\bar{\varrho}_y^{(1)} - \bar{\varrho}_y^{(0)})(\langle \varrho_x^2 \rangle_t - \langle \varrho_x \rangle_t^2)$. Clearly

$$|c_i| \leqslant \ell_2^d \gamma^\alpha \tag{5.60}$$

with some $\alpha > 0$.

Applying the Laplace method one has

$$\langle \varrho_x^{k+1} \rangle_t - \langle \varrho_x^k \rangle_t \langle \varrho_x \rangle_t = \frac{k\varrho_*(t)^k}{\beta(\partial^2/\partial\varrho_x^2)\, \tilde{\mathscr{F}}_{\gamma,\beta,\lambda}(\varrho_x \mid t)|_{\varrho_*(t)}} + O(\ell_2^{-2d}) \qquad (5.61)$$

where again $\varrho_*(t)$ is the minimum of $\tilde{\mathscr{F}}_{\gamma,\beta,\lambda}(\varrho_x \mid t)$. Since

$$\beta \frac{\partial^2}{\partial \varrho_x^2} \tilde{\mathscr{F}}_{\gamma,\beta,\lambda}(\varrho_x \mid t) = \ell_2^d \varrho_x^{-1} + \beta 2c_2 + \beta 6c_3\varrho_x$$

$$+ \beta 12c_4\varrho_x^2 + \beta t(2c_5 + 6c_6\varrho_x)(\bar{\varrho}_y^{(1)} - \bar{\varrho}_y^{(0)}) \qquad (5.62)$$

and because of (5.60) all corrections above do not destroy the arguments of the previous subsection.

As the next step we incorporate in our calculation $\Delta H^{(1)}_{\gamma,\lambda}(\varrho \mid \bar{q}^{(\Lambda^c)})$. According to (5.34) this is the same type of a polynomial correction which was just discussed. Thus the same arguments work. The necessary smallness, like in (5.60), of coefficient $W(D)$ is a consequence of the smallness of the statistical weights of links (5.22)–(5.23) and the definition of $W(D)$ (see also (5.30)).

Finally, to treat $\sum_{i=2,3} \Delta H^{(i)}_{\gamma,\lambda}(\varrho \mid \bar{q}^{(\Lambda^c)})$ we simply observe that for any boundary condition $\bar{\varrho}$ given on $\mathbb{Z}^d_{\ell_2} \setminus x$

$$R(\nu(\varrho_x \mid \bar{\varrho}), \nu_{2,3}(\varrho_x \mid \bar{\varrho})) < \gamma^d \qquad (5.63)$$

where $\nu(\varrho_x \mid \bar{\varrho})$ is the conditional Gibbs distribution given by the whole Hamiltonian (5.43) while $\nu_{2,3}(\varrho_x \mid \bar{\varrho})$ is a similar distribution given by the Hamiltonian (5.43) without $\sum_{i=2,3} \Delta H^{(i)}_{\gamma,\lambda}(\varrho \mid \bar{q}^{(\Lambda^c)})$. This estimate is obvious in view of definition (5.44) and bounds (5.31) and (5.37). One can comment that the contribution of contours or long polymers to the free energy of the auxiliary model is too small to affect anything at all. ∎

6. PROPERTIES OF AUXILIARY MODEL

In this section we use the Dobrushin uniqueness result established for the auxiliary model in the previous section to prove Statements 3.3 and 3.5. We begin with a construction which is necessary for the proof of both statements. Namely, given a phase σ and a particle configuration $\bar{q}$ from the ground state ensemble of the phase σ we derive an appropriate representation for the logarithm of partition function (3.37). Lemma 5.2 gives

$$Z^{\Lambda}_{\gamma,\beta,\lambda}(\Lambda \mid \bar{q}^{(\Lambda^c)}) = \sum_{\varrho^{(\ell_2)}(\Lambda)} \exp\left(-\beta\tilde{\mathscr{F}}_{\gamma,\beta,\lambda}(\varrho^{(\ell_2)}(\Lambda) \mid \bar{q}^{(\Lambda^c)}) - \sum_{i=1}^{3} \Delta H^{(i)}_{\gamma,\lambda}(\varrho^{(\ell_2)}(\Lambda) \mid \bar{q}^{(\Lambda^c)})\right) \tag{6.1}$$

Below we consider density configurations, lattice volumes, etc. related to the scale ℓ_2 and in most cases we omit the superscript (ℓ_2) from the notations. On few occasions when we need other scales we specify them explicitly.

In Section 4 we studied in detail the minimizers of $\mathscr{F}_{\gamma,\beta,\lambda}(\varrho(\Lambda) \mid \bar{q}^{(\Lambda^c)})$ for bounded Λ and $\Lambda = \mathbb{Z}^d_{\ell_2}$. Now we need to study minimizers for

$$\tilde{H}_{\gamma,\lambda}(\varrho(\Lambda) \mid \bar{q}^{(\Lambda^c)}) = \beta\tilde{\mathscr{F}}_{\gamma,\beta,\lambda}(\varrho(\Lambda) \mid \bar{q}^{(\Lambda^c)}) + \Delta H^{(1)}_{\gamma,\lambda}(\varrho(\Lambda) \mid \bar{q}^{(\Lambda^c)}) \tag{6.2}$$

where both $\bar{q}^{(\Lambda^c)}$ and $\varrho(\Lambda)$ belong to the ground state ensemble of the phase σ. This is a finite range translation-invariant Hamiltonian similar to $\mathscr{F}_{\gamma,\beta,\lambda}(\varrho(\Lambda) \mid \bar{q}^{(\Lambda^c)})$ but not allowing a simple representation of the type (4.35). For a bounded Λ the existence of at least one minimizer of $\tilde{H}_{\gamma,\lambda}(\varrho(\Lambda) \mid \bar{q}^{(\Lambda^c)})$ follows from the compactness of its domain, $(\varrho_{\beta,\sigma} - \zeta, \varrho_{\beta,\sigma} + \zeta)^{\Lambda}$. This domain is convex and the minimizer is unique because the function $\tilde{H}_{\gamma,\lambda}(\varrho(\Lambda) \mid \bar{q}^{(\Lambda^c)})$ (of the finite number of variables ϱ_x, $x \in \Lambda$) is also convex. This is a consequence of estimate (3.17) from which it is not hard to see that for any $\bar{q}^{(\Lambda^c)}$ and $\varrho(\Lambda)$ the Hessian matrix of $\tilde{H}_{\gamma,\lambda}(\cdot \mid \bar{q}^{(\Lambda^c)})$ calculated at $\varrho(\Lambda)$ is positive definite with the mass bounded from 0 independently of $\bar{q}^{(\Lambda^c)}$ and $\varrho(\Lambda)$. Depending on the context we denote the unique minimizer of $\tilde{H}_{\gamma,\lambda}(\varrho(\Lambda) \mid \bar{q}^{(\Lambda^c)})$ by $\hat{\varrho}$, $\hat{\varrho}(\Lambda)$ or $\hat{\varrho}^{\bar{q}^{(\Lambda^c)}}$.

To clarify the structure and convexity of $\tilde{H}_{\gamma,\lambda}(\varrho(\Lambda) \mid \bar{q}^{(\Lambda^c)})$ we introduce

$$\Delta\tilde{H}_{\gamma,\lambda}(\varrho(\Lambda) \mid \bar{q}^{(\Lambda^c)}) = \tilde{H}_{\gamma,\lambda}(\varrho(\Lambda) \mid \bar{q}^{(\Lambda^c)}) - \tilde{H}_{\gamma,\lambda}(\hat{\varrho}(\Lambda) \mid \bar{q}^{(\Lambda^c)}) \tag{6.3}$$

It is not hard to check that

$$\Delta\tilde{H}_{\gamma,\lambda}(\varrho(\Lambda) \mid \bar{q}^{(\Lambda^c)}) = \ell_2^d \sum_{x \in [\Lambda]^{(\ell_2)}} \sum_{n=2}^{\infty} \frac{1}{n(n-1)\,\hat{\varrho}_x^{n-1}} \Delta\varrho_x^n + \ell_2^d \sum_{D \not\subset [\Lambda^c]^{(\ell_2)}} W(D) \sum_{X_1(D),X_2(D)} \prod_{x \in X_1(D)} \hat{\varrho}_x(\Lambda) \prod_{x \in X_2(D)} \Delta\varrho_x \tag{6.4}$$

Here $\Delta\varrho_x = \varrho_x - \hat{\varrho}_x(\Lambda)$ and D is a connected family of links with endpoints forming a set $X(D) \subset \mathbb{Z}^d_{\ell_2}$. The number of links in D is less than $c_{(5.31)}$. The subsets $X_1(D)$ and $X_2(D)$ form a partition of $X(D)$ with at least two elements in $X_2(D)$. The second internal sum is taken over all partitions of that type. For $W(D)$ one has the estimate

$$\sum_{\substack{D:\, X(D) \ni x \\ D \neq L^{(2)},\, L^{(4)}}} |W(D)|\, \varrho_{\max}^{|X(D)|} \leqslant \gamma^{\alpha_2/3} \tag{6.5}$$

as follows from (5.30). Also

$$|\Delta\varrho_x| \leqslant 2\zeta \tag{6.6}$$

The quadratic part of (6.4) is

$$\frac{1}{2!}\ell_2^d \sum_{x \in [\Lambda]^{(\ell_2)}} \frac{1}{\hat{\varrho}} \Delta\varrho_x^2 - \frac{1}{2!}\ell_2^d \sum_{x_1, x_2:\, x_1 \cup x_2 \notin [\Lambda^c]^{(\ell_2)}} \tilde{I}^{(2)}(x_1, x_2 \mid \hat{\varrho}(\Lambda))\, \Delta\varrho_{x_1} \Delta\varrho_{x_2} \tag{6.7}$$

with

$$\frac{1}{\hat{\varrho}_{x_1}(\Lambda)} - \sum_{x_2} \tilde{I}^{(2)}(x_1, x_2 \mid \hat{\varrho}(\Lambda)) \geqslant m(\beta) > 0 \tag{6.8}$$

for any $x_1 \in [\Lambda]^{(\ell_2)}$. Here $\tilde{I}^{(2)}(x_1, x_2 \mid \hat{\varrho}(\Lambda))$ is an analogue of (4.41) and satisfies the same estimate (4.42) for γ small enough. (By construction $|I^{(2)}(x_1, x_2 \mid \hat{\varrho}(\Lambda)) - \tilde{I}^{(2)}(x_1, x_2 \mid \hat{\varrho}(\Lambda))| \leqslant c\gamma^{\alpha_2} I^{(2)}(x_1, x_2))$. Moreover, in view of (6.5) and assuming that ζ is chosen to be small enough with respect to $m(\beta)$

$$\Delta\tilde{H}_{\gamma,\lambda}(\varrho(\Lambda) \mid \bar{q}^{(\Lambda^c)}) = \frac{1}{2!}\ell_2^d \sum_{x \in [\Lambda]^{(\ell_2)}} \frac{m(\beta)}{2} \Delta\varrho_x^2 + U(\hat{\varrho}(\Lambda), \Delta\varrho(\Lambda)) \tag{6.9}$$

with positive convex $U(\hat{\varrho}(\Lambda), \Delta\varrho(\Lambda))$ having minimum at $\Delta\varrho(\Lambda) \equiv 0$.

Along with the minimizers of $\tilde{H}_{\gamma,\lambda}(\varrho(\Lambda) \mid \bar{q}^{(\Lambda^c)})$ for bounded regions Λ we also need a global minimizer, i.e., an analogue of $\varrho_{\beta,\sigma}$. To find such a minimizer and to show its uniqueness (for given σ) we consider a sufficiently large (with respect to the range of $\tilde{H}_{\gamma,\lambda}(\cdot)$) periodic box Λ and the corresponding Hamiltonian $\tilde{H}_{\gamma,\lambda}(\varrho(\Lambda)$. All convexity considerations remain true for $\tilde{H}_{\gamma,\lambda}(\varrho(\Lambda))$ such that $\tilde{H}_{\gamma,\lambda}(\varrho(\Lambda))$ has a unique minimizer $\hat{\varrho}(\Lambda)$.

Because of the translation invariance of $\tilde{H}_{\gamma,\lambda}(\varrho(\Lambda))$ this minimizer is a constant configuration $\hat{\varrho}_x(\Lambda) \equiv \hat{\varrho}$, $x \in \Lambda$. (Otherwise one is able to construct another minimizers by space translations.) Since $\tilde{H}_{\gamma,\lambda}(\varrho(\Lambda))$ is of finite range we observe that $\hat{\varrho}$ is independent on Λ for all sufficiently large periodic Λ and hence $\hat{\varrho}$ is the global minimizer we are looking for. Note that the specific energy $\tilde{h}_{\gamma,\beta,\lambda}(s) = \lim_{\Lambda \to \mathbb{Z}^d_{\ell_2}} |\Lambda|^{-1} \tilde{H}_{\gamma,\lambda}(\varrho(\Lambda))$ of any constant density configuration $\varrho(\Lambda) \equiv s$ is just a finite sum and $\hat{\varrho}$ is nothing but the value of s at which $\tilde{h}_{\gamma,\beta,\lambda}(s)$ achieves its minimum. Thus given σ such a minimum is unique for $s \in (\varrho_{\beta,\sigma} - \zeta, \varrho_{\beta,\sigma} + \zeta)$ and we have different values $\hat{\varrho}_{\beta,\lambda,-}$ and $\hat{\varrho}_{\beta,\lambda,+}$ corresponding to $\sigma = -1$ and $\sigma = +1$.

Denote by $\hat{\lambda}(\beta,\gamma)$ the value of λ at which $\tilde{h}_{\gamma,\beta,\lambda}(\hat{\varrho}_{\beta,\lambda,-}) = \tilde{h}_{\gamma,\beta,\lambda}(\hat{\varrho}_{\beta,\lambda,+})$ and set $\hat{\varrho}_{\beta,\sigma} = \hat{\varrho}_{\beta,\hat{\lambda}(\beta),\sigma}$. The important consequences of the representation of $\hat{\varrho}_{\beta,\lambda,\sigma}$ via $\tilde{h}_{\gamma,\beta,\lambda}(s)$ are the existence of $\hat{\lambda}(\beta,\gamma)$ and the estimates

$$|\varrho_{\beta,\sigma} - \hat{\varrho}_{\beta,\lambda,\sigma}| \leqslant c\gamma^{\alpha_1/2} \tag{6.10}$$

and

$$|\lambda(\beta) - \hat{\lambda}(\beta,\gamma)| \leqslant c\gamma^{\alpha_1} \tag{6.11}$$

Indeed, let $h_{\gamma,\beta,\lambda}(s)$ be the specific energy of $\varrho \equiv s$ calculated via the Hamiltonian $\tilde{H}_{\gamma,\lambda}(\varrho(\Lambda)) + \Delta H^{(2)}_{\gamma,\lambda}(\varrho(\Lambda))$ with periodic Λ. Then

$$|\tilde{h}_{\gamma,\beta,\lambda}(s) - h_{\gamma,\beta,\lambda}(s)| \leqslant c\gamma^{4d} \tag{6.12}$$

as follows from (5.31).

Another way to calculate $h_{\gamma,\beta,\lambda}(s)$ is the approach of Section 4. Here the starting point is the partition function (5.15) with fixed density configuration $\varrho^{(\ell_2)} \equiv s$. Then one approximate this partition function by using the density configurations defined with respect to the scale $\ell_1 \ll \ell_2$. (Note that the density configuration which is constant in the scale ℓ_2 is not necessarily a constant one in the finer scale ℓ_1.) As the first step of the approximation one shifts all particles to the centers of the corresponding boxes $C^{(\ell_1)}$ for the price of the error (in the value of $\tilde{H}_{\gamma,\lambda}(\varrho^{(\ell_1)}(\Lambda)) + \Delta H^{(2)}_{\gamma,\lambda}(\varrho^{(\ell_1)}(\Lambda)))$ which in absolute value does not exceed $c\gamma^{\alpha_1}|\Lambda|$ (see Lemma 4.5). Lemmas 4.6 and 4.7 then show that up to the same error the Hamiltonian $\tilde{H}_{\gamma,\lambda}(\varrho^{(\ell_1)}(\Lambda)) + \Delta H^{(2)}_{\gamma,\lambda}(\varrho^{(\ell_1)}(\Lambda))$ can be approximated by (4.26).

Now the convexity considerations (see (3.6)) imply that among all density configurations $\varrho^{(\ell_1)}(\Lambda)$ such that $\varrho^{(\ell_2)}(\varrho^{(\ell_1)}(\Lambda)) \equiv s$ the configuration

$\varrho^{(\ell_1)}(\Lambda) \equiv s$ has the minimal value of the mean field functional (4.26). Hence (see (4.19) and above) counting only the contribution of $\varrho^{(\ell_1)}(\Lambda) \equiv s$ produces the error which again in absolute value does not exceed $c\gamma^{\alpha_1} |\Lambda|$ (in fact it is much smaller). The specific energy of the constant density configuration $\varrho^{(\ell_1)} \equiv s$ calculated via (4.26) is nothing but the mean field specific energy $F_{\beta, \lambda}(s)$ and therefore

$$|F_{\beta, \lambda}(s) - h_{\gamma, \beta, \lambda}(s)| \leqslant c\gamma^{\alpha_1} \tag{6.13}$$

Since $\tilde{h}_{\gamma, \beta, \lambda}(s) = F_{\beta, \lambda}(s) + \sum_{k=2}^{K} a_k s^k$ with $|a_k| < c\gamma^{\alpha_2}$, i.e., $\tilde{h}_{\gamma, \beta, \lambda}(s)$ is a small polynomial perturbation of $F_{\beta, \lambda}(s)$, and

$$|F_{\beta, \lambda}(s) - \tilde{h}_{\gamma, \beta, \lambda}(s)| \leqslant c\gamma^{\alpha_1} \tag{6.14}$$

as follows from (6.12) and (6.13), one immediately obtains (6.10) and (6.11). (The existence of $\hat{\lambda}(\beta, \gamma)$ follows from an elementary linear analysis and continuity of $\tilde{h}_{\gamma, \beta, \lambda}(s)$.)

Finally observe that a straightforward modification of the arguments of Lemma 4.8 shows that $|\hat{\varrho}_{\beta, \lambda, \sigma} - \hat{\varrho}^{\bar{q}^{(\Lambda^c)}}|$ satisfy (4.36).

With $\hat{\varrho}_{\beta, \lambda, \sigma}$ and $\hat{\varrho}^{\bar{q}^{(\Lambda^c)}}$ being properly defined we construct now a convenient representation for the log of the partition function (6.1) using the following interpolation trick. For $t \in [0, 1]$ and $s \in [0, 1]$ introduce an interpolated Hamiltonian

$$\begin{aligned} \Delta H_{\gamma, \lambda}(\Delta\varrho(\Lambda) \mid \bar{q}^{(\Lambda^c)}; t, s) &= \ell_2^d \sum_{x \in [\Lambda]^{(\ell_2)}} \sum_{n=2}^{\infty} \frac{1}{n(n-1)\, \hat{\varrho}_x^{n-1}} \Delta\varrho_x^n \\ &\quad + t\ell_2^d \sum_{D \not\subset [\Lambda^c]^{(\ell_2)}} W(D) \sum_{\substack{X_1(D),\, x \in X_1(D) \\ X_2(D)}} \prod \hat{\varrho}_x(\Lambda) \prod_{x \in X_2(D)} \Delta\varrho_x \\ &\quad + s \sum_{i=2,3} \Delta H_{\gamma, \lambda}^{(i)}(\varrho(\Lambda) \mid \bar{q}^{(\Lambda^c)}) \end{aligned} \tag{6.15}$$

The Hamiltonian $\Delta H_{\gamma, \lambda}(\Delta\varrho(\Lambda) \mid \bar{q}^{(\Lambda^c)}; t, s)$ satisfies the Dobrushin uniqueness condition similarly to $\Delta H_{\gamma, \lambda}(\Delta\varrho(\Lambda) \mid \bar{q}^{(\Lambda^c)}; 1, 1)$. We denote by $\langle \cdot \rangle_{t, s, \Lambda, \bar{q}^{(\Lambda^c)}}$ the expectation with respect to the corresponding Gibbs measure in the finite domain Λ with the boundary condition $\bar{q}^{(\Lambda^c)}$ and we denote by $\langle \cdot \rangle_{t, s}$ the expectation with respect to the corresponding unique limit Gibbs distribution.

From (6.1) and (6.15) by direct calculation one obtains

$$\log Z^{\Lambda}_{\gamma,\beta,\lambda}(\Lambda \mid \bar{q}^{(\Lambda^c)})$$

$$= -\tilde{H}_{\gamma,\lambda}(\hat{\varrho}(\Lambda) \mid \bar{q}^{(\Lambda^c)}) + \ell_2^d \sum_{D \not\subset [\Lambda^c]^{(\ell_2)}} W(D) \sum_{\substack{X_1(D),\\ X_2(D)}} \prod_{x \in X_1(D)} \hat{\varrho}_x$$

$$\times \int_0^1 dt \left\langle \prod_{x \in X_2(D)} \Delta\varrho_x \right\rangle_{t,0,\Lambda,\bar{q}^{(\Lambda^c)}}$$

$$+ \sum_{\tau \in \Lambda:\, L(\tau) \geqslant c_{(5.31)}} W(\tau) \int_0^1 ds \langle P(\tau, \Delta\varrho)\rangle_{1,s,\Lambda,\bar{q}^{(\Lambda^c)}}$$

$$+ \sum_{\xi \in \Lambda} \int_0^1 ds \langle W(\xi, \Delta\varrho)\rangle_{1,s,\Lambda,\bar{q}^{(\Lambda^c)}} \tag{6.16}$$

Here $\hat{\varrho}(\Lambda) = \hat{\varrho}^{\bar{q}^{(\Lambda^c)}}$ is the minimizer of $\tilde{H}_{\gamma,\lambda}(\cdot \mid \bar{q}^{(\Lambda^c)})$ and notations $P(\tau, \Delta\varrho)$ and $W(\xi, \Delta\varrho)$ are used instead of $P(\tau)$ and $W(\xi)$ to underline the density configuration $\hat{\varrho}(\Lambda) + \Delta\varrho(\Lambda)$ with respect to which these quantities are calculated. We use two parameters s and t instead of a single one for the technical transparency. In models $\langle \cdot \rangle_{t,0,\Lambda,\bar{q}^{(\Lambda^c)}}$, i.e., with $s = 0$, the range of interaction is finite so it is simpler to control decay of correlations.

Namely, a standard consequence of the Dobrushin uniqueness theorem [D2] is the estimate

$$|\langle \Delta\varrho_x\rangle_{t,0,\Lambda,\bar{q}^{(\Lambda^c)}} - \langle \Delta\varrho_x\rangle_{t,0}| \leqslant 2\zeta \sum_{n=2}^{\infty} \sum_{\substack{y_1 = x\\ y_2,\ldots,y_{n-1} \in [\Lambda]^{(\ell_2)}\\ y_n \in [\Lambda^c]^{(\ell_2)}}} \prod_{i=1}^{n-1} r_{y_i y_{i+1}} \tag{6.17}$$

where r_{xy} are defined by (5.56) with nonessential modification discussed below (5.62). Similarly

$$\left| \left\langle \prod_{x \in X} \Delta\varrho_x \right\rangle_{t,0,\Lambda,\bar{q}^{(\Lambda^c)}} - \langle \Delta\varrho_x\rangle_{t,0} \right|$$

$$\leqslant (2\zeta)^{|X|} \sum_{n=2}^{\infty} \sum_{\substack{y_1 \in X\\ y_2,\ldots,y_{n-1} \in [\Lambda]^{(\ell_2)}\\ y_n \in [\Lambda^c]^{(\ell_2)}}} \prod_{i=1}^{n-1} r_{y_i y_{i+1}} \tag{6.18}$$

Now if follows from definition (5.56) of r_{xy} and estimate (3.17) that

$$|\langle \Delta\varrho_x\rangle_{t,0,\Lambda,\bar{q}^{(\Lambda^c)}} - \langle \Delta\varrho_x\rangle_{t,0}| \leqslant \frac{c\zeta}{1 - a(\beta)} \left(\frac{1 + a(\beta)}{2}\right)^{[\gamma\, \mathrm{dist}(x, \Lambda^c)]} \tag{6.19}$$

and

$$\left|\left\langle \prod_{x\in X} \Delta\varrho_x \right\rangle_{t,0,\Lambda,\bar{q}^{(\Lambda^c)}} - \langle \Delta\varrho_x\rangle_{t,0}\right| \leqslant |X|(2\zeta)^{|X|}\frac{c\zeta}{1-a(\beta)}\left(\frac{1+a(\beta)}{2}\right)^{[\gamma\,\mathrm{dist}(X,\Lambda^c)]} \tag{6.20}$$

which are similar to (4.36). Also given x

$$\sum_{n=2}^{\infty} \sum_{\substack{y_1=x\\ y_2,\dots,y_n\in\mathbb{Z}^d_{\ell_2}}} \prod_{i=1}^{n-1} r_{y_i y_{i+1}} \leqslant \frac{2}{1-a(\beta)} \tag{6.21}$$

In fact is easily follows from the structure and smallness of $|W(\tau)|$ and $|W(\xi)|$ expressed by (5.31) and (5.37) that the same exponential decay of correlations takes place for models with $s\neq 0$ which include the infinite range part $s\sum_{i=2,3}\Delta H^{(i)}_{\gamma,\lambda}(\varrho(\Lambda)\mid \bar{q}^{(\Lambda^c)})$ of the Hamiltonian (6.15)

The final representation of $\log Z^A_{\gamma,\beta,\lambda}(\Lambda\mid\bar{q}^{(\Lambda^c)})$ is

$$\begin{aligned}
\log Z^A_{\gamma,\beta,\lambda}&(\Lambda\mid\bar{q}^{(\Lambda^c)})\\
&= -\tilde{H}_{\gamma,\lambda}(\hat{\varrho}(\Lambda)\mid\bar{q}^{(\Lambda^c)})\\
&\quad -\ell_2^d \sum_{D\not\subset[\Lambda^c]^{(\ell_2)}} W(D) \sum_{\substack{X_1(D),\\ X_2(D)}} \prod_{x\in X_1(D)} \hat{\varrho}_x \int_0^1 dt \left\langle \prod_{x\in X_2(D)} \Delta\varrho_x\right\rangle_{t,0}\\
&\quad - \sum_{\tau\notin\Lambda^c:\,L(\tau)\geqslant c} W(\tau)\int_0^1 ds\langle P(\tau,\Delta\varrho)\rangle_{1,s}\\
&\quad - \sum_{\xi\notin\Lambda^c}\int_0^1 ds\langle W(\xi,\Delta\varrho)\rangle_{1,s}\\
&\quad +\ell_2^d \sum_{D\not\subset[\Lambda^c]^{(\ell_2)}} W(D) \sum_{\substack{X_1(D),\\ X_2(D)}} \prod_{x\in X_1(D)} \hat{\varrho}_x \int_0^1 dt\\
&\quad \times\left(\left\langle \prod_{x\in X_2(D)} \Delta\varrho_x\right\rangle_{t,0,\Lambda,\bar{q}^{(\Lambda^c)}} - \left\langle \prod_{x\in X_2(D)} \Delta\varrho_x\right\rangle_{t,0}\right)\\
&\quad + \sum_{\tau\notin\Lambda^c:\,L(\tau)\geqslant c} W(\tau)\int_0^1 ds(\langle P(\tau,\Delta\varrho)\rangle_{1,s,\Lambda,\bar{q}^{(\Lambda^c)}} - \langle P(\tau,\Delta\varrho)\rangle_{1,s})\\
&\quad + \sum_{\xi\notin\Lambda^c}\int_0^1 ds(\langle W(\xi,\Delta\varrho)\rangle_{1,s,\Lambda,\bar{q}^{(\Lambda^c)}} - \langle W(\xi,\Delta\varrho)\rangle_{1,s})
\end{aligned} \tag{6.22}$$

Here speaking about $\langle f(\Delta\varrho_x)\rangle_{t,s,\Lambda,\bar{q}^{(\Lambda^c)}}$ we assume that $\Delta\varrho_x=\varrho_x-\hat{\varrho}_x(\Lambda)$ while for $\langle f(\Delta\varrho_x)\rangle_{t,s}$ we mean $\Delta\varrho_x=\varrho_x-\hat{\varrho}_{\beta,\lambda,\sigma}$ such that $\langle f(\Delta\varrho_x)\rangle_{t,s}$ is truly independent on any boundary conditions.

Proof of Statement 3.3. From the interpolation trick we have a representation for the metastable free energy

$$\begin{aligned}
f^{\Lambda}_{\sigma,\lambda,\gamma} &= \ell_2^d\hat{\varrho}_{\beta,\lambda,\sigma}(\log\hat{\varrho}_{\beta,\lambda,\sigma}-1)-\ell_2^d\lambda\hat{\varrho}_{\beta,\lambda,\sigma}-\ell_2^d\frac{1}{2!}\sum_{x\in\mathbb{Z}^d_{\ell_2}}J^{(2)}_\gamma(0,x)\,\hat{\varrho}^2_{\beta,\lambda,\sigma}\\
&+\ell_2^d\frac{1}{4!}\sum_{x_1,x_2,x_3\in\mathbb{Z}^d_{\ell_2}}J^{(4)}_\gamma(0,x_1,\dots,x_3)\,\hat{\varrho}^4_{\beta,\lambda,\sigma}\\
&-\ell_2^d\sum_{\substack{D:\,X(D)\ni 0\\ D\neq L^{(2)},L^{(4)}}}\frac{W(D)}{|X(D)|}\hat{\varrho}^{|X(D)|}_{\beta,\lambda,\sigma}\\
&-\sum_{\tau:\,X(\tau)\ni 0}\frac{W(\tau)}{|X(\tau)|}P(\tau,\hat{\varrho}_{\beta,\lambda,\sigma})-\sum_{\xi:\,X(\xi)\ni 0}\frac{W(\xi,\hat{\varrho}_{\beta,\lambda,\sigma})}{|X(\xi)|}\\
&-\ell_2^d\sum_{\substack{D:\,X(D)\ni 0\\ |X_2(D)|\geqslant 2}}\frac{W(D)}{|X(D)|}\sum_{\substack{X_1(D),\\X_2(D)}}\hat{\varrho}^{|X_1(D)|}_{\beta,\lambda,\sigma}\int_0^1 dt\left\langle\prod_{x\in X_2(D)}\Delta\varrho_x\right\rangle_{t,0}\\
&-\sum_{\tau:\,X(\tau)\ni 0}\frac{W(\tau)}{|X(\tau)|}\int_0^1 ds\langle P(\tau,\Delta\varrho)\rangle_{1,s}\\
&-\sum_{\xi:\,X(\xi)\ni 0}\int_0^1 ds\,\frac{\langle W(\xi,\Delta\varrho)\rangle_{1,s}}{|X(\xi)|}
\end{aligned}\tag{6.23}$$

Here the definition of $X(\tau)$ and $X(\xi)$ is similar to that of $X(D)$ and $\Delta\varrho$ is defined with respect to $\hat{\varrho}_{\beta,\lambda,\sigma}$.

The difference $g_\sigma(\Lambda\mid\bar{q}^{(\Lambda^c)})$ between RHS of (6.22) and $f^{\Lambda}_{\sigma,\lambda,\gamma}|\Lambda|$ can be estimated as

$$|g_\sigma(\Lambda\mid\bar{q}^{(\Lambda^c)})|\leqslant c\ell_2^d\frac{|\bar{\Lambda}\cap\bar{\Lambda}^c|}{\ell_2^{d-1}}\gamma^{-\alpha_2}\leqslant c\gamma^{-1}|\bar{\Lambda}\cap\bar{\Lambda}^c|\tag{6.24}$$

Here $|\bar{\Lambda}\cap\bar{\Lambda}^c|$ is the $(d-1)$-dimensional volume (area) of hypersurface separating Λ and Λ^c and $|\bar{\Lambda}\cap\bar{\Lambda}^c|\,\ell_2^{-d+1}\gamma^{-\alpha_2}$ is the number of lattice points in $[\delta\Lambda]^{(\ell_2)}$. The first factor $c\ell_2^d$ is the upper estimate for the sum of the absolute values of the error terms associated with given site $x\in[\partial\Lambda]^{(\ell_2)}$. To be more precise, the contribution to $g_\sigma(\Lambda\mid\bar{q}^{(\Lambda^c)})$ is given by terms in (6.22) and (6.23) crossing the boundary of Λ directly or indirectly. Directly crossing terms are D's with $X(D)$ intersecting both Λ and Λ^c,

τ's with $X(\tau)$ intersecting both Λ and Λ^c, ξ's with $X(\xi)$ intersecting both Λ and Λ^c and links with endpoints in both Λ and Λ^c. Indirectly intersection terms come from estimate (6.18) for the difference terms in the last three lines of (6.22). Graphically those terms can be represented by chains of r's joining something inside Λ, say some $D \in \Lambda$, with the sites outside Λ. It is important for us that geometrically any of the terms crossing the boundary passes through some site $x \in \partial \Lambda$. This site x may be: one of the endpoints of the link, contributing itself or as a part of D, τ or ξ, or the endpoint of some r_{xy} or it may belong to the support of some contour which is a part of some polymer ξ. For all these involved weighted objects we always keep the property that the sum of the absolute values of the statistical weights of all objects passing through given point is less than absolute constant. Multiplying this constant by the factor ℓ_2^d, which depending on the notations is present explicitly or implicitly in front of the sums just discussed, we reproduce (6.24). We need now some additional work to improve on (6.24).

The simplest observation is that using (5.43) and (5.37) one obtains that the sum over the objects containing τ with $L(\tau) > c_{(5.31)}$ or ξ as a constituting element is less than $c\gamma^{4d}$.

Another source of smallness are various expectations $\langle \prod_x \Delta\varrho_x \rangle$ in the corresponding terms. In particular we check that given x and for any $k \geqslant 1$

$$\langle |\Delta\varrho_x|^k \rangle \leqslant c(\ell_2^d)^{-k(5/12)} \tag{6.25}$$

where the expectation means any one of those contributing to (6.22) and (6.23). The power $-\frac{5}{12}$ is taken for the definiteness only and it can be replaced by $-\frac{1}{2}+\varepsilon$. We prove (6.25) adapting to our situation relatively abstract Lemma 8.1 in the spirit of [R2]. The correspondence between current notations and those of Section 8 is given by the following list of analogous objects: ϕ_x and $\Delta\varrho_x$, κ and ℓ_2^d, a and 2ζ, M and $2\zeta\ell_2^d$, R and $c_{(5.31)}\gamma^{-1}$, J_{xy} and $\hat{\varrho}\bar{I}^{(2)}(x, y \mid \hat{\varrho})$ and so on. The bound (6.25) is an easy consequence of Lemma 8.1 (which deals the finite range interactions) as the infinite radius part, $s \sum_{i=2,3} \Delta H^{(i)}_{\gamma,\lambda}(\varrho^{(\ell_2)} \mid \bar{q}^{(\Lambda^c)})$, of (6.15) is so small that it can not increase the expectation of $|\Delta\varrho_x|$ more than in $1 + c\gamma^{4d}$ times. Note also that in (6.22) and (6.23) we have terms $\langle |\Delta\varrho_x|^k \rangle$ only with $k \geqslant 2$.

In a similar way applying estimate (8.12) to $|\langle \prod_{x \in X_2(D)} \Delta\varrho_x \rangle_{\iota, 0, \Lambda, \bar{q}^{(\Lambda^c)}} - \langle \prod_{x \in X_2(D)} \Delta\varrho_x \rangle_{\iota, 0}|$ we obtain improved (6.20). Indeed, the RHS of (6.18) is nothing but the upper bound for the Vasserstein distance and a square root of this estimate produces an exponential bound just with twice smaller exponent than in (6.20).

These improvements reduce by the same factor, $(\ell_2^d)^{-2(5/12)}$, the estimates of the cross boundary terms coming from lines 2 and 5 in (6.22) and line 7 in (6.23).

As an immediate consequence of these improvements we can see that the contribution, $e(\lambda)$, to the difference $\ell_2^{-d}(f^A_{+,\lambda,\gamma} - f^A_{-,\lambda,\gamma})$ coming from the non energy terms (lines 7, 8, 9 in (6.23)) is a continuous function of λ and $|e(\lambda)| \leqslant c(\ell_2^d)^{-2(5/12)}$. The energy parts (lines 1–6 in (6.23)) of $\ell_2^{-d} f^A_{+,\lambda,\gamma}$ and $\ell_2^{-d} f^A_{-,\lambda,\gamma}$ coincide with each other at $\lambda = \hat{\lambda}(\beta, \gamma)$ and they contain explicitly terms which are linear in λ. Hence shiftig $\hat{\lambda}(\beta, \gamma)$ by at most $c(\ell_2^d)^{-2(5/12)}$ one can find the solution, $\lambda = \lambda(\beta, \gamma)$ of (3.42) which proves Statement 3.3. ∎

A useful consequence of our improvements is the estimate

$$|g_\sigma(\Lambda \mid \bar{q}^{(\Lambda^c)}) + (\tilde{H}_{\gamma,\lambda}(\hat{\varrho}_{\beta,\lambda,\gamma}(\Lambda) \mid \hat{\varrho}_{\beta,\lambda,\gamma}(\Lambda^c)) - \tfrac{1}{2}\tilde{U}_\gamma(\hat{\varrho}_{\beta,\lambda,\gamma}(\Lambda^c) \mid \hat{\varrho}_{\beta,\lambda,\gamma}(\Lambda^c))$$
$$- \tilde{H}_{\gamma,\lambda}(\hat{\varrho}(\Lambda) \mid \bar{q}^{(\Lambda^c)}))| \leqslant c\gamma^{-1}(\ell_2^d)^{-2(5/12)} |\bar{\Lambda} \cap \bar{\Lambda}^c| \leqslant \gamma^{1/2} |\bar{\Lambda} \cap \bar{\Lambda}^c| \qquad (6.26)$$

where $\tilde{U}_\gamma(\hat{\varrho}_{\beta,\lambda,\gamma}(\Lambda) \mid \hat{\varrho}_{\beta,\lambda,\gamma}(\Lambda^c))$ is defined as in (2.11) but in terms of the Hamiltonian $\tilde{H}(\cdot)$. (We remind that $|\Lambda \cap \Lambda^c|$ is the $(d-1)$-dimensional volume of the hypersurface separating closed domains Λ and Λ^c.) Observe that $\tilde{H}_{\gamma,\lambda}(\hat{\varrho}_{\beta,\lambda,\gamma}(\Lambda) \mid \hat{\varrho}_{\beta,\lambda,\gamma}(\Lambda^c)) - \tilde{H}_{\gamma,\lambda}(\hat{\varrho}(\Lambda) \mid \bar{q}^{(\Lambda^c)})$ is rather small while the compensating term $-\frac{1}{2}\tilde{U}_\gamma(\hat{\varrho}_{\beta,\lambda,\gamma}(\Lambda) \mid \hat{\varrho}_{\beta,\lambda,\gamma}(\Lambda^c))$ is of order $c\gamma^{-1} |\Lambda \cap \Lambda^c|$.

Proof of Statement 3.5. To check estimate (3.45) consider $\mathrm{Int}^{\neq}(\Gamma) = \bigcup_{m:\, \sigma_m(\Gamma) \neq \sigma(\Gamma)} \mathrm{Int}_m(\Gamma)$ and introduce a strip $S(\Gamma) = \partial^{(\ell_3/2)} \mathrm{Int}^{\neq}(\Gamma) \setminus \tilde{\delta}^{\neq}(\Gamma)$, where $\tilde{\delta}^{\neq}(\Gamma)$ is defined similarly to $\tilde{\delta}^{=}(\Gamma)$ (see (4.57) and below). First we rewrite the numerator of (3.39) as

$$\int_{\mathscr{Q}(\mathrm{Supp}(\Gamma) \setminus S(\Gamma))} dq\, \mathbb{1}_{\eta(q) = \eta^\Gamma}\, e^{-\beta H_{\gamma,\lambda}(q \mid \bar{q}^{(\delta^{=}(\Gamma))})}\, Z^A_{\gamma,\beta,\lambda}(\mathrm{Int}^{\neq}(\Gamma) \cup S(\Gamma) \mid q^{(\tilde{\delta}^{\neq}(\Gamma))}) \qquad (6.27)$$

Observe that for any q contributing to (6.27) the restriction of q to $\tilde{\delta}^{\neq}(\Gamma)$ belongs to the ground state ensemble of the phase $-\sigma(\Gamma)$. Given such $q^{(\tilde{\delta}^{\neq}(\Gamma))}$ denote by $\hat{\varrho}(\mathrm{Int}^{\neq}(\Gamma) \cup S(\Gamma)) = \hat{\varrho}^{q^{(\tilde{\delta}^{\neq}(\Gamma))}}$ the minimizer of (6.2). Then using representation (6.22) and definition (6.23) one can see that

$$\begin{aligned}
&Z^A_{\gamma,\beta,\lambda}(\mathrm{Int}^{\neq}(\Gamma) \cup S(\Gamma) \mid q^{(\tilde{\delta}^{\neq}(\Gamma))}) \\
&\quad = Z^A_{\gamma,\beta,\lambda}(S(\Gamma) \mid q^{(\tilde{\delta}^{\neq}(\Gamma))}, \hat{\varrho}^{q^{(\tilde{\delta}^{\neq}(\Gamma))}}(\mathrm{Int}^{\neq}(\Gamma)))\, \exp(\ell_2^{-d} f^A_{-\sigma(\Gamma),\lambda,\gamma} |\mathrm{Int}^{\neq}(\Gamma)| \\
&\qquad - \tfrac{1}{2} U(\hat{\varrho}^{q^{(\tilde{\delta}^{\neq}(\Gamma))}}(\mathrm{Int}^{\neq}(\Gamma)) \mid \hat{\varrho}^{q^{(\tilde{\delta}^{\neq}(\Gamma))}}(\partial\, \mathrm{Int}^{\neq}(\Gamma))) \\
&\qquad - (\tilde{H}_{\gamma,\lambda}(\hat{\varrho}^{q^{(\tilde{\delta}^{\neq}(\Gamma))}}(\mathrm{Int}^{\neq}(\Gamma)) \mid \hat{\varrho}^{q^{(\tilde{\delta}^{\neq}(\Gamma))}}(\partial\, \mathrm{Int}^{\neq}(\Gamma))) \\
&\qquad - \tilde{H}_{\gamma,\lambda}(\hat{\varrho}_{\beta,\lambda,\sigma}(\mathrm{Int}^{\neq}(\Gamma)) \mid \hat{\varrho}_{\beta,\lambda,\sigma}(\partial\, \mathrm{Int}^{\neq}(\Gamma)))) \\
&\qquad + \tilde{g}_{-\sigma(\Gamma)}(\mathrm{Int}^{\neq}(\Gamma) \mid \hat{\varrho}^{q^{(\tilde{\delta}^{\neq}(\Gamma))}}(\partial\, \mathrm{Int}^{\neq}(\Gamma)))) \qquad (6.28)
\end{aligned}$$

The term $\tilde{g}_{-\sigma(\Gamma)}(\mathrm{Int}^{\neq}(\Gamma) \mid \hat{\varrho}^{q^{(\tilde{\delta}^{\neq}(\Gamma))}}(\partial\,\mathrm{Int}^{\neq}(\Gamma)))$ is the sum over "polymer type" terms cross the boundary of $\mathrm{Int}^{\neq}(\Gamma)$ and it can be estimated (see (6.26)) by

$$|\tilde{g}_{-\sigma(\Gamma)}(\mathrm{Int}^{\neq}(\Gamma) \mid \hat{\varrho}^{q^{(\tilde{\delta}^{\neq}(\Gamma))}}(\partial\,\mathrm{Int}^{\neq}(\Gamma)))| \leqslant c\gamma^{1/2}\,|\overline{\mathrm{Int}^{\neq}(\Gamma)} \cap \overline{\mathrm{Int}^{\neq}(\Gamma)^{c}}| \tag{6.29}$$

On the other hand the strip $S(\Gamma)$ is so wide that in the γ^{-1} neighborhood of $\mathrm{Int}^{\neq}(\Gamma) \cap \mathrm{Int}^{\neq}(\Gamma)^{c}$ the difference $|\hat{\varrho}^{q^{(\tilde{\delta}^{\neq}(\Gamma))}} - \hat{\varrho}_{\beta,\lambda,\sigma}|$ is exponentially small as follows from a straightforward analogue of (4.36). Thus

$$\begin{aligned} Z^{A}_{\gamma,\beta,\lambda}&(\mathrm{Int}^{\neq}(\Gamma) \cup S(\Gamma) \mid q^{(\tilde{\delta}^{\neq}(\Gamma))}) \\ &= Z^{A}_{\gamma,\beta,\lambda}(S(\Gamma) \mid q^{(\tilde{\delta}^{\neq}(\Gamma))}, \hat{\varrho}_{\beta,\lambda,\sigma}(\mathrm{Int}^{\neq}(\Gamma)))\exp(\ell_{2}^{-d} f^{A}_{-\sigma(\Gamma),\lambda,\gamma}\,|\mathrm{Int}^{\neq}(\Gamma)| \\ &\quad - \tfrac{1}{2}U(\hat{\varrho}_{\beta,\lambda,\sigma}(\mathrm{Int}^{\neq}(\Gamma)) \mid \hat{\varrho}_{\beta,\lambda,\sigma}(\partial\,\mathrm{Int}^{\neq}(\Gamma))) + \bar{g}_{-\sigma(\Gamma)}(\mathrm{Int}^{\neq}(\Gamma))) \end{aligned} \tag{6.30}$$

where

$$|\bar{g}_{-\sigma(\Gamma)}(\mathrm{Int}^{\neq}(\Gamma))| \leqslant c\gamma^{1/2}\,|\overline{\mathrm{Int}^{\neq}(\Gamma)} \cap \overline{\mathrm{Int}^{\neq}(\Gamma)^{c}}| \tag{6.31}$$

Here $\bar{g}_{-\sigma(\Gamma)}(\mathrm{Int}^{\neq}(\Gamma))$ collects the contribution of $\tilde{g}_{-\sigma(\Gamma)}(\mathrm{Int}^{\neq}(\Gamma) \mid \hat{\varrho}^{q^{(\tilde{\delta}^{\neq}(\Gamma))}}(\partial\,\mathrm{Int}^{\neq}(\Gamma)))$ together with corrections due to replacement of $U(\hat{\varrho}^{q^{(\tilde{\delta}^{\neq}(\Gamma))}}(\mathrm{Int}^{\neq}(\Gamma)) \mid \hat{\varrho}^{q^{(\tilde{\delta}^{\neq}(\Gamma))}}(\partial\,\mathrm{Int}^{\neq}(\Gamma)))$ by $U(\hat{\varrho}_{\beta,\lambda,\sigma}(\mathrm{Int}^{\neq}(\Gamma)) \mid \hat{\varrho}_{\beta,\lambda,\sigma}(\partial\,\mathrm{Int}^{\neq}(\Gamma)))$ and the estimate of $\tilde{H}_{\gamma,\lambda}(\hat{\varrho}^{q^{(\tilde{\delta}^{\neq}(\Gamma))}}(\mathrm{Int}^{\neq}(\Gamma)) \mid \hat{\varrho}^{q^{(\tilde{\delta}^{\neq}(\Gamma))}}(\partial\,\mathrm{Int}^{\neq}(\Gamma))) - \tilde{H}_{\gamma,\lambda}(\hat{\varrho}_{\beta,\lambda,\sigma}\mathrm{Int}^{\neq}(\Gamma)) \mid \hat{\varrho}_{\beta,\lambda,\sigma}(\partial\,\mathrm{Int}^{\neq}(\Gamma)))$.

Using (6.10) we obtain from (6.30) our last estimate

$$\begin{aligned} Z^{A}_{\gamma,\beta,\lambda}&(\mathrm{Int}^{\neq}(\Gamma) \cup S(\Gamma) \mid q^{\tilde{\delta}^{\neq}(\Gamma))}) \\ &= Z^{A}_{\gamma,\beta,\lambda}(S(\Gamma) \mid q^{(\tilde{\delta}^{\neq}(\Gamma))}, \varrho_{\beta,\lambda,\sigma}(\mathrm{Int}^{\neq}(\Gamma)))\exp(\ell_{2}^{-d} f^{A}_{-\sigma(\Gamma),\lambda,\gamma}\,|\mathrm{Int}^{\neq}(\Gamma)| \\ &\quad - \tfrac{1}{2}U(\varrho_{\beta,\lambda,\sigma}(\mathrm{Int}^{\neq}(\Gamma)) \mid \varrho_{\beta,\lambda,\sigma}(\partial\,\mathrm{Int}^{\neq}(\Gamma))) + \bar{g}_{-\sigma(\Gamma)}(\mathrm{Int}^{\neq}(\Gamma)) + \Delta \end{aligned} \tag{6.32}$$

where $|\Delta| \leqslant c\gamma^{1/4}\,|S(\Gamma)$.

A similar estimate is true for the denominator of (3.39) which after integration over $q^{(\tilde{\delta}^{\neq}(\Gamma))}$ implies (3.45). ▮

7. APPENDIX. POLYMER EXPANSION THEOREM

Consider a finite or countable set Θ the elements of which are called (abstract) *diagrams* and denoted θ, θ', etc. Fix some reflexive and symmetric relation on $\Theta \times \Theta$. A pair $\theta, \theta' \in \Theta \times \Theta$ is called incompatible ($\theta \nsim \theta'$)

if it satisfies given relation and compatible ($\theta \sim \theta'$) in the opposite case. A collection $\{\theta_j\}$ is called a *compatible collection of diagrams* if any two its elements are compatible. Every diagram θ is assigned a complex-valued *statistical weight* denoted by $w(\theta)$, and for any finite $\Lambda \subseteq \Theta$ an (abstract) *partition function* is defined as

$$Z(\Lambda) = \sum_{\{\theta_j\} \subseteq \Lambda} \prod_j w(\theta_j) \tag{7.1}$$

where the sum is extended to all compatible collections of diagrams $\theta_i \in \Lambda$. The empty collection is compatible by definition, and it is included in $Z(\Lambda)$ with statistical weight 1.

A polymer $\pi = [\theta_i^{\varepsilon_i}]$ is an (unordered) finite collection of different diagrams $\theta_i \in \Theta$ taken with positive integer multiplicities ε_i, such that for every pair θ', $\theta'' \in \pi$ there exists a sequence $\theta' = \theta_{i_1}, \theta_{i_2},..., \theta_{i_s} = \theta'' \in \pi$ with $\theta_{i_j} \nsim \theta_{i_{j+1}}$, $j = 1, 2,..., s-1$. The notation $\pi \subseteq \Lambda$ means that $\theta_i \in \Lambda$ for every $\theta_i \in \pi$.

With every polymer π we associate an (abstract) graph $G(\pi)$ which consists of $\sum_i \varepsilon_i$ vertices labeled by the diagrams from π and edges joining every two vertices labeled by incompatible diagrams. It follows from the definition of $G(\pi)$ that it is connected and we denote by $r(\pi)$ the quantity

$$r(\pi) = \prod_i (\varepsilon_i!)^{-1} \sum_{G' \subset G(\pi)} (-1)^{|G'|} \tag{7.2}$$

where the sum is taken over all connected subgraphs G' of $G(\pi)$ containing all of $\sum_i \varepsilon_i$ vertices and $|G'|$ denotes the number of edges in G'. For any $\theta \in \pi$ we denote by $\varepsilon(\theta, \pi)$ the multiplicity of θ in the polymer π.

The polymer expansion theorem below is a modification of results of [Se] and [KP] proven in [MSu]. See also [D3] for similar results.

Theorem 7.1. Suppose that there exists a function $a(\theta)$: $\Theta \mapsto \mathbf{R}^+$ such that for any diagram θ

$$\sum_{\theta' : \theta' \nsim \theta} |w(\theta')| \, e^{a(\theta')} \leqslant a(\theta) \tag{7.3}$$

Then for any finite Λ,

$$\log Z(\Lambda) = \sum_{\pi \subseteq \Lambda} w(\pi) \tag{7.4}$$

where the statistical weight of a polymer $\pi = [\theta_i^{\varepsilon_i}]$ equals

$$w(\pi) = r(\pi) \prod_i w(\theta_i)^{\varepsilon_i} \tag{7.5}$$

Moreover, the series (7.4) for $\log Z(\Lambda)$ is absolutely convergent in view of the estimate

$$\sum_{\pi:\, \pi \ni \theta} \varepsilon(\theta, \pi)\, |w(\pi)| \leqslant |w(\theta)|\, e^{a(\theta)} \tag{7.6}$$

which is true for any diagram θ.

Corollary 7.2. For any function $b(\theta)\colon \Theta \mapsto \mathbb{R}^+$ consider modified statistical weights of diagrams

$$\tilde{w}(\theta) = w(\theta)\, e^{b(\theta)} \tag{7.7}$$

and suppose that still

$$\sum_{\theta':\, \theta' \nsim \theta} |\tilde{w}(\theta')|\, e^{a(\theta')} \leqslant a(\theta) \tag{7.8}$$

Then for any family, Π, of polymers such that any $\pi \in \Pi$ contains given diagram θ

$$\sum_{\pi \in \Pi} \varepsilon(\theta, \pi)\, |w(\pi)| \leqslant |\tilde{w}(\theta)|\, e^{a(\theta)} \left[\min_{\pi \in \Pi} \left(\prod_{\theta' \in \pi} e^{b(\theta')} \right) \right]^{-1} \tag{7.9}$$

8. APPENDIX. A TECHNICAL LEMMA

Consider a spin model on the lattice $\mathbb{Z}^d$ given by the formal Hamiltonian

$$H(\phi) = \sum_x \phi_x^2 + \tfrac{1}{2} \sum_{x \neq y} J_{xy} \phi_x \phi_y + \sum_{A \in \mathscr{A}} K_A \phi_A \tag{8.1}$$

Here the spin variable ϕ_x takes M discrete values, including 0, from the bounded interval $[-a, a]$. The last sum runs over some family $\mathscr{A}$ of set $A = \{x_1, \ldots, x_{|A|} \in \mathbb{Z}^d\}$ containing $|A|$ not necessarily different sites x and $\phi_A = \prod_{x \in A} \phi_x$. The interaction, J_{xy} and K_A, is of finite range $R < \infty$, i.e., $J_{xy} = 0$ if $\operatorname{dist}(x, y) \geqslant R$ and $K_A = 0$ if $\operatorname{diam}(A) \geqslant R$. Suppose that

$$\sum_{y \neq x} |J_{xy}| = 1 - \alpha, \qquad 0 < \alpha < 1 \tag{8.2}$$

and

$$\sum_{A \ni x} a^{|A|}\, |K_A| \ll \alpha \tag{8.3}$$

and denote by $\mu(\phi(\Lambda))$ the corresponding Gibbs distribution in the finite domain $\Lambda \subset \mathbb{Z}^d$ with zero ($\equiv$empty) boundary condition in Λ^c.

Lemma 8.1. For any $c > 1$ and sufficiently large inverse temperature

$$\kappa > \frac{M+R}{c} \tag{8.4}$$

one has

$$\mu(|\phi_x| \geqslant \kappa^{-5/12}) \leqslant e^{-c\kappa^{1/6}}, \qquad \forall x \in \Lambda \tag{8.5}$$

Proof. for an arbitrary configuration $\phi(\Lambda)$ define *spots* $S(\phi(\Lambda))$ as R-connected components of site $x \in \Lambda$ with $|\phi_x| \geqslant \kappa^{-5/12}$. Taking $x \in S$ set

$$h(x, \phi(\Lambda)) = \sum_{y \neq x} J_{xy} \phi_x \phi_y \mathbb{1}_{|\phi_x| \geqslant |\phi_y|} + \sum_{A \in \mathscr{A}:\, A \ni x} K_A \phi_A \mathbb{1}_{|\phi_x| \geqslant \max_{y \in A, y \neq x} |\phi_y|} \tag{8.6}$$

Then

$$H(\phi(S) \mid \phi(\Lambda \backslash S)) = \sum_{x \in S} h(x, \phi(\Lambda)) \tag{8.7}$$

and

$$h(x, \phi(\Lambda)) \geqslant \frac{\alpha}{2} \phi_x^2 \tag{8.8}$$

Here in (8.7)–(8.8) we used (8.2), (8.3) and the fact that $|\phi_x| \geqslant |\phi_y|$ for any $x \in S$ and $y \in \partial^{(R)} S$.

Now the probability that $\phi(\Lambda)$ contains a spot S with the fixed value of $\phi(S)$ can be estimated as

$$\begin{aligned}
\mu(\phi(S)) &= \frac{\sum_{\phi(\Lambda\backslash S)} \exp(-\kappa H(\phi(S) \mid \phi(\Lambda\backslash S)) - \kappa H(\phi(\Lambda\backslash S)))}{\sum_{\phi(\Lambda)} \exp(-\kappa H(\phi(\Lambda)))} \\
&\leqslant \exp\left(-\kappa \frac{\alpha}{2} \sum_{x \in S} \phi_x^2\right) \frac{\sum_{\phi(\Lambda\backslash S)} \exp(-\kappa H(\phi(\Lambda\backslash S)))}{\sum_{\phi(\Lambda):\, \phi(S) \equiv 0} \exp(-\kappa H(\phi(\Lambda)))} \\
&\leqslant \exp\left(-\frac{\alpha}{2} \kappa^{1/6} |S|\right) \frac{\sum_{\phi(\Lambda\backslash S)} \exp(-\kappa H(\phi(\Lambda\backslash S)))}{\sum_{\phi(\Lambda):\, \phi(S) \equiv 0} \exp(-\kappa H(\phi(\Lambda)))} \\
&= \exp\left(-\frac{\alpha}{2} \kappa^{1/6} |S|\right)
\end{aligned} \tag{8.9}$$

Hence the probability of the spot S

$$\mu(S)=\sum_{\phi(S)}\mu(\phi(S))\leqslant(c_{(8.4)}\kappa)^{|S|}\exp\left(-\frac{\alpha}{2}\kappa^{1/6}|S|\right)$$

$$\leqslant\exp\left(-\frac{\alpha}{4}\kappa^{1/6}|S|\right)\tag{8.10}$$

where the last inequality is true for κ large enough.

Finally for any $x\in\Lambda$

$$\begin{aligned}\mu(|\phi_x|\geqslant\kappa^{-5/12})&=\sum_{S\ni x}\mu(S)\\&\leqslant\sum_{|S|=1}^{\infty}(c_{(8.4)}\kappa)^{|S|}\exp\left(-\frac{\alpha}{4}\kappa^{1/6}|S|\right)\\&\leqslant\sum_{|S|=1}^{\infty}\exp\left(-\frac{\alpha}{8}\kappa^{1/6}|S|\right)\\&\leqslant e^{-c\kappa^{1/6}}\end{aligned}\tag{8.11}$$

where again κ is large enough. ∎

Let now $\mu^{(0)}(\phi_x,\phi_y)$ and $\mu^{(1)}(\phi_x,\phi_y)$ be a distribution at sites $x, y\in\Lambda$ of conditional Gibbs measures in Λ with two arbitrary boundary conditions $\bar{\phi}^{(0)}(\Lambda^c)$ and $\bar{\phi}^{(1)}(\Lambda^c)$ respectively. The Vasserstein distance $R(\mu^{(0)},\mu^{(1)})$ between $\mu^{(0)}$ and $\mu^{(1)}$ is given by some coupling $\mu(\phi_x^{(0)},\phi_y^{(0)};\phi_x^{(1)},\phi_y^{(1)})$. Denote by $\langle\cdot\rangle^{(0)}$, $\langle\cdot\rangle^{(1)}$ and $\langle\cdot\rangle$ the expectation with respect to $\mu^{(0)}$, $\mu^{(1)}$ and μ and suppose that $2a<1$. Then

$$\begin{aligned}&|\langle\phi_x,\phi_y\rangle^{(0)}-\langle\phi_x,\phi_y\rangle^{(1)}|\\&\quad=\langle(\phi_x^{(0)}-\phi_x^{(1)})\,\phi_y^{(0)}+(\phi_y^{(0)}-\phi_y^{(1)})\,\phi_x^{(1)}\rangle\\&\quad\leqslant\langle(\phi_x^{(0)}-\phi_x^{(1)})^2\rangle^{1/2}\langle(\phi_y^{(0)})^2\rangle^{1/2}+\langle(\phi_y^{(0)}-\phi_y^{(1)})^2\rangle^{1/2}\langle(\phi_x^{(1)})^2\rangle^{1/2}\\&\quad\leqslant\langle|\phi_x^{(0)}-\phi_x^{(1)}|\rangle^{1/2}\langle(\phi_y^{(0)})^2\rangle^{1/2}+\langle|\phi_y^{(0)}-\phi_y^{(1)}|\rangle^{1/2}\langle(\phi_x^{(1)})^2\rangle^{1/2}\\&\quad\leqslant\max((\langle\phi_x^2\rangle^{(0)})^{1/2},(\langle\phi_y^2\rangle^{(1)})^{1/2})\\&\qquad\times\max(R(\mu^{(0)}(\phi_x),\mu^{(1)}(\phi_x))^{1/2},R(\mu^{(0)}(\phi_y),\mu^{(1)}(\phi_y))^{1/2})\end{aligned}\tag{8.12}$$

Similarly one can estimate via the Vasserstein distance and the second moment the difference $|\langle\prod_{x\in X}\phi_x\rangle^{(0)}-\langle\prod_{x\in X}\phi_x\rangle^{(1)}|$ for any $X\subset\Lambda$.

ACKNOWLEDGMENTS

The work of J.L.L. and A.M. was supported by NSF Grant DMR 95-23266 and AFOSR Grant 4-26435.

REFERENCES

[Bax] R. J. Baxter, *Exactly Solved Models in Statistical Mechanics* (Academic Press, London/New York, 1982).

[BKL] J. Bricmont, K. Kuroda, and J. L. Lebowitz, First order phase transitions in lattice and continuous systems: Extension of Pirogov–Sinai theory, *Comm. Math. Phys.* **101**:501–538 (1985).

[BP] T. Bodineau and E. Presutti, Phase diagram of Ising systems with additional long range forces, *Comm. Math. Phys.* **189**:287–298, (1996).

[BZ] A. Bovier and M. Zahradnik, The low temperature phase of Kac–Ising models, *J. Stat. Phys.*, to appear.

[CCK] J. T. Chayes, L. Chayes, and R. Kotecky, The analysis of the Widom–Rowlinson model by stochastic geometric methods, *Comm. Math. Phys.* **172**:551–569 (1995).

[COPP] M. Cassandro, E. Olivieri, A. Pellegrinotti, and E. Presutti, Existence and uniqueness of DLR measures for unbounded spin systems, *Z. Wahr. verv. Geb.* **41**:313–334 (1978).

[CP] M. Cassandro and E. Presutti, Phase transitions in Ising systems with long but finite range interactions, *Markov Processes and Related Fields* **2**:241–262 (1996).

[D1] R. L. Dobrushin, Existence of phase transition in two- and three-dimensional Ising models, *Th. Prob. Appl.* **10**:193–313 (1965).

[D2] R. L. Dobrushin, Prescribing a system of random variables by conditional distributions, *Th. Prob. Appl.* **15**:456–458 (1970).

[D3] R. L. Dobrushin, Estimates of semi-invariants for the Ising model at low temperatures, topics in statistical and theoretical physics, *Amer. Math. Soc. Transl. (2)* **177**:59–81 (1996).

[DS] E. I. Dinaburg and Ya. G. Sinai, Contour models with interaction and their applications, *Sel. Math. Sov.* **7**:291–315 (1988).

[DZ] R. L. Dobrushin and M. Zahradnik, Phase diagrams for continuous spin models. Extension of Pirogov–Sinai theory, in *Mathematical Problems of Statistical Mechanics and Dynamics*, R. L. Dobrushin, ed. (Kluwer Academic Publishers, Dordrecht, Boston, 1986), pp. 1–123.

[F] M. V. Fedoryuk, *Asymptotic: Integrals and Series* (Nauka, Moscow, 1987).

[FF] B. U. Felderhof and M. E. Fisher, Phase transitions in one-dimensional cluster-interaction fluids. IA. Thermodynamics, IB. Critical behavior, II. Simple logarithmic model, *Annals of Phys.* **58**:176–216, 217–267, 268–280 (1970).

[GH] H.-O. Georgii and Haggstrom, Phase transitions in continuum Potts models, *Comm. Math. Phys.* **181**:507–528 (1996).

[G] R. B. Griffiths, Peierls' proof of spontaneous magnetization in a two-dimensional Ising ferromagnet, *Phys. Rev. A* **136**:437–439 (1964).

[J] K. Johansson, On separation of phases in one-dimensional gases, *Comm. Math. Phys.* **169**:521–561 (1995).

[vK] N. G. van Kampen, Condensation of a classical gas with long-range attraction, *Phys. Rev. A* **135**:362 (1964).

[KP] R. Kotecky and D. Preiss, Cluster expansion for abstract polymer models, *Comm. Math. Phys.* **103**:491–498 (1986).

[KUH] M. Kac, G. Uhlenbeck, and P. C. Hemmer, On the Van der Waals theory of vapor–liquid equilibrium, *J. Mat. Phys.* **4**:216–228, 229–247 (1963); **5**:60–74 (1964).

[LL] J. L. Lebowitz and E. H. Lieb, Phase transition in a continuum classical system with finite interactions, *Phys. Lett. A* **39**:98–100 (1972).

[LMPI] J. L. Lebowitz, A. E. Mazel, and E. Presutti, Rigorous proof of a liquid–vapor phase transition in a continuum particle system, *Phys. Rev. Lett.* **80**:4701–4704 (1998).

[LMPII] J. L. Lebowitz, A. E. Mazel, and E. Presutti, in preparation.

[LP] J. L. Lebowitz and O. Penrose, Rigorous treatment of the Van der Waals–Maxwell theory of the liquid–vapor transition, *J. Mat. Phys.* **7**:98–113 (1966).

[MS] A. E. Mazel and Yu. M. Suhov, Ground states of boson quantum lattice model, *Amer. Math. Soc. Transl. (2)* **171**:185–226 (1996).

[O] L. Onsager, Crystal statistics I. A two-dimensional model with an order–disorder transition, *Phys. Rev.* **65**:117–149 (1944).

[P] R. Peierls, On Ising's model of ferromagnetism, *Proc. Camb. Phil. Soc.* **32**:477–481 (1936).

[Pr] E. Presutti, Notes on phase transition in the continuum, Lectures at IHP, Paris, June–July 1998, preprint.

[PS] S. A. Pirogov and Ya. G. Sinai, Phase diagrams of classical lattice systems, *Theor. Math. Phys.* **25**:358–369, 1185–1192 (1975).

[R1] D. Ruelle, Existence of a phase transition in a continuous classical system, *Phys. Rev. Lett.* **27**:1040–1041 (1971).

[R2] D. Ruelle, Probability estimates for continuous spin systems, *Comm. Math. Phys.* **50**:189–194 (1976).

[S] Ya. G. Sinai, *Theory of Phase Transitions* (Academia Kiado and Pergamon Press, Budapest, London, 1982).

[Se] E. Seiler, Gauge theories as a problem of constructive quantum field theory and statistical mechanics, *Lect. Notes in Physics*, Vol. 159 (Springer-Verlag, Berlin, 1982).

[WR] B. Widom and J. S. Rowlinson, New model for the study of liquid–vapor phase transitions, *J. Chem. Phys.* **52**:1670–1684 (1970).

[Z] M. Zahradnik, An alternate version of Pirogov–Sinai theory, *Comm. Math. Phys.* **93**:559–581 (1984).

Journal of Statistical Physics, Vol. 95, Nos. 1/2, 1999

A Gallavotti–Cohen-Type Symmetry in the Large Deviation Functional for Stochastic Dynamics

Joel L. Lebowitz[1] and Herbert Spohn[2]

Received November 9, 1998; final December 16, 1998

We extend the work of Kurchan on the Gallavotti–Cohen fluctuation theorem, which yields a symmetry property of the large deviation function, to general Markov processes. These include jump processes describing the evolution of stochastic lattice gases driven in the bulk or through particle reservoirs, general diffusive processes in physical and/or velocity space, as well as Hamiltonian systems with stochastic boundary conditions. For dynamics satisfying local detailed balance we establish a link between the average of the action functional in the fluctuation theorem and the macroscopic entropy production. This gives, in the linear regime, an alternative derivation of the Green–Kubo formula and the Onsager reciprocity relations. In the nonlinear regime consequences of the new symmetry are harder to come by and the large deviation functional difficult to compute. For the asymmetric simple exclusion process the latter is determined explicitly using the Bethe ansatz in the limit of large N.

KEY WORDS: Fluctuation theorem; current fluctuations; asymmetric exclusion process.

1. INTRODUCTION

The study of stationary nonequilibrium states (SNS) of macroscopic systems evolving according to classical (or even quantum) mechanics, which are kept out of equilibrium by contact with thermal reservoirs has a long history. The interactions of the system with the reservoirs are usually modeled by the addition of stochastic boundary terms to the deterministic evolution equations describing the isolated system.[1-7] Fully deterministic evolutions of systems coupled to infinitely extended reservoirs

[1] Department of Mathematics and Physics, Rutgers University, Hill Center, New Brunswick, New Jersey 08903; e-mail: lebowitz@math.rutgers.edu.

[2] Zentrum Mathematik and Physik Department, TU München, D-80290 München, Germany; e-mail: spohn@mathematik.tu-muenchen.de.

0022-4715/99/0400-0333$16.00/0

heave also been investigated,[8–10] as have fully stochastic models.[11–14] A more recent development has been the study of SNS of systems evolving via deterministic "thermostatted" dynamics.[15–17] In this scheme the SNS is maintained by deterministic external driving forces which are balanced by a "friction term" suitably chosen so that the system evolves on a compact surface (generally one of constant energy) in phase space. The resulting dynamics no longer conserves phase space volume, which is contracted on the average.[19] Consequently, the SNS is singular with respect to the Liouville (volume) measure induced on the surface.

While the jury is still out on what has or can be achieved by such a dynamical systems approach, there are some intriguing theoretical results which have emerged.[20, 21] In particular Gallavotti and Cohen[22] motivated by results of computer simulations[23] discovered that under suitable assumptions these SNS satisfy a certain symmetry, which they doobed a "fluctuation theorem." Assuming that the dynamics satisfies *time reversal invariance* and is sufficiently chaotic, so that the SNS is given by an SRB measure, they prove that the probability distribution for the phase space contraction averaged along a trajectory over the time span τ has, for large τ, a highly non-obvious symmetry, whose specific form will be given below. Near equilibrium this fluctuation theorem implies Onsager reciprocity and the Einstein relations.[24] We refer to refs. 20, 22, and 25 for an exhaustive discussion of the fluctuation theorem and of the "chaotic hypothesis," which implies the validity of the said symmetry for more general SNS. Transient fluctuations in the phase space volume contraction over the time interval $(0, t)$ starting from an initial state uniform on the (kinetic) energy surface were considered by Evans and Searles.[26]

In a recent work Kurchan showed that, with proper definitions, the fluctuation theorem is valid also for certain diffusion processes.[27] In the present work we extend Kurchan's results and show that for Markov processes the fluctuation theorem holds in great generality. Our proof is based on the Perron–Frobenius theorem and goes much beyond the cases considered in ref. 27. While the assumed stochasticity of the dynamics should provide "for free" a suitable modification of the chaotic hypothesis it is far from obvious how to find the quantity that plays the role of the phase space contraction. In analogy to the thermostatted case it is reasonable to expect that it will have some relation to the production of Gibbs entropy. In fact, this comes out from our general construction under the additional assumption of local detailed balance. This condition seems to play, in stochastic dynamics, the same role as time reversal invariance in deterministic Gaussian thermostatted dynamics. When local detailed balance is satisfied, then the action functional is proportional to the current of the conserved quantity, e.g., in the case of bulk driven lattice gases it is the bulk particle current. If local detailed balance is not satisfied, then the action functional can still be defined but it does not allow for such a direct physical interpretation.

Near equilibrium the fluctuation theorem, generalized to systems with several currents, yields the Onsager symmetry and the usual Kubo formulae for the linear transport coefficients. These give a relation between linear response and current fluctuations in equilibrium. In this sense the fluctuation theorem, which holds also far from equilibrium, can be thought of as a natural extension, in the spirit of GC, of the fluctuation-dissipation theorem which holds for systems close to (local) equilibrium. In fact, when the large deviation functional is approximately quadratic over a sufficiently large range of the driving field, the fluctuation theorem does indeed yield such an approximate extension. We do not, however, have a quantitative criteria of when this approximation is valid. Even aside from this extension it would be of great interest to have an experimentally verifiable consequence of the novel symmetry predicted by the fluctuation theorem for the distribution of large deviations in current carrying systems. This seems difficult to obtain for macroscopic systems since the time required for observing a large deviation in such a system far exceeds any observation time.

The physical meaning of the fluctuation theorem can be understood best through the application to simple models. We therefore determine here the action functional for which such a fluctuation theorem holds in various examples: bulk and/or boundary driven lattice gases, the Fokker–Planck equation with driving forces and a spatially varying temperature, and Hamiltonian particle systems driven by boundary reservoirs. In general, the rate function appearing in the fluctuation theorem cannot be computed explicitly and one has to rely on numerical simulations.[23, 28] In stochastic models the task is easier as long as we stick to noninteracting particles. We discuss one example of interacting particles, where the rate function can be computed via the Bethe ansatz. When the size of the system goes to infinity the rate function becomes singular. This is analogous to the behavior of thermodynamic potentials at a phase transition. This is done in the Appendix which may be read independently of the rest of the paper.

Our analysis extends directly to discrete time stochastic processes, i.e., to probabilistic cellular automata. In this situation it is natural to think of the GC fluctuation theorem as a symmetry of discrete space-time Gibbs measures, as has been pointed out in ref. 28 and at a more general level by C. Maes.[29] In fact, the framework of space-time Gibbs measures used by Maes contains as special cases both the stochastic dynamics studied here and the thermostatted systems satisfying the chaotic hypothesis. In the latter case the space-time Gibbs measure is constructed through a Markov partition of phase space.

2. THE FLUCTUATION THEOREM FOR JUMP PROCESSES

We start with the fluctuation theorem for a stochastic time evolution governed by a master equation on a state space with a finite number of points. This minimizes technical complications and at the same time provides a blue-print for stochastic systems with Langevin type dynamics. In this section we investigate the abstract structure, specific examples and applications will be discussed in later sections.

2.1. Action Functional, Large Deviations

We consider a continuous time Markov jump process with finite state space $\mathscr{S}$. Points in $\mathscr{S}$ are denoted by σ. The jump process is determined by the rates, $k(\sigma, \sigma') \geqslant 0$, for jumping from σ to σ'. More precisely, if the system is in the state σ it waits a random time $t \geqslant 0$ distributed according to the exponential law $r(\sigma)\, e^{-r(\sigma)\, t}\, dt$, $r(\sigma) = \sum_{\sigma'} k(\sigma, \sigma')$, and then jumps to σ' with probability $r(\sigma)^{-1}\, k(\sigma, \sigma')$, etc.. The generator acting on functions $f\colon \mathscr{S} \to \mathbb{R}$ is given by

$$Lf(\sigma) = \sum_{\sigma'} k(\sigma, \sigma')[f(\sigma') - f(\sigma)] \tag{2.1}$$

This means, if $\mu(\sigma, t)$ is the probability distribution of σ at time t, then the rate of change of the average of f, $\langle f \rangle_{\mu(t)} = \sum_\sigma \mu(\sigma, t)\, f(\sigma)$, is given by

$$\frac{d}{dt} \langle f \rangle_{\mu(t)} = \langle Lf \rangle_{\mu(t)} \tag{2.2}$$

Taking f to be a Kronecker delta at σ (2.2) corresponds to the master equation

$$\frac{\partial \mu(\sigma, t)}{\partial t} = \sum_{\sigma'} k(\sigma', \sigma)\, \mu(\sigma', t) - r(\sigma)\, \mu(\sigma, t) = L^* \mu(\sigma, t) \tag{2.3}$$

for the evolution of $\mu(\sigma, t)$, where L^* denotes the adjoint of L.

We assume that if $k(\sigma, \sigma') > 0$, then also $k(\sigma', \sigma) > 0$, and that from every σ all other $\sigma' \in \mathscr{S}$ can be reached by a succession of steps with non-zero rates. It follows then from the general theory of finite state Markov processes that there exists a unique stationary measure, μ_s, which satisfies $L^* \mu_s(\sigma) = 0$ and is strictly positive, $\mu_s(\sigma) > 0$. Starting from any initial state σ_0, μ_s is approached exponentially fast in time.

We could as well take the state space $\mathscr{S}$ to be countable or replace $\mathscr{S}$ by $\mathbb{R}^n$. In the latter case one prescribes the rates $k(x, x')\, dx'$ for a jump from x to the volume element $x'+dx'$. Setting $r(x)=\int k(x, x')\,dx'$, the waiting time at x is exponential with rate $r(x)$ and the jump probability to $x'+dx'$ is $r(x)^{-1}\,k(x, x')\,dx'$. Formally the difference is only in notation. However, now one has to make sure that the jump process has a unique invariant measure.

Let $s\to\sigma_s$ be a trajectory (or history) of the jump process and define for it the quantity

$$W(t, \{\sigma_s, 0\leqslant s\leqslant t\})=\int_0^t \sum_{\sigma, \sigma'} w_{\sigma, \sigma'}(s)\,ds \tag{2.4}$$

with $w_{\sigma, \sigma'}(s)$ a sequence of δ-functions, located exactly at those times s when σ_s jumps from σ to σ', with weight

$$w(\sigma, \sigma')=\log k(\sigma, \sigma')-\log k(\sigma', \sigma) \tag{2.5}$$

This means, if the trajectory σ_s, $0\leqslant s\leqslant t$, visits in succession the states $\sigma_0, \sigma_1,..., \sigma_n$, where σ_0 is the state at time 0 and σ_n the one at time t, then

$$W(t, \{\sigma_s, 0\leqslant s\leqslant t\})=\log\left[\frac{k(\sigma_0, \sigma_1)}{k(\sigma_1, \sigma_0)}\cdots\frac{k(\sigma_{n-1}, \sigma_n)}{k(\sigma_n, \sigma_{n-1})}\right] \tag{2.6}$$

(Note that at an allowed transition $k(\sigma, \sigma')>0$ and, by assumption, also $k(\sigma', \sigma)>0$).

For lack of a better name we call $W(t)$, and the similar quantities to be defined below, an *action functional*. (We will generally not indicate explicitly the dependence of $W(t)$ on the stochastic trajectory $\{\sigma_s, 0\leqslant s\leqslant t\}$).

Let $\langle\cdot\rangle$ denote the expectation over all trajectories in the stationary process, i.e., starting in the steady state $\mu_s(\sigma)$. We consider the generating function $\langle\exp[-\lambda W(t)]\rangle$ of $W(t)$ and define

$$\lim_{t\to\infty} -\frac{1}{t}\log\langle e^{-\lambda W(t)}\rangle=e(\lambda) \tag{2.7}$$

The *fluctuation theorem* we shall prove states that the limit (2.7) exists with $e(\lambda)$ convex downwards, and that

$$e(\lambda)=e(1-\lambda) \tag{2.8}$$

By a general result, analogous to the equivalence of ensembles in equilibrium statistical mechanics, (2.7) implies a large deviation property for the probability distribution $p_t(w)$ of $W(t)/t$, i.e., for large t

$$p_t(w)\cong e^{-t\hat{e}(w)} \tag{2.9}$$

where $\hat{e}$ is the Legendre transform of e. $\hat{e}(w)$ is convex up, $\hat{e}(w) \geqslant 0$, $\hat{e}(w_0) = 0$ for $w_0 = \lim_{t\to\infty} \langle (1/t)\, W(t) \rangle$, and by (2.8) $\hat{e}$ satisfies

$$\hat{e}(w) - \hat{e}(-w) = -w \tag{2.10}$$

i.e., the odd part of $\hat{e}$ is linear with slope $-1/2$.

To be more precise, from (2.7) one concludes that for every interval I

$$\lim_{t\to\infty} -\frac{1}{t} \log \text{Prob}\left(\frac{1}{t}\, W(t) \in I\right) = \min_{w\in I} \hat{e}(w) \tag{2.11}$$

with

$$\begin{aligned} \hat{e}(w) &= \max_{\lambda} \{e(\lambda) - \lambda w\} \\ &= \max_{\lambda} \{e(1-\lambda) - \lambda w\} = \max_{\lambda} \{e(\lambda) - (1-\lambda)\, w\} \\ &= \hat{e}(-w) - w \end{aligned} \tag{2.12}$$

Equation (2.10) is of the same form as the GC fluctuation theorem, with the phase space contraction integrated along an orbit of the thermostatted dynamics in GC replaced by (2.5) summed over the jumps along the stochastic trajectory. We note that the convexity of $e(\lambda)$ and (2.8) imply that $\langle w_0 \rangle = (de(\lambda)/d\lambda)_{|\lambda=0} \geqslant 0$. This is analogous to the result[19] that the mean phase space volume contraction in the stationary state is nonnegative.

To prove (2.8) we first define

$$g(\sigma, t) = \mathbb{E}_\sigma[e^{-\lambda W(t)}] \tag{2.13}$$

as the expectation value of $e^{-\lambda W(t)}$ conditioned on the system being in state σ at time $t = 0$. We then have

$$\begin{aligned} \frac{d}{dt}\, g(\sigma, t) &= \sum_{\sigma'} k(\sigma, \sigma')\, e^{-\lambda w(\sigma, \sigma')} g(\sigma', t) - r(\sigma)\, g(\sigma, t) \\ &= \sum_{\sigma'} k(\sigma, \sigma')^{1-\lambda}\, k(\sigma', \sigma)^{\lambda}\, g(\sigma', t) - r(\sigma)\, g(\sigma, t) \\ &= L_\lambda g(\sigma, t) \end{aligned} \tag{2.14}$$

Equation (2.14) is to be solved subject to the initial condition $g(\sigma, 0) = 1$. Therefore

$$\langle e^{-\lambda W(t)} \rangle = \sum_{\sigma} \mu_s(\sigma)\, g(\sigma, t) = \sum_{\sigma, \sigma'} \mu_s(\sigma) (e^{L_\lambda t})_{\sigma\sigma'} \tag{2.15}$$

Clearly $(e^{L_\lambda t})_{\sigma\sigma'} > 0$ by our assumption on $k(\sigma, \sigma')$ above. The Perron–Frobenius theorem states that L_λ has a unique maximal eigenvector f_λ which is characterized by $f_\lambda(\sigma) > 0$ and satisfies $L_\lambda f_\lambda(\sigma) = -e(\lambda) f_\lambda(\sigma)$ with real maximal $-e(\lambda)$. This implies the existence of the limit in (2.7). Furthermore L_λ and L_λ^* have the same maximal eigenvalue. Using now the definition (2.16) we see that

$$L_\lambda^* = L_{1-\lambda} \tag{2.16}$$

Hence the maximal eigenvector $\bar{f}_\lambda$ of L_λ^* satisfies

$$L_\lambda^* \bar{f}_\lambda(\sigma) = -e(\lambda) \bar{f}_\lambda(\sigma) = L_{1-\lambda} \bar{f}_\lambda(\sigma) \tag{2.17}$$

Since $L_{1-\lambda} f_{1-\lambda}(\sigma) = -e(1-\lambda) f_{1-\lambda}(\sigma)$ and since $\bar{f}_\lambda > 0$, we conclude by uniqueness that $\bar{f}_\lambda = f_{1-\lambda}$ and $e(\lambda) = e(1-\lambda)$. We may add that the same conclusion could be inferred from the Donsker–Varadhan theory of large deviations.[30, 31]

2.2. Time Reversal

For a Markov chain with transition probability $p(\sigma, \sigma')$ from σ to σ' the history $\{\sigma\} = \{\sigma_0, ..., \sigma_n\}$ has the probability

$$P(\{\sigma\}) = \mu_s(\sigma_0)\, p(\sigma_0, \sigma_1) \cdots p(\sigma_{n-1}, \sigma_n) \tag{2.18}$$

when starting in the stationary measure μ_s. In analogy to (2.6) we define the action functional as

$$W(n, \{\sigma\}) = -\sum_{j=1}^{n} \log[p(\sigma_j, \sigma_{j-1})/p(\sigma_{j-1}, \sigma_j)] \tag{2.19}$$

If we now denote the time reversed history of $\{\sigma\}$ by $R\{\sigma\} = \{\sigma_n, ..., \sigma_0\}$, then (2.19) can be rewritten as

$$\begin{aligned} W(n, \{\sigma\}) &= -\log[P(R\{\sigma\})/P(\{\sigma\})] + \log[\mu_s(\sigma_n)/\mu_s(\sigma_0)] \\ &= \bar{W}(n, \{\sigma\}) + \log[\mu_s(\sigma_n)/\mu_s(\sigma_0)] \end{aligned} \tag{2.20}$$

One now immediately verifies that $\langle e^{-\lambda \bar{W}(n)} \rangle = \langle e^{-(1-\lambda)\bar{W}(n)} \rangle$, where the average is with respect to $P(\{\sigma\})$. Thus for $\bar{W}(n)$ the fluctuation theorem holds even for finite n.

The form (2.20) is analogous to the form which the GC functional takes for thermostatted systems when the latter deterministic dynamical evolution is described via Markov partitions and it shows explicitly the role of time reversal in the original GC theorem. This observation forms the basis of the analysis of C. Maes in ref. 29, where he studies the fluctuation theorem for space-time Gibbs measures and also discusses symmetry transformations different from time reversal.

For the time continuous jump process a path $\{\sigma_s, 0 \leqslant s \leqslant t\}$ is time reversed as $\{\sigma_{t-s}, 0 \leqslant s \leqslant t\}$. The time reversed process has the jump rates $k^R(\sigma, \sigma') = k(\sigma', \sigma)\, \mu_s(\sigma')/\mu_s(\sigma)$ whereas the inverse waiting times, $r(\sigma)$, and the steady state, $\mu_s(\sigma)$, remain unmodified. If $P_{[0,t]}$ denotes the path measure of the stationary process in the time window $[0, t]$ and $P^R_{[0,t]}$ the one of the corresponding time reversed process, then $P^R_{[0,t]}$ has a density relative to $P_{[0,t]}$ and

$$P^R_{[0,t]} = e^{-W(t)}(\mu_s(\sigma_t)/\mu_s(\sigma_0))\, P_{[0,t]} \tag{2.21}$$

with $W(t)$ from (2.4). Thus, up to boundary terms, the action functional equals $-\log(dP^R_{[[0,t]}/dP_{[0,t]})$ with dP^R/dP denoting the Radon–Nikodym derivative.

(2.21) remains meaningful for a stochastic dynamics which is *not Markov* and it can thus be used as a definition of the action functional in such a more general context.

One may wonder whether time-reversal could be replaced by some other transformation on path space. An obvious candidate is an internal symmetry $S\colon \mathscr{S} \to \mathscr{S}$ such that $S \circ S = 1$. Using this transformation in (2.21) the GC action functional becomes

$$W(t) = -\int_0^t ds \log[r(S\sigma_s)/r(\sigma_s)] \tag{2.22}$$

up to boundary terms. From the examples we have studied it seems that only time-reversal leads to an action functional which has a simple physical interpretation.

2.3. Sum of Several Generators

We can generalize the above analysis to the case where the generator in (2.1) is the sum of several generators: $L = \sum_{j=1}^m L^{(j)}$ with

$$L^{(j)}f(\sigma) = \sum_{\sigma'} k^{(j)}(\sigma, \sigma')[f(\sigma') - f(\sigma)] \tag{2.23}$$

We require that for transitions between a given pair σ, σ' at most one of the rates $k^{(j)}$ is different from zero. It follows then from our assumptions on $k(\sigma, \sigma')$ that if $k^{(j)}(\sigma, \sigma') > 0$, then also $k^{(j)}(\sigma', \sigma) > 0$. With this decomposition we define, in complete analogy with (2.4)–(2.6), the jth action functional

$$W^{(j)}(t, \{\sigma_s, 0 \leqslant s \leqslant t\}) = \int_0^t \sum_{\sigma, \sigma'} w^{(j)}_{\sigma, \sigma'}(s)\, ds \tag{2.24}$$

and the logarithm of the generating function

$$e(\lambda_1,..., \lambda_m) = \lim_{t \to \infty} -\frac{1}{t} \log \left\langle \exp\left[-\sum_{j=1}^{m} \lambda_j W^{(j)}(t)\right]\right\rangle \tag{2.25}$$

Repeating the arguments of Section 2.1 we conclude that the fluctuation theorem now takes the form

$$e(\lambda_1,..., \lambda_m) = e(1-\lambda_1,..., 1-\lambda_m) \tag{2.26}$$

The corresponding analog of (2.10) is

$$\hat{e}(w_1,..., w_n) - \hat{e}(-w_1,..., -w_m) = -\sum_{j=1}^{m} w_j \tag{2.27}$$

We shall use the general form (2.26) in Section 8 to prove the Onsager relations but note here that there are many ways of splitting L into a sum of $L^{(j)}$'s. In fact we can choose a different $L^{(j)}$ for every pair of points (σ, σ') for which $k(\sigma, \sigma') \neq 0$. In all cases setting $\lambda_j = \lambda$ for all j we recover (2.8) and (2.10) for $W = \sum_{j=1}^{m} W^{(j)}$. We do not expect in general that the distribution of $(1/t)\, W^{(j)}(t)$ separately satisfies any fluctuation theorem. On the other hand we can consider, as a degenerate case of (2.23), a situation in which our system is composed of several independent systems with generators $L^{(j)}$. In that case, of course, each $L^{(j)}$ satisfies the symmetry relation, $e_j(\lambda_j) = e_j(1-\lambda_j)$, $\hat{e}_j(w_j) - \hat{e}_j(-w_j) = -w_j$, and $e(\lambda_1,..., \lambda_m) = \sum_{j=1}^{m} e_j(\lambda_j)$, $\hat{e}(w_1,..., w_m) = \sum_{j=1}^{m} \hat{e}_j(w_j)$.

2.4. Entropy Production

For the probability distribution $\mu(\sigma, t)$ at time t the Gibbs entropy is, as usual, given by

$$S_G(\mu(t)) = -\sum_{\sigma} \mu(\sigma, t) \log \mu(\sigma, t) \tag{2.28}$$

Using (2.3) the rate of change of S_G can be written in the form

$$\frac{d}{dt} S_G(\mu(t)) = R(\mu(t)) - A(\mu(t)) \tag{2.29}$$

where

$$R(\mu(t)) = \frac{1}{2} \sum_{\sigma, \sigma'} [k(\sigma, \sigma')\, \mu(\sigma, t) - k(\sigma', \sigma)\, \mu(\sigma', t)] \log \left[\frac{\mu(\sigma, t)\, k(\sigma, \sigma')}{\mu(\sigma', t)\, k(\sigma', \sigma)} \right] \geqslant 0 \tag{2.30}$$

and

$$A(\mu(t)) = \langle I \rangle_{\mu(t)}, \qquad I(\sigma) = \sum_{\sigma'} k(\sigma, \sigma') \log \left[\frac{k(\sigma, \sigma')}{k(\sigma', \sigma)} \right] \tag{2.31}$$

(When the generator is a sum, $L = \sum_{j=1}^{m} L^{(j)}$, as in the previous section, then the rate of change in Gibbs entropy is also a sum over j and consequently $R = \sum_{j=1}^{m} R^{(j)}$, $A = \sum_{j=1}^{m} A^{(j)}$).

The breakup in (2.30) has the property that R is non-negative, while A is linear in μ. This suggests that we identify R as the entropy produced by the stochastic jumps and A with the entropy flow which can have either sign. Defining now the "integrand" in A, $I(\sigma)$, as a "microscopic" entropy flux we note that the expectation value of $W(t)$, starting with some initial distribution $\mu(\sigma, 0)$, is given by

$$\langle W(t) \rangle_{\mu(0)} = \int_0^t ds \langle I \rangle_{\mu(s)} \tag{2.32}$$

We can therefore say that the average action equals the entropy flow integrated over the time span t. There is of course some arbitrariness in this identification. We could for example add some positive linear term to R and subtract it from A. Our choice of splitting is however quite natural in the examples discussed below.

When the system is in the steady state μ_s, $(d/dt)\, S_G(\mu_s) = 0$ and the entropy flow balances the entropy production, i.e.,

$$R(\mu_s) = A(\mu_s) \tag{2.33}$$

Thus

$$\langle t^{-1} W(t) \rangle = R(\mu_s) = A(\mu_s) \tag{2.34}$$

and hence the average action equals the rate of entropy production in the steady state. This suggests to identify W with the microscopic production of Gibbs entropy, a role played by the phase space contraction in the context of thermostatted systems; see also Section 7.

We remark that $R(\mu_s)=A(\mu_s)=0$ in case $k(\sigma,\sigma')$ satisfies detailed balance with respect to the invariant measure $\mu_s(\sigma)\sim e^{-V(\sigma)}$, i.e.,

$$e^{-V(\sigma)}k(\sigma,\sigma')=e^{-V(\sigma')}k(\sigma',\sigma) \tag{2.35}$$

Then each term in the sum defining $R(\mu_s)$ in (2.30) vanishes. This implies by (2.29) that $A(\mu_s)=\langle I\rangle_{\mu_s}=0$. The action functional (2.6) now becomes a telescoping sum with

$$W(t)=V(\sigma_0)-V(\sigma_t) \tag{2.36}$$

It then follows from (2.7) that $e(\lambda)=0$ and the fluctuation theorem becomes empty. More generally if

$$k(\sigma,\sigma')=k_0(\sigma,\sigma')\,k_1(\sigma,\sigma') \tag{2.37}$$

and $k_0(\sigma,\sigma')$ satisfies detailed balance while k_1 is a nonequilibrium driving "force," then the $\log k_0$ terms of $W(t)$ sum to a boundary value as before and only the weight corresponding to k_1 contributes in $W(t)$. We will encounter such a situation for driven lattice gases.

3. BULK DRIVEN LATTICE GASES

An important example of an SNS with spatial structure is a stochastic lattice gas where particles jump at random to neighboring lattice sites. We envisage two mechanisms for how the conserved particle density is driven away from equilibrium. The first one, studied in this section, is a global driving force which corresponds to driven diffusive systems (DDS). In the second one, considered in the next section, particles are injected/removed at the boundaries whereas the bulk is governed by reversible dynamics satisfying detailed balance with respect to an equilibrium stationary measure. In both cases the full dynamics does not satisfy detailed balance with respect to the stationary measure μ_s.

We consider particles with exclusion hopping on a regular lattice $\mathbb{Z}^d$. The configuration is specified by the occupation variables η_x, $x\in\mathbb{Z}^d$, taking the values 0 (empty) and 1 (occupied). Particles on different sites may interact by making the jump rate of a given particle depend on the configurations on nearby sites. As explained in Section 2, the fluctuation theorem holds in general. For notational simplicity we restrict ourselves however to the one-dimensional lattice, $d=1$.

We first study particles subject to a uniform drive. This will set up a steady state current in a finite system of particles jumping on a ring, $x=1,..., \ell$, with $x=\ell+1$ identified with $x=1$. A particle configuration is denoted by $\eta=(\eta_1,..., \eta_\ell)$. We denote by $c_{xx+1}(\eta)$ the exchange rate for the pair of sites $(x, x+1)$, i.e., if $\eta_x=1$ (0) and $\eta_{x+1}=0$ (1), then $c_{xx+1}(\eta)$ is the rate at which the particle at x $(x+1)$ jumps to the empty site at $x+1$ (x).

In equilibrium the distribution of particles is determined by the Gibbs measure $\mu_{eq} \sim \exp[-\beta H]$ where β is the inverse temperature and $H(\eta)$ is the energy function. The rates c_{xx+1} satisfy the condition of detailed balance with respect to μ_{eq} whenever

$$e^{-\beta H(\eta)} c_{xx+1}(\eta) = e^{-\beta H(\eta^{xx+1})} c_{xx+1}(\eta^{xx+1}) \tag{3.1}$$

Here η^{xx+1} denotes the configuration η with the occupations at $x, x+1$ interchanged. A simple example of $H(\eta)$ is the energy of a system with nearest neighbor interactions

$$H(\eta) = -J \sum_{x=1}^{\ell} \eta_x \eta_{x+1} \tag{3.2}$$

but such an explicit form is not used here. Rates satisfying (3.1) lead the system to equilibrium. In order to drive the lattice gas by a uniform external force field of strength F we add to $H(\eta)$ the linear term $-\sum_x Fx\eta_x$. While this term does not respect the boundary conditions, energy differences always do. Therefore the rates for a uniform drive are assumed to satisfy

$$c^F_{xx+1}(\eta) = c^F_{xx+1}(\eta^{xx+1}) \exp[-\beta(H(\eta^{xx+1}) - H(\eta) - F(\eta_x - \eta_{x+1}))] \tag{3.3}$$

We have called this condition local detailed balance,[11, 13] because locally a particle feels a linear potential. It is only through the periodic boundary condition, that the steady state becomes a nonequilibrium one. For a closed system the particles would pile up at the right (for $F>0$).

To state the fluctuation theorem we merely have to apply the results of Section 2. The logarithmic ratio in (2.5) is

$$-\beta(H(\eta^{xx+1}) - H(\eta)) + \beta F(\eta_x - \eta_{x+1}) \tag{3.4}$$

with η the configuration before and η^{xx+1} the configuration after the jump. The energy difference summed over the jumps is telescoping and yields a pure surface term as in (2.36) Therefore the relevant part of the action functional is given by

$$W(t) = \beta F \int_0^t \sum_{x=1}^{\ell} J_{x,x+1}(s)\, ds \tag{3.5}$$

$J_{x,x+1}(s)$ is the actual particle current across the bond $(x, x+1)$, i.e., for a given history $\{\eta(s), 0 \leqslant s \leqslant t\}$, $J_{x,x+1}(s)$ is a sequence of δ-functions located at the times of jump through the bond $(x, x+1)$ with weight $+1$ (-1) if the jump is from x $(x+1)$ to $x+1$ (x). The action functional is βF times the total current integrated over the time span t, which equals the signed number of all jumps up to time t.

The entropy production $R(\mu)$ can be read off from (2.30). If we set $\mu(\eta) = g(\eta)\, Z^{-1} e^{-\beta H(\eta)}$ and denote by $\langle \cdot \rangle_{eq}$ the average over $Z^{-1} e^{-\beta H(\eta)}$, then

$$\begin{aligned} R(\mu) = \frac{1}{2} \sum_{x=1}^{\ell} \langle c^F_{xx+1}(\eta)[e^{-\beta F(\eta_x - \eta_{x+1})} g(\eta^{xx+1}) - g(\eta)] \\ \times [\log e^{-\beta F(\eta_x - \eta_{x+1})} g(\eta^{xx+1}) - \log g(\eta)] \rangle_{eq} \end{aligned} \tag{3.6}$$

We clearly have $R(\mu) > 0$ unless $F = 0$.[32] The identification of $R(\mu)$ as entropy production is further supported by taking μ to be a state of local equilibrium. In the limit of slow density variations R coincides with the phenomenological entropy production based on the nonlinear diffusion equation as a macroscopic equation for the density.[32]

We emphasize that the action functional is equal to the total current only if the condition (3.3) of local detailed balance holds. To illustrate this point we consider nearest and next nearest neighbor jumps. Local detailed balance for next nearest neighbor jumps means

$$c^F_{xx+2}(\eta) = c^F_{xx+2}(\eta^{xx+2}) \exp[-\beta(H(\eta^{xx+2}) - H(\eta) - 2F(\eta_x - \eta_{x+2}))] \tag{3.7}$$

As before, $W(t)$ constructed according to the rule (2.5) equals the total current. There are many ways to violate (3.7). Just as an example, if in (3.7) we replace $2F$ by αF, then the action functional is $W(t) = \beta F \int dt \times \sum_x (J_{x,x+1}(t) + \alpha J_{x,x+2}(t))$ which equals the particle current only if $\alpha = 2$. Thus, while the fluctuation theorem holds in great generality, the specific physical interpretation is linked to local detailed balance. We could also in this case consider the generator as the sum of two generators, one for nearest and one for next nearest neighbor jumps, and have more general fluctuation theorems as in Section 2.3. The physical interpretation would however remain obscure.

4. BOUNDARY-DRIVEN LATTICE GASES

We use the same set-up as in the previous section. Only now the box consisting of sites $\{1, 2,..., \ell\}$ is not periodic, and there is no driving force. Instead particles are injected and removed at the left $(x=1)$ and right $(x=\ell)$ boundary. The generator, L, for the full dynamics is naturally decomposed as a sum of three pieces,

$$L = L_b + L_1 + L_\ell \tag{4.1}$$

L_b is the dynamics inside the box satisfying detailed balance with respect to $e^{-\beta H}$, cf. (3.1) with $x = 1,..., \ell - 1$. L_b conserves the number of particles. L_i models the particle source/sink at the right and left boundaries and has the form

$$L_i f(\eta) = c_i(\eta)[f(\eta^i) - f(\eta)], \qquad i = 1, \ell \tag{4.2}$$

$c_i(\eta)$ is the rate for the transition η_i to $1 - \eta_i$ while the remaining configuration is untouched. The configuration after this transition is denoted by η^i. As before we will require the rate $c_i(\eta)$ to satisfy local detailed balance in the form[13]

$$c_i(\eta) = c_i(\eta^i) \exp[-\beta(H(\eta^i) - H(\eta)) + \beta\gamma_i(1 - 2\eta_i)] \tag{4.3}$$

where local refers now to the boundary points which are coupled to reservoirs with chemical potentials γ_1 and γ_ℓ. Thus if we consider the dynamics generated only by $L_b + L_i$, then the dynamics will satisfy detailed balance with respect to the stationary distribution $Z^{-1} \exp[-\beta H(\eta) + \beta\gamma_i N(\eta)]$ (achieved in the long time limit) with $N(\eta) = \sum_{x=1}^{\ell} \eta_x$ the number of particles. The same will be true for the full L in (4.1), if $\gamma_1 = \gamma_\ell$. If on the other hand $\gamma_1 \neq \gamma_\ell$, then the sources are unbalanced and there will be in the steady state a net flux of particles through the system going from the reservoir with high to the reservoir with low chemical potential, i.e., from left to right if $\gamma_1 > \gamma_\ell$.

We now apply the results of Section 2 (with $\lambda_j = \lambda$). The logarithmic ratio (2.5) reads

$$-\beta(H(\eta^{xx+1}) - H(\eta)) - \beta \sum_{i=1,\ell} (H(\eta^i) - H(\eta)) + \sum_{i=1,\ell} \beta\gamma_i(N(\eta^i) - N(\eta)) \tag{4.4}$$

where η is the configuration before and η^{xx+1}, resp. η^i, the configuration after the jump. Again the energy differences sum to a surface term. Let $J_i(t)$ be the boundary current, $i=1, \ell$. It is a sequence of delta functions located at times when η_i jumps to $1-\eta_i$ with weight 1 (-1) for the transition from 0 to 1 (from 1 to 0). When the boundary current J_1 is positive there is a net flux of particles from the ith reservoir into the system. Then

$$W(t)=\beta\int_0^t (\gamma_1 J_1(s)+\gamma_\ell J_\ell(s))\,ds \tag{4.5}$$

Because of particle conservation there are many equivalent expressions for $W(t)$ in the sense that they differ only by surface terms and have therefore the same rate function. Let us denote by $N_{[x,y]}(t)$, $x\leqslant y$, the number of particles in the interval $[x, y]$ at time t. By the conservation law for the number of particles we have

$$N_{[x,y]}(t)-N_{[x,y]}(0)=\int_0^t (J_{x-1,x}(s)-J_{y,y+1}(s))\,ds \tag{4.6}$$

Since $N_{[x,y]}(t)\leqslant |y-x|+1$, $\int_0^t J_{x-1,x}(s)\,ds$ and $\int_0^t J_{y,y+1}(s)\,ds$ must have the same large deviations. If we include in the mass balance also the boundary currents, then we can replace in (4.5) $J_1(t)$ by any $J_{x,x+1}(t)$ and $J_\ell(t)$ by any $-J_{x,x+1}(t)$ without modifying the rate function $\hat{e}(w)$. While this gives many equivalent choices, the simplest one is perhaps

$$W_x(t)=\beta(\gamma_1-\gamma_\ell)\int_0^t J_{x,x+1}(s)\,ds=\beta(\gamma_1-\gamma_\ell)\,\tilde{W}_x(t) \tag{4.7}$$

$W_x(t)$ differs from $W(t)$ only by a surface term. The action functional $\tilde{W}_x(t)$ is the time integrated particle current across the bond $(x, x+1)$. The generating function for $\tilde{W}_x(t)$ satisfies the fluctuation theorem in the form $\tilde{e}(\lambda)=\tilde{e}(\beta(\gamma_1-\gamma_\ell)-\lambda)$, with $\tilde{e}$ independent of the choice of x.

The conservation law (4.6) can be used also in the driven lattice gas of the previous section to produce equivalent action functionals. For example $W(t)$ of (3.5) is equivalent to

$$\bar{W}(t)=\ell\beta F\int_0^t J_{x,x+1}(s)\,ds \tag{4.8}$$

However the symmetric form (3.5) seems to be more accessible to analytic computations, as can be seen from the examples described in the Appendix.

5. FLUCTUATION THEOREM FOR DIFFUSION PROCESSES

We consider next a general diffusion process $x_t \in \mathbb{R}^n$. It is specified by a drift vector $c(x)$ (a vector field on $\mathbb{R}^n$) and a positive definite diffusion matrix $a(x) > 0$ (an $n \times n$ matrix valued function on $\mathbb{R}^n$). For later convenience we write the generator as

$$L = \tfrac{1}{2}\nabla \cdot (a\nabla) + c \cdot \nabla \tag{5.1}$$

with ∇ representing differentiation with respect to x. x_t is the solution of the stochastic differential equation

$$dx_t = (c + \tfrac{1}{2}(\nabla \cdot a))(x_t)\, dt + \sqrt{a}(x_t)\, db(t) \tag{5.2}$$

with $db(t)/dt$ standard white noise. We use Ito's convention for stochastic differentials.

To construct an action functional for (5.2) such that it satisfies the fluctuation theorem, we follow the same strategy as for Markov jump processes: We add to the generator L an operator linear in λ such that $L_\lambda = L^*_{1-\lambda}$. More systematically, we consider the process x^R_t defined as time-reversal of x_t and their Radon–Nikodym derivative, compare with (2.21). By either argument the action functional $W(t)$ for diffusion processes turns out to be given by

$$W(t) = 2\int_0^t (a^{-1}c(x_s)) \cdot dx_s + \int_0^t a\nabla(a^{-1}c)(x_s)\, ds \tag{5.3}$$

We may think of $W(t)$ as the limit of the symmetrized (Stratonovich) approximation,

$$W(t) = \lim_{\varepsilon \to 0} \varepsilon \sum_{j=0}^{t/\varepsilon} (a^{-1}c(x_{(j+1)\varepsilon}) + a^{-1}c(x_{j\varepsilon})) \cdot (x_{(j+1)\varepsilon} - x_{j\varepsilon}) \tag{5.4}$$

If a is diagonal and independent of x, then $a_{ii}/2$ corresponds physically to the temperature β^{-1}. In this case $W(t)/\beta$ equals the work done by the force c on the system during the time span t, in accordance with the physical meaning of the action functional for lattice gases.

As before, we expect that the equilibrium forces do not show up in the large deviations of $W(t)$. For diffusion processes detailed balance with respect to $e^{-U(x)}$ means that

$$c = -\tfrac{1}{2}a\nabla U \tag{5.5}$$

Inserting into (5.3), by Ito's lemma,[33]

$$-\int_0^t \nabla U(x_s)\cdot dx_s - \tfrac{1}{2}\int_0^t a\nabla\nabla U(x_s)\, ds = U(x_0) - U(x_t) \tag{5.6}$$

which is indeed a pure surface term. Only driving forces make a contribution to $W(t)$ proportional to t.

Since $a > 0$ by assumption, we are in the uniformly elliptic case, where the transition probability, e^{Lt}, has a density and $(e^{Lt})_{x,x'} > 0$ for $t > 0$. We require that the drift c is sufficiently confining. Then x_t has a unique stationary measure with density $\mu_s(x) > 0$, satisfying $\mu_s e^{Lt} = \mu_s$.

Let us denote by $\langle \cdot \rangle$ expectation with respect to the stationary process x_t (starting in the invariant measure μ_s). Then

$$\lim_{t\to\infty} -\frac{1}{t}\log\langle e^{-\lambda W(t)}\rangle = e(\lambda) \tag{5.7}$$

and the fluctuation theorem

$$e(\lambda) = e(1-\lambda) \tag{5.8}$$

should hold.

To prove (5.8), as for jump processes, we first define the function $g(x, t) = \mathbb{E}_x(\exp[-\lambda W(t)])$ as the expectation with respect to the stationary diffusion process x_t conditioned to start at x. Then $g(x, t)$ satisfies

$$\frac{\partial}{\partial t} g(x, t) = L_\lambda g(x, t) \tag{5.9}$$

with

$$L_\lambda = \tfrac{1}{2}\nabla\cdot(a\cdot\nabla) + c\cdot\nabla - 2\lambda c\cdot\nabla - \lambda(\nabla\cdot c) - 2\lambda(1-\lambda)\, c\cdot a^{-1}c \tag{5.10}$$

where the last two terms act as multiplication operators. We conclude that

$$\langle e^{-\lambda W(t)}\rangle = \iint \mu_s(x)(e^{L_\lambda t})_{x,x'}\, dx\, dx' \tag{5.11}$$

On the other hand, from (5.10),

$$L_\lambda^* = L_{1-\lambda} \tag{5.12}$$

which implies (5.8) by Perron–Frobenius.

Equations (5.10) and (5.11) can also be viewed as an application of the Cameron–Martin–Girsanov formula.[33] The path measure of the process generated by L_λ has a density relative to the path measure of the process generated by L. We write this density as $\exp[R]$. Then, according to ref. 33, Section 6.4,

$$\begin{aligned}
R &= -2\lambda \int a^{-1}c(x_t)\cdot dx_t + 2\lambda \int (a^{-1}c)\cdot(c+\tfrac{1}{2}\nabla a)(x_t)\,dt \\
&\quad - 2\lambda^2 \int c\cdot a^{-1}c(x_t)\,dt - 2\lambda(1-\lambda)\int c\cdot a^{-1}c(x_t)\,dt - \lambda\int \nabla\cdot c(x_t)\,dt \\
&= -2\lambda \int a^{-1}c(x_t)\cdot dx_t - \lambda \int a\nabla(a^{-1}c)(x_t)\,dt \\
&= -\lambda W(t) \qquad (5.13)
\end{aligned}$$

where we used the identity

$$a(\alpha)\frac{d}{d\alpha}a^{-1}(\alpha) = -\left(\frac{d}{d\alpha}a(\alpha)\right)a^{-1}(\alpha) \qquad (5.14)$$

valid for a parameter-dependent matrix $a(\alpha)$.

From the argument of Section 2, we conclude that $W(t)/t$ has large deviations as stated in (2.11) with a rate function satisfying the symmetry (2.10).

6. PARTICLE UNDER MECHANICAL AND THERMAL DRIVE

In the stochastic differential equations of Section 5 the diffusion acts everywhere on the configuration space $\mathbb{R}^n$. This is appropriate for strongly overdamped stochastic systems such as those considered in time-dependent Ginzburg–Landau theories. In many physical situations, however, stochasticity is assumed to act only in velocity space. Then the matrix a in (5.1) has zero eigenvalues and, the results of Section 5 are not directly applicable. To understand the required modifications we consider, as an example, a mechanical particle of mass m with position x in the d-dimensional torus T^d and velocity v in $\mathbb{R}^d$ subject to noise and friction. The particle has the mechanical energy $H = \frac{1}{2}mv^2 + U(x)$, where U is some periodic potential. In equilibrium the stationary distribution is given by $\exp[-\beta H(x, v)]$ with β^{-1} the temperature. We envision two mechanisms to drive the system out of equilibrium: (i) There is a mechanical driving force $F(x)$. The standard example is a constant electric field E, i.e.,

$F(x) = E$. If F has a periodic potential part, it may be added to U but as we will see this has no effect on the large deviation of the action functional. (ii) There is a nonuniform temperature $\beta(x)^{-1}$. The case of a periodic variation in β (and constant friction) has been studied recently in the context of Brownian motors[34, 35] with the at first sight surprising result that the steady state maintains a nonzero current. The dynamics of the particle is governed by the stochastic differential equation

$$m \frac{d^2}{dt^2} x_t = -\nabla U(x_t) + F(x_t) - m\gamma(x_t)\, v_t + (2m\gamma(x_t)/\beta(x_t))^{1/2}\, \xi(t) \tag{6.1}$$

Here $\gamma(x) > 0$ is the friction coefficient, $\beta(x) > 0$ the inverse temperature and $\xi(t)$ standard white noise. The generator of the corresponding Fokker–Planck equation reads

$$L = v \cdot \nabla_x + \left(-\frac{1}{m} \nabla U + \frac{1}{m} F - \gamma(x)\, v \right) \cdot \nabla_v + (\gamma(x)/m\beta(x))\, \nabla_v^2 \tag{6.2}$$

Using either a modified generator or time-reversal the action functional for the fluctuation theorem turns outer to be

$$W(t) = \int_0^t \beta(x_s)\, F(x_s) \cdot dx_s + \int_0^t H(x_s, v_s)\, d\beta(x_s) \tag{6.3}$$

for x_s, v_s a solution to (6.1). The first term is $\beta \times$(work done by F) and the second term represents $\beta \times$(work due to thermal gradients) integrated along the trajectory x_s, $0 \leqslant s \leqslant t$. If $F = -\nabla V$, then up to a surface term $\int_0^t \beta(x_s)\, F(x_s) \cdot dx_s = \int_0^t V(x_s)\, d\beta(x_s)$ which, in (6.3), should be added to the energy $H(x, v)$.

Let $\langle \cdot \rangle$ denote the stationary average for (6.1). Then

$$\lim_{t \to \infty} -\frac{1}{t} \log \langle e^{-\lambda W(t)} \rangle = e(\lambda) \tag{6.4}$$

and $-e(\lambda)$ is the maximal eigenvalue of

$$L_\lambda = L - \lambda(\beta(x)\, F(x) \cdot v + H(x, v)\, v \cdot \nabla_x \beta(x)) \tag{6.5}$$

the last two terms being considered as multiplication operators. The corresponding maximal eigenvector is denoted by $f_\lambda(x, v) > 0$. Let R be the velocity reversal operator, $Rf(x, v) = f(x, -v)$. Then

$$Re^{-\beta(x) H} L_\lambda e^{\beta(x) H} R^{-1} Re^{-\beta(x) H} f_\lambda(x, v) = -e(\lambda)\, Re^{-\beta(x) H} f_\lambda(x, v) \tag{6.6}$$

By a straightforward computation

$$Re^{-\beta(x)H}L_\lambda e^{\beta(x)H}R^{-1}=L^*_{1-\lambda} \tag{6.7}$$

Therefore

$$L^*_{1-\lambda}Re^{-\beta H}f_\lambda=-e(\lambda)\,Re^{-\beta H}f_\lambda=-e(1-\lambda)\,Re^{-\beta H}f_\lambda \tag{6.8}$$

by Perron–Frobenius, since $Re^{-\beta(x)H}f_\lambda(x,v)>0$. We conclude that the fluctuation theorem holds and $e(\lambda)=e(1-\lambda)$.

As for lattice gases, we expect that the action functional $W(t)$ is linked to the entropy balance. To spell out the details we consider the time-dependent probability density $\rho_t(x,v)$ of (x_t,v_t). It satisfies $(\partial/\partial t)\,\rho_t=L^*\rho_t$. As usual the Gibbs entropy is given by $S_G(\rho_t)=-\int\rho_t\log\rho_t\,dx\,dv$ and changes in time as

$$\frac{d}{dt}S_G(\rho_t)=\sigma(\rho_t)+j_s(\rho_t) \tag{6.9}$$

The first term

$$\sigma(\rho_t)=\int\gamma\frac{1}{\rho_t}\Big[\sqrt{m\beta}\;v\rho_t+\frac{1}{\sqrt{m\beta}}\nabla_v\rho_t\Big]^2\,dx\,dv \tag{6.10}$$

is positive definite. We identify it as the entropy production in the system. The remainder reads

$$j_s(\rho_t)=\frac{d}{dt}\int\rho_t\beta H\,dx\,dv-\int\rho_t\beta F\cdot v\,dx\,dv-\int\rho_t Hv\cdot\nabla_x\beta\,dx\,dv \tag{6.11}$$

We regard it as the entropy flow from the system to the mechanical and heat reservoirs.

According to the rules used before, we construct $W(t)$ by integrating the linear functional defining the entropy flow along a stochastic trajectory. Using (6.11) this yields

$$\int_0^t\Big\{-\frac{d}{ds}\beta H(x_s,v_s)+\beta F(x_s)\cdot v_s+H(x_s,v_s)\,v_s\cdot\nabla\beta(x_s)\Big\}\,ds \tag{6.12}$$

Since the first summand is a surface term, we recover (6.3).

7. STOCHASTIC AND THERMOSTATTING HEAT RESERVOIRS

Physical systems with stationary flows of heat, momentum, etc., are usually modeled by coupling the system to thermal reservoirs represented by stochastic forces acting near the boundaries.[1–7] Away from the boundary the time evolution obeys the same laws as an isolated system. To be specific, let us consider one such model of heat flow through a classical system of N particles of mass m with position q_j, momentum p_j, $j = 1,..., N$, inside the slab $\Lambda = [-\ell - a, \ell + a] \times [0, \ell]^2$. Let the boxes $\Lambda_- = [-\ell - a, -\ell] \times [0, \ell]^2$, $\Lambda_+ = [\ell, \ell + a] \times [0, \ell]^2$ be the left and right boundary zones. The indicator functions of these sets are denoted by $\chi_\pm$. The particles interact through a short range pair potential V and are confined to the slab by the wall potential V_w. The equations of motion are

$$\begin{aligned} \frac{d}{dt} q_j &= \frac{1}{m} p_j \\ \frac{d}{dt} p_j &= F_j + F_w(q_j) + \sum_{\delta = \pm} \chi_\delta(q_j)(-\gamma p_j + (2m\gamma/\beta_\delta)^{1/2}\, \xi_j(t)) \end{aligned} \tag{7.1}$$

where $q_j \in \Lambda$, $p_j \in \mathbb{R}^3$, F_j is the force acting on the jth particle, $F_j = -\sum_{i \neq j=1}^N \nabla V(q_j - q_i)$, $F_w = -\nabla V_w$ is the force from the wall. $\{\xi_j(t)\}$ are a collection of independent white noises. The friction and the stochastic forces operate only when the particle is in the boundary layers acting there like thermal reservoirs at inverse temperature β_δ. If $\beta_+ = \beta = \beta_-$, then the steady state is $Z^{-1} \exp[-\beta H]$, $H = \sum_{j=1}^N ((1/2m)\, p_j^2 + V_w(q_j)) + \frac{1}{2} \sum_{i,j=1}^N V(q_i - q_j)$. If $\beta_+ \neq \beta_-$, there is, in the steady state, a constant heat flux from the higher to the lower temperature reservoir. (This is a variation of the model used to describe heat flow in a crystal where the Langevin forces act on the particles in the end layers[3, 10]). Note that, since phase space is unbounded, even the existence of the SNS requires considerable effort, compare with ref. 36.

We now follow the method explained in Section 6. For the similarity transformation we use

$$\exp\left[-\sum_{j=1}^N \beta(q_j) \left\{\frac{1}{2m} p_j^2 + V_w(q_j) + \frac{1}{2} \sum_{i=1}^N V(q_i - q_j)\right\}\right] \tag{7.2}$$

The local inverse temperature $\beta(q) = \beta_\delta$ for $q \in \Lambda_\delta$ and $\beta(q)$ interpolates smoothly otherwise. For each choice of $\beta(q)$ we obtain a distinct action functional $W(t)$. By local conservation of energy, these differ only through boundary terms. For later purposes it is convenient to choose a step interpolation as $\beta(q) = \beta_-$ for $q_1 < 0$ and $\beta(q) = \beta_+$ for $q_1 \geqslant 0$. Then the action functional is

$$W(t) = (\beta_+ - \beta_-) \int_0^t ds\, J_0(s) \tag{7.3}$$

with J_0 the energy current through the plane $\{q_1=0\}$,

$$J_0=\sum_{j=1}^{N}\frac{1}{m}p_j\left\{\frac{1}{2m}p_j^2+\frac{1}{2}\sum_{i\neq j=1}^{N}V(q_i-q_j)\right\}\delta(q_{j1})$$
$$+\frac{1}{2}\sum_{i,j=1}^{N}\frac{1}{2m}[((p_j+p_i)\cdot F(q_j-q_i))(q_j-q_i)]$$
$$\times\int_0^1 d\lambda\,\delta(\lambda q_{j1}+(1-\lambda)\,q_{i1}) \tag{7.4}$$

where q_{j1} is the x-component of the jth particle position vector.

As an advantageous numerical alternative deterministic thermostatting reservoirs have been developed.(16) We mention here a model for thermal reservoirs originally proposed by refs. 37 and 38 and further studied in refs. 22 and 39. The equations of motions are Newtonian in the bulk. At the boundary layers Λ_+ and Λ_- additional friction terms are added which are constructed in such a way that the total kinetic energy of the particles in Λ_+, resp. Λ_-, is kept constant. The particular boundary temperature is then fixed through the initial conditions. The equations of motion read

$$\frac{d}{dt}q_j=\frac{1}{m}p_j$$
$$\frac{d}{dt}p_j=F_j+F_w(q_j)+\sum_{\delta=\pm}\chi_\delta(q_j)\,\alpha_\delta p_j \tag{7.5}$$

The "friction" coefficients α_+, α_- are given by

$$\alpha_\delta=\left[\sum_{j=1}^{N}\chi_\delta(q_j)\,p_j^2\right]^{-1}\sum_{j=1}^{N}\left\{\frac{1}{m}p_j\cdot\nabla\chi_\delta(q_j)\frac{1}{2m}p_j^2\right.$$
$$\left.+\frac{1}{m}\chi_\delta(q_j)(F_j+F_w)\cdot p_j\right\},\qquad \delta=\pm 1 \tag{7.6}$$

For thermostatted systems the action functional is the phase space contraction integrated along a trajectory of the dynamical system, which in our case equals

$$3N\sigma=3N_+\alpha_++3N_-\alpha_-+\mathcal{O}(1) \tag{7.7}$$

with $N_\pm$ the number of particles in $\Lambda_\pm$. The fluctuation theorem holds provided the dynamics is sufficiently hyperbolic for the SNS to be described by an SRB measure. This is not expected to be true in general.

It is then an assumption, embodied in the chaotic hypothesis, that the fluctuation theorem remain valid for realistic physical systems. The numerical simulation in ref. 40 gives strong support for the validity of the fluctuation theorem in one such model. There the two heat reservoirs are linked through an anharmonic chain, rather than a fluid. Further numerical support comes from a study of shear flow for hard disks in a box[41] where momentum but no energy is transferred to the system at opposite boundaries. The fluctuation theorem is well verified,[42] although it seems to be difficult to get beyond the quadratic approximation.

The action functional defined with the phase space contraction (7.7) appears to be rather different than the one of (7.3). However the conservation law can be used again to transform a boundary flux to an interior flux. Let H_1 be the total energy of particles in the left half box $[-a-\ell, 0] \times [0, \ell]^2$. We have

$$\begin{aligned} \frac{d}{dt} H_1(t) &= -J_0(t) - \left[\sum_{j=1}^{N} \chi_-(q_j)\, p_j^2 \right] \alpha_- \\ &= -J_0(t) - \frac{1}{\beta_-} 3N_- \alpha_- \end{aligned} \tag{7.8}$$

since $\sum_{j=1}^{N} \chi_-(q_j)\, p_j^2$ is a constant of motion and initially fixed to $3N_-/\beta_-$. We conclude that

$$\int_0^t ds\, 3N\sigma(s) = -H(t) + H(0) + (\beta_+ - \beta_-) \int_0^t ds\, J_0(s) \tag{7.9}$$

Up to a boundary term the phase space contraction is just the energy flux across the plane $\{q_1 = 0\}$, the same quantity which has been obtained for stochastic reservoirs, compare with (4.7).

It is tempting to assume that for stochastic and thermostatted boundaries, the rate functions of the fluctuation theorem are identical. This requires however a very strong form of equivalence of ensembles, in the sense that in the interior of the system even the large deviations of the current are in the steady state independent of the mechanism, by which the boundaries are cooled/heated. This appears to be the case for the model studied in ref. 40. On the other hand recent simulations ref. 43 suggest that this equivalence fails for the model described in refs. 41–42.

8. GREEN–KUBO FORMULA, ONSAGER RECIPROCITY

As observed in ref. 24 the fluctuation theorem yields the Green–Kubo formula at equilibrium, see also refs. 39, 44, and 45 for discussions of linear response for chaotic thermostatted systems. We follow the derivation in ref. 24 for the stochastic models considered here. For the sake of concreteness we consider a k-species lattice gas on a d-dimensional lattice. The lattice gas is driven by an external field, cf. Section 3, and satisfies local detailed balance. For each species the driving field is a d-component vector. It is convenient to regard the driving field as the m-component vector $\vec{F}=(F_1,..., F_m)$, $m=d\times k$. There are m different types of jump, labeled by the species and lattice directions. Therefore the generator of the lattice gas is the sum

$$L=\sum_{j=1}^{m} L^{(j)} \tag{8.1}$$

where each $L^{(j)}$ satisfies the conditions of Section 2.3, compare with (2.23). To each $L^{(j)}$ there is associated the current $J_j(t)$.

We define

$$e(\vec{F}; \lambda_1,..., \lambda_m)=\lim_{t\to\infty} -\frac{1}{t}\log\left\langle \exp\left[-\beta\sum_{j=1}^{m}\int_0^t \lambda_j J_j(s)\,ds\right]\right\rangle_{\vec{F}} \tag{8.2}$$

where we have, in the action, absorbed F_j into the λ_j's and indicated explicitly the dependence of the stationary state and the dynamics on $\vec{F}$. The fluctuation theorem now takes the form

$$e(\vec{F}; \lambda_1,..., \lambda_m)=e(\vec{F}; F_1-\lambda_1,..., F_m-\lambda_m) \tag{8.3}$$

We have $\langle J_j(t)\rangle_0=0$, since $\vec{F}=0$ corresponds to equilibrium. Differentiating (8.2) at $\lambda=0$, we obtain

$$\frac{\partial}{\partial\lambda_j} e(\vec{F}; 0)=\beta\langle F_j(t)\rangle_{\vec{F}} \tag{8.4}$$

the average jth current.

Now the linear response in the average current to a small driving field is given by

$$\frac{\partial}{\partial F_j}\frac{1}{\beta}\frac{\partial}{\partial\lambda_i} e(0; 0)=\sigma_{ij} \tag{8.5}$$

Differentiating (8.3) with respect to $\vec{F}$ and $\vec{\lambda}$ at $\vec{F}=\vec{\lambda}=0$ we conclude that

$$\beta\sigma_{ij} = -\beta\sigma_{ij} + \beta^2\hat{\sigma}_{ij} \tag{8.6}$$

and therefore

$$\sigma_{ij} = \tfrac{1}{2}\beta\hat{\sigma}_{ij} \tag{8.7}$$

where

$$\hat{\sigma}_{ij} = -\frac{1}{\beta^2}\frac{\partial}{\partial\lambda_i}\frac{\partial}{\partial\lambda_j}e(0;0) = \int_{-\infty}^{\infty} dt \langle J_i(t)\, J_j(0)\rangle_0 \tag{8.8}$$

This is the standard Einstein–Green–Kubo relation between linear response in the current and the time-integrated current-current correlation, which satisfies the Onsager relation $\hat{\sigma}_{ij} = \hat{\sigma}_{ji}$.

We remark that the derivation in ref. 24 and here of the Onsager symmetry (8.7) differs considerably from the standard computation based on linear response. We refer to ref. 32 for an exposition of the latter. There one differentiates, at finite volume, the steady state current $\langle j_{\vec{F}}\rangle_{\vec{F}}$ at $\vec{F}=0$, which by definition equals σ_{ij}. This gives two terms: one from differentiating the current function $j_{\vec{F}}$ and one from differentiating the steady state $\langle\cdot\rangle_{\vec{F}}$. Their sum is then (8.8), because $\langle J_i(t)\, J_j(0)\rangle_0$ has a δ-peak at $t=0$, whose weight is the first term, and a smooth piece, whose time integral equals the second term.

9. CONCLUSIONS

Within the framework of stochastic dynamics the fluctuation theorem holds in great generality. For finite state spaces, like stochastic lattice gases, we have given a proof. For the nondegenerate diffusion processes of Section 5 we could provide sufficient conditions on the driving forces and the diffusion matrix which ensure the validity of (5.8). The situation is more delicate for degenerate diffusion processes, in particular for deterministic bulk systems driven by stochastic boundaries. Since the noise vanishes on large parts of the phase space, even the convergence to the steady state is not obvious. To our knowledge there are only a few rigorous investigations of SNS of this type.[3–5, 7, 10] In refs. 5 and 7 classical point particles are considered. At a collision with the wall they are reflected with a Maxwellian distribution at the local temperature of the wall. Existence and uniqueness of the invariant measure is established provided the forces are repulsive and have a range of the size of the box. This latter property prevents pockets in phase space which never see the wall. In ref. 3 explicit

SNS measures are obtained for a harmonic chain coupled to stochastic reservoirs specified by Langevin forces at the ends of the chain. While in ref. 10 an anharmonic chain is coupled to a free field leading to similar boundary conditions. For this system existence and uniqueness of the invariant measure is established in under a Hörmander condition. We expect that for these systems the GC fluctuation theorem can be proved. Thermostatted boundaries are beyond our mathematical abilities at the present time.

Viewed from the physics point of view the fluctuation theorem is a consequence of time-reversal. For a given stationary stochastic process one considers the density of the path measure of the time-reversed process relative to the original one. The logarithm of this density is the action functional which satisfies the fluctuation theorem. At this level of generality it is somewhat unexpected that in concrete models, which satisfy the condition of local detailed balance, the action functional has a direct physical interpretation. By separating the rate of change of Gibbs entropy into a production and flow term we identify the observable whose average defines the entropy flow. The action functional is then this observable integrated along a stochastic trajectory. Generically it is the current, multiplied by the driving force, corresponding to the conserved field. The GC fluctuation theorem is then a symmetry property for the large deviations in the current. Such large deviations are not readily observed in physical systems, because they refer to exponentially small probabilities. Also in numerical experiments special efforts are needed.[23, 42, 28] Even worse, the action functional is extensive and the probabilities are exponentially small also in N. Thus it would be desirable to have a local quantity satisfying (at least approximately) the fluctuation theorem but this does not appear to be the case for the systems investigated numerically thus far.[42, 43]

There is one prediction of the fluctuation theorem which is worthwhile to emphasize. We use the convention of the previous section where the symmetry reads $e(\lambda)=e(F-\lambda)$, i.e., e is even relative to $\lambda=F/2$, and we may expand as $e(\lambda)=e_0+\frac{1}{2}e_2\lambda(F-\lambda)+\cdots$. Assuming that in the interval $-\delta<\lambda<F+\delta$ with some small $\delta>0$ the quadratic approximation is valid, then by differentiating at $\lambda=0$ we obtain $\langle j\rangle_F=e_2F/2$, where $e_2=\int_{-\infty}^{\infty}dt(\langle J(t)\,J(0)\rangle_F-\langle J(t)\rangle_F\langle J(0)\rangle_F)$ will generally depend on F. The nonlinear response in the current is then linked to the time-integral over the current-current correlation. The knowledge of $e(\lambda)$ thus provides information about, the range of validity of the Einstein relation and linear response theory, see also ref. 23. We should note, however, that even when $e(\lambda)$ looks approximately Gaussian the "best fit" Gaussian may not correspond to the curvature at the top, i.e., we might have $e(\lambda)=e_0+\frac{1}{2}\bar{e}_2\lambda(F-\lambda)$ with $\bar{e}_2\neq e_2$.[54]

APPENDIX: LARGE DEVIATION FUNCTION FOR THE ASYMMETRIC EXCLUSION PROCESS

To determine the large deviation function in any of the models we have discussed is like computing a free energy in equilibrium statistical mechanics. Therefore it is not surprising that even in the stochastic framework explicit results are available only for a few models. We discuss here the large deviation function for the asymmetric simple exclusion process in one dimension where particles random walk with an average drift and interact through the constraint that there can be at most one particle per lattice site. As a warm up we consider this model without constraint, i.e., independent biased random walks on a lattice. It would be of interest to have also an example for the boundary driven lattice gases of Section 4. Besides independent particles, the simplest model is symmetric exclusion in the bulk and boundary reservoirs as discussed in Section 4. Its large deviation function, for $\tilde{W}_x(t)$, satisfies then $e(\lambda) = e(\beta(\gamma_1 - \gamma_\ell) - \lambda)$, cf. (4.7). Thus in the limit $\gamma_1 \to \infty$, i.e., at the left boundary particles are only injected, the GC fluctuation theorem predicts $\lim_{t\to\infty}(-1/t)\log \mathrm{Prob}(\{\tilde{W}_x < at\}) = 0$ for $a < 0$.

We first consider N independent particles, i.e., no exclusion, on a ring of size ℓ driven by a uniform force of strength F. Since particles are independent, it suffices to study the one-particle problem. For N particles we merely have to multiply by N at the end. It is convenient to set $\beta F/2 = E$. The generator for the dynamics reads

$$L_E f(x) = (e^E/2\cosh E)\, f(x+1) + (e^{-E}/2\cosh E)\, f(x-1) - f(x) \qquad \text{(A.1)}$$

The elementary jump time is normalized such that for $E \to \infty$ the jump rate to the right is one. $W(t)$ is now the total number of signed jumps up to time t. To compute the generating function $\langle e^{-\lambda W(t)}\rangle$ for large t we need the maximal eigenvalue of

$$L_{E,\lambda} f(x) = (e^{E-\lambda}/2\cosh E)\, f(x+1) + (e^{-E+\lambda}/2\cosh E)\, f(x-1) - f(x) \qquad \text{(A.2)}$$

Clearly $f = 1$ solves $L_{F,\lambda} f = -e(\lambda) f$ with

$$e(\lambda) = 1 - (\cosh(E-\lambda)/\cosh E) \qquad \text{(A.3)}$$

Its Legendre transform is given by

$$\hat{e}(w) = -Ew + 1 + w \operatorname{argsinh}(w \cosh E) - \sqrt{w^2 + (\cosh E)^{-2}} \qquad \text{(A.4)}$$

$\hat{e}$ vanishes at the average current $j = \tanh E$. On general grounds we know that $e(\lambda)$ is convex down, $e(0)=0$, $e'(0) = -j$, the average current, $e(\lambda) = e(2E-\lambda)$ and that $\hat{e}(w)$ is convex up, $\hat{e}(w) \geqslant 0$, $\hat{e}(w_0)=0$ for $w_0 = j = (1/t)\langle W(t)\rangle$, $\hat{e}(w) - \hat{e}(-w) = -2Ew$.

Note that in the limit $E \to \infty$, $\hat{e}$ degenerates to $\hat{e}(w) = 1 - w + w \log w$ for $w \geqslant 0$ and $\hat{e}(w) = \infty$ for $w < 0$, which merely reflects that left jumps have probability zero for $E = \infty$.

We expect more structure for particles interacting through hard core exclusion. This corresponds to $H(\eta)=0$ in the notation of Section 3. The jump rates with a uniform driving field are then

$$c^E_{xx+1}(\eta) = (e^E/2 \cosh E)\, \eta_x(1-\eta_{x+1}) + (e^{-E}/2 \cosh E)(1-\eta_x)\, \eta_{x+1} \tag{A.5}$$

We normalized the rates such that for $E \to \infty$ a particle jumps with unit rate to the right. (A.5) are the jump rates of the asymmetric simple exclusion process (ASEP).

The action functional $W(t)$ is the total current integrated over the time span t. Expectations, $\langle \cdot \rangle$, are taken in the stationary process at a given number of particles N, $1 \leqslant N \leqslant \ell - 1$. For the ASEP this means that at time $t=0$ all allowed configurations have equal weight. Then

$$\lim_{t \to \infty} -\frac{1}{t} \log \langle e^{-\lambda W(t)} \rangle = e_E(\lambda, \ell, N) \tag{A.6}$$

and $-e_E$ is the maximal eigenvalue of L_λ with

$$\begin{aligned} L_\lambda f(\eta) = (2 \cosh E)^{-1} \sum_{x=1}^{\ell} \{ &(e^{E-\lambda} \eta_x(1-\eta_{x+1}) + e^{-(E-\lambda)}(1-\eta_x)\, \eta_{x+1}) \\ &\times f(\eta^{xx+1}) - (e^E \eta_x(1-\eta_{x+1}) + e^{-E}(1-\eta_x)\, \eta_{x+1})\, f(\eta) \} \end{aligned} \tag{A.7}$$

The fluctuation theorem reads

$$e_E(\lambda) = e_E(2E - \lambda) \tag{A.8}$$

which can be seen also directly from (A.7).

L_λ is a linear operator with a structure which is similar to that of the hamiltonian for a quantum spin chain. To make contact with that model we rewrite L_λ in terms of the Pauli spin $\frac{1}{2}$ matrices $\vec{\sigma}_x = (\sigma^1_x, \sigma^2_x, \sigma^3_x)$

associated to every lattice site x. We identify $\eta_x = 1$ with spin up, $\sigma_x^3 = 1$, and $\eta_x = 0$ with spin down, $\sigma_x^3 = -1$. Then

$$-L_\lambda = H_\lambda = \sum_{x=1}^{\ell} \{(1/4)(1 - \sigma^3{}_x \sigma^3_{x+1}) - (e^{E-\lambda}/2 \cosh E)\, \sigma_x^+ \sigma_{x+1}^- \\ - (e^{-E+\lambda}/2 \cosh E)\, \sigma_x^- \sigma_{x+1}^+\} \tag{A.9}$$

where $\sigma_x^\pm = \frac{1}{2}(\sigma_x^1 \pm i\sigma_x^2)$ are the spin raising and lowering operators. We note that $H_\lambda^* \neq H_\lambda$, unless $E = \lambda$, and H_λ does not have directly a quantum mechanical interpretation, except through analytic continuation in λ.

To determine the ground state energy of H_λ, the only technique available is the Bethe ansatz. Fortunately, H_λ is covered by the famous paper of Sutherland and Yang.[46] However, the analysis of the nonlinear Bethe equations still require a considerable effort, which in our section of parameter space has been carried out only recently. We use the results by D. Kim.[47] He is in fact interested in the energy gap, but also gives $e_E(\lambda, \ell, N)$ to leading order in N in the limit $\ell \to \infty$, $N \to \infty$, $\rho = N/\ell$ fixed. We expect that also order 1 terms could be obtained, but such an analysis is certainly beyond the scope of our paper. In the limiting case $E \to \infty$, the Bethe equations simplify. This has been exploited recently by Derrida *et al.*[48, 49] They determine $e_\infty(\lambda, \ell, N)$ fairly explicitly and discuss the order 1 corrections to the infinite volume limit.

To leading order in N, $N/\ell = \rho$, we have

$$e_E(\lambda, \ell, N) \cong \ell \bar{e}_E(\lambda, \rho) \tag{A.10}$$

i.e., e_E is extensive. $\bar{e}_E$ has a flat piece, $\bar{e}_E(\lambda, \rho) = 0$ for $0 \leqslant \lambda \leqslant 2E$. At $\lambda = 0$, $\bar{e}_E$ has the slope $\rho(1-\rho) \tanh E$ and the asymptotics[47]

$$\bar{e}_E(\lambda, \rho) = \rho(1-\rho)(\tanh E)\, \lambda \\ - \frac{1}{20}\left(\frac{3\pi}{2}\right)^{2/3} (4\rho(1-\rho))^{4/3} (\tanh E)\, |\lambda|^{5/3} (1 + \mathcal{O}(|\lambda|^{2/3})) \tag{A.11}$$

for $\lambda < 0$. By symmetry $\bar{e}_E$ has the same asymptotics at $\lambda = 2E$. For $E \to \infty$, the right zero $e_E(\lambda, \ell, N)$ moves to ∞ and the flat piece of $e_E(\lambda, \ell, N)$ never levels off, in agreement with refs. 48 and 49.

$\bar{e}'(0)$ is just the average current which is $\rho(1-\rho) \tanh E$. $\bar{e}''(0)$ is formally the integral over the total current-current correlation and therefore the mobility of the lattice gas.[32] In our case $\bar{e}''(0) = \infty$, which reflects the fact that in the ASEP density fluctuations propagate superdiffusively.[50]

At $E=0$ (A.6) gives the large deviations in the total current at equilibrium, i.e., for the symmetric simple exclusion process. The fluctuation theorem reduces then to $e_{0,\ell,N}(\lambda)=e_{0,\ell,N}(-\lambda)$, which also follows from time reversibility. $\bar{e}_0(\lambda,\rho)$ has no longer a flat piece. For small λ, $\bar{e}_0(\lambda,\rho)=-\frac{1}{2}\lambda^2\rho(1-\rho)$. Thus the mobility $\sigma(\rho)=\rho(1-\rho)$ and the diffusion constant, which is σ divided by the compressibility, $D(\rho)=1$, independent of the density in agreement with the known bulk diffusion of the symmetric simple exclusion process. The more precise asymptotics[47] is given by

$$\bar{e}_0(\lambda,\rho)=-\tfrac{1}{2}\rho(1-\rho)\,\lambda^2-\tfrac{1}{20}2^{1/3}(2\pi)^{2/3}\,(\rho(1-\rho))^{4/3}\,|\lambda|^{8/3}\,(1+\mathcal{O}(|\lambda|^{4/3}) \tag{A.12}$$

On a formal level the "Burnett coefficients" correspond to higher derivatives of $\bar{e}_E(\lambda)$ at $\lambda=0$.[51] (A.12) shows that for the symmetric simple exclusion process the Burnett coefficients are infinite.

Just as free energies, the large deviation function may become singular in the infinite volume limit and this happens already for the first nontrivial example. To understand the origin of this behavior, it is necessary to analyze those configurations which give the main contribution to $\langle e^{-\lambda W(t)}\rangle$. If $\lambda\to\infty$ only left jumps are permitted and similarly for $\lambda\to-\infty$ only right jumps are permitted, essentially independent of the driving field E, $E>0$ for the sake of discussion. However, in the intermediate regime $0\leqslant\lambda\leqslant 2E$, the particles have two opposing instructions. Should they follow E or λ? In fact, neither. As can be inferrred from the maximal eigenvector, they just stick and form one big cluster.[52] The clustering is seen most easily at $\lambda=E$, where H_λ is the ferromagnetic, anisotropic Heisenberg model and the maximal eigenvector its ground state. At our parameters, H_λ is in the ferromagnetic phase. Since the magnetization is fixed, typical configurations condense into one large cluster. In our units the corresponding ground state energy is of order 1 independent of ℓ. In parentheses we remark that in the context of chaotic thermostatted systems configurations typical for the large deviation functional are studied in ref. 53.

It is instructive to reconsider our result from the point of view of the probability distribution of $W(t)/t$, which is obtained from the Legendre transform of $\bar{e}_E(\lambda)$. To make it well defined, we note that, for large N, $e_E(\lambda)$ smoothly interpolates between $e_E(0)=0=e_E(2E)$, such that $e'_E(0)=Nj(\rho)$, $e'_E(2E)=-Nj(\rho)$, $j(\rho)=\rho(1-\rho)\,tanhE$, and $e_E(E)=\mathcal{O}(1)$. Thereby we obtain, for $E\geqslant 0$,

$$\hat{\bar{e}}_E(w,\rho)=\begin{cases}0 & \text{for}\quad 0\leqslant w\leqslant j(\rho)\\ -2Ew & \text{for}\quad -j(\rho)\leqslant w\leqslant 0\end{cases} \tag{A.13}$$

For w slightly larger than $j(\rho)$ we have $\hat{\hat{e}}_E(w, \rho) \simeq (w - j(\rho))^{5/2}$ and $\hat{\hat{e}}_E(w, \rho) \simeq w \log w$ in the limit of large w. For $w < -j(\rho)$ the values of $\hat{\hat{e}}_E$ are fixed by the fluctuation theorem.

To understand the N-dependence of $\hat{e}_E(w, \ell, N)/\ell$, we note that for $W(t)/Nt$ to have a value larger than $j(\rho)$ we have to speed up all j particles. For it to have a value smaller than $j(\rho)$ it suffices to slow down a single particle, since the other particles pile up behind. Typical configurations consist of essentially a single cluster. The large deviation rate is $\mathcal{O}(1/N)$ which on our scale corresponds to $\hat{\hat{e}}_E = 0$. Such a mechanism works only for $(W(t)/Nt) \geqslant 0$. To have a negative total current, again all N particles are forced to move, now opposite to E. The linear decrease of $\hat{e}_E(w < N, \ell)/\ell$ with corrections of $\mathcal{O}(1/N)$ for $-j(\rho) \leqslant w \leqslant 0$ follows from the fluctuation theorem. In the limit $E \to \infty$, $\hat{e}_E(w, \ell, N) = \infty$ for $w < 0$. This just reflects the fact that the underlying process does not allow for histories with negative currents.

From our discussion of $\hat{\hat{e}}$ we conclude that the non-smooth behavior of $\bar{e}(\lambda)$ for large N comes from blocking through slow particles, which suggests that, if we go to higher spatial dimension or soften the hard core, a smooth large deviation function may be recovered. But this remains to be seen. The symmetric case, $E = 0$, warns against too fast a conjecture.

ACKNOWLEDGMENTS

We are grateful to F. Bonetto, G. Gallavotti, C. Maes, and D. Ruelle for instructive discussions. The research was supported by NSF Grant NSF-DMR 95-23266 and AFOSR Grant F49620-98-1-0207. We also acknowledge DIMACS and its supporting agencies, the NSF under contract STC-91-19999 and the NJ Commission on Science and Technology.

REFERENCES

1. P. G. Bergmann and J. L. Lebowitz, New approach to nonequlibrium processes, *Phys. Rev.* **99**:578 (1955).
2. J. L. Lebowitz and P. G. Bergmann, Irreversible Gibbsian ensembles, *Ann. Phys.* **1**:1 (1957).
3. Z. Rieder, J. L. Lebowitz, and E. Lieb, Properties of a harmonic crystal in a stationary nonequilibrium state, *J. Math. Phys.* **8**:1073 (1967).
4. A. J. O'Conner and J. L. Lebowitz, Heat conduction and sound transmission in isotopically disordered harmonic crystals, *Journ. Math. Phys.* **15**:629 (1974).
5. S. Goldstein, J. L. Lebowitz, and E. Presutti, Mechanical system with stochastic boundaries, in *Random Fields*, Vol. I, J. Fritz, J. L. Lebowitz, and D. Szász, eds. (North-Holland, Amsterdam, 1979).
6. J. L. Lebowitz, Exact results in nonequilibrium statistical mechanics: Where do we stand?, *Prog. Theor. Physics, Supplement* **64**:35 (1978).

7. S. Goldstein, C. Kipnis, and N. Ianiro, Stationary states for a mechanical system with stochastic boundary conditions, *J. Stat. Phys.* **41**:915 (1985).
8. H. Spohn, and J. L. Lebowitz, Stationary nonequilibrium states of infinite harmonic systems, *Comm. Math. Phys.* **54**:97 (1977).
9. S. Goldstein, J. L. Lebowitz, and K. Ravishankar, Approach to equilibrium in models of a system in contact with a heat bath, *J. Stat. Phys.* **43**:303 (1986).
10. J.-P. Eckmann, C.-A. Pillet, and L. Rey-Bellet, Non-equilibrium statistical mechanics of anharmonic chains coupled to two heat baths at different temperatures, preprint 1998, Texas Archive for Mathematical Physics.
11. S. Katz, J. L. Lebowitz, and H. Spohn, Stationary nonequilibrium states for stochastic lattice gas models of ionic superconductors, *J. Stat. Phys.* **34**:497 (1984).
12. H. Spohn, Long range correlations for stochastic lattice gases in a nonequilibrium steady state, *J. Phys. A* **16**:4275 (1983).
13. G. Eyink, J. L. Lebowitz, and H. Spohn, Microscopic origin of hydrodynamic behavior: Entropy production and the steady state, in *Chaos, Soviet-American Perspectives in Nonlinear Science*, Hg. D. K Campbell, (American Institute of Physics, 1990), p. 367.
14. G. Eyink, J. L. Lebowitz, and H. Spohn, Hydrodynamics, fluctuations, and large deviations outside local equilibrium, *J. Stat. Phys.* **83**:385 (1996).
15. W.G. Hoover, *Molecular Dynamics*, Lecture Notes in Physics, Vol. 258 (Springer, Heidelberg, 1986).
16. D. J. Evans and G. P. Morriss, *Statistical Mechanics of Nonequilibrium Fluids* (Academic Press, London, 1990).
17. N. J. Chernov, G. L. Eyink, J. L. Lebowitz, and Ya. G. Sinai, Derivation of Ohm's law in a determinisitic mechanical model, *Phys. Rev. Lett.* **70**:2209 (1993).
18. N. J. Chernov, G. L. Eyink, J. L. Lebowitz, and Ya. G. Sinai, Steady-state electrical conduction in the periodic Lorentz gas, *Comm. Math. Phys.* **154**:569 (1993).
19. D. Ruelle, Positivity of entropy production in nonequilibrium statistical mechanics, *J. Stat. Phys.* **85**:1 (1996).
20. G. Gallavotti, Chaotic dynamics, fluctuations, non-equilibrium ensembles, *Chaos* **8**:384 (1998).
21. D. Ruelle, New theoretical ideas in nonequilibrum statistical mechanics, Lecture Notes (Rutgers University, fall 1997).
22. G. Gallavotti, and E. G. D. Cohen, Dynamical ensembles in stationary states, *J. Stat. Phys.* **80**:931 (1995).
23. D. J. Evans, E. G. D. Cohen, and G. P. Morriss, Probability of second law violations in steady flows, *Phys. Rev. Lett.* **71**:2401 (1993).
24. G. Gallavotti, Extension of Onsager's reciprocity to large fields and the chaotic hypothesis, *Phys. Rev. Lett.* **77**:4334 (1996).
25. G. Gallavotti, New methods in nonequilibrium gases and fluids, *Proceedings of the conference "Let's face chaos through nonlinear dynamics"*, University of Maribor, 24 june–5 july 1996, M. Robnik, ed. Open Systems and Information Dynamics, Vol. 5, 1998, to be published. Archived as: chao-dyn 9610018.
26. D. J. Evans and D. J. Searles, Equilibrium microstates which generate second law violating steady states, *Phys. Rev. E* **50**:1645 (1994).
27. J. Kurchan, Fluctuation theorem for stochastic dynamics, *J. Phys. A.: Math. Gen.* **31**:3719 (1998).
28. F. Bonetto, G. Gallavotti, and P. Garrido, Chaotic principle: an experimental test, *Physica D* **105**:226 (1997).
29. C. Maes, The fluctuation theorem as a Gibbs property, preprint, 1998.
30. S. R. S. Varadhan, *Large Deviations and Applications* (SIAM, Philadelphia, 1984).

31. J.-D. Deuschel and D. W. Strook, *Large Deviations* (Academic Press, San Diego, 1989).
32. H. Spohn, *Large Scale Dynamics of Interacting Particles* (Springer, Heidelberg, 1991).
33. D. Stroock and S. R. S. Varadhan, *Multidimensional Diffusion Processes* (Springer, Berlin, 1979).
34. M. Büttiker, Transport as a consequence of state-dependent diffusion, *Z. Physik* **68**:161 (1987).
35. Ya. M. Blanter and M. Büttiker, Rectification of fluctuations in an underdamped ratchet, preprint 1998.
36. J.-P. Eckmann, C.-A. Pillet, and L. Rey-Bellet, Entropy production in non-linear, thermally driven Hamiltonian systems, *J. Stat. Phys.*, to appear; Nonequilibrium statistical mechanics of anharmonic chains coupled to two heat baths at different temperatures, *Commun. Math. Phys.*, to appear.
37. B. L. Holian, W. G. Hoover, and H. A. Posch, Resolution of Loschmidt's paradox: The origin of irreversible behavior in reversible atomistic dynamics, *Phys. Rev. Lett.* **59**:10 (1987).
38. H. A. Posch and W. G. Hoover, Non equilibrium molecular dynamics of a classical fluid, in *Molecular Liquids: New Perspectives in Physics and Chemistry*, J. Teixeira-Dias, ed. (Kluwer Academic Publishers, 1992), p. 527.
39. G. Gallavotti, Chaotic hypothesis: Onsager reciprocity and fluctuation dissipation theorem, *J. Stat. Phys.* **84**:899 (1996).
40. S. Lipri, R. Livi, and A. Politi, Energy transport in anharmonic lattice close and far from equilibrium, preprint, archived in xxx.lanl.gov.cond-mat #9709195.
41. N. I. Chernov and J. L. Lebowitz, Stationary nonequilibrium states in boundary-driven Hamiltonian systems: Shear flow, *J. Stat. Phys.* **86**:953 (1997).
42. F. Bonetto, N. I. Chernov, and J. L. Lebowitz, (Global and local) fluctuations of phase space contraction in deterministic stationary non-equilibrium, *Chaos* **8**:823–833 (1998).
43. F. Bonetto and J. L. Lebowitz (work in progress).
44. G. Gallavotti and D. Ruelle, SNOB states and nonequilibrium statistical mechanics close to equilibrium, *Comm. Math. Phys.* **190**:279 (1997).
45. D. Ruelle, Differentiation of SRB states, *Comm. Math. Phys.* **187**:227 (1997).
46. B. Suthertand, C. N. Yang, and C. P. Yang, Exact solution of a model of two-dimensional ferroelectric in an arbitrary external electric field, *Phys. Rev. Lett.* **19**:588 (1967).
47. D. Kim, Bethe ansatz solution for crossover scaling functions of the asymmetric XXZ chain and the KPZ-type growth model, *Phys. Rev. E* **52**:3512 (1995).
48. B. Derrida and J. L. Lebowitz, Exact large deviation function in the asymmetric exclusion process, *Phys. Rev. Lett.* **80**:209 (1998).
49. B. Derrida and C. Appert, Universal large deviation function of the Kardar–Parisi–Zhang equation in one dimension, preprint, 1998.
50. H. van Beijeren, R. Kutner, and H. Spohn, Excess noise for driven diffusive systems, *Phys. Rev. Lett.* **54**:2026 (1985).
51. H. van Beijeren, Transport properties of stochastic Lorentz models, *Rev. Mod. Phys.* **54**:195 (1982).
52. C. N. Yang and C. P. Yang, Ground state energy of a Heisenberg-Ising lattice, *Phys. Rev.* **147**:303 (1966).
53. G. Gallavotti, Fluctuation patterns and conditional reversibility in nonequilibrium systems, *Ann. Institut H. Poincaré*, in print, and chao-dyn@xyz.lanl.gov #9703007.
54. G. Gallavotti, private communication.

VOLUME 87, NUMBER 15 PHYSICAL REVIEW LETTERS 8 OCTOBER 2001

Free Energy Functional for Nonequilibrium Systems: An Exactly Solvable Case

B. Derrida,[1,*] J. L. Lebowitz,[2,3,†] and E. R. Speer[2,‡]

[1]*Laboratoire de Physique Statistique, 24 rue Lhomond, 75231 Paris Cedex 05, France*
[2]*Department of Mathematics, Rutgers University, New Brunswick, New Jersey 08903*
[3]*Department of Physics, Rutgers University, New Brunswick, New Jersey 08903*
(Received 30 April 2001; published 19 September 2001)

We consider the steady state of an open system in which there is a flux of matter between two reservoirs at different chemical potentials. For a large system of size N, the probability of any macroscopic density profile $\rho(x)$ is $\exp[-N\mathcal{F}(\{\rho\})]$; $\mathcal{F}$ thus generalizes to nonequilibrium systems the notion of free energy density for equilibrium systems. Our exact expression for $\mathcal{F}$ is a nonlocal functional of ρ, which yields the macroscopically long range correlations in the nonequilibrium steady state previously predicted by fluctuating hydrodynamics and observed experimentally.

DOI: 10.1103/PhysRevLett.87.150601 PACS numbers: 05.70.Ln, 05.40.–a, 82.20.–w

The extension of the central object of equilibrium statistical mechanics, entropy or free energy, to nonequilibrium systems in which there is a transport of matter or energy has been the holy grail of nonequilibrium statistical mechanics since the time of Boltzmann. An important step in that direction was taken by Onsager and Machlup [1] for linear deviations from equilibrium, and there have been many further extensions to time dependent evolutions starting and remaining in local equilibrium [2–6]. The extension to stationary nonequilibrium states, for which one has no *a priori* control of how close the system is to local equilibrium, has, however, remained difficult. One knows from approximate theories such as fluctuating hydrodynamics [7] that such states exhibit long range correlations totally absent from equilibrium systems (even at the critical point). These correlations have been measured experimentally in a fluid with a steady heat current [8]. Their derivation from a well-defined macroscopic functional valid beyond local equilibrium is clearly an essential step in understanding pattern formation in more general stationary nonequilibrium states such as the Bénard system. We report here what we believe is the first exact derivation of such a functional for a nonequilibrium model which is relatively simple but exhibits the realistic feature of macroscopically long range correlations.

For an equilibrium system, such as a lattice gas, in a unit cube containing L^d sites with spacing $1/L$ (or a similar continuum system), at temperature T and chemical potential ν, the probability of observing a specified macroscopic density profile, with density $\rho(x)$ at macroscopic position x in the unit cube, is given by $P(\{\rho(x)\}) \sim \exp[-L^d\mathcal{F}_{\rm eq}(\{\rho\})]$. Here, $\mathcal{F}_{\rm eq}(\{\rho\}) = \int\{f[\rho(x)] - f[\bar\rho(x)]\}\,dx$, where the integration is over the unit cube, with $f(\rho)$ the grand canonical free energy density and $\bar\rho$ the equilibrium density profile, obtained by minimizing $\int f[\rho(x)]\,dx$ [9]. The profile $\bar\rho(x)$ will be independent of x unless there is an external potential. We have suppressed the dependence of f and $\mathcal{F}_{\rm eq}$ on the constant temperature T, and assume that neither ρ nor $\bar\rho$ passes through a phase transition region at this temperature.

In this Letter, we generalize the expression for $\mathcal{F}_{\rm eq}$ to the case of a system maintained, by contact with two boundary reservoirs at unequal chemical potentials ν_0 and ν_1, in a stationary nonequilibrium state with a constant particle flux. We consider perhaps the simplest such system, the one-dimensional symmetric simple exclusion process on a lattice of N sites with open boundaries [10]. Each site i, $i = 1,\dots,N$, is either empty ($\tau_i = 0$) or occupied by a single particle ($\tau_i = 1$). Each particle independently attempts to jump to its right neighboring site, and to its left neighboring site, in each case at rate 1. It succeeds if the target site is empty; otherwise nothing happens. At the boundary sites, 1 and N, particles are added or removed: A particle is added to site 1, when the site is empty, at rate α, and removed, when the site is occupied, at rate γ; similarly particles are added to site N at rate δ and removed at rate β. We can think of sites $i = 0$ and $i = N + 1$ as occupied with probabilities $\rho_0 = \exp\nu_0/(1 + \exp\nu_0) = \alpha/(\alpha + \gamma)$ and $\rho_1 = \exp\nu_1/(1 + \exp\nu_1) = \delta/(\beta + \delta)$, independent of τ. We will assume for definiteness that $\rho_0 > \rho_1$, but, of course, due to the left-right symmetry, this is not a restriction.

The probabilities of the microscopic configurations in the steady state may be calculated through the so-called matrix method [11]. Here, we ask for the probability of seeing a specified macroscopic density profile $\rho(x)$, where $0 \le x \le 1$ and $0 \le \rho(x) \le 1$. By definition, this is the sum of the probabilities of all microscopic configurations consistent with the given profile. We shall not give a precise definition (see [3,9]) of "consistent" here; roughly speaking, we include all configurations τ such that for any y, z with $0 \le y < z \le 1$, $|\frac{1}{N}\sum_{i=yN}^{zN}\tau_i - \int_y^z \rho(x)\,dx| < \delta_N$, with $\delta_N \to 0$ as $N \to \infty$.

Our main result is a parametric formula for this probability: for large N, $P[\{\rho(x)\}] \sim \exp[-N\mathcal{F}(\{\rho\})]$, with

 0031-9007/01/87(15)/150601(4)$15.00

$$\mathcal{F}(\{\rho\}) \equiv \int_0^1 dx \left\{ \rho(x) \log\left(\frac{\rho(x)}{F(x)}\right) + [1 - \rho(x)] \log\left(\frac{1 - \rho(x)}{1 - F(x)}\right) + \log\left(\frac{F'(x)}{\rho_1 - \rho_0}\right) \right\}. \quad (1)$$

Here, F is an auxiliary function determined by $\rho(x)$: It is the decreasing solution of the differential equation

$$\rho(x) = F(x) + \frac{F(x)[1 - F(x)]F''(x)}{F'(x)^2}, \quad (2)$$

satisfying the boundary conditions,

$$F(0) = \rho_0, \qquad F(1) = \rho_1. \quad (3)$$

It can be shown that such a solution exists and is (at least when $\rho_1 > 0$ and $\rho_0 < 1$) unique [12].

We note that, if one looks for a monotone function F, satisfying the constraint (3), for which the expression $\mathcal{G}(\{\rho\}, \{F\})$ given by the right-hand side of (1) is stationary, one obtains (2) as a Euler-Lagrange equation:

$$\frac{\delta \mathcal{G}(\{\rho\}, \{F\})}{\delta F(x)} = 0. \quad (4)$$

We can in fact prove that this stationary point is a maximum [12] when $0 < \rho_1 < \rho_0 < 1$ and expect this to be true in general:

$$\mathcal{F}(\{\rho\}) = \sup_F \mathcal{G}(\{\rho\}, \{F\}). \quad (5)$$

Before explaining the main steps of the derivation of (1)–(3), let us comment on some of their consequences.

(a) By its very nature $\mathcal{F}$ satisfies $\mathcal{F}(\{\rho\}) \geq 0$, with equality only for $\rho(x) = \bar{\rho}(x)$, where $\bar{\rho}(x) = \rho_0(1 - x) + \rho_1 x$ is the profile obtained with probability one in the limit $N \to \infty$. Any other profile will have $\mathcal{F}(\{\rho\}) > 0$ and, thus, for large N, exponentially small probability. When $\rho_0 = 1$ or $\rho_1 = 0$, there are some profiles for which $\mathcal{F} = +\infty$; their probability is superexponentially small in N. For examples, see (b) below and [12].

(b) For a constant profile $\rho(x) = r$, F satisfies $F' = AF^r(1 - F)^{1-r}$, where A is fixed by (3), and

$$\mathcal{F}(\{\rho\}) = \log\left[\int_{\rho_0}^{\rho_1} \left(\frac{r}{z}\right)^r \left(\frac{1 - r}{1 - z}\right)^{1-r} \frac{dz}{\rho_1 - \rho_0}\right]. \quad (6)$$

We see that $\mathcal{F}(\{\rho\}) = \infty$ if $r = 1$ and $\rho_1 = 0$, or $r = 0$ and $\rho_0 = 1$.

(c) Using (4), one finds immediately that

$$\frac{\delta \mathcal{F}}{\delta \rho(x)} = \log\left[\frac{\rho(x)}{F(x)} \frac{1 - F(x)}{1 - \rho(x)}\right]. \quad (7)$$

This expression can then be used to find optimal profiles subject to various constraints. For example, setting (7) to zero implies that $\rho(x) = F(x)$, leading to the most likely profile $\bar{\rho}(x)$ given in (a).

(d) If we minimize $\mathcal{F}$ subject to the constraint of a fixed mean density $\int_0^1 \rho(x)\, dx$, the right-hand side of (7) becomes an arbitrary constant, and together with (2) one obtains that the most likely profile is exponential: $\rho(x) = A_1 \exp(\theta x) + A_2$, the constants being determined by the value of the mean density and the boundary conditions (3). (This exponential form, which is the stationary solution of a diffusion equation with drift, was first suggested to us, for some special cases, by Errico Presutti).

Similarly, if we impose a fixed mean density in k nonoverlapping intervals, with no other constraints, the optimal profile is exponential inside these intervals, linear outside, and in general not continuous at the end points of the intervals.

(e) When the chemical potentials of the two reservoirs are equal, i.e., $\rho_0 = \rho_1$, the system is in true equilibrium with $\bar{\rho}(x) = \rho_1$. Equations (1)–(3) have a well-defined limit for $\rho_1 \nearrow \rho_0$, with $F(x) = \rho_0 + (\rho_1 - \rho_0)x + O[(\rho_1 - \rho_0)^2]$. It is also natural to consider a local equilibrium Gibbs measure corresponding to a spatially varying chemical potential [2,9] which is adjusted to maintain the same optimal profile $\bar{\rho}(x)$. For this system, the large deviation functional (free energy) is just $\mathcal{F}_{\text{eq}}(\{\rho\})$, which has the explicit form,

$$\mathcal{F}_{\text{eq}}(\{\rho\}) = \int \left\{ \rho(x) \log \frac{\rho(x)}{\bar{\rho}(x)} + [1 - \rho(x)] \log \frac{[1 - \rho(x)]}{[1 - \bar{\rho}(x)]} \right\} dx. \quad (8)$$

From (5), $\mathcal{F}(\{\rho\}) \geq \mathcal{G}(\{\rho\}, \{\bar{\rho}\}) = \mathcal{F}_{\text{eq}}(\{\rho\})$, so that expressions (1) and (8) are in general different,

$$\mathcal{F}(\{\rho\}) \geq \mathcal{F}_{\text{eq}}(\{\rho\}), \quad (9)$$

with equality only for $\rho(x) = \bar{\rho}(x)$ or $\rho_0 = \rho_1$.

(f) Using the fact that the exponential is the optimal profile for the case of a fixed mean density in the entire interval, we may compute the distribution of M, the total number of particles in the system, in the steady state for large N. We find that the fluctuations of M predicted by (1) are reduced in comparison to those in a system in local equilibrium (8) with the same $\bar{\rho}$:

$$\lim_{N\to\infty} N^{-1}[\langle M^2 \rangle_{\text{SNS}} - \langle M \rangle^2] = \lim_{N\to\infty} N^{-1}[\langle M^2 \rangle_{\text{eq}} - \langle M \rangle^2] - \frac{(\rho_1 - \rho_0)^2}{12}. \quad (10)$$

We may also obtain (10) by expanding $\rho(x)$ about $\bar{\rho}(x)$ in (1). The result agrees with that obtained in [13] directly from the microscopic model and from fluctuating hydrodynamics [7].

Derivation.—Let us now sketch the derivation of (1)–(3). The probability of a configuration $\tau = \{\tau_1, \ldots, \tau_N\}$ in the steady state of our model is given by [11]

$$P_N(\tau) = \frac{\omega_N(\tau)}{\langle W|(D + E)^N|V\rangle}, \quad (11)$$

where the weights $\omega_N(\tau)$ are given by

$$\omega_N(\tau) = \langle W|\Pi_{i=1}^N[\tau_i D + (1 - \tau_i)E]|V\rangle, \quad (12)$$

and the matrices D and E and the vectors $|V\rangle$ and $\langle W|$ satisfy

$$DE - ED = D + E, \tag{13}$$

$$(\beta D - \delta E)|V\rangle = |V\rangle, \qquad \langle W|(\alpha E - \gamma D) = \langle W|. \tag{14}$$

Although proving that (11)–(14) do give the weights in the steady state is rather easy [11], there is not so far a simple physical interpretation of the matrices D and E or of the vectors $|V\rangle$ or $\langle W|$.

To obtain the probability $P_{N_1,\ldots,N_n}(M_1, M_2, \ldots, M_n)$ that M_1 particles are located on the first N_1 sites, M_2 particles on the next N_2 sites, etc., we first calculate the sum $\Omega_{N_1,\ldots,N_n}(M_1, M_2, \ldots, M_n)$ of the weights of all the corresponding configurations. The key to obtaining (1) is that the following generating function, which plays the role of the grand-canonical-pressure partition function, can be computed exactly:

$$Z(\lambda_1, \ldots, \lambda_n; \mu_1, \ldots, \mu_n)$$
$$\equiv \sum \frac{\mu_1^{N_1}}{N_1!} \cdots \frac{\mu_n^{N_n}}{N_n!} \lambda_1^{M_1} \cdots \lambda_n^{M_n} \frac{\Omega_{N_1,\ldots,N_n}(M_1, \ldots, M_n)}{\langle W|V\rangle}$$
$$= \frac{\langle W|e^{\mu_1\lambda_1 D + \mu_1 E} \cdots e^{\mu_n\lambda_n D + \mu_n E}|V\rangle}{\langle W|V\rangle}, \tag{15}$$

where the sum is over all $N_i, M_i \geq 0$ and the parameters μ_i and λ_i are conjugate to the N_i and M_i. To do the calculation, we used repeatedly the following identity, which follows from (13),

$$e^{xD+yE} = \left(\frac{(x-y)e^y}{xe^y - ye^x}\right)^E \left(\frac{(x-y)e^x}{xe^y - ye^x}\right)^D, \tag{16}$$

and the result is that

$$Z = \left(\frac{\rho_0 - \rho_1}{g}\right)^{a+b} \exp\left[a \sum_{i=1}^n \mu_i(1 - \lambda_i)\right], \tag{17}$$

where $a = (\alpha + \gamma)^{-1}$, $b = (\beta + \delta)^{-1}$, and

$$g = \left[-\rho_1 + \rho_0 e^{\sum_{i=1}^n \mu_i(1-\lambda_i)} + \sum_{i=1}^n \frac{1}{\lambda_i - 1}(e^{\mu_i(1-\lambda_i)} - 1)e^{\sum_{j>i}\mu_j(1-\lambda_j)}\right]. \tag{18}$$

All the rest of the derivation consists in extracting from the exact expression (17) the behavior of Ω and P for large N_i's.

First (17) gives the normalization factor in (11),

$$\frac{\langle W|(D+E)^N|V\rangle}{\langle W|V\rangle} = \frac{\Gamma(a+b+N)}{\Gamma(a+b)(\rho_0 - \rho_1)^N}, \tag{19}$$

so that $P_{N_1,\ldots,N_n}(M_1, M_2, \ldots, M_n)$ is given simply by $\Omega_{N_1,\ldots,N_n}(M_1, M_2, \ldots, M_n)$ divided by (19).

Now the large N_i behavior of $\Omega_{N_1,\ldots,N_n}$ controls the position and the nature of the singularities of Z closest to the origin $\mu_i = \lambda_i = 0$; conversely, this relation can be inverted [12] to determine the asymptotic behavior of $\Omega_{N_1,\ldots,N_n}$ from the equation $g = 0$ of the surface on which Z is singular, just as the growth of the coefficients of a power series of one variable is determined by the singularity nearest the origin. In particular, if one sets $N_j = Ny_j$ and $M_j = N_j r_j$ then, for large N and fixed y_i, r_i,

$$\frac{\log P_{N_1,\ldots,N_n}(M_1, \ldots, M_n)}{N} \simeq \log(\rho_0 - \rho_1) - \sum_{j=1}^n y_j(\log\mu_j/y_j + r_j \log\lambda_j), \tag{20}$$

where y_j and r_j are related to the parameters $\mu_1, \ldots, \mu_n$ and $\lambda_1, \ldots, \lambda_n$ by

$$y_j = \frac{\frac{\partial g}{\partial \log\mu_j}}{\sum_{i=1}^n \frac{\partial g}{\partial \log\mu_i}}, \qquad r_j = \frac{\frac{\partial g}{\partial \log\lambda_j}}{\frac{\partial g}{\partial \log\mu_j}}, \tag{21}$$

with all derivatives calculated on the manifold $g = 0$.

Equations (20) and (21) determine $\mathcal{F}$ in a parametric form. As the μ_j's and the λ_j's vary, they give the sizes of the boxes y_j, their particle densities r_j, and the corresponding probabilities. Note that the parameters a and b do not appear in (18) and therefore in the equation $g = 0$; so for large N, only ρ_0 and ρ_1 remain relevant.

This parametric form can be simplified by replacing the role of the μ_i and λ_i by a single sequence of parameters G_i. Let us define the constant C by $C = \sum_{i=1}^n \partial g/\partial \log\mu_i$ and the sequence G_i by

$$G_i = C^{-1}e^{\sum_{j=i}^n \mu_j(1-\lambda_j)}, \qquad G_{n+1} = C^{-1}. \tag{22}$$

It follows from (22) that $\mu_i = \log(G_i/G_{i+1})/(1 - \lambda_i)$ and

$$\frac{1}{\lambda_i - 1} = \frac{1}{G_{i+1}} \frac{y_i}{\log(G_i/G_{i+1})} - \rho_0 + \sum_{j=1}^i \left(\frac{1}{G_j} - \frac{1}{G_{j+1}}\right)\frac{y_j}{\log(G_j/G_{j+1})}. \tag{23}$$

The condition that $g = 0$ becomes

$$\rho_0 - \rho_1 = \sum_{j=1}^n \left(\frac{1}{G_j} - \frac{1}{G_{j+1}}\right)\frac{y_j}{\log(G_j/G_{j+1})}, \tag{24}$$

which can be thought of as an equation which determines G_{n+1} in terms of $G_1, \ldots, G_n$.

Once the λ_i's are known, one gets for the r_i's and the large deviation function,

$$r_i = -\frac{\lambda_i}{1 - \lambda_i} - \frac{\lambda_i}{(1-\lambda_i)^2}\frac{G_i - G_{i+1}}{y_i}, \tag{25}$$

and

$$\frac{\log[P_{N_1,\ldots,N_n}(M_1, \ldots, M_n)]}{N} = -\sum_{i=1}^n \left\{y_i \log\left[\frac{\log(\frac{G_i}{G_{i+1}})}{y_i(\rho_0 - \rho_1)}\right] - y_i \log(1 - \lambda_i) + y_i r_i \log(\lambda_i)\right\}. \tag{26}$$

To take the limit $N \to \infty$ followed by $n \to \infty$, $y_i \to 0$, we introduce a continuous variable x, $0 \leq x \leq 1$, and let $x_i = y_1 + y_2 + \cdots + y_i$. All the discrete sequences can now be thought of as functions of $x_1, \ldots, x_n$ with $G_i \equiv G(x_i)$, $\lambda_i \equiv \lambda(x_i)$, and $r_i \equiv \rho(x_i)$, $i = 1, \ldots, n$. Then, extending G, λ, and ρ to be smooth functions of x, so that $G_{i+1} - G_i \simeq y_i G'(x_i)$, one finds that (23) and (24) become

$$\frac{1}{\lambda(x) - 1} = -\frac{1}{G'(x)} - \rho_0 - \int_0^x \frac{du}{G(u)}, \qquad (27)$$

$$\rho_1 = \rho_0 + \int_0^1 \frac{du}{G(u)}. \qquad (28)$$

At this stage it is more convenient to replace $G(x)$ by the function $F(x)$ defined by $F(x) = \rho_0 + \int_0^x G(u)^{-1}\, du$. The expression (27) for $\lambda(x)$ becomes then

$$\frac{1}{\lambda(x) - 1} = \frac{F'(x)^2}{F''(x)} - F(x). \qquad (29)$$

Using the above relations, we may rewrite (25) and (26) in terms of F, obtaining (1) and (2); the boundary conditions (3) come from (28). The monotonicity of F follows from the uniqueness of the sign of G [see (22)].

Possible extensions.—It would be nice to give a physical interpretation of the Euler-Lagrange equations (4) and (5) directly from a macroscopic or semimacroscopic approach. Such an approach might allow an extension to the large deviation function for time-dependent profiles and permit one to unify various results recently obtained for the large deviation functions of time-dependent profiles or currents [6,14,15].

The matrix method [with modified rules (13) and (14)] also applies [11] to the asymmetric simple exclusion process, in which particles jump to the left and right at different rates. We are at present trying to generalize our calculations to this case. We have already calculated the probability of a specified mean density ρ (total number of particles) for the special case in which particles jump only to their right and with only input, with rate $\alpha > 1$, at the left and output, with rate $\beta > 1$, at the right:

$$N^{-1} \log P(\rho) \sim -2[\rho \log 2\rho + (1 - \rho)\log 2(1 - \rho)]. \qquad (30)$$

For a Bernoulli measure at uniform density $1/2$, the result would be the same except for the factor 2 in front of the whole expression, which here again expresses the fact that fluctuations are reduced by long range correlations.

In conclusion, for the simple model we studied here, the large deviation function $\mathcal{F}$ (which extends the notion of free energy to nonequilibrium systems) is a nonlocal functional (1)–(3) of the density ρ. This implies that the probability of a given profile is not the product of probabilities for, say, the left and right halves of the system, in contrast to the situation for equilibrium systems. Since long range correlations are expected for general stationary nonequilibrium states and have even been measured experimentally [8], the nonlocal nature of $\mathcal{F}(\{\rho\})$ is presumably also general, in contrast with the conjecture [4,5] that $\mathcal{F}$ is a local function of the hydrodynamic variables.

We thank E. Presutti for very helpful discussions. The work of J. L. Lebowitz was supported by NSF Grant No. DMR-9813268, AFOSR Grant No. F49620/0154, DIMACS and its supporting agencies, and NATO Grant No. PST.CLG.976552. J. L. L. and E. R. Speer acknowledge the hospitality of the I.H.E.S. in the spring of 2000, where this work was begun.

*Email address: derrida@lps.ens.fr
†Email address: lebowitz@math.rutgers.edu
‡Email address: speer@math.rutgers.edu

[1] L. Onsager and S. Machlup, Phys. Rev. **91**, 1505 (1953); **91**, 1512 (1953).
[2] C. Kipnis, S. Olla, and S. R. S. Varadhan, Commun. Pure Appl. Math. **42**, 115 (1989).
[3] C. Kipnis and C. Landim, *Scaling Limits of Interacting Particle Systems* (Springer-Verlag, Berlin, 1999).
[4] R. Graham, in *Order and Fluctuations in Equilibrium and Nonequilibrium Statistical Mechanics,* edited by G. Nicolis, G. Dewel, and J. Turner (Wiley, New York, 1981).
[5] G. Eyink, J. Stat. Phys. **61**, 533 (1990).
[6] L. Bertini *et al.*, Phys. Rev. Lett. **87**, 040601 (2001).
[7] J. R. Dorfman, T. R. Kirkpatrick, and J. V. Sengers, Annu. Rev. Phys. Chem. **45**, 213 (1994).
[8] W. B. Li *et al.*, Phys. Rev. Lett. **81**, 5580 (1998).
[9] H. Spohn, *Large Scale Dynamics of Interacting Particles* (Springer-Verlag, Berlin, 1991).
[10] G. Eyink, J. L. Lebowitz, and H. Spohn, Commun. Math. Phys. **132**, 253 (1990); **140**, 119 (1991).
[11] B. Derrida, M. R. Evans, V. Hakim, and V. Pasquier, J. Phys. A **26**, 1493 (1993); F. H. L. Essler and V. Rittenberg, J. Phys. A **29**, 3375 (1996); N. Rajewsky and M. Schreckenberg, Physica (Amsterdam) **245A**, 139 (1997); T. Sasamoto, J. Phys. A **32**, 7109 (1999); R. A. Blythe, M. R. Evans, F. Colaiori, and F. H. L. Essler, J. Phys. A **33**, 2313 (2000).
[12] B. Derrida, J. L. Lebowitz, and E. R. Speer (to be published).
[13] H. Spohn, J. Phys. A **16**, 4275 (1983).
[14] B. Derrida and J. L. Lebowitz, Phys. Rev. Lett. **80**, 209 (1998).
[15] M. Prähofer and H. Spohn, Phys. Rev. Lett. **84**, 4882 (2000).

Available online at www.sciencedirect.com
SCIENCE DIRECT®
ELSEVIER
Physica D 193 (2004) 53–66
PHYSICA D
www.elsevier.com/locate/physd

On the (Boltzmann) entropy of non-equilibrium systems

S. Goldstein, Joel L. Lebowitz*

Departments of Mathematics and Physics, Rutgers University, Piscataway, NJ 08854, USA

Abstract

Boltzmann defined the entropy of a macroscopic system in a macrostate M as the log of the volume of phase space (number of microstates) corresponding to M. This agrees with the thermodynamic entropy of Clausius when M specifies the locally conserved quantities of a system in local thermal equilibrium (LTE). Here we discuss Boltzmann's entropy, involving an appropriate choice of macro-variables, for systems not in LTE. We generalize the formulas of Boltzmann for dilute gases and of Resibois for hard sphere fluids and show that for macro-variables satisfying any deterministic autonomous evolution equation arising from the microscopic dynamics the corresponding Boltzmann entropy must satisfy an $\mathcal{H}$-theorem.

PACS: 05.20.−y; 05.20.Dd; 05.20.Jj; 05.70.Ln

Keywords: Non-equilibrium entropy; $\mathcal{H}$-theorem; Phase space volume; Macroscopic evolution; Microscopic dynamics

1. Introduction

Thermodynamics associates to isolated equilibrium macroscopic systems with specified thermodynamic parameters M an additive, macroscopically well defined, entropy $S(M)$. The second law of thermodynamics then asserts that in any temporal change occurring in such isolated systems (due e.g. to the relaxation of some constraint) the new equilibrium state, with parameters M', must satisfy $S(M') \geq S(M)$. $S(M)$ as well as the second law naturally generalize to an entropy $S_{\text{loc.eq.}}(M) = S_{\text{loc.eq.}}(\{M(\mathbf{x})\})$ for systems in local thermal equilibrium (LTE) with particle, momentum and energy densities varying slowly (on a microscopic scale) in space and time. $S_{\text{loc.eq.}}(M)$ increases with time when $M(\mathbf{x})$ evolves to $M_t(\mathbf{x})$ according to macroscopic hydrodynamical equations [1,2]. This is reviewed briefly in Section 2.

We then discuss in Section 3 Boltzmann's microscopic interpretation of $S(M)$ as the log of the volume of phase space associated to M. This not only provides a formula for computing $S(M)$ microscopically, but also explains the origin of the time-asymmetric second law in the time-symmetric microscopic laws [3,4]. It shows in particular that if there is a deterministic autonomous equation describing the time evolution of a macrostate M_t of an isolated system, be it hydrodynamic or kinetic, e.g. the Boltzmann equation, it must give an $S(M_t)$ which is monotone non-decreasing in t.

* Corresponding author.
E-mail addresses: oldstein@math.rutgers.edu (S. Goldstein), lebowitz@sakharov.rutgers.edu (J.L. Lebowitz).

0167-2789/$ – see front matter © 2004 Published by Elsevier B.V.
doi:10.1016/j.physd.2004.01.008

Boltzmann's macroscopic formulation leads naturally to a formula for the entropy of dilute gases which may be far from LTE. For such systems the macrostate M may be specified by $f(\mathbf{x}, \mathbf{v})$, the density of gas particles in the six-dimensional one-particle phase space. Boltzmann showed that this entropy, $S_{\text{gas}}(f_t)$, increases with time when f_t evolves according to the Boltzmann equation ($\mathcal{H}$-theorem) [3–5]. This is discussed in Section 4.

In Section 5 we give a formula for $S(f, E)$, the log of the phase space volume of a general system whose macrostate is specified by $f(\mathbf{x}, \mathbf{v})$ and the total energy E. This reduces to $S_{\text{gas}}(f)$ for a dilute gas and to $S_{\text{hs}}(f)$ for a system of hard spheres. $S_{\text{hs}}(f)$ was found by Resibois [6] to satisfy an $\mathcal{H}$-theorem when f evolves according to the (modified) Enskog equation for a system of hard spheres. We note that the general argument given by Boltzmann for the origin of the second law suggests that $S(f_t, E)$ should be monotone in time even if f_t does not satisfy an autonomous evolution equation. This is discussed further in Section 6.

Section 7 consists of some remarks comparing and contrasting Boltzmann's definition of the entropy of a macroscopic system with other definitions of entropy. We raise, but do not resolve, the question of the appropriate choice of macrostates for general non-equilibrium systems.

The article is written in an informal style describing the ideas and facts (not necessarily in the right historical order) we think important for understanding the notion of the entropy of a macroscopic system, made up of a very large number of atoms or molecules. We restrict ourselves to isolated classical systems, assume familiarity with the basic notions of thermodynamics and statistical mechanics, and omit many details (including units, boundary conditions, etc.). We refer the interested reader to [7] and references therein.

2. Clausius' macroscopic entropy

Rudolf Clausius' 1865 paper [8] concludes with his celebrated "two fundamental theorems of the mechanical theory of heat": (1) the energy of the universe is constant and (2) the entropy of the universe tends to a maximum. These express in succinct form what is generally referred to as the first and second law of thermodynamics; see [3,4] for their interesting history.

The first law needs no elaboration. The existence of a conserved energy for isolated systems goes back to Newton for mechanical systems. The experiments of Joule then showed that thermal phenomena are subject to the same mechanical laws.

The second law, on the other hand, which contains the newly coined word entropy, does need elaboration. Let us quote Lars Onsager [9]: "The second law of thermodynamics forbids perpetual motion of the second kind and implies the existence of a definable entropy for any system in a state that can be reached by a succession of reversible processes. These "thermodynamic" states are typically defined as states of "equilibrium" under specified restraints on composition, energy, and external boundary conditions, in the sense that no spontaneous change can occur in the system as long as the constraints remain fixed." The implicit "restraints" exclude chemical or nuclear reactions which would change the species present, etc.

As put in the textbooks, e.g. [1]: given an equilibrium system with energy E and mole (or particle) numbers $\mathbf{N}$ in a spatial region V, with a volume which we shall also denote by V, there exists a function $S(E, \mathbf{N}, V)$ such that in a reversible process

$$\mathrm{d}S = \frac{[\mathrm{d}E + p\,\mathrm{d}V - \sum \mu_j\,\mathrm{d}N_j]}{T}, \tag{1}$$

where T is the absolute temperature, p the pressure and μ_j the chemical potential of species j. The terms in the square bracket just give the amount of heat added to the system in a reversible process.

Thermodynamics further states that the entropy of two isolated macroscopic systems, each in equilibrium, with their own energies, mole numbers and volumes, is the sum of their individual entropies, i.e.

$$S_{1,2}(E_1, \mathbf{N}_1, V_1 \cdot E_2, \mathbf{N}_2, V_2) = S_1(E_1, \mathbf{N}_1, V_1) + S_2(E_2, \mathbf{N}_2, V_2). \tag{2}$$

Suppose now that these two systems are permitted to interact and exchange energy over some period of time after which they are again isolated. Then, according to the first law, their new energies E_1' and E_2' will satisfy $E_1' + E_2' = E_1 + E_2$. If we now wait until each of the systems comes to equilibrium then, according to the second law, their combined new entropy must satisfy the inequality,

$$S_{1,2}' = S_1(E_1', \mathbf{N}_1, V_1) + S_2(E_2', \mathbf{N}_2, V_2) \geq S_{1,2} = S_1(E_1, \mathbf{N}_1, V_1) + S_2(E_2, \mathbf{N}_2, V_2). \tag{3}$$

Similar inequalities hold when relaxing other constraints. As an extreme example imagine that initially system 1 contained a mixture of hydrogen and oxygen at a low temperature T_1, with an implicit constraint prohibiting their chemical reaction, while system 2 was at a high temperature T_2. The final state could now be very hot steam in system 1, with a temperature T_1' higher than the temperature T_2' in system 2, e.g. $T_1' > T_2' > T_2 > T_1$.

In all cases, *if the two systems have come to a joint equilibrium* at the end of this period of interaction, with possibly new $\mathbf{N}'$, then E_1' and E_2' must be such that the new entropy satisfies

$$S_1'(E_1', \mathbf{N}_1', V_1) + S_2'(E_2', \mathbf{N}_2', V_2) = S_{1,2}' = \sup_{U_1, U_2} \{S_1(U_1, \mathbf{N}_1', V_1) + S_2(U_2, \mathbf{N}_2', V_2)\} \tag{4}$$

subject only to energy conservation, $U_1 + U_2 = E_1 + E_2$. This implies

$$T_1' \equiv \left[\frac{\partial S_1}{\partial E_1'}(E_1', \mathbf{N}_1', V_1)\right]^{-1} = T_2' \equiv \left[\frac{\partial S_2(E_2', \mathbf{N}_2', V_2)}{\partial E_2'}\right]. \tag{5}$$

Similar relations hold when there can be an exchange of matter or volume between the systems.

Using the fact that for macroscopic systems surface areas (multiplied by a suitable microscopic length) and surface energies are negligible compared to the corresponding bulk quantities, the additivity of the entropy, expressed by (2), also gives extensivity. That is, for systems uniform in the bulk,

$$S(E, \mathbf{N}, V) = Vs(e, \mathbf{n}), \tag{6}$$

where $e = E/V$ and $\mathbf{n} = \mathbf{N}/V$.

The thermodynamic entropy can now be extended to a system in LTE [1,2]: a system in a volume V which can be considered, to a good approximation, as being locally in equilibrium with energy density $e(\mathbf{x})$, particle density $n(\mathbf{x})$ and hydrodynamic velocity $\mathbf{u}(\mathbf{x})$: for a precise definition see [10]. For such systems we can, by extension of (6), write

$$S_{\text{loc.eq.}}(n, \mathbf{u}, e) = \int_V s\left(e(\mathbf{x}) - \frac{1}{2}mn(\mathbf{x})\mathbf{u}^2(\mathbf{x}), n(\mathbf{x})\right) \mathrm{d}\mathbf{x}, \tag{7}$$

where m is the mass (per mole) and we consider just one component for simplicity [1–12]. (The "mechanical" energy associated with $\mathbf{u}(\mathbf{x})$ does not contribute to S until it is dissipated and we always take $\int_V n(\mathbf{x})u(\mathbf{x})\,\mathrm{d}\mathbf{x} = 0$.)

Eq. (7) clearly agrees with (6) when the system is in true equilibrium, $\mathbf{u} = 0$, and e and n are independent of $\mathbf{x}$. Furthermore, if a LTE state evolves in time according to macroscopic equations then $S_{\text{loc.eq.}}$ must increase (or at least not decrease) as a function of t. Consider for example an isolated system in LTE (with $\mathbf{u} = 0$ and n constant) with an energy density profile $e_0(\mathbf{x})$ and corresponding temperature profile $T_0(\mathbf{x})$ for which we have, using an extension of (1) to continuous time,

$$\frac{\partial s}{\partial t} = -\frac{\nabla \cdot \mathbf{J}}{T} = -\nabla \cdot \left(\frac{\mathbf{J}}{T}\right) + \mathbf{J} \cdot \nabla \left(\frac{1}{T}\right), \tag{8}$$

where $\mathbf{J}$ is the heat flux, $\mathbf{J}/T$ is the entropy flux, and $\mathbf{J}\cdot\nabla(1/T)$ is the entropy production. According to Fourier's law, $\mathbf{J} = -\mathcal{K}(T)\nabla T$ so (8) can be written in the more familiar form as

$$C_V(T)\frac{\partial}{\partial t}T(\mathbf{x},t) = \nabla\cdot[\mathcal{K}(T)\nabla T], \tag{9}$$

where $C_V(T) = T^{-1}(\mathrm{d}s(T)/\mathrm{d}T)$ is the specific heat and $\mathcal{K}$ is the heat conductivity. Eq. (9) is to be solved, for $t > 0$, subject to no heat flux, or $\nabla T = 0$, at the surface of V. Integrating (8) or (9) then yields

$$\frac{\mathrm{d}S_{\text{loc.eq.}}}{\mathrm{d}t} = \frac{\mathrm{d}}{\mathrm{d}t}\int_V s\,\mathrm{d}\mathbf{x} = \int_V \mathbf{J}\cdot\left(\nabla\frac{1}{T}\right)\mathrm{d}\mathbf{x} = \int_V \mathcal{K}T^2\left(\nabla\frac{1}{T}\right)^2\mathrm{d}\mathbf{x} \geq 0. \tag{10}$$

As $t\to\infty$, $T(\mathbf{x},t)\to\bar{T}$, for all $\mathbf{x}\in V$, with $\bar{T}$ determined by energy conservation and $S_{\text{loc.eq.}}$ approaches its maximum (equilibrium) value $Vs(\bar{e},n)$.

The second law thus manifests itself for LTE by the requirement that the entropy production $\mathbf{J}\cdot\nabla(1/T)$ be non-negative, i.e. that $\mathcal{K}(T)\geq 0$. The formula for entropy production generalizes to other macroscopic equations [2,12], e.g. to the compressible Navier–Stokes equations which describe the time evolution of $n(\mathbf{x},t)$, $\mathbf{u}(\mathbf{x},t)$ and $e(\mathbf{x},t)$. The Euler equations, on the other hand, conserve $S_{\text{loc.eq.}}$, in the absence of singularities of the flow. They do not give an approach to equilibrium of an isolated system and thus do not provide a description valid for times over which such an approach takes place. In the presence of shocks, however, the solutions are non-unique and the requirement that $S_{\text{loc.eq.}}$ increase picks out the correct evolution.

3. Boltzmann's microscopic entropy

There are many ways to stretch the notion of LTE and apply the second law to processes taking place in systems which are clearly very far from equilibrium, e.g. living organisms [3,9]. These ad hoc extensions work quite well in the hands of seasoned practitioners [9] but are far from systematic. It would certainly be desirable to find systematic ways for defining and calculating the entropy, expressed as a function of the appropriate macroscopic variables of systems that are not in LTE. This entropy would be monotone in time and coincide with $S_{\text{loc.eq.}}$ for a system in LTE.

This is exactly what was accomplished by Boltzmann's microscopic interpretation of the macroscopic Clausius equilibrium entropy $S(M)$. This interpretation provides a formula for the computation of $S(M)$ from the microscopic Hamiltonian. Even more importantly, it explains the origin of the time-asymmetric second law in the time-reversible dynamics of the atoms and molecules which are the microscopic constituents of macroscopic matter, and shows its applicability to systems not in LTE. We will be very brief here [1–14].

A very good summary of Boltzmann's accomplishment is given by this quote from Einstein [13]: "On the basis of the kinetic theory of gases Boltzmann had discovered that, aside from a constant factor, entropy is equivalent to the logarithm of the "probability" of the [macro]state under consideration. Through this insight he recognized the nature of the course of events which, in the sense of thermodynamics, are "irreversible". Seen from the molecular-mechanical point of view, however, all courses of events are reversible. If one calls a molecular-theoretically defined state a microscopically described one, or, more briefly, microstate, then an immensely large number (Z) of states belong to a macroscopic condition. Z then is a measure of the probability of a chosen macrostate. This idea appears to be of outstanding importance also because of the fact that its usefulness is not limited to microscopic description on the basis of mechanics."

Let us make Einstein's remarks more explicit by considering a classical system of N particles in a box V. Its microstate X is given by a point in the $6N$ dimensional phase space which specifies everything about the system, e.g. the energy given by its Hamiltonian $H(X)$, etc. When N is very large a more appropriate coarse-grained description of the system is provided by its macrostate M. We can specify M, for example, by dividing V into J cubes Δ_k, $J \ll N$, so that each cube contains a very large number of particles, and specifying coarse-grained values of the energy, momentum and number of particles in each Δ_k. Let Γ_M be the region of the phase space consisting of all microstates consistent with M, i.e. the set of all X such that the appropriate phase space function $M(X) = M$. Let $|\Gamma_M|$ be the volume of Γ_M in appropriate units (this is Einstein's Z for a classical system). Boltzmann defined the entropy of a macroscopic system with microstate X by

$$S_B(X) = k \log|\Gamma_{M(X)}|. \tag{11}$$

Boltzmann then showed for a gas in an equilibrium macrostate, M_{eq}, which, for the above choice of M, corresponds to a uniform density of the macro-variables in V, that S_B agrees (to leading order in N) with the thermodynamic entropy of Clausius. The same is true for LTE states, i.e. if $M(X) = \{n(\mathbf{x}), \mathbf{u}(\mathbf{x}), e(\mathbf{x})\}$ then $S_B(M) = k \log|\Gamma_{M(X)}| = S_{loc.eq.}(n, \mathbf{u}, e)$. This means that if the entropies $S_B(M) > S_B(M')$ differ by a macroscopic amount, the ratio of their corresponding phase space volumes is exponentially large in N. Thus if the system contains one mole of material, the ratio of $|\Gamma_{M_{eq}}|$ to $|\Gamma_M|$ for a macrostate M in which all the particles are in the left half of the box is of order $\exp[10^{20}]$. This is far larger than the ratio of the volume of the known universe to the volume of one proton.

Boltzmann then argued that given this disparity in sizes of Γ_M for different M's, the time evolved $M_t = M(X_t)$ will be such that $|\Gamma_M(X_t)|$ and thus $S_B(X_t)$ will *typically* increase in accord with the second law. By "typically" we mean that for any Γ_M (of the kind described above) the relative volume of the set of microstates X in Γ_M for which the second law is violated by a macroscopic amount, i.e. by an amount proportional to N, during any fixed time period (not bigger than the age of the universe), goes to zero rapidly (exponentially) in the number of atoms and molecules in the system.

In fact let us consider the case where M_t satisfies an autonomous deterministic evolution, e.g. Eq. (9). This means that if that evolution carries $M_{t_1} \to M_{t_2}$, then the microscopic dynamics Φ_t carries $\Gamma_{M_{t_1}}$ inside $\Gamma_{M_{t_2}}$, i.e. $\phi_{t_2-t_1}\Gamma_{M_{t_1}} \subset \Gamma_{M_{t_2}}$, with negligible error. Now the fact that phase space volume is conserved by the Hamiltonian time evolution (Liouville's theorem) implies that $|\Gamma_{M_{t_1}}| \leq |\Gamma_{M_{t_2}}|$ and thus by (11) that $S_B(M_{t_2}) \geq S_B(M_{t_1})$ for $t_2 \geq t_1$. We have thus derived an "$\mathcal{H}$-theorem" for any deterministic evolution of the macro-variables arising from the microscopic dynamics. The explicit form for the rate of change of $S_B(M_t)$ (including strict positivity) depends on the detailed macroscopic evolution equation. The fact that $\Gamma_{M_{eq}}$ essentially coincides for large N with the whole energy surface $H(X) = E$ also *explains* the evolution towards and persistence of equilibrium in an isolated macroscopic system.

The emergence of definite time-asymmetric behavior in the observed evolution of macroscopic systems, despite the total absence of such asymmetry in the microscopic dynamics, is thus a consequence of the great disparity between microscopic and macroscopic scales, together with the fact (or very reasonable assumption) that what we observe in nature is typical behavior, corresponding to typical initial conditions [7].

4. Going beyond LTE: dilute gases

As is clear from Eq. (11), the choice of the macro-variables M is essential for the computation of $S_B(X)$. For equilibrium systems or those in LTE these are specified by thermodynamics although there is a large leeway in choosing the sizes of the boxes Δ_k, if we consider only leading terms in N. They are the locally conserved and hence microscopically slowly varying quantities—precisely those for which one has hydrodynamic type autonomous equations.

To obtain useful quantitative information from the second law for systems not in LTE one has to find appropriate macro-variables M for the system under consideration, e.g. those which satisfy autonomous time evolution equations, and for which one can compute $S_B(M)$. A paradigmatic example of where this has been achieved is a dilute gas. Following Boltzmann, we refine the M considered in the last section for a system of N particles in a box V. This is done by noting that the microstate $X = \{\mathbf{r}_i, \mathbf{v}_i\}$, $i = 1, \ldots, N$, can be considered as a set of N points in six-dimensional μ-space, somewhat analogous to positions $\{\mathbf{r}_i\}$ in $V \subset \mathbb{R}^3$. We may then divide up this μ-space space into $\tilde{J}$ cells $\tilde{\Delta}_\alpha$, centered on $(\mathbf{r}_\alpha, \mathbf{v}_\alpha)$, of volume $|\tilde{\Delta}_\alpha|$. A macrostate $\tilde{M}$ is then specified by the (coarse-grained) number of particles in each $\tilde{\Delta}_\alpha$,

$$\tilde{M} = \{N_\alpha\}, \quad \alpha = 1, \ldots, \tilde{J} \ll N. \tag{12}$$

For dilute gases one can *neglect*, for typical configurations, the existence of interactions between the particles, although of course they still play a role in the dynamics now described by a succession of collisions between pairs of particles [3–5]. Under these conditions the coarse-grained energy of the system in the state $\tilde{M}$ is given by

$$\frac{1}{2}m \sum_\alpha N_\alpha \mathbf{v}_\alpha^2 = E, \tag{13}$$

$$\sum N_\alpha = N. \tag{14}$$

We do not therefore need to specify the energy separately and the phase space volume associated with such an $\tilde{M}$ is then readily computed to be [4b]

$$|\Gamma_{\tilde{M}}| = \prod_\alpha (N_\alpha!)^{-1} |\tilde{\Delta}_\alpha|^{N_\alpha}, \tag{15}$$

where we do not distinguish between configurations in which particle labels are interchanged. For large enough N and a judicious choice of the $\{\tilde{\Delta}_\alpha\}$ we can, for almost all X consistent with (13) and (14), use Stirling's formula in (15) and obtain

$$S_B(\tilde{M}) \sim -k \left\{ \sum_\alpha \left(\frac{N_\alpha}{|\tilde{\Delta}_\alpha|} \log \frac{N_\alpha}{|\tilde{\Delta}_\alpha|} \right) |\tilde{\Delta}_\alpha| - N \right\}. \tag{16}$$

Using $\tilde{M}$ we can associate with a typical X a coarse-grained density $f_X \sim N_\alpha / |\tilde{\Delta}_\alpha|$ in μ-space, i.e. such that $N_\alpha = \int_{\tilde{\Delta}_\alpha} \mathrm{d}\mathbf{x}\, \mathrm{d}\mathbf{v}\, f_X(\mathbf{x}, \mathbf{v})$. Eq. (16) then shows that, up to a constant (depending on N), the Boltzmann entropy $S_B(X)$ is given by the negative of Boltzmann's $\mathcal{H}$-function,

$$S_{\mathrm{gas}}(f) = -k \int_V \mathrm{d}\mathbf{x} \int_{\mathbb{R}^3} \mathrm{d}\mathbf{v}\, f(\mathbf{x}, \mathbf{v}) \log f(\mathbf{x}, \mathbf{v}), \tag{17}$$

where $f = f_X$. (We shall drop the subscript X unless we want to emphasize that f is associated with a given microstate X.) The maximum of $S_{\mathrm{gas}}(f)$ over all f which satisfy the conditions,

$$\int_V \mathrm{d}\mathbf{x} \int_{\mathbb{R}^3} \mathrm{d}\mathbf{v}\, f(\mathbf{x}, \mathbf{v}) = N, \tag{18}$$

$$\int_V \mathrm{d}\mathbf{x} \int_{\mathbb{R}^3} \mathrm{d}\mathbf{v}\, \frac{1}{2} m\mathbf{v}^2 f(\mathbf{x}, \mathbf{v}) = E \tag{19}$$

is given by the equilibrium distribution

$$f_{\mathrm{eq}} = \frac{N}{V} \left(\frac{2\pi kT}{m} \right)^{-3/2} \exp\left[-\frac{m\mathbf{v}^2}{2kT} \right], \tag{20}$$

where $kT = 2/3(E/N)$. f_{eq} coincides of course with the density $f_X(\mathbf{x}, \mathbf{v})$ obtained for a typical microstate X on the energy surface $H(X) = E$ when N is macroscopic (with deviations going to zero as $N \to \infty$).

When $f \neq f_{\text{eq}}$ then f and consequently $S_{\text{gas}}(f)$ will change in time. The microscopic version of the second law, discussed in Section 3, now says that typical $X \in \Gamma_{\tilde{M}}$ at the initial time $t = 0$, will have an $\tilde{M}_t = \tilde{M}(X_t)$ with the property that $S_{\text{B}}(\tilde{M}(X_t)) \geq S_{\text{B}}(\tilde{M}(X_{t'}))$, for $t \geq t'$. This means that $f_{X_t}(\mathbf{x}, \mathbf{v}) = f_t(\mathbf{x}, \mathbf{v})$ has to be such that $S_{\text{gas}}(f_t) \geq S_{\text{gas}}(f_{t'})$, for $t \geq t'$. This is exactly what happens for a dilute gas for which the time evolution of $f_t(\mathbf{x}, \mathbf{v})$ is well described by the Boltzmann equation (BE) [3–5] which we shall not write out here: see [14] for a rigorous derivation of the BE under suitable conditions. As shown by Boltzmann, in his famous $\mathcal{H}$-theorem, it indeed follows from the BE that $(\mathrm{d}/\mathrm{d}t)S_{\text{gas}}(f_t) \geq 0$, with equality holding only when $f(\mathbf{x}, \mathbf{v})$ is a local Maxwellian $\hat{f}(\mathbf{v}; n, \mathbf{u}, T)$ [5]

$$\hat{f} = n(\mathbf{x}) \left[\frac{2\pi k T(\mathbf{x})}{m}\right]^{-3/2} \exp\left\{\frac{-m[\mathbf{v} - \mathbf{u}(\mathbf{x})]^2}{2kT(\mathbf{x})}\right\}. \tag{21}$$

Eq. (21) defines LTE for a dilute gas with the $n(\mathbf{x})$, $\mathbf{u}(\mathbf{x})$ and $e(\mathbf{x})$ obtained from f in the usual way, $n = \int_{\mathbb{R}^3} f \, \mathrm{d}\mathbf{v}$, $\mathbf{u}(\mathbf{x}) = \int \mathbf{v} f \, \mathrm{d}\mathbf{v}/n$, $kT(\mathbf{x}) = (2/3)[e(\mathbf{x}) - (1/2)mn(\mathbf{x})\mathbf{u}^2(x)]/n(\mathbf{x})$. When (21) is substituted into (17) we obtain

$$S_{\text{gas}}(\hat{f}) = \int_V \mathrm{d}\mathbf{x} \, s_{\text{gas}}(e, n) \tag{22}$$

with

$$s_{\text{gas}}(e, n) = k\{\tfrac{3}{2} n \log(kT) - n(\log n - 1)\} + \text{const.}, \tag{23}$$

the Clausius entropy density given in (7), for a gas in LTE. Since $\hat{f}$ is not stationary unless n, e and $\mathbf{u}$ are uniform in the whole box, i.e. $\hat{f} = f_{\text{eq}}$, it is expected and partially proven [5,15], that starting with an initial $f_0(\mathbf{x}, \mathbf{v})$, which can be far from a local Maxwellian, $f_t(\mathbf{x}, \mathbf{v})$ will "rapidly" approach an f which is close to $\hat{f}(\mathbf{v}; n, \mathbf{u}, T)$ and will stay close to it while the local variables n, $\mathbf{u}$ and e change on a slower time scale. As the gradients become smaller this evolution will be hydrodynamic, i.e. n, $\mathbf{u}$, e will evolve according to the compressible Navier–Stokes equations, which will then bring the gas to equilibrium with $S_{\text{loc.eq.}}$ increasing with time.

Note that f satisfies the requirements for macro-variables discussed in the beginning of this section so that $S_{\text{gas}}(f)$ is indeed a useful entropy functional. The non-decrease of $S_{\text{gas}}(f_t)$ for f_t a solution for the BE is, as already noted, a consequence of Boltzmann's interpretation of the second law. As put by Boltzmann: "In one respect we have even generalized the entropy principle here, in that we have been able to define the entropy in a gas that is not in a stationary state" [4b, p. 75].

It is important to distinguish between $f_{X_t}(\mathbf{x}, \mathbf{v})$ and another object with the same name, the marginal one-particle (probability) distribution $F_1(\mathbf{x}, \mathbf{v}, t)$ obtained from an N-particle ensemble density evolving according to the Liouville equation. We mention here an instructive example in which $F_1(\mathbf{x}, \mathbf{v}, 0) = f_{X_0}(\mathbf{x}, \mathbf{v})$ but $F_1(\mathbf{x}, \mathbf{v}, t) \neq f_{X_t}(\mathbf{x}, \mathbf{v})$ so that $F_1(\mathbf{x}, \mathbf{v}, t)$ does not give an appropriate description of the macrostate of the system. Consider a macroscopic system of N non-interacting point particles, moving among a periodic array of scatterers in a macroscopic volume V [7,10]. Starting with a non-uniform initial density $f_{X_0}(\mathbf{x}, \mathbf{v})$ the time evolved $f_{X_t}(\mathbf{x}, \mathbf{v})$ will approach f which depends only on $|\mathbf{v}|$ and which will have a larger $S_{\text{gas}}(f)$. Since, however, $F_1(\mathbf{x}, \mathbf{v}, t)$ evolves according to the one-particle Liouville equation, $\iint F_1 \log F_1 \, \mathrm{d}\mathbf{x} \, \mathrm{d}\mathbf{v}$ remains constant in time. What is crucial here is that what one might have regarded as the obvious evolution equation for f_{X_t} for this system, namely the one-particle Liouville equation, in fact does not describe the evolution of f_{X_t} for times after which $F_1(\mathbf{x}, \mathbf{v}, t)$ has developed structure on the microscopic scale. (We note that when the periodicity of the scatterers is on the microscopic scale then for macroscopic times the spatial density profile $n_{X_t}(\mathbf{x})$ will satisfy a diffusion equation [7,10].)

5. The Boltzmann entropy of dense fluids not in LTE

As already noted, it is very important for the microscopic derivation of the second law and ipso facto for the increase of $S_{\text{gas}}(f_{X_t})$ that the initial microstate X_0 of the system under consideration be typical of Γ_{M_0}. Consider now the case when the interaction potential energy between the particles is not negligible so that (19) is just the kinetic energy K rather than the total energy E. The region $\Gamma_{\tilde{M}}$ will then include phase points with widely differing total energies. The set of microstates X of a system with a specified energy, $H(X) = E$, will then correspond to a small fraction of $\Gamma_{\tilde{M}(X)}$ unless E is such that almost all the points in $\Gamma_{\tilde{M}(X)}$ are also in the energy shell around E.

This illuminates the example considered by Jaynes [16]: one starts with a non-equilibrium system with energy E in which the kinetic energy K is larger than what it would be for an equilibrium system with the same energy E. The system will evolve towards equilibrium with a Maxwellian velocity distribution at temperature $T_\infty = (2/3k)K_E$, where K_E is the equilibrium kinetic energy corresponding to E. Assuming that f_0 is a Maxwellian with temperature T_0 we will have $T_\infty < T_0$ and so $S_{\text{gas}}(f_\infty) < S_{\text{gas}}(f_0)$. However, since the microstates corresponding to the initial situation just described are not typical for the macrostates Γ_{f_0}, this example does not contradict the typical second law behavior of $S(f_t)$.

To properly describe non-LTE macrostates for dense fluids let us suppose now that both $f(\mathbf{x}, \mathbf{v})$ and E are used to specify the macrostate $\tilde{M}$ for a fluid with Hamiltonian H

$$H = \sum \frac{1}{2} m \mathbf{v}_i^2 + \sum \phi(\mathbf{r}_i - \mathbf{r}_j) \tag{24}$$

in which interactions are important. We expect then that the evolution of a typical initial $X_0 \in \Gamma_{M_0}$, with $H(X_0) = E$ and $f_{X_0}(\mathbf{x}, \mathbf{v}) = f_0(\mathbf{x}, \mathbf{v})$, will indeed be such that $S_{\text{B}}(\tilde{M}(X_t)) = S_{\text{B}}(f_t, E)$ will increase with time in an actual system *even* if there is no autonomous evolution law for f_t.

To compute $S_{\text{B}}(f, E) = \log|\Gamma_{f,E}|$, let us consider first the case where the macrostate $\tilde{M}$ is specified by both $f(\mathbf{x}, \mathbf{v})$ and the local energy density $e(\mathbf{x})$. The particle, kinetic and potential energy densities $n(\mathbf{x})$, $K(\mathbf{x})$ and $\Phi(\mathbf{x})$ are determined from f and e,

$$n(\mathbf{x}) = \int_{\mathbb{R}^3} f(\mathbf{x}, \mathbf{v})\, \mathrm{d}\mathbf{v}, \qquad K(\mathbf{x}) = \frac{1}{2} m \int \mathbf{v}^2 f\, \mathrm{d}\mathbf{v}, \qquad \Phi(\mathbf{x}) = e(\mathbf{x}) - K(\mathbf{x}). \tag{25}$$

It is easy to see that the entropy corresponding to M can be split into momentum space and configuration space contributions

$$S_{\text{B}}(f, e) = S^{(\text{m})}(f) + S^{(\text{c})}(n, \Phi). \tag{26}$$

The momentum contribution $S^{(\text{m})}$ can be readily computed along the lines of formulas (15)–(17),

$$S^{(\text{m})}(f) = -\int_V \mathrm{d}\mathbf{x} \int_{\mathbb{R}^3} \mathrm{d}\mathbf{v}\, f(\mathbf{x}, \mathbf{v}) \log\left[\frac{f(\mathbf{x}, \mathbf{v})}{n(\mathbf{x})}\right] \tag{27}$$

while $S^{(\text{c})}(n, \Phi)$ is the configurational local equilibrium entropy corresponding to the density $n(\mathbf{x})$ and the potential energy density $\Phi(\mathbf{x})$. $S^{(\text{c})}$ is clearly the same as the configurational part of $S_{\text{loc.eq.}}$ computed at the energy density $e'(\mathbf{x})$ that corresponds to $\Phi(\mathbf{x})$ in LTE, i.e.

$$S^{(\text{c})}(n, \Phi) = S_{\text{loc.eq.}}(n, 0, e') - \frac{3}{2} k \int_V n(\mathbf{x}) \log T'(\mathbf{x})\, \mathrm{d}\mathbf{x} \tag{28}$$

where $T'(\mathbf{x})$ is the temperature corresponding to $e'(\mathbf{x})$. $S_{\text{loc.eq.}}$ is defined in (7) using the equilibrium $s(e, n)$ and the subtracted kinetic term corresponds to the first term on the right side of (23).

We can now obtain $S_{\text{B}}(f, E)$ by taking the sup of $S_{\text{B}}(f, e)$ over all energy densities $e(\mathbf{x})$ such that $\int_V e(\mathbf{x})\, \mathrm{d}\mathbf{x} = E$.

An alternative way to compute $S_B(f, E)$ is to note that

$$S_B(f, E) = S^{(m)}(f) + S^{(c)}(n, \Phi_{tot.}), \tag{29}$$

where $S^{(c)}(n, \Phi_{tot.})$ is the (configurational) entropy associated with the macro-variables n and $\Phi_{tot.}$ for

$$\Phi_{tot.} = E - \int K(\mathbf{x})\, d\mathbf{x}. \tag{30}$$

We now observe that $S^{(c)}(n, \Phi_{tot.})$ must agree with the configurational entropy of a system with a Hamiltonian

$$H' = \frac{m}{2}\sum \mathbf{v}_i^2 + \sum \phi(\mathbf{r}_i - \mathbf{r}_j) + \sum u(\mathbf{r}_i) \tag{31}$$

in equilibrium at energy E', *when* we choose E' and $u(\mathbf{x})$ in such a way that the equilibrium density is equal to $n(\mathbf{x})$ and the total (coarse-grained) internal potential energy has the value $\Phi_{tot.}$, i.e. $\sum \phi(\mathbf{r}_i - \mathbf{r}_j) = \Phi_{tot.}$, where $\Phi_{tot.}$ is given by (30). Letting then $S(E', N, V; u)$ be the equilibrium entropy of the system (31) we have

$$S^{(c)}(n, \Phi_{tot.}) = S(E', N, V; u) - \tfrac{3}{2}Nk\log(kT'), \tag{32}$$

where $T' = (2/3k)(K_{E'}/N)$ is the temperature corresponding to the energy E' and the right-hand side is the configurational part of the equilibrium entropy of system (30) at energy E'. (To actually compute $S(E', N, V; u)$ and find the appropriate $u(\mathbf{x})$ for a specified $n(\mathbf{x})$ and Φ we would have to use the canonical or grand canonical ensembles and the corresponding Gibbs entropies, see Section 6.)

These considerations simplify for a system of hard spheres where the interactions do not contribute to the energy, i.e. $E = K$. The phase space domain for a specified f is then just a direct product of the configurational and momentum space regions. The configuration space itself is modified from what it is for a non-interacting system to exclude all microstates X such that the distance between any pair of particles is less than a, the hard sphere diameter. Eq. (32) now assumes the form

$$S_{hs}(f) = S^{(m)}(f) + S_{hs}^{(c)}(n), \tag{33}$$

where E is specified by f and $S_{hs}^{(c)}(n)$ is the configurational part of the entropy of an equilibrium system of hard spheres kept at a non-uniform density $n(\mathbf{x}) = \int_{\mathbb{R}^3} f(\mathbf{x}, \mathbf{v})\, d\mathbf{v}$ by some external potential $u(\mathbf{x})$.

The entropy function $S_{hs}(f)$ in (33) is the same as that used by Resibois [6] (in a different form) to obtain an $\mathcal{H}$-theorem for the modified Enskog equation (MEE). It is generally believed that the MEE, which is a heuristic extension of the BE, accurately describes the evolution of $f_t(\mathbf{x}, \mathbf{v})$ for a moderately dense hard sphere fluid, say $na^3 \leq 0.1$, where a is the hard sphere diameter [5,6]. The remarkable result, proved by Resibois [6], is that when $f_t(\mathbf{x}, \mathbf{v})$ evolves according to the MEE then

$$\frac{d}{dt} S_{hs}(f_t) \geq 0 \tag{34}$$

with equality holding only when $f = f_{eq}$. (This is actually a stronger statement about entropy increase than what is given by the BE where collisions alone do not change $S_{gas}(f)$ when f is a local Maxwellian. The reason for this difference is the non-locality of the collisions in the MEE, see [6]. We still have, however, $S_{hs}(f) = S_{loq.eq.}(e, \mathbf{u}, n)$ when $f = \hat{f}(\mathbf{v}; n, \mathbf{u}, T)$ and the hydrodynamic variables change slowly in space.)

Resibois was driven to an expression for S_{hs} equivalent to (33) by the structure of the MEE and argued that it should have an intrinsic significance. Our identification of $S_{hs}(f)$ with the Boltzmann entropy $S_B(f, E)$, i.e. as the log of the phase space volume for such a non-equilibrium system, completely justifies Resibois' intuition. It further shows the necessity of the modification of the original Enskog equation: as we have pointed out repeatedly, any deterministic evolution equation arising from the microscopic dynamics for macro-variables such as f must be such that the corresponding Boltzmann entropy satisfies an $\mathcal{H}$-theorem for that equation. The unmodified Enskog equation apparently does not have that property [5,6].

Let us make the statement about the increase of $S(f, E)$ a bit more concrete. Define

$$T_\Phi(n) = \left[\frac{\partial s^{(c)}(n, \Phi)}{\partial \Phi}\right]^{-1}, \tag{35}$$

where $s^{(c)}(n, \Phi)$ is the configurational entropy per unit volume for a system with Hamiltonian (24) at uniform particle density n and uniform potential energy density Φ. $T_\Phi(n)$ is the inverse of the function $\Phi(n, T)$ relating Φ to the temperature T for this system. Similarly

$$\frac{\partial s^{(c)}(n, \Phi)}{\partial n} = \frac{-\mu_\Phi(n, T_\Phi)}{T_\Phi} = \frac{-\mu(n, T_\Phi)}{T_\Phi} - \frac{3}{2}k \log T_\Phi + \text{const.}, \tag{36}$$

where $\mu(n, T)$ is the chemical potential of such an equilibrium system. (Of course for an equilibrium system $T_\Phi = T_K = (m/3)\langle(\mathbf{v} - \mathbf{u})^2\rangle/k$ where T_K is the kinetic temperature of the system.) Our definitions are motivated by (1), (11) and the structure of H in (24).

Turning now to non-equilibrium systems with macrostates given by $f(\mathbf{x}, \mathbf{v})$ and $e(\mathbf{x})$ we have, for $\Phi = \Phi(\mathbf{x})$,

$$S^{(c)}(n, \Phi) = \int_V \mathrm{d}\mathbf{x}\, s^{(c)}(n(\mathbf{x}), \Phi(\mathbf{x})). \tag{37}$$

To obtain $S^{(c)}(n, \Phi_{\text{tot.}})$ we have to take the sup of $S^{(c)}(n, \Phi)$ over all $\Phi(\mathbf{x})$ such that

$$\int_V \Phi(\mathbf{x})\, \mathrm{d}\mathbf{x} = \Phi_{\text{tot.}}. \tag{38}$$

Using (35) this yields immediately, as might be expected, that the sup is achieved when $T_{\Phi(\mathbf{x})}(n(\mathbf{x})) = \text{const.}$, which we shall call $T_{\Phi_{\text{tot.}}}$ since its value is determined by the requirement (38).

We are now able, after some manipulations, to obtain

$$\begin{aligned}\frac{\mathrm{d}S(f_t, E)}{\mathrm{d}t} &= -k \iint \left(\log \frac{f_t}{\tilde{f}_t}\right) \frac{\partial f_t}{\partial t} \mathrm{d}\mathbf{x}\, \mathrm{d}\mathbf{v} - \int \mathrm{d}\mathbf{x} \left[\frac{\mu_\Phi(n, T_{\Phi_{\text{tot.}}})}{T_{\Phi_{\text{tot.}}}} + \frac{\mu_K(n, T_K)}{T_K}\right] \frac{\partial n(x, t)}{\partial t} \\ &\quad + \int \mathrm{d}\mathbf{x}\, T_K^{-1} \frac{\partial}{\partial t} K_{\mathbf{u}}(\mathbf{x}, t) + T_{\Phi_{\text{tot.}}}^{-1} \frac{\mathrm{d}\Phi_{\text{tot.}}}{\mathrm{d}t},\end{aligned} \tag{39}$$

where

$$\mu_K = -\frac{3}{2}k \log T_K(\mathbf{x}, t) + \text{const.}, \qquad T_K(\mathbf{x}, t) = \frac{2}{3k} \frac{K_{\mathbf{u}}(\mathbf{x}, t)}{n(\mathbf{x}, t)}, \tag{40}$$

with

$$K_{\mathbf{u}}(\mathbf{x}, t) = \frac{m}{2} \int (\mathbf{v} - \mathbf{u}(\mathbf{x}, t))^2 f(\mathbf{x}, \mathbf{v}, t)\, \mathrm{d}\mathbf{v} = K(\mathbf{x}, t) - \frac{m}{2} n(\mathbf{x}, t) \mathbf{u}^2(\mathbf{x}, t)$$

and where

$$\tilde{f}_t = n(\mathbf{x}, t)(2\pi m T_K)^{-3/2} \exp\left[-\frac{m}{2} \frac{(\mathbf{v} - \mathbf{u}(\mathbf{x}, t))^2}{k T_K}\right] \tag{41}$$

is a local Maxwellian with parameters T_K, n, K and $\mathbf{u}$ computed from f_t. The first term in (39) corresponds to changes in the entropy due to the redistribution of velocities, while the other terms are as expected from thermodynamic considerations.

Eq. (39) simplifies for a spatially uniform system for which $\mathbf{u} = 0$, $(\partial n/\partial t) = 0$ and T_K is independent of $\mathbf{x}$. If, furthermore, f_t happens to be a Maxwellian with a temperature T_K (a case considered by Jaynes [16]), then (at that instant)

$$\frac{\mathrm{d}S(f_t, E)}{\mathrm{d}t} = \frac{\mathrm{d}K}{\mathrm{d}t}[T_K^{-1} - T_{\Phi_{\mathrm{tot.}}}^{-1}], \tag{42}$$

where $K = E - \Phi_{\mathrm{tot.}}$ is the total kinetic energy of the system. Our assertion is then that $\mathrm{d}K/\mathrm{d}t$ has the same sign as $(T_{\Phi_{\mathrm{tot.}}} - T_K)$, which is certainly expected even in the absence of any deterministic equation for f.

6. Other kinetic equations

The Boltzmann and modified Enskog equations are appropriate for systems in which the interaction between the particles can be represented by a succession of uncorrelated binary encounters or collisions. The extension of these equation to dense, non-hard sphere fluids or to mixtures of hard spheres is not straightforward [17]. We are not aware of any results about $\mathcal{H}$-theorems for such equations.

For systems with dominant long range interactions, such as plasmas, the time evolution of f_t (where f has several components) is determined in suitable regimes of temperature and density by a Vlasov equation combined with Boltzmann, Balescu-Lenard, or Landau collision terms [18]. The Vlasov term describes in a mean-field way the long range interaction between the particles. These interactions contribute to the energy, which is determined entirely by f, but not directly to the entropy of the system, i.e. the entropy continues to be given (at the level of approximation considered) by $S_{\mathrm{gas}}(f)$ in (17). When the short-range collisions are neglected, the smooth solutions of the Vlasov equation, like those of the Euler equations, leave the entropy unchanged. The inclusion of the short-range collision terms then provides an $\mathcal{H}$-theorem for $S_{\mathrm{gas}}(f_t)$. This is analogous to what happens when the viscosity terms are added to the Euler equations to yield the Navier–Stokes equations.

Going beyond deterministic equations, there must also be an $\mathcal{H}$-theorem when the macro-variables undergo a stochastic Markovian evolution, since, in the thermodynamic limit, the probability of a transition to lower entropy is of much smaller order than the (order unity) probabilities describing the Markov process. However, we know of no examples of such macro-variables. (The small-scale stochastic correction to the deterministic evolution of macro-variables will of course fail to obey an $\mathcal{H}$-theorem, since the probability of these fluctuations is of the same order as the exponential of the entropy changes.)

7. Concluding remarks

The entropy of a macroscopic state, defined by Eqs. (1) and (6), is clearly a property of an individual macroscopic system specified by macro-variables M. We neither have nor need ensembles to observe the time asymmetric evolution of the color profile of a glass of water in which we dissolve a capsule of purple ink. The appropriate choice of M for this process is clearly that corresponding to dividing the glass into a suitable large number of little cubes and specifying the coarse-grained fraction of ink molecules in each cube as was done in Section 3. The exact number of little cubes, as long as it is still small compared to the number of ink molecules, will not affect S_{B} to leading order in the number of molecules. To this order S_{B} will coincide with $S_{\mathrm{loc.eq.}}$. The evolution of M_t will be given by the solution to a diffusion equation and $S_{\mathrm{B}}(M_t)$ will satisfy the second law.

As argued in Section 3, this behavior can be understood fully from Boltzmann's microscopic interpretation of entropy. This direct explanatory connection between Boltzmann's entropy and the observed behavior of individual macroscopic systems seems lacking in other definitions of entropy in which probability distributions are a key ingredient. The best known of these is the Gibbs entropy,

$$S_{\mathrm{G}}(\rho) = -k \int_{\Gamma} \rho \log \rho \, \mathrm{d}X, \tag{43}$$

where $\rho(X)$ is some given probability (ensemble) density. Clearly if $\rho = \rho_M$

$$\rho_M(X) = \begin{cases} |\Gamma_M|^{-1} & \text{if } X \in \Gamma_M, \\ 0 & \text{otherwise,} \end{cases} \tag{44}$$

then

$$S_{\mathrm{G}}(\rho_M) = k \log |\Gamma_M| = S_{\mathrm{B}}(M). \tag{45}$$

For a system specified by the thermodynamic variables $M(X) = (E, N, V)$, ρ_M corresponds to the microcanonical ensemble. Using the equivalence of equilibrium ensembles for macroscopic systems, Eq. (45) holds also for the canonical and grandcanonical ensembles, in the thermodynamic limit. In fact S_{G} is of paramount importance both in the mathematical foundations and practical applications of equilibrium statistical mechanics. On the other hand, as is very well known, S_{G}, unlike S_{B}, does not change in time for an isolated system evolving under Hamiltonian dynamics [$S_{\mathrm{G}}(t) = S_{\mathrm{G}}(\rho_t)$, where $\rho_t(X)$ is $\rho_0(X_{-t})$]. It is therefore inappropriate, we believe, to use S_{G} or quantities like it in "derivations" of the second law without explicitly considering typical behavior. Attempts to remedy this through the use of coarse-grained ensembles may be useful mathematically, but conceptually they are just variations on the Boltzmann entropy [19].

The Boltzmann entropy itself is, as indicated by the use of macro-variables for its very formulation, meaningful only for macroscopic systems. For such systems one can speak of the behavior of the macro-variables M as arising from the evolution of a typical microstate in Γ_{M_0}. It might still be true that a system containing just a few particles exhibits ergodicity, mixing, positive Lyapunov exponents, etc.; this is true e.g. for a particle moving among fixed convex scatterers on a torus (Sinai billiard). But the physical system of one such particle will not exhibit any time asymmetric behavior, corresponding to the diffusion of the purple ink in the glass of water. Unlike the glass of inky water or a very large number of particles moving among such scatterers (see the discussion at the end of Section 4) a film of the particle's motion run backwards will look the same as one run forward.

The situation is different when one considers open systems, e.g. systems in contact with thermal reservoirs [19,20]. The time evolution of the microstate X of such a system is then no longer given by a Hamiltonian since the system is not isolated and $S_{\mathrm{G}}(\rho_t)$ need no longer be constant in time. It is in fact reasonable in some cases to treat X as a random variable evolving via a stochastic Markovian dynamics. It is then easy to show that when the Markov process has a stationary density $\bar{\rho}(X)$ then the relative entropy

$$S_{\mathrm{G}}(\rho_t|\bar{\rho}) = -k \int_{\Gamma} \rho_t \log\left(\frac{\rho_t}{\bar{\rho}}\right) \mathrm{d}X \tag{46}$$

increases monotonically in time.

A different situation, of current interest, in which $S_{\mathrm{G}}(\rho)$ is not constant is that of a closed system evolving under a deterministic non-Hamiltonian thermostated dynamics [19,20]. Starting with an initial density $\rho(X, 0)$, uniform (or absolutely continuous) with respect to the appropriate Lebesgue measure, the dynamics leads to $\rho(X, t) \to \bar{\rho}$, as $t \to \infty$, with $\bar{\rho}$ singular with respect to Lebesgue measure ($\bar{\rho}$ is generally an SRB measure), and $S_{\mathrm{G}}(\rho) \to -\infty$ decreasing with time in such a way that

$$\frac{\mathrm{d}}{\mathrm{d}t} S_{\mathrm{G}}(\rho) \to -\sigma, \tag{47}$$

with $\sigma > 0$. This σ is frequently interpreted as an "entropy production" and indeed has, in some cases, a form similar to the hydrodynamic entropy production in a stationary open system in which there are currents. The connection between the steady states of systems evolving under the invented thermostated dynamics and those obtained from more realistic models for which the stationary $\bar{\rho}$ is not singular is still not entirely clear. The same is true for the correct identification of entropy in such systems [17–23].

It should be noted that the word entropy is used very widely in contexts other than that of macroscopic physical systems. There is the Shannon information entropy—which is formally similar to S_G, but is designed for situations which have apparently nothing to do with thermodynamics—the Kolmogorov–Sinai and topological entropies of dynamical systems, etc. These entropies are clearly very useful and clearly different from the Clausius and Boltzmann entropies. Surprisingly, there are frequently some unexpected deep connections between these different entropies which are very interesting [24–26].

Let us finally discuss the choice of appropriate macro-variables M in terms of which to describe a particular non-equilibrium system of interest in some microstate X which is far from LTE. We do not want to choose too coarse a description: one for which X is not typical of $\Gamma_{M(X)}$. This occurred in the example of Jaynes discussed in Section 5; the problem could be remedied there by introducing more refined macrostates, given not just by f but by f and E. We also do not want to use a description more detailed than is relevant to macroscopic behavior.

What we are after is a useful minimal description via macro-variables M, adapted to the situation (corresponding to a microstate X) under consideration. This is always achieved when M is such that (1) M_{X_t} typically obeys an autonomous deterministic evolution law, such as those corresponding to hydrodynamics, the BE, or the MEE and (2) X is typical in this sense of $\Gamma_{M(X)}$.

Acknowledgements

We want to thank Raffaello Esposito, Giovanni Gallavotti, Michael Kiessling, Christian Maes, David Ruelle and Roderich Tumulka for useful discussions. J.L. also wants to thank Harry Swinney and Constantino Tsallis for their hospitality at the International Workshop on Anomalous Distributions, Nonlinear Dynamics and Nonextensivity, Santa Fe, November 2002, which prompted this article. Research supported by NSF Grant DMS 01-279-26 and AFOSR Grant AF 49620-01-1-0154.

References

[1] H.B. Callen, Thermodynamics and an Introduction to Thermostatistics, Wiley, New York, 1960.

[2] S.R. de Groot, P. Mazur, Nonequilibrium Thermodynamics, North-Holland, Amsterdam, 1959.

[3] For a historical survey see: S.G. Brush, The Kind of Motion We Call Heat, North-Holland, Amsterdam, 1976; M. Klein, The development of Boltzmann's statistical ideas, in: E.G.D. Cohen, W. Thirring (Eds.), The Boltzmann Equation, Acta Physica Austriaca Suppl. X, Vienna, 1973, p. 53.

[4] (a) For a collection of basic papers from the second half of the nineteenth century on the subject (in English Trans.), see: S.G. Brush, Kinetic Theory, vols. 1 and 2, Pergamon, Elmsford, NY, 1965–1966; The apparently little known 1874 article by W. Thomson (Lord Kelvin), The kinetic theory of the dissipation of energy, Proceedings of the Royal Society of Edinburgh 8, 1874, pp. 325–328, reprinted there, is highly recommended;
(b) L. Boltzmann, Vorlesungen über Gastheorie, 2 vols., Barth, Leipzig, 1896, 1898 (English Trans.: S.G. Brush, Lectures on Gas Theory, Cambridge University Press, London, 1964, reprinted Dover, 1995).

[5] C. Cercignani, The Boltzmann Equation and Its Applications, Springer-Verlag, Berlin, 1988.

[6] P. Resibois, $\mathcal{H}$-theorem for the (modified) nonlinear Enskog equation, J. Stat. Phys. 19 (1978) 593.

[7] (a) J.L. Lebowitz, Boltzmann's entropy and time's arrow, Phys. Today 46 (1993) 32–38; Microscopic reversibility and macroscopic behavior: physical explanations and mathematical derivations, in: J.J. Brey, J. Marro, J.M. Rubí, M. San Miguel (Eds.), Proceedings of the Sitges Conference on 25 Years of Non-Equilibrium Statistical Mechanics, Barcelona, Spain, 1994, Lecture Notes in Physics, Springer-Verlag,

Berlin, 1995; Microscopic origins of irreversible macroscopic behavior, Physica A 263 (1999) 516–527;
(b) S. Goldstein, Boltzmann's approach to statistical mechanics, in: J. Bricmont, et al. (Eds.), Chance in Physics: Foundations and Perspectives, Lecture Notes in Physics 574, Springer-Verlag, Berlin, 2001. cond-mat/0105242.
[8] R. Clausius, The nature of the motion we call heat, Ann. Phys. [2] 125 (1865) 353, reprinted in [4a].
[9] L. Onsager, Thermodynamics and some molecular aspects of biology, in: G.C. Quarton, et al. (Eds.), The Neurosciences: A Study Program, Rockefeller University Press, New York, 1967, pp. 75–79.
[10] H. Spohn, Large Scale Dynamics of Interacting Particle Systems, Springer-Verlag, Berlin, 1991.
[11] L. Landau, S. Lifshitz, Fluid Mechanics, Butterworth-Heinemann, 1987.
[12] G. Gallavotti, Foundations of Fluid Dynamics, Springer-Verlag, Berlin, 2002.
[13] A. Einstein, Autobiographical notes, in: P.A. Schlipp (Ed.), Albert Einstein, Philosopher–Scientist, The Library of Living Philosophers, 1949, p. 43 (sixth printing 1995).
[14] O. Lanford, Time evolution of large classical systems, in: J. Moser (Ed.), Dynamical Systems, Theory and Applications, Lecture Notes in Physics 38, Springer-Verlag, Berlin, 1975; Physica A 106 (1981) 70.
[15] E. Carlen, M. Carvalho, Entropy production estimates for Boltzmann equation with physically realistic collision kernels, J. Stat. Phys. 74 (1994) 743–782;
L. Desvillettes, C. Villani, On the trend to global equilibrium for spatially inhomogeneous kinetic systems: the Boltzmann equation, Preprint.
[16] E.T. Jaynes, Violation of Boltzmann's H theorem in real gases, Phys. Rev. A 4 (1971) 747–750.
[17] J.A. McLennan, Introduction to Non-equilibrium Statistical Mechanics, Prentice-Hall, Englewood Cliffs, NJ, 1989.
[18] R. Balescu, Transport Processes in Plasmas, North-Holland, Amsterdam, 1988.
[19] C. Maes, K. Netocny, Time-reversal and entropy, J. Stat. Phys. 110 (2003) 269–310.
[20] F. Bonetto, J.L. Lebowitz, L. Rey-Bellet, Fourier's law: a challenge to theorists, in: A. Fokas, A. Grigoryan, T. Kibble, B. Zegarlinski (Eds.), Mathematical Physics 2000, Imperial College Press, London, 2000, pp. 128–150.
[21] G. Gallavotti, E.G.D. Cohen, Dynamical ensembles in nonequilibrium statistical mechanics, Phys. Rev. Lett. 95 (74) (1995) 2694;
G. Gallavotti, Nonequilibrium thermodynamics, 2002, Preprint.
[22] D. Ruelle, Smooth dynamics and new theoretical ideas in nonequilibrium statistical mechanics, J. Stat. Phys. 95 (1999) 393–468; Extending the definition of entropy to nonequilibrium steady states, Proc. Natl. Acad. Sci. U.S.A. 100 (2003) 3054–3058. math phys archive 03-98.
[23] N. Chernov, J.L. Lebowitz, Stationary nonequilibrium states in boundary driven Hamiltonian systems: shear flow, J. Stat. Phys. 86 (1997) 953–990. cond-mat/9703097, Texas 97–113.
[24] Ya.G. Sinai, Introduction to Ergodic Theory, Lecture 14, Princeton University Press, Princeton, NJ, 1976.
[25] G. Gallavotti, Statistical Mechanics: A Short Treatise, Springer-Verlag, Berlin, 1999.
[26] Ya. Sinai (Ed.), Dynamical Systems, Ergodic Theory and Applications, Springer-Verlag, Berlin, 2000.

PRL 96, 050403 (2006) PHYSICAL REVIEW LETTERS week ending 10 FEBRUARY 2006

Canonical Typicality

Sheldon Goldstein,[1,*] Joel L. Lebowitz,[1,†] Roderich Tumulka,[2,‡] and Nino Zanghì[3,§]

[1]*Departments of Mathematics and Physics, Hill Center, Rutgers, The State University of New Jersey, 110 Frelinghuysen Road, Piscataway, New Jersey 08854-8019, USA*

[2]*Mathematisches Institut, Eberhard-Karls-Universität, Auf der Morgenstelle 10, 72076 Tübingen, Germany*

[3]*Dipartimento di Fisica dell'Università di Genova and INFN sezione di Genova, Via Dodecaneso 33, 16146 Genova, Italy*

(Received 4 November 2005; published 8 February 2006)

It is well known that a system S weakly coupled to a heat bath B is described by the canonical ensemble when the composite $S + B$ is described by the microcanonical ensemble corresponding to a suitable energy shell. This is true for both classical distributions on the phase space and quantum density matrices. Here we show that a much stronger statement holds for quantum systems. Even if the state of the composite corresponds to a single wave function rather than a mixture, the reduced density matrix of the system is canonical, for the overwhelming majority of wave functions in the subspace corresponding to the energy interval encompassed by the microcanonical ensemble. This clarifies, expands, and justifies remarks made by Schrödinger in 1952.

DOI: 10.1103/PhysRevLett.96.050403 PACS numbers: 05.30.Ch, 03.65.−w, 05.20.Gg

A quantum system in thermal equilibrium at inverse temperature β is described by the canonical density matrix

$$\rho_\beta = \frac{1}{Z}\exp(-\beta H^{(S)}), \tag{1}$$

where $H^{(S)}$ is the system Hamiltonian and

$$Z = \mathrm{tr}\exp(-\beta H^{(S)}).$$

The usual justification for (1) is that it is the reduced density matrix of the system when it is weakly coupled to a heat bath, and the composite system is described by the microcanonical density matrix at a suitable total energy E.

More explicitly, one assumes that it is permissible to neglect the relatively small interaction between the system and the bath, so that the total Hamiltonian of the composite $S + B$ is given by

$$H = H^{(S+B)} = H^{(S)} + H^{(B)}. \tag{2}$$

The composite $S + B$ is then assumed to be represented by a microcanonical ensemble in some energy interval $[E, E + \delta]$, where δ is small on the macroscopic scale, $\delta \ll E$, but large enough for the interval to contain very many eigenvalues. The corresponding microcanonical density matrix is

$$\rho_{E,\delta} = (\dim \mathcal{H}_{[E,E+\delta]})^{-1} P_{\mathcal{H}_{[E,E+\delta]}}, \tag{3}$$

with $P_{\mathcal{H}_{[E,E+\delta]}}$ the projection to $\mathcal{H}_{[E,E+\delta]}$, the spectral subspace for H associated with energies in the interval $[E, E + \delta]$ in the Hilbert space $\mathcal{H} = \mathcal{H}^{(S+B)} = \mathcal{H}^{(S)} \otimes \mathcal{H}^{(B)}$. One readily proves (see below) that in the thermodynamic limit, when the size, i.e., the number of components N of the heat bath, goes to infinity while $E/N \to e$, the reduced density matrix of the system S,

$$\rho^{(S)} = \mathrm{tr}^{(B)}\rho_{E,\delta}, \tag{4}$$

is equal to ρ_β. Here $\mathrm{tr}^{(B)}$ denotes the partial trace over $\mathcal{H}^{(B)}$ and $\beta = \beta(e)$.

In this Letter we show how this result can be substantially strengthened: we prove that, in the thermodynamic limit, the reduced density matrices of the overwhelming majority of the wave functions of $S + B$ are canonical. We call this statement *canonical typicality.* As a consequence of canonical typicality, it follows that the canonical ensemble is even more inevitable in quantum mechanics—arising as it does there without the invocation of any genuine randomness—than it is classically. Results in this direction were first obtained by Schrödinger [1], and later by Gemmer and Mahler [2]; related results have been obtained by Tasaki [3] (see later).

Typicality in quantum mechanics, as well as in classical mechanics, involves a probability distribution on the possible microstates of the system, the distribution in terms of which "overwhelming majority" is to be understood. In classical mechanics these microstates are points in the appropriate phase space and the distribution is then a measure on this phase space. To define typicality for quantum systems we shall take the microstates to be wave functions, i.e., points on the unit sphere of $\mathcal{H}$ (up to a phase). Even with this identification it may not be clear which distribution is appropriate for the (composite) system described by the density matrix $\rho_{E,\delta}$. Here we take that to be the probability distribution proposed long ago by Schrödinger [1,4] and Bloch [5]: it is the (normalized) uniform (surface area) measure $u_{E,\delta}$ on the unit sphere in the subspace $\mathcal{H}_{[E,E+\delta]}$, i.e., the uniform probability distribution over all normalized wave functions Ψ with energies in $[E, E + \delta]$. If we expand Ψ in terms of energy eigenfunctions $|E_\alpha\rangle$ of H, $\Psi = \sum c_\alpha |E_\alpha\rangle$, where the sum is restricted to levels with energies in the interval $[E, E + \delta]$, then $u_{E,\delta}$ corresponds to the uniform distribution on the surface of the sphere $\sum |c_\alpha|^2 = 1$. This measure was shown in [5] to be, in a well-defined sense, the most appropriate distribution corresponding to the microcanonical density matrix, i.e., such that

$$\rho_{E,\delta} = \int u_{E,\delta}(d\Psi)|\Psi\rangle\langle\Psi|. \quad (5)$$

Let ρ^{Ψ} denote the reduced density matrix of the system, given that $S + B$ is in a pure state $\Psi \in \mathcal{H}_{[E,E+\delta]}$,

$$\rho^{\Psi} = \mathrm{tr}^{(B)}|\Psi\rangle\langle\Psi|. \quad (6)$$

We may then ask this: for which wave functions Ψ is ρ^{Ψ} (approximately) of the canonical form (1)? We make the standard assumption that both $H^{(S)}$ and $H^{(B)}$ have pure point spectrum and are bounded from below. In $\mathcal{H}^{(S)}$ ($\mathcal{H}^{(B)}$) we choose an eigenbasis of $H^{(S)}$ ($H^{(B)}$), denoted $|E_1^{(S)}\rangle, |E_2^{(S)}\rangle, \ldots$ ($|E_1^{(B)}\rangle, |E_2^{(B)}\rangle, \ldots$), with corresponding eigenvalues $E_1^{(S)} \leq E_2^{(S)} \leq \cdots$ ($E_1^{(B)} \leq E_2^{(B)} \leq \cdots$).

First, we note the following: saying that for the majority of Ψ's, ρ^{Ψ} is close to ρ_β is equivalent to saying that if Ψ is *randomly* chosen with distribution $u_{E,\delta}$ then with overwhelming *probability* ρ^{Ψ} is close to ρ_β. From now on we will thus regard Ψ as a (Hilbert-space-valued) random variable.

We begin by recalling the standard derivation of the canonical ensemble from the microcanonical. One has for the reduced density matrix $\rho^{(S)}$ of the system that

$$\rho^{(S)} = (\dim \mathcal{H}_{[E,E+\delta]})^{-1} \sum_i \dim(\mathcal{H}_i^{(B)}) |E_i^{(S)}\rangle\langle E_i^{(S)}|, \quad (7)$$

where

$$\mathcal{H}_i^{(B)} = \mathcal{H}^{(B)}_{[E-E_i^{(S)}, E-E_i^{(S)}+\delta]} \quad (8)$$

is the spectral subspace for $H^{(B)}$ associated with energies in the interval $[E - E_i^{(S)}, E - E_i^{(S)} + \delta]$. It is then more or less clear, and can be rigorously proven under suitable conditions, that when the bath is sufficiently large $\rho^{(S)} \approx \rho_\beta$, with $\beta = dS(E)/dE$ where $S(E)$ is the bath's entropy. This follows from the basic fact that $S(E) \approx \log \dim(\mathcal{H}^{(B)}_{[E,E+\delta]})$, so that

$$\dim(\mathcal{H}_i^{(B)}) \approx e^{S(E-E_i^{(S)})} \approx e^{S(E)-\beta E_i^{(S)}} \sim e^{-\beta E_i^{(S)}}. \quad (9)$$

More precisely, one proves that $\rho^{(S)} \to \rho_\beta$ in the thermodynamic limit with $\beta = ds(e)/de$, where $s(e) = \lim[S(E)/N]$ and $e = \lim[E/N]$.

Thus to demonstrate canonical typicality it suffices to establish that (7) holds, at least approximately, when $\rho^{(S)}$ is replaced by ρ^{Ψ} for typical $\Psi \in \mathcal{H}_{[E,E+\delta]}$. A key step of our argument is to note that the uniformly distributed random vector Ψ can always be regarded as arising by normalization

$$\Psi = \frac{\Phi}{\| \Phi \|} \quad (10)$$

from a Gaussian random vector $\Phi \in \mathcal{H}_{[E,E+\delta]}$ with mean zero and covariance given by the identity operator on $\mathcal{H}_{[E,E+\delta]}$. This means that in the decomposition

$$\Phi = \sum_i \sum_j C_{ij} |E_i^{(S)}\rangle |E_j^{(B)}\rangle, \quad (11)$$

the real and imaginary parts, $\mathrm{Re}C_{ij}$ and $\mathrm{Im}C_{ij}$, of the coefficients are independent real Gaussian random variables with mean zero and variance $1/2$ for those i and j for which $E_i^{(S)} + E_j^{(B)} \in [E, E+\delta]$ (and $C_{ij} = 0$ otherwise). We obtain from (11) that

$$\Phi = \sum_i |E_i^{(S)}\rangle |\Phi_i\rangle \quad (12)$$

with

$$|\Phi_i\rangle = \sum_{j \in I_i} C_{ij} |E_j^{(B)}\rangle, \quad (13)$$

where I_i is the set of bath levels j such that $E_j^{(B)} \in [E - E_i^{(S)}, E - E_i^{(S)} + \delta]$, whence the reduced density matrix (6) is of the form

$$\rho^{\Psi} = \frac{1}{\| \Phi \|^2} \mathrm{tr}^{(B)} |\Phi\rangle\langle\Phi| = \frac{1}{\| \Phi \|^2} \sum_{i,i'} \langle \Phi_i | \Phi_{i'} \rangle |E_i^{(S)}\rangle\langle E_{i'}^{(S)}|. \quad (14)$$

Now, if δ were so small that the system's energy spacings $\Delta E_i^{(S)} = E_{i+1}^{(S)} - E_i^{(S)}$ are all greater than δ, then the relevant energy intervals I_i for the heat bath would be pairwise disjoint and the Φ_i pairwise orthogonal,

$$\langle \Phi_i | \Phi_{i'} \rangle = \delta_{ii'} \| \Phi_i \|^2 . \quad (15)$$

We argue now that Eq. (15) will continue to hold, at least approximately, even without the above assumption on $\Delta E_i^{(S)}$. This is so because, when I_i and $I_{i'}$ have significant overlap, the contributions to Φ_i and $\Phi_{i'}$ corresponding to the sum over the terms in (13) belonging to both I_i and $I_{i'}$ will typically be approximately orthogonal. To see this, note that these sums form two independent random vectors, with uniformly distributed directions, in a high-dimensional space. As such, they are, with probability approaching unity, nearly orthogonal since the expected value of the absolute square of the dot product of two independent random unit vectors in an n-dimensional space is, by symmetry, $1/n$.

From the representation (13) we have that

$$\| \Phi_i \|^2 = \sum_j |C_{ij}|^2, \quad (16)$$

so that $\| \Phi_i \|^2$ is the sum of the $N_i = \dim(\mathcal{H}_i^{(B)})$ independent, identically distributed random variables $|C_{ij}|^2$ with mean 1. It thus follows, by the law of large numbers, that typically

$$\| \Phi_i \|^2 \approx \dim(\mathcal{H}_i^{(B)}). \tag{17}$$

We thus have for the reduced density matrix (6), using (14), that typically

$$\rho^\Psi \approx (\dim \mathcal{H}_{[E,E+\delta]})^{-1} \sum_i \dim(\mathcal{H}_i^{(B)}) |E_i^{(S)}\rangle\langle E_i^{(S)}|, \tag{18}$$

which is what we needed to show.

Concerning (18), we remark that it follows merely from the fact that the reduced density matrix ρ^Ψ does not depend upon Ψ for typical $\Psi \in \mathcal{H}_{E,\delta}$, that whenever the reduced microcanonical density matrix $\rho^{(S)} \approx \rho_\beta$, the same is true for ρ^Ψ for typical Ψ: applying the partial trace $\mathrm{tr}^{(B)}$ to (5), one obtains that $\rho^{(S)} = \int u_{E,\delta}(d\Psi)\rho^\Psi \approx \rho^\Psi$, for typical Ψ.

Some essential parts of the argument we have presented here have already been described by Schrödinger in an appendix, written in 1952, to his book on statistical thermodynamics [1]. We note that Schrödinger in [1] made the assumption that "in a big system ... the amplitude-squares ... are on the average [in time] equal for ... eigenfunctions belonging to the same [energy] eigenvalue." He uses this assumption for the combined system $S + B$ to derive "exactly the same canonical distribution between the amplitude-squares, as is in the customary treatment said to indicate the probability of the system being on this or that energy level." However, Schrödinger neither connects the assumption with typicality, nor his conclusion with the reduced density matrix ρ^Ψ, which he does not even mention. His concern is rather with showing that one need not regard a system in thermal equilibrium as being in an energy eigenstate. As he states in his preface, "To ascribe to every system always one of its sharp energy values is an indefensible attitude."

Tasaki [3] has studied, as we do here, the reduced density matrix of a system coupled to a heat bath when $S + B$ is described by a wave function Ψ. He shows that for a special form of the coupling Hamiltonian the long-time average of $\rho^{\Psi(t)}$ is canonical. He then argues in a heuristic way that also for typical large times, $\rho^{\Psi(t)} \approx \rho_\beta$. [In this argument, there is a hidden typicality assumption on the initial wave function $\Psi(0)$, namely, that all of the energy expansion coefficients are small, which is true of most wave functions.] His argument does not yield our stronger statement that $\rho^{\Psi(t)} \approx \rho_\beta$ even at $t = 0$ for typical wave functions. Tasaki also includes some examples that he studied rigorously, concerning energy eigenstates of the composite or states that are initially product states $\Psi^{(S)} \otimes \Psi^{(B)}$ with $\Psi^{(S)}$ an eigenstate of $H^{(S)}$.

More recently, Gemmer and Mahler [2] have established canonical typicality under the assumption of very large degeneracy for energy eigenstates by computing appropriate Hilbert space volumes.

We appreciate the hospitality that some of us have enjoyed, while working on this Letter, at the Institut des Hautes Études Scientifiques (Bures-sur-Yvette, France), the Mathematics Department of Rutgers University, and the Dipartimento di Fisica of Università di Genova. The work of S. G. was supported in part by NSF Grant No. DMS-0504504, and that of J. L. by NSF Grant No. DMR 01-279-26 and AFOSR Grant No. AF 49620-01-1-0154. The work of R. T. was supported by INFN and by the European Commission through its 6th Framework Program "Structuring the European Research Area" and Contract No. RITA-CT-2004-505493 for the provision of Transnational Access implemented as Specific Support Action. The work of N. Z. was supported by INFN.

Note added.—In a very recent paper, Popescu, Short, and Winter have established canonical typicality under great generality by invoking Levy's Lemma [6].

*Electronic address: oldstein@math.rutgers.edu
†Electronic address: lebowitz@math.rutgers.edu
‡Electronic address: tumulka@everest.mathematik.uni-tuebingen.de
§Electronic address: zanghi@ge.infn.it

[1] E. Schrödinger, *Statistical Thermodynamics* (Cambridge University Press, Cambridge, England, 1952).
[2] J. Gemmer and G. Mahler, Eur. Phys. J. B **31**, 249 (2003).
[3] H. Tasaki, Phys. Rev. Lett. **80**, 1373 (1998).
[4] E. Schrödinger, Ann. Phys. (Berlin) **83**, 956 (1927).
[5] John Dirk Walecka, *Fundamentals of Statistical Mechanics. Manuscript and Notes of Felix Bloch* (Stanford University Press, Stanford, CA, 1989).
[6] S. Popescu, A. J. Short, and A. Winter, quant-ph/0511225.

PHYSICAL REVIEW E **81**, 011109 (2010)

Approach to thermal equilibrium of macroscopic quantum systems

Sheldon Goldstein,[1,*] Joel L. Lebowitz,[1,†] Christian Mastrodonato,[2,‡] Roderich Tumulka,[3,§] and Nino Zanghì[2,∥]
[1]*Departments of Mathematics and Physics, Rutgers University, 110 Frelinghuysen Road, Piscataway, New Jersey 08854-8019, USA*
[2]*Dipartimento di Fisica, Università di Genova and INFN Sezione di Genova, Via Dodecaneso 33, 16146 Genova, Italy*
[3]*Department of Mathematics, Rutgers University, 110 Frelinghuysen Road, Piscataway, New Jersey 08854-8019, USA*
(Received 9 November 2009; published 7 January 2010)

We consider an isolated macroscopic quantum system. Let $\mathcal{H}$ be a microcanonical "energy shell," i.e., a subspace of the system's Hilbert space spanned by the (finitely) many energy eigenstates with energies between E and $E+\delta E$. The thermal equilibrium macrostate at energy E corresponds to a subspace $\mathcal{H}_{eq}$ of $\mathcal{H}$ such that $\dim \mathcal{H}_{eq}/\dim \mathcal{H}$ is close to 1. We say that a system with state vector $\psi \in \mathcal{H}$ is in thermal equilibrium if ψ is "close" to $\mathcal{H}_{eq}$. We show that for "typical" Hamiltonians with given eigenvalues, all initial state vectors ψ_0 evolve in such a way that ψ_t is in thermal equilibrium for most times t. This result is closely related to von Neumann's quantum ergodic theorem of 1929.

DOI: 10.1103/PhysRevE.81.011109 PACS number(s): 05.30.−d, 03.65.−w

I. INTRODUCTION

If a hot brick is brought in contact with a cold brick, and the two bricks are otherwise isolated, then energy will flow from the hot to the cold brick until their temperatures become equal, i.e., the system equilibrates. Since the bricks ultimately consist of electrons and nuclei, they form a quantum system with a huge number ($>10^{20}$) of particles; this is an example of an *isolated macroscopic quantum system.*

From a microscopic point of view the state of the system at time t is described by a vector

$$\psi(t) = e^{-iHt}\psi(0) \quad (1)$$

in the system's Hilbert space or a density matrix

$$\rho(t) = e^{-iHt}\rho(0)e^{iHt}, \quad (2)$$

where H is the Hamiltonian of the isolated system and we have set $\hbar=1$. In this paper we prove a theorem asserting that for a sufficiently large quantum system with a typical Hamiltonian and an *arbitrary initial state* $\psi(0)$, the system's state $\psi(t)$ spends most of the time, in the long run, in thermal equilibrium. (Of course, before the system even reaches thermal equilibrium there could be a waiting time longer than the present age of the universe.) This implies the same behavior for an arbitrary $\rho(0)$.

This behavior of isolated macroscopic quantum systems is an instance of a phenomenon we call *normal typicality* [1], a version of which is expressed in von Neumann's quantum ergodic theorem [2]. However, our result falls outside the scope of von Neumann's theorem because of the technical assumptions made in that theorem. Our result also differs from the related results in [3–8], which use different notions of when a system is in an equilibrium state. In particular they do not regard the thermal equilibrium of an isolated macroscopic system as corresponding to its wave function being close to a subspace $\mathcal{H}_{eq}$ of Hilbert space. See Sec. VI for further discussion.

The rest of this paper is organized as follows. In the remainder of Sec. I, we define more precisely what we mean by thermal equilibrium. In Sec. II we outline the problem and our result, Theorem 1. In Sec. III we prove the key estimate for the proof of Theorem 1. In Sec. IV we describe examples of exceptional Hamiltonians, illustrating how a system can fail to ever approach thermal equilibrium. In Sec. V we compare our result to the situation with classical systems. In Sec. VI we discuss related works.

A. Equilibrium subspace

Let $\mathcal{H}_{\text{total}}$ be the Hilbert space of a macroscopic system in a box Λ, and let H be its Hamiltonian. Let $\{\phi_\alpha\}$ be an orthonormal basis of $\mathcal{H}_{\text{total}}$ consisting of eigenvectors of H with eigenvalues E_α. Consider an energy interval $[E, E+\delta E]$, where δE is small on the macroscopic scale but large enough for the interval $[E, E+\delta E]$ to contain very many eigenvalues. Let $\mathcal{H} \subseteq \mathcal{H}_{\text{total}}$ be the corresponding subspace,

$$\mathcal{H} = \text{span}\{\phi_\alpha : E_\alpha \in [E, E+\delta E]\}. \quad (3)$$

A subspace such as $\mathcal{H}$ is often called a *microcanonical energy shell.* Let D be the dimension of $\mathcal{H}$, i.e., the number of energy levels, including multiplicities, between E and $E+\delta E$. In the following we consider only quantum states ψ that lie in $\mathcal{H}$, i.e., of the form

$$\psi = \sum c_\alpha \phi_\alpha \quad (4)$$

with $c_\alpha \neq 0$ only for α such that $E_\alpha \in [E, E+\delta E]$.

According to the analysis of von Neumann [2,9] and others (cf. [10]), the macroscopic (coarse-grained) observables in a macroscopic quantum system can be naturally "rounded" to form a set of commuting operators,

$$\{M_i\}_{i=1,\ldots,k}. \quad (5)$$

The operators are defined on $\mathcal{H}_{\text{total}}$, but since we can take them to include (and thus commute with) a coarse-grained

*oldstein@math.rutgers.edu
†lebowitz@math.rutgers.edu
‡christian.mastrodonato@ge.infn.it
§tumulka@math.rutgers.edu
∥zanghi@ge.infn.it

Hamiltonian, we can (and will) take them to commute with the projection to $\mathcal{H}$, and thus to map $\mathcal{H}$ to itself. We write $\nu=(m_1,\ldots,m_k)$ for a list of eigenvalues m_i of the restriction of M_i to $\mathcal{H}$, and $\mathcal{H}_\nu$ for the joint eigenspace. Such a set of operators generates an orthogonal decomposition of the Hilbert space

$$\mathcal{H} = \bigoplus_\nu \mathcal{H}_\nu, \tag{6}$$

where each $\mathcal{H}_\nu$, called a macrospace, represents a macrostate of the system. The dimension of $\mathcal{H}_\nu$ is denoted by d_ν; note that $\Sigma_\nu d_\nu = D$. If any $\mathcal{H}_\nu$ has a dimension of zero, we remove it from the family $\{\mathcal{H}_\nu\}$. In practice, $d_\nu \gg 1$, since we are considering a macroscopic system with coarse-grained observables.

It can be shown in many cases, and is expected to be true generally, that among the macrospaces $\mathcal{H}_\nu$ there is a particular macrospace $\mathcal{H}_{eq}$, the one corresponding to thermal equilibrium, such that

$$d_{eq}/D \approx 1, \tag{7}$$

indeed with the difference $1-d_{eq}/D$ exponentially small in the number of particles.[1] This implies, in particular, that each of the macro-observables M_i is "nearly constant" on the energy shell $\mathcal{H}$ in the sense that one of its eigenvalues has multiplicity at least $d_{eq}\approx D$. We say that a system with quantum state ψ (with $\|\psi\|=1$) is *in thermal equilibrium* if ψ is very close (in the Hilbert space norm) to $\mathcal{H}_{eq}$, i.e., if

$$\langle\psi|P_{eq}|\psi\rangle \approx 1, \tag{8}$$

where P_{eq} is the projection operator to $\mathcal{H}_{eq}$. The corresponding relation for density matrices is

$$\mathrm{Tr}(P_{eq}\rho) \approx 1. \tag{9}$$

Condition (8) implies that a quantum measurement of the macroscopic observable M_i on a system with wave function ψ will yield, with probability close to 1, the "equilibrium" value of M_i. Likewise, a joint measurement of $M_1,\ldots,M_k$ will yield, with probability close to 1, their equilibrium values.

Let $\mu(d\psi)$ be the uniform measure on the unit sphere in $\mathcal{H}$ [12,13]. It follows from Eq. (7) that most ψ relative to μ are in thermal equilibrium. Indeed,

$$\int \langle\psi|P_{eq}|\psi\rangle\mu(d\psi) = \frac{1}{D}\mathrm{Tr}\, P_{eq} = \frac{d_{eq}}{D} \approx 1. \tag{10}$$

Since the quantity $\langle\psi|P_{eq}|\psi\rangle$ is bounded from above by 1, most ψ must satisfy Eq. (8).[2]

B. Examples of equilibrium subspaces

To illustrate the decomposition into macrostates, we describe two examples. As example 1, consider a system composed of two identical subsystems designated 1 and 2, e.g., the bricks mentioned in the beginning of this paper, with Hilbert space $\mathcal{H}_{\text{total}}=\mathcal{H}_1\otimes\mathcal{H}_2$. The Hamiltonian of the total system is

$$H = H_1 + H_2 + \lambda V, \tag{11}$$

where H_1 and H_2 are the Hamiltonians of subsystems 1 and 2, respectively, and λV is a small interaction between the two subsystems. We assume that H_1, H_2, and H are positive operators. Let $\mathcal{H}$ be spanned by the eigenfunctions of H with energies between E and $E+\delta E$.

In this example, we consider just a single macro-observable M, which is a projected and coarse-grained version of H_1/E, i.e., of the fraction of the energy that is contained in subsystem 1 alone. We cannot take M to simply equal to H_1/E because H_1 is defined on $\mathcal{H}_{\text{total}}$, not $\mathcal{H}$, and will generically not map $\mathcal{H}$ to itself, while we would like M to be an operator on $\mathcal{H}$. To obtain an operator on $\mathcal{H}$, let P be the projection $\mathcal{H}_{\text{total}}\to\mathcal{H}$ and set

$$H_1' = PH_1P \tag{12}$$

(more precisely, H_1' is PH_1 restricted to $\mathcal{H}$). Note that H_1' is a positive operator, but might have eigenvalues greater than E. Now define[3]

$$M = f(H_1'/E) \tag{13}$$

with the coarse-graining function

$$f(x) = \begin{cases} 0 & \text{if } x < 0.01 \\ 0.02 & \text{if } x \in [0.01, 0.03) \\ 0.04 & \text{if } x \in [0.03, 0.05) \\ \text{etc.} & \ldots . \end{cases} \tag{14}$$

$\mathcal{H}_\nu$ are the eigenspaces of M; clearly, $\oplus_\nu\mathcal{H}_\nu=\mathcal{H}$. If, as we assume, λV is small, then we expect $\mathcal{H}_{0.5}=\mathcal{H}_{eq}$ to have the overwhelming majority of dimensions. In a thorough treatment we would need to prove this claim, as well as that H_1' is not too different from H_1, but we do not give such a treatment here.

As example 2, consider N bosons (fermions) in a box $\Lambda=[0,L]^3\subseteq\mathbb{R}^3$; i.e., $\mathcal{H}_{\text{total}}$ consists of the square-integrable (anti)symmetric functions on Λ^N. Let the Hamiltonian be

[1]This dominance of the equilibrium state can be expressed in terms of the (Boltzmann) entropy S_ν of a macroscopic system in the macrostate ν, be it the equilibrium state or some other (see [11]), defined as $S_\nu=k_B\log d_\nu$, where k_B is the Boltzmann constant: d_{eq}/D being close to 1 just expresses the fact that the entropy of the equilibrium state is close to the microcanonical entropy S_{mc}, i.e., $S_{eq}=k_B\log d_{eq}\approx k_B\log D=S_{mc}$.

[2]It should in fact be true for a large class of observables A on $\mathcal{H}$ that, for most ψ relative to μ, $\langle\psi|A|\psi\rangle\approx\mathrm{Tr}(\rho_{mc}A)$, where ρ_{mc} is the microcanonical density matrix, i.e., $1/D$ times the identity on $\mathcal{H}$. This is relevant to the various results on thermalization described in Sec. VI.

[3]Recall that the application of a function f to a self-adjoint matrix A is defined to be $f(A)=\Sigma f(a_\alpha)|\varphi_\alpha\rangle\langle\varphi_\alpha|$ if the spectral decomposition of A reads $A=\Sigma a_\alpha|\varphi_\alpha\rangle\langle\varphi_\alpha|$.

$$H = -\frac{1}{2m}\sum_{i=1}^{N}\nabla_i^2 + \sum_{i<j} v(|\boldsymbol{q}_i - \boldsymbol{q}_j|), \tag{15}$$

where the Laplacian ∇_i^2 has Dirichlet boundary conditions, $v(r)$ is a given pair potential, and $\boldsymbol{q}_i$ is the triple of position coordinates of the ith particle. Let $\mathcal{H}$ again be spanned by the eigenfunctions with energies between E and $E+\delta E$.

In this example, we consider again a single macro-observable M, based on the operator N_{left} for the number of particles in the left half of the box Λ,

$$N_{\text{left}}\psi(\boldsymbol{q}_1,\ldots,\boldsymbol{q}_N) = \#\{i:\boldsymbol{q}_i \in [0,L/2]\times[0,L]^2\}\psi(\boldsymbol{q}_1,\ldots,\boldsymbol{q}_N). \tag{16}$$

Note that the spectrum of N_{left} consists of the $N+1$ eigenvalues $0,1,2,\ldots,N$. To obtain an operator on $\mathcal{H}$, let P be the projection $\mathcal{H}_{\text{total}}\to\mathcal{H}$ and set $N'_{\text{left}}=PN_{\text{left}}P$. Note that the spectrum of N'_{left} is still contained in $[0,N]$. Now define $M=f(N'_{\text{left}}/N)$ with the coarse-graining function (14). We expect that for large N, the eigenspace with an eigenvalue of 0.5, $\mathcal{H}_{eq}=\mathcal{H}_{0.5}$, has the overwhelming majority of dimensions (and that $N'_{\text{left}}\approx N_{\text{left}}$).

II. FORMULATION OF PROBLEM AND RESULTS

Our goal is to show that, for typical macroscopic quantum systems,

$$\langle\psi(t)|P_{eq}|\psi(t)\rangle \approx 1 \quad \text{for most } t. \tag{17}$$

To see this, we compute the time average of $\langle\psi(t)|P_{eq}|\psi(t)\rangle$. We denote the time average of a time-dependent quantity $f(t)$ by an overbar,

$$\overline{f(t)} = \lim_{T\to\infty}\frac{1}{T}\int_0^T dt\, f(t). \tag{18}$$

Since $\langle\psi(t)|P_{eq}|\psi(t)\rangle$ is always a real number between 0 and 1, it follows that if its time average is close to 1 then it must be close to 1 most of the time. Moreover, for μ-most $\psi(0)$, where μ is the uniform measure on the unit sphere of $\mathcal{H}$, $\psi(t)$ is in thermal equilibrium most of the time. This result follows from Fubini's theorem (which implies that taking the μ average commutes with taking the time average) and the unitary invariance of μ,

$$\int\overline{\langle\psi(t)|P_{eq}|\psi(t)\rangle}\mu(d\psi) = \int\overline{\langle\psi|e^{iHt}P_{eq}e^{-iHt}|\psi\rangle}\mu(d\psi)$$

$$= \int\langle\psi|P_{eq}|\psi\rangle\mu(d\psi) \approx 1. \tag{19}$$

That is, the ensemble average of the time average is near 1, so, for μ-most $\psi(0)$, the time average must be near 1, which implies our claim above. So the interesting question is about the behavior of exceptional $\psi(0)$, e.g., of systems which are not in thermal equilibrium at $t=0$. Do they ever go to thermal equilibrium?

As we will show, for many Hamiltonians statement (17) holds in fact for *all* $\psi(0)\in\mathcal{H}$. From now on, let H denote the restriction of the Hamiltonian to $\mathcal{H}$, and let $\phi_1,\ldots,\phi_D$ be an orthonormal basis of $\mathcal{H}$ consisting of eigenvectors of the Hamiltonian H with eigenvalues $E_1,\ldots,E_D$. If

$$\psi(0) = \sum_{\alpha=1}^{D} c_\alpha\phi_\alpha, \quad c_\alpha = \langle\phi_\alpha|\psi(0)\rangle, \tag{20}$$

then

$$\psi(t) = \sum_{\alpha=1}^{D} e^{-iE_\alpha t}c_\alpha\phi_\alpha. \tag{21}$$

Thus,

$$\overline{\langle\psi(t)|P_{eq}|\psi(t)\rangle} = \sum_{\alpha,\beta=1}^{D}\overline{e^{i(E_\alpha-E_\beta)t}}c_\alpha^* c_\beta\langle\phi_\alpha|P_{eq}|\phi_\beta\rangle. \tag{22}$$

If H is nondegenerate (which is the generic case) then $E_\alpha - E_\beta$ vanishes only for $\alpha=\beta$, so the time averaged exponential is $\delta_{\alpha\beta}$, and

$$\overline{\langle\psi(t)|P_{eq}|\psi(t)\rangle} = \sum_{\alpha=1}^{D}|c_\alpha|^2\langle\phi_\alpha|P_{eq}|\phi_\alpha\rangle. \tag{23}$$

Thus, for the system to be in thermal equilibrium most of the time it is necessary and sufficient that the right-hand side of Eq. (23) is close to 1.

Now if an energy eigenstate ϕ_α is not itself in thermal equilibrium then when $\psi(0)=\phi_\alpha$ the system is never in thermal equilibrium since this state is stationary. Conversely, if we have that

$$\langle\phi_\alpha|P_{eq}|\phi_\alpha\rangle \approx 1 \quad \text{for all } \alpha, \tag{24}$$

then the system will be in thermal equilibrium most of the time for all $\psi(0)$. This follows directly from Eq. (23) since the right-hand side of Eq. (23) is an average of $\langle\phi_\alpha|P_{eq}|\phi_\alpha\rangle$. We show below that Eq. (24) is true for "most" Hamiltonians and, thus, for most Hamiltonians it is the case that every wave function spends most of the time in thermal equilibrium.

A. Main result

The measure of most we use is the following: for any given D (distinct) energy values $E_1,\ldots,E_D$, we consider the uniform distribution μ_{Ham} over all Hamiltonians with these eigenvalues. Choosing H at random with distribution μ_{Ham} is equivalent to choosing the eigenbasis $\{\phi_\alpha\}$ according to the uniform distribution μ_{ONB} over all orthonormal bases of $\mathcal{H}$, and setting $H=\sum_\alpha E_\alpha|\phi_\alpha\rangle\langle\phi_\alpha|$. The measure μ_{ONB} can be defined as follows: choosing a random basis according to μ_{ONB} amounts to choosing ϕ_1 according to the uniform distribution over the unit sphere in $\mathcal{H}$, then ϕ_2 according to the uniform distribution over the unit sphere in the orthogonal complement of ϕ_1, etc. Alternatively, μ_{ONB} can be defined in terms of the Haar measure $\mu_{\mathcal{U}(D)}$ on the group $\mathcal{U}(D)$ of unitary $D\times D$ matrices: any given orthonormal basis $\{\chi_\alpha\}$ of $\mathcal{H}$ defines a one-to-one correspondence between $\mathcal{U}(D)$ and the set of all orthonormal bases of $\mathcal{H}$, associating with the matrix $U=(U_{\alpha\beta})\in\mathcal{U}(D)$ the basis

$$\phi_\alpha = \sum_{\beta=1}^{D} U_{\alpha\beta}\chi_\beta. \quad (25)$$

The image of the Haar measure under this correspondence is in fact independent of the choice of $\{\chi_\beta\}$ (because of the invariance of the Haar measure under right multiplication) and is μ_{ONB}.

Put differently, the ensemble μ_{Ham} of Hamiltonians can be obtained by starting from a given Hamiltonian H_0 on $\mathcal{H}$ (with distinct eigenvalues $E_1,\ldots,E_D$) and setting

$$H = UH_0U^{-1} \quad (26)$$

with U a random unitary matrix chosen according to the Haar measure. Note that, while considering different possible Hamiltonians H in $\mathcal{H}$, we keep $\mathcal{H}_{eq}$ fixed, although in practice it would often be reasonable to select $\mathcal{H}_{eq}$ in a way that depends on H (as we did in the examples of Sec. I B).

For our purpose it is convenient to choose the basis $\{\chi_\alpha\}$ in such a way that the first d_{eq} basis vectors lie in $\mathcal{H}_{eq}$ and the other ones are orthogonal to $\mathcal{H}_{eq}$. Then, we have that

$$\langle\phi_\alpha|P_{eq}|\phi_\alpha\rangle = \sum_{\beta=1}^{d_{eq}} |U_{\alpha\beta}|^2 \quad (27)$$

with $U_{\alpha\beta}$ the unitary matrix satisfying Eq. (25).

We will show first, in Lemma 1, that for every $0<\varepsilon<1$, if D is sufficiently large and d_{eq}/D sufficiently close to 1, then most orthonormal bases $\{\phi_\alpha\}$ are such that

$$\langle\phi_\alpha|P_{eq}|\phi_\alpha\rangle > 1-\varepsilon \quad \text{for all } \alpha. \quad (28)$$

This inequality is a precise version of Eq. (24). How close to 1 should d_{eq}/D be? The fact that the average of $\langle\psi|P_{eq}|\psi\rangle$ over all wave functions ψ on the unit sphere of $\mathcal{H}$ equals d_{eq}/D, mentioned already in Eq. (10), implies that Eq. (28) cannot be true for most orthonormal bases if $d_{eq}/D \leq 1-\varepsilon$. To have enough wiggling room, we require that

$$\frac{d_{eq}}{D} > 1 - \frac{\varepsilon}{2}. \quad (29)$$

We will show then, in Theorem 1, that for every (arbitrarily small) $0<\eta<1$ and for sufficiently large D, most H are such that for every initial wave function $\psi(0)\in\mathcal{H}$ with $\|\psi(0)\|=1$, the system will spend most of the time in thermal equilibrium with accuracy $1-\eta$, where we say that a system with wave function ψ is in thermal equilibrium with accuracy $1-\eta$ if

$$\langle\psi|P_{eq}|\psi\rangle > 1-\eta. \quad (30)$$

This inequality is a precise version of Eq. (8). In order to have no more exceptions in time than the fraction $0<\delta'<1$, we need to set ε in Eqs. (28) and (29) equal to $\eta\delta'$.

Lemma 1. Let $\mu_{\mathcal{U}(D)}$ denote the Haar measure on $\mathcal{U}(D)$, and

$$\mathcal{S}_\varepsilon := \left\{ U \in \mathcal{U}(D) \,\middle|\, \forall\, \alpha: \sum_{\beta=1}^{d_{eq}} |U_{\alpha\beta}|^2 > 1-\varepsilon \right\}. \quad (31)$$

Then for all $0<\varepsilon<1$ and $0<\delta<1$, there exists $D_0 = D_0(\varepsilon,\delta)>0$ such that

$$\text{if } D>D_0, \quad d_{eq}>(1-\varepsilon/2)D \quad \text{then } \mu_{\mathcal{U}(D)}(\mathcal{S}_\varepsilon) \geq 1-\delta. \quad (32)$$

The proof of Lemma 1 is given in Sec. III. It also shows that D_0 can, for example, be chosen to be

$$D_0(\varepsilon,\delta) = \max[10^3\varepsilon^{-2}\log(4/\delta), 10^6\varepsilon^{-4}]. \quad (33)$$

Here and throughout this paper, log denotes the natural logarithm. From Eq. (27), we obtain

Theorem 1. For all $\eta,\delta,\delta'\in(0,1)$, all integers $D>D_0(\eta\delta',\delta)$ and all integers $d_{eq}>(1-\eta\delta'/2)D$ the following is true: let $\mathcal{H}$ be a Hilbert space of dimension D, let $\mathcal{H}_{eq}$ be a subspace of dimension d_{eq}, let P_{eq} denote the projection to $\mathcal{H}_{eq}$, let $E_1,\ldots,E_D$ be pairwise distinct but otherwise arbitrary, and choose a Hamiltonian at random with eigenvalues E_α and an eigenbasis ϕ_α that is uniformly distributed. Then, with probability at least $1-\delta$, every initial quantum state will spend $(1-\delta')$-most of the time in thermal equilibrium as defined in Eq. (30), i.e.,

$$\liminf_{T\to\infty} \frac{1}{T}|\{0<t<T: \langle\psi(t)|P_{eq}|\psi(t)\rangle > 1-\eta\}| \geq 1-\delta', \quad (34)$$

where $|M|$ denotes the size (Lebesgue measure) of the set M.

Proof. It follows from Lemma 1 that under the hypotheses of Theorem 1,

$$\overline{\langle\psi(t)|P_{eq}|\psi(t)\rangle} \geq 1-\eta\delta'$$

with probability at least $1-\delta$. Thus, since $\eta\delta' \geq \overline{1-\langle\psi(t)|P_{eq}|\psi(t)\rangle} \geq \eta\tilde{\delta}$, where $\tilde{\delta}$ is the lim $\sup_{T\to\infty}$ of the fraction of the time in $(0,T)$ for which $\langle\psi(t)|P_{eq}|\psi(t)\rangle \leq 1-\eta$, it follows that $\tilde{\delta}\leq\delta'$. ■

B. Remarks

Normal typicality. Theorem 1 can be strengthened; with the same sense of "most" as in Theorem 1, we have that for most Hamiltonians and for all $\psi(0)$,

$$\langle\psi(t)|P_\nu|\psi(t)\rangle \approx \frac{\dim \mathcal{H}_\nu}{\dim \mathcal{H}}, \quad \text{for all } \nu, \quad (35)$$

for most t. For $\nu=eq$, this implies that $\langle\psi(t)|P_{eq}|\psi(t)\rangle\approx 1$. This stronger statement we have called normal typicality [1]. A version of normal typicality was proven by von Neumann [2]. However, because of the technical assumptions he made, von Neumann's result, while much more difficult, does not quite cover the simple result of this paper.

Typicality and probability. When we express that something is true for most H or most ψ relative to some normalized measure μ, it is often convenient to use the language of probability theory and speak of a random H or ψ chosen with distribution μ. However, by this we do not mean to imply that the actual H or ψ in a concrete physical situation is random, nor that one would obtain, in repetitions of the experiment or in a class of similar experiments, different H's or ψ's whose empirical distribution is close to μ. That would be a misinterpretation of the measure μ, one that suggests the question whether perhaps the actual distribution in reality could be nonuniform. This question misses the point, as there need not be any actual distribution in reality. Rather, Theorem 1 means that the set of "bad" Hamiltonians has very small measure $\mu_{\rm Ham}$.

Consequences for example 2. From Lemma 1 it follows for example 2 that typical Hamiltonians of the form (26) with H_0 given by the right-hand side of Eq. (15) are such that all eigenfunctions are close to $\mathcal{H}_{0.5}$; this fact in turn strongly suggests (although we have not proved this) that the eigenfunctions are essentially concentrated on those configurations that have approximately 50% of the particles in the left half and 50% in the right half of the box.

Equilibrium statistical mechanics. Theorem 1 implies that, for typical H, every $\psi(0)\in\mathcal{H}$ is such that for most t,

$$\langle\psi(t)|M_i|\psi(t)\rangle \approx \mathrm{Tr}(\rho_{mc}M_i), \tag{36}$$

where ρ_{mc} is the standard microcanonical density matrix (i.e., $1/D$ times the projection $\mathcal{H}_{\rm total}\to\mathcal{H}$), for all macro-observables M_i as described in Sec. I A. This justifies replacing $|\psi(t)\rangle\langle\psi(t)|$ with ρ_{mc} as far as macro-observables in equilibrium are concerned. However, this does not—by itself—justify the use of ρ_{mc} for observables A not among $\{M_i\}$. For example, consider a microscopic observable A that is not "nearly constant" on the energy shell $\mathcal{H}$. Then, standard equilibrium statistical mechanics tells us to use ρ_{mc} for the expected value of A in equilibrium. We believe that this is in fact correct for most such observables, but it is not covered by Theorem 1. Results concerning many such observables are described in Sec. VI. These results, according to which, in an appropriate sense,

$$\langle\psi(t)|A|\psi(t)\rangle \approx \mathrm{Tr}(\rho_{mc}A) \tag{37}$$

for suitable A and $\psi(0)$, are valid only in quantum mechanics. The justification of the broad use of ρ_{mc} in classical statistical mechanics relies on rather different sorts of results requiring different kinds of considerations.

III. PROOF OF LEMMA 1

Proof. Let us write $\mathbb{P}$ for the Haar measure $\mu_{\mathcal{U}(D)}$, and let

$$p := \mathbb{P}\left(\bigcap_{\alpha=1}^{D}\left\{\sum_{\beta=1}^{d_{eq}}|U_{\alpha\beta}|^2 > 1-\varepsilon\right\}\right). \tag{38}$$

Observe that

$$p = 1-\mathbb{P}\left(\bigcup_{\alpha=1}^{D}\left\{\sum_{\beta=1}^{d_{eq}}|U_{\alpha\beta}|^2 \le 1-\varepsilon\right\}\right) \tag{39}$$

$$\ge 1 - D\max_{\alpha}\mathbb{P}\left\{\sum_{\beta=1}^{d_{eq}}|U_{\alpha\beta}|^2 \le 1-\varepsilon\right\}. \tag{40}$$

Since $U=(U_{\alpha\beta})$ is a random unitary matrix with Haar distribution, its αth column is a random unit vector $\vec{U}:=(U_{\alpha\beta})_\beta$ whose distribution is uniform over the unit sphere of $\mathbb{C}^D$ (i.e., the distribution is, up to a normalizing constant, the surface area measure). Therefore, the probability in the last line does not, in fact, depend on α, and so the step of taking the maximum over α can be omitted.

A random unit vector such as $\vec{U}$ can be thought of as arising from a random Gaussian vector $\vec{G}$ by normalization: let G_β for $\beta=1,\ldots,D$ be independent complex Gaussian random variables with mean zero and variance $\mathbb{E}|G_\beta|^2=1/D$; i.e., Re G_β and Im G_β are independent real Gaussian random variables with mean zero and variance $1/2D$. Then the distribution of $\vec{G}=(G_1,\ldots,G_D)$ is symmetric under rotations from $\mathcal{U}(D)$, and thus

$$\frac{\vec{G}}{\|\vec{G}\|} = \vec{U} \quad \text{in distribution,} \quad \text{with } \|\vec{G}\|^2 = \sum_{\beta=1}^{D}|G_\beta|^2. \tag{41}$$

We thus have that

$$p \ge 1 - D\,\mathbb{P}\left\{\sum_{\beta=1}^{d_{eq}}\frac{|G_\beta|^2}{\|\vec{G}\|^2} \le 1-\varepsilon\right\}. \tag{42}$$

To estimate the probability on the right-hand side of Eq. (42), we introduce three different events:

$$A(\eta') := \{|\|\vec{G}\|^2-1| < \eta'\}, \tag{43}$$

$$B(\eta'') := \left\{(1-\eta'')\frac{d_{eq}}{D} < \sum_{\beta=1}^{d_{eq}}|G_\beta|^2 < (1+\eta'')\frac{d_{eq}}{D}\right\}, \tag{44}$$

$$C(\eta''') := \left\{(1-\eta''')\frac{d_{eq}}{D} < \sum_{\beta=1}^{d_{eq}}\frac{|G_\beta|^2}{\|\vec{G}\|^2} < (1+\eta''')\frac{d_{eq}}{D}\right\}. \tag{45}$$

Let us now assume that

$$\frac{d_{eq}}{D} > 1-\frac{\varepsilon}{2}. \tag{46}$$

We then have that

$$(1-\varepsilon/2)\frac{d_{eq}}{D} > 1-\varepsilon+\frac{\varepsilon^2}{4} > 1-\varepsilon, \tag{47}$$

so that

$$C(\varepsilon/2) \subseteq \left\{(1-\varepsilon/2)\frac{d_{eq}}{D} < \sum_{\beta=1}^{d_{eq}}\frac{|G_\beta|^2}{\|\vec{G}\|^2}\right\}$$

$$\subseteq \left\{1-\varepsilon < \sum_{\beta=1}^{d_{eq}}\frac{|G_\beta|^2}{\|\vec{G}\|^2}\right\}, \tag{48}$$

and thus

$$p \ge 1 - D\,\mathbb{P}(C^c(\varepsilon/2)), \tag{49}$$

where the superscript c means complement. Our goal is to find a good upper bound for $\mathbb{P}(C^c(\varepsilon/2))$.

If the event $A(\eta')$ occurs for $0<\eta'<\frac{1}{2}$ then

$$1-\eta' < \frac{1}{\|\vec{G}\|^2} < 1+2\eta', \tag{50}$$

and, consequently, if $A(\eta')\cap B(\eta'')$ occurs then

$$\frac{d_{eq}}{D}(1-\eta')(1-\eta'') < \frac{\sum_{\beta=1}^{d_{eq}}|G_\beta|^2}{\|\vec{G}\|^2} < \frac{d_{eq}}{D}(1+2\eta')(1+\eta''). \tag{51}$$

It is now easy to see that $A(\eta')\cap B(\eta'')\subseteq C(2\eta'+\eta''+2\eta'\eta'')$, so if we choose $\eta'=\eta''=\varepsilon/8$ we obtain that

$$A\left(\frac{\varepsilon}{8}\right)\cap B\left(\frac{\varepsilon}{8}\right)\subseteq C\left(\frac{3}{8}\varepsilon+\frac{1}{32}\varepsilon^2\right)\subseteq C(\varepsilon/2) \quad \text{for } 0<\varepsilon<1. \tag{52}$$

We thus have the following upper bound:

$$\mathbb{P}(C^c(\varepsilon/2)) \le \mathbb{P}(A^c(\varepsilon/8)) + \mathbb{P}(B^c(\varepsilon/8)). \tag{53}$$

To find an estimate of $\mathbb{P}(A(\varepsilon/8))$ and $\mathbb{P}(B(\varepsilon/8))$ we use the *large deviations principle*. It is convenient to use a slightly stronger version of this principle than usual (see Sec. 2.2.1 of [14]), which states that for a sequence of N independent and identically distributed (i.i.d.) random variables X_i,

$$\mathbb{P}\left(\left|\sum_{i=1}^{N}\frac{X_i}{N}-\mathbb{E}(X_1)\right|>\delta\right)\le 2e^{-NI[\mathbb{E}(X_1)+\delta]}, \tag{54}$$

where $I(x)$ is the *rate function* [14] associated with the distribution of X_i, defined to be

$$I(x)=\sup_{\theta>0}(\theta x-\log \mathbb{E}e^{\theta X_i}). \tag{55}$$

In our case, where X_i will be the square of a standard normal random variable, the rate function is

$$I(x)=\tfrac{1}{2}(x-1-\log x) \quad \forall x>1, \tag{56}$$

as a simple calculation shows.

To estimate $\mathbb{P}(A(\varepsilon/8))$, set

$$N=2D,\quad X_\beta=2D(\mathrm{Re}\,G_\beta)^2,\quad X_{D+\beta}=2D(\mathrm{Im}\,G_\beta)^2 \quad \text{for } \beta=1,\ldots,D. \tag{57}$$

Thus, for $i=1,\ldots,2D$, X_i are i.i.d. variables with mean $\mathbb{E}X_i=2D\mathbb{E}(\mathrm{Re}\,G_i)^2=1$; we thus obtain

$$\mathbb{P}(A^c(\varepsilon/8))=\mathbb{P}\{|\|\vec{G}\|^2-1|>\varepsilon/8\} \tag{58}$$

$$=\mathbb{P}\left\{\left|\sum_{\beta=1}^{D}|G_\beta|^2-1\right|>\varepsilon/8\right\} \tag{59}$$

$$=\mathbb{P}\left\{\left|\sum_{i=1}^{2D}\frac{X_i}{2D}-1\right|>\varepsilon/8\right\} \tag{60}$$

$$\le 2e^{-2DI(1+\varepsilon/8)} \tag{61}$$

$$=2e^{-D[\varepsilon/8-\log(1+\varepsilon/8)]} \tag{62}$$

$$\le 2\exp\left(-\frac{D\varepsilon^2}{192}\right). \tag{63}$$

In the last step we have used $\log(1+x)\le x-x^2/3$ for $0<x<1/2$.

We use a completely analogous argument for B, setting

$$N=2d_{eq},\quad X_\beta=2D(\mathrm{Re}\,G_\beta)^2,$$

$$X_{D+\beta}=2D(\mathrm{Im}\,G_\beta)^2, \quad \text{for } \beta=1,\ldots,d_{eq}, \tag{64}$$

and obtain that

$$\mathbb{P}(B^c(\varepsilon/8))=\mathbb{P}\left\{\left|\sum_{\beta=1}^{d_{eq}}|G_\beta|^2-\frac{d_{eq}}{D}\right|/\frac{d_{eq}}{D}>\varepsilon/8\right\} \tag{65}$$

$$=\mathbb{P}\left\{\left|\sum_{i=1}^{2d_{eq}}\frac{X_i}{2d_{eq}}-1\right|>\varepsilon/8\right\} \tag{66}$$

$$\le 2\exp\left(-\frac{d_{eq}\varepsilon^2}{192}\right). \tag{67}$$

From Eqs. (53), (63), and (67) it follows that

$$\mathbb{P}(C^c(\varepsilon/2))\le 2\exp\left(-\frac{d_{eq}\varepsilon^2}{192}\right)+2\exp\left(-\frac{D\varepsilon^2}{192}\right)$$

$$\le 4\exp\left(-\frac{D\varepsilon^2}{384}\right), \tag{68}$$

where we have used $d_{eq}>D/2$. Therefore, by Eq. (49),

$$p\ge 1-4D\exp\left(-\frac{D\varepsilon^2}{384}\right). \tag{69}$$

The last term converges to zero as $D\to\infty$, so there exists $D_0>0$ such that for all $D>D_0$,

$$p\ge 1-\delta, \tag{70}$$

which is what we wanted to show. In order to check this for D_0 specified in Eq. (33) right after Lemma 1, note that the desired relation

$$4D\exp\left(-\frac{D\varepsilon^2}{384}\right)\le\delta \tag{71}$$

is equivalent to

$$D\left(\frac{\varepsilon^2}{384}-\frac{\log D}{D}\right)\ge\log(4/\delta). \tag{72}$$

Thus, it suffices that $D>10^3\varepsilon^{-2}\log(4/\delta)$ and

$$\frac{\log D}{D}<10^{-3}\varepsilon^2. \tag{73}$$

Since $\log D<\sqrt{D}$ for all positive numbers D, condition (73) will be satisfied if $\sqrt{D}>10^3\varepsilon^{-2}$, i.e., if $D>10^6\varepsilon^{-4}$. ■

IV. EXAMPLES OF SYSTEMS THAT DO NOT APPROACH THERMAL EQUILIBRIUM

We shall now present examples of atypical behavior, namely, examples of bad Hamiltonians, i.e., Hamiltonians for which not all wave functions approach thermal equilibrium [or, equivalently, for which Eq. (24) is not satisfied]. According to Theorem 1, bad Hamiltonians form a very small subset of the set of all Hamiltonians. Of course, to establish that Eq. (24) holds for a particular Hamiltonian can be a formidable challenge. Moreover, the small subset might include all standard many-body Hamiltonians (e.g., all those which are a sum of kinetic and potential energies). But there is no *a priori* reason to believe that this should be the case.

The first example consists of two noninteracting subsystems. This can be expressed in the framework provided by example 1 in Sec. I B with the Hamiltonian $H=H_1+H_2+\lambda V$ by setting $\lambda=0$. Let $\{\phi_i^1\}$ be an orthonormal basis of $\mathcal{H}_1$ consisting of eigenvectors of H_1 with eigenvalues E_i^1, and let $\{\phi_j^2\}$ be one of $\mathcal{H}_2$ consisting of eigenvectors of H_2 with eigenvalues E_j^2. Clearly, for $\lambda=0$ not every wave function will approach thermal equilibrium. After all, in this case, $\phi_i^1\otimes\phi_j^2$ forms an eigenbasis of H, while

$$\mathcal{H}=\mathrm{span}\{\phi_i^1\otimes\phi_j^2 : E_i^1+E_j^2\in[E,E+\delta E]\}, \tag{74}$$

$$\mathcal{H}_{eq}=\mathrm{span}\{\phi_i^1\otimes\phi_j^2 : E_i^1\in[0.49E,0.51E),\quad E_i^1+E_j^2\in[E,E+\delta E]\}. \tag{75}$$

Thus, any $\phi_i^1\otimes\phi_j^2$ such that $E_i^1+E_j^2\in[E,E+\delta E]$ but, say, $E_i^1<0.49E$, will be an example of an element of $\mathcal{H}$ that is orthogonal to $\mathcal{H}_{eq}$ and, as it is an eigenfunction of H, forever remains orthogonal to $\mathcal{H}_{eq}$.

As another example, we conjecture that some wave functions will fail to approach thermal equilibrium also when λ is nonzero but sufficiently small. We prove this now for a slightly simplified setting, corresponding to the following modification of example 1 of Sec. I B. For the usual energy interval $[E,E+\delta E]$, let $\mathcal{H}$ be, independently of λ, given by Eq. (74); and, instead of $H_1+H_2+\lambda V$, let H be given by

$$H=H(\lambda)=P(H_1+H_2+\lambda V)P, \tag{76}$$

where P is the projection to $\mathcal{H}$. Then H defines a time evolution on $\mathcal{H}$ that depends on λ. (Note that $\mathcal{H}$ is still an "energy shell" for all sufficiently small λ, as all nonzero eigenvalues of $H(\lambda)$ are still contained in an interval just slightly larger than $[E,E+\delta E]$, and the corresponding eigenvectors lie in $\mathcal{H}$.) Let $\mathcal{H}_{eq}$ for $\lambda\neq 0$ also be given by Eq. (75). Again, choose one particular ϕ_i^1 and one particular ϕ_j^2 (independently of λ), so that $E_i^1+E_j^2\in[E,E+\delta E]$ and $E_i^1<0.49E$, and consider as the initial state of the system again

$$\psi(t=0)=\phi_i^1\otimes\phi_j^2, \tag{77}$$

which evolves to

$$\psi(\lambda,t)=e^{-iH(\lambda)t}\phi_i^1\otimes\phi_j^2. \tag{78}$$

Suppose for simplicity that $H(\lambda=0)=H_1+H_2$ is nondegenerate.[4] Then, according to standard results of perturbation theory [15], also $H(\lambda)$, regarded as an operator on $\mathcal{H}$, is nondegenerate for all $\lambda\in(-\lambda_0,\lambda_0)$ for some $\lambda_0>0$; moreover, its eigenvalues $E(\lambda)$ depend continuously (even analytically) on λ, and so do the eigenspaces. In particular, it is possible to choose for every $\lambda\in(-\lambda_0,\lambda_0)$ a normalized eigenstate $\phi(\lambda)\in\mathcal{H}$ of $H(\lambda)$ with eigenvalue $E(\lambda)$ in such a way that $\phi(\lambda)$ and $E(\lambda)$ depend continuously on λ, and $\phi(\lambda=0)=\phi_i^1\otimes\phi_j^2$.

We are now ready to show that, for sufficiently small $\lambda>0$,

$$\langle\psi(\lambda,t)|P_{eq}|\psi(\lambda,t)\rangle\approx 0 \tag{79}$$

for all t; that is, $\psi(\lambda,t)$ is nearly orthogonal to $\mathcal{H}_{eq}$ for all t, and thus is never in thermal equilibrium. To see this, note first that since $\phi(0)\approx\phi(\lambda)$ for sufficiently small λ, and since $e^{-iH(\lambda)t}$ is unitary, also

$$e^{-iH(\lambda)t}\phi(0)\approx e^{-iH(\lambda)t}\phi(\lambda) \tag{80}$$

(with error independent of t). Since the right-hand side equals

$$e^{-iE(\lambda)t}\phi(\lambda)\approx e^{-iE(\lambda)t}\phi(0), \tag{81}$$

we have that

$$\langle e^{-iH(\lambda)t}\phi(0)|P_{eq}|e^{-iH(\lambda)t}\phi(0)\rangle\approx\langle\phi(0)|P_{eq}|\phi(0)\rangle=0. \tag{82}$$

This proves Eq. (79) with an error bound independent of t that tends to zero as $\lambda\to 0$.

Another example of bad Hamiltonians is provided by the phenomenon of *Anderson localization* (see in particular [16,17]): certain physically relevant Hamiltonians possess some eigenfunctions ϕ_α that have a spatial energy density function that is macroscopically nonuniform whereas wave functions in $\mathcal{H}_{eq}$ should have macroscopically uniform energy density over the entire available volume. Thus, some eigenfunctions are not close to $\mathcal{H}_{eq}$, violating Eq. (24).

V. COMPARISON WITH CLASSICAL MECHANICS

In classical mechanics, one would expect as well that a macroscopic system spends most of the time in the long run in thermal equilibrium. Let us define what thermal equilibrium means in classical mechanics. [We defined it for quantum systems in Eq. (8).] We denote a point in phase space by $X=(\mathbf{q}_1,\ldots,\mathbf{q}_N,\mathbf{p}_1,\ldots,\mathbf{p}_N)$. Instead of the orthogonal decomposition of $\mathcal{H}$ into subspaces $\mathcal{H}_\nu$ we consider a partition of an energy shell Γ in phase space, $\Gamma=\{X:E\le H(X)\le E+\delta E\}$, into regions Γ_ν corresponding to different macrostates ν, i.e., if the microstate X of the system is in Γ_ν then the macrostate

[4]Since this requires that no eigenvalue difference of H_1, $E_i^1-E_{i'}^1$, coincides with an eigenvalue difference of H_2, $E_j^2-E_{j'}^2$, we need to relax our earlier assumption that systems 1 and 2 be identical; so, let them be almost identical, with slightly different eigenvalues, and let H_1 and H_2 each be nondegenerate.

of the system is ν. It has been shown [18] for realistic systems with large N that one of the regions Γ_ν, corresponding to the macrostate of thermal equilibrium and denoted Γ_{eq}, is such that, in terms of the (uniform or Liouville) phase-space volume measure μ on Γ,

$$\frac{\mu(\Gamma_{eq})}{\mu(\Gamma)} \approx 1. \tag{83}$$

Although the subspaces $\mathcal{H}_\nu$ play a role roughly analogous to the regions Γ_ν, a basic difference between the classical and the quantum cases is that while every classical phase point in Γ belongs to one and only one Γ_ν, and thus is in one macrostate, a quantum state ψ need not lie in any one $\mathcal{H}_\nu$, but can be a nontrivial superposition of vectors in different macrostates. (Indeed, almost all ψ do not lie in any one $\mathcal{H}_\nu$. That is why we defined being in thermal equilibrium in terms of ψ lying in a neighborhood of $\mathcal{H}_{eq}$, rather than lying in $\mathcal{H}_{eq}$ itself.)

The time evolution of the microstate X is given by the solution of the Hamiltonian equations of motion, which sends X (at time zero) to X_t (at time t), $t \in \mathbb{R}$. We expect that for realistic systems with a sufficiently large number N of constituents and for every macrostate ν, most initial phase points $X \in \Gamma_\nu$ will be such that X_t spends most of the time in the set Γ_{eq}. This statement follows if the system is ergodic,[5] but in fact is much weaker than ergodicity. Theorem 1 is parallel to this statement in that it implies, for typical Hamiltonians, that initial states [here, $\psi(0)$] out of thermal equilibrium will spend most of the time in thermal equilibrium; it is different in that it applies, for typical Hamiltonians, to *all*—rather than most—initial states $\psi(0)$.

VI. COMPARISON WITH THE LITERATURE

von Neumann [2] proved, as his "quantum ergodic theorem," a precise version of normal typicality (defined in Sec. II B); his proof requires much more effort, and more refined methods, than our proof of Theorem 1. However, his theorem assumes that the dimension d_ν of each macrospace $\mathcal{H}_\nu$ is much smaller than the full dimension D, and thus does not apply to the situation considered in this paper, in which one of the macrospaces, $\mathcal{H}_{eq}$, has the majority of dimensions. The reason von Neumann treated the more difficult case of small d_ν but left out the easier and particularly interesting case of the thermal equilibrium macrostate is that he had in mind a notion of thermal equilibrium different from ours. He thought of a thermal equilibrium wave function ψ, not as one in (or close to) a particular $\mathcal{H}_\nu$, but as one with $\|P_\nu\psi\|^2 \approx d_\nu/D$ for every ν, i.e., one for which $|\psi\rangle\langle\psi| \approx \rho_{mc}$ in a suitable coarse-grained sense. Because of this different focus, he did not consider the situation presented here. We also note that von Neumann's quantum ergodic theorem makes an assumption on H that we do not need in our Theorem 1; this

[5]A classical system is ergodic if and only if the time evolved microstate X_t spends, in the long run, a fraction of time in each (measurable) set $B \subseteq \Gamma$ that is equal to $\mu(B)/\mu(\Gamma)$ for μ-almost all X.

assumption, known as a "no resonances" [8,10] or "nondegenerate energy gaps" [4] condition, asserts that

$$E_\alpha - E_\beta \neq E_{\alpha'} - E_{\beta'} \quad \text{unless} \quad \begin{cases} \text{either } \alpha = \alpha', \ \beta = \beta' \\ \text{or } \alpha = \beta, \ \alpha' = \beta'. \end{cases} \tag{84}$$

The Schnirelman theorem [19] states that, in the semiclassical limit and under suitable hypotheses, the Wigner distribution corresponding to an eigenstate ϕ_α becomes the microcanonical measure. That is, ϕ_α has a property resembling thermal equilibrium, similar to our condition (24) expressing that all eigenstates are in thermal equilibrium. Srednicki [7] observed other thermal equilibrium properties in energy eigenstates of example systems, a phenomenon he referred to as "eigenstate thermalization."

The results of [4,5,8] also concern conditions under which a quantum system will spend most of the time in thermal equilibrium. For the sake of comparison, their results, as well as ours, can be described in a unified way as follows. Let us say that a system with initial wave function $\psi(0)$ *equilibrates* relative to a class $\mathcal{A}$ of observables if for most times τ,

$$\langle\psi(\tau)|A|\psi(\tau)\rangle \approx \mathrm{Tr}(\overline{|\psi(t)\rangle\langle\psi(t)|}A) \quad \text{for all } A \in \mathcal{A}. \tag{85}$$

We then say that the system *thermalizes* relative to $\mathcal{A}$ if it equilibrates and, moreover,

$$\mathrm{Tr}(\overline{|\psi(t)\rangle\langle\psi(t)|}A) \approx \mathrm{Tr}(\rho_{mc}A) \quad \text{for all } A \in \mathcal{A}, \tag{86}$$

with ρ_{mc} the microcanonical density matrix (in our notation, $1/D$ times the projection P to $\mathcal{H}$). With these definitions, the results of [4,5,8] can be formulated by saying that, under suitable hypotheses on H and $\psi(0)$ and for large enough D, a system will equilibrate, or even thermalize, relative to a suitable class $\mathcal{A}$.

Our result is also of this form. We have just one operator in $\mathcal{A}$, namely P_{eq}. We establish thermalization for arbitrary $\psi(0)$ assuming that H is nondegenerate and satisfies $\langle\phi_\alpha|P_{eq}|\phi_\alpha\rangle \approx 1$ for all α, which (as we show) is typically true.

von Neumann's quantum ergodic theorem [2] establishes thermalization for a family $\mathcal{A}$ of commuting observables; $\mathcal{A}$ is the algebra generated by $\{M_1, \ldots, M_k\}$ in the notation of Sec. I A. He assumed that the dimensions of the joint eigenspaces $\mathcal{H}_\nu$ are not too small and not too large and that H obeys Eq. (84), he made an assumption about the relation between H and the subspaces $\mathcal{H}_\nu$ that he showed is typically true, and he admitted arbitrary $\psi(0)$. See [1] for further discussion. Rigol *et al.* [6] numerically simulated an example system and concluded that it thermalizes relative to a certain class $\mathcal{A}$ consisting of commuting observables.

Tasaki [8] as well as Linden *et al.* [4] considered a system coupled to a heat bath, $\mathcal{H}_{\rm total} = \mathcal{H}_{\rm sys} \otimes \mathcal{H}_{\rm bath}$, and took $\mathcal{A}$ to contain all operators of the form $A_{\rm sys} \otimes 1_{\rm bath}$. Tasaki considered a rather special class of Hamiltonians and established thermalization assuming that

$$\max_{\alpha}|\langle\phi_\alpha|\psi(0)\rangle|^2 \ll 1, \tag{87}$$

a condition that implies that many eigenstates of H contribute to $\psi(0)$ appreciably and that can (more or less) equivalently be rewritten as

$$\sum_{\alpha}|\langle\phi_\alpha|\psi(0)\rangle|^4 \ll 1. \tag{88}$$

Under assumption (88) on $\psi(0)$, Linden *et al.* established equilibration for H satisfying Eq. (84). They also established a result in the direction of thermalization under the additional hypothesis that the dimension of the energy shell of the bath is much greater than dim $\mathcal{H}_{sys}$.

Reimann's mathematical result [5] can be described in the above scheme as follows. Let $\mathcal{A}$ be the set of all observables A with (possibly degenerate) eigenvalues between 0 and 1 such that the absolute difference between any two eigenvalues is at least (say) 10^{-1000}. He established equilibration for H satisfying Eq. (84), assuming that $\psi(0)$ satisfies Eq. (88).

ACKNOWLEDGMENTS

We thank Matthias Birkner (LMU München), Peter Reimann (Bielefeld), Anthony Short (Cambridge), Avraham Soffer (Rutgers), and Eugene Speer (Rutgers) for helpful discussions. S.G. was supported in part by National Science Foundation (Grant No. DMS-0504504). N.Z. is supported in part by Istituto Nazionale di Fisica Nucleare. J.L.L. and C.M. are supported in part by NSF (Grant No. DMR 08-02120) and by AFOSR (Grant No. AF-FA 09550-07).

[1] S. Goldstein, J. Lebowitz, C. Mastrodonato, R. Tumulka, and N. Zanghì, e-print arXiv:0907.0108
[2] J. von Neumann, Z. Phys. **57**, 30 (1929).
[3] J. M. Deutsch, Phys. Rev. A **43**, 2046 (1991).
[4] N. Linden, S. Popescu, A. J. Short, and A. Winter, Phys. Rev. E **79**, 061103 (2009).
[5] P. Reimann, Phys. Rev. Lett. **101**, 190403 (2008).
[6] M. Rigol, V. Dunjko, and M. Olshanii, Nature (London) **452**, 854 (2008).
[7] M. Srednicki, Phys. Rev. E **50**, 888 (1994).
[8] H. Tasaki, Phys. Rev. Lett. **80**, 1373 (1998).
[9] J. von Neumann, *Mathematical Foundation of Quantum Mechanics* (Princeton University Press, Princeton, NJ, 1955); translation of J. von Neumann, *Mathematische Grundlagen der Quantenmechanik* (Springer-Verlag, Berlin, 1932).
[10] R. Jancel, *Foundations of Classical and Quantum Statistical Mechanics* (Pergamon, Oxford, 1969); translation of R. Jancel, *Les Fondements de la Mécanique Statistique Classique e Quantique* (Gauthier-Villars, Paris, 1963).
[11] J. Lebowitz, in *Boltzmann's Legacy*, edited by G. Gallavotti, W. L. Reiter, and J. Yngvason (European Mathematical Society, Zurich, 2008), pp. 63–88.
[12] E. Schrödinger, *Statistical Thermodynamics*, 2nd ed. (Cambridge University Press, Cambridge, England, 1952).
[13] J. D. Walecka, *Fundamentals of Statistical Mechanics: Manuscript and Notes of Felix Bloch* (Stanford University Press, Stanford, CA, 1989).
[14] A. Dembo and O. Zeitouni, *Large Deviations Techniques and Applications*, 2nd ed. (Springer, New York, 1998).
[15] T. Kato, *A Short Introduction to Perturbation Theory for Linear Operators* (Springer-Verlag, New York, 1982).
[16] P. W. Anderson, Phys. Rev. **109**, 1492 (1958).
[17] V. Oganesyan and D. A. Huse, Phys. Rev. B **75**, 155111 (2007).
[18] O. E. Lanford, in Lecture Notes in Physics Vol. 2, edited by A. Lenard (Springer-Verlag, Berlin, 1973), pp. 1–113.
[19] Y. Colin de Verdière, Commun. Math. Phys. **102**, 497 (1985).

Commun. Math. Phys. 363, 97–110 (2018)
Digital Object Identifier (DOI) https://doi.org/10.1007/s00220-018-3226-5

Communications in **Mathematical Physics**

Generalized Stealthy Hyperuniform Processes: Maximal Rigidity and the Bounded Holes Conjecture

Subhroshekhar Ghosh[1], **Joel L. Lebowitz**[2]

[1] National University of Singapore, Singapore, Singapore. E-mail: subhrowork@gmail.com
[2] Rutgers University, New Brunswick, USA. E-mail: lebowitz@math.rutgers.edu

Received: 21 July 2017 / Accepted: 15 February 2018
Published online: 24 August 2018 –

Abstract: We study translation invariant stochastic processes on $\mathbb{R}^d$ or $\mathbb{Z}^d$ whose diffraction spectrum or structure function $S(k)$, i.e. the Fourier transform of the *truncated* total pair correlation function, vanishes on an open set U in the wave space. A key family of such processes are "stealthy" hyperuniform point processes, for which the origin $k = 0$ is in U; these are of much current physical interest. We show that all such processes exhibit the following remarkable maximal rigidity: namely, the configuration outside a bounded region determines, with probability 1, the exact value (or the exact locations of the points) of the process inside the region. In particular, such processes are completely determined by their tail. In the 1D discrete setting (i.e. $\mathbb{Z}$-valued processes on $\mathbb{Z}$), this can also be seen as a consequence of a recent theorem of Borichev et al. (Spectra of stationary processes on Z, vol. 261. Birkhäuser, Cham, 2017); in higher dimensions or in the continuum, such a phenomenon seems novel. For stealthy hyperuniform point processes, we prove the Zhang–Stillinger–Torquato conjecture (Zhang et al. in Soft Matter 13:6197–6207, 2017. arXiv:1705.04415) that such processes have bounded holes (empty regions), with a universal bound that depends inversely on the size of U.

1. Introduction

In recent years, a special class of hyperuniform particle systems, known as stealthy hyperuniform (henceforth abbreviated as SH) systems, have attracted considerable attention [ToZS, ZhST-2, ZhST-1, ZhST-3, ChDZCT, To-2]. These systems are characterized by the structure function $S(k)$ vanishing in a neighbourhood of $k = 0$. A natural generalization of SH point processes is to consider point processes, or random fields, with a gap in the spectrum on an open set U which may not include the origin. We shall denote these processes as Generalized Stealthy (henceforth abbreviated as GS) processes. For precise definitions we refer the reader to Sect. 2.

The nomenclature "stealthy", as well as the physical interest in SH particle systems, stems from the fact that such systems are optically transparent (invisible) for wave vectors k in the gap U. Numerical and experimental investigations have been carried out regarding how to construct SH particle systems. These systems cannot be equilibrium systems, with tempered potentials, at finite temperatures. They may, however, be ground states of such systems, e.g. the zero temperature states of classical systems, or they can be generated as non-equilibrium states.

SH systems are an extension of hyperuniform (superhomogeneous) particle systems. Hyperuniform systems, which have been studied extensively both in the physics and the mathematics literature [AiMa, MaY, GoLSp, GhL, ToSt, MaST, JiLHMCT, JiT, FITS, BuFN, HaMS, DeSMRPRJTB, HeL, HeCL], have reduced fluctuations: the variance of the particle number in a domain D in $\mathbb{R}^d$ or $\mathbb{Z}^d$ grows slower than the volume of D. The significance of hyperuniform materials, and in particular SH systems, lies in the fact that they embody properties of both crystalline and disordered or random systems (see [To-1, GhL] and the references therein).

For translation invariant systems, on which we shall focus here, an equivalent characterization of hyperuniformity can be obtained by looking at its structure function. Hyperuniformity then boils down to the vanishing of the structure function $S(k)$ at $k = 0$. SH systems, therefore, involve a specific manner in which this vanishing of the structure function takes place.

In particle systems theory, an important object of investigation is the event that there is a large "hole" in the particle configuration, by a "hole" we generally mean a ball, with centre at the origin, that is devoid of any particle. The rate of decay of the hole probability encodes important structural information about the particle system at hand. Heuristically speaking, the faster this rate of decay the stronger is the crystalline (or lattice-like) behaviour of the particle system.

The examination of hole probabilities for stealthy hyperuniform processes at the numerical level brings up intriguing challenges. In particular, numerical studies led to the conjecture that the hole sizes for stealthy hyperuniform processes are, in fact, uniformly bounded. Furthermore, it was conjectured that the upper bound on the hole size depends only on the size of the gap for the structure function of the process. In the article [ZhST-4], Zhang, Stillinger and Torquato provide numerical evidence in support of these remarkable conjectural properties of stealthy hyperuniform processes. Note that this is definitely not true for all hyperuniform systems, as exemplified by the one component Coulomb systems, in particular by the Ginibre ensemble [JaLMa, Gi].

In this paper, we carry out a rigorous mathematical analysis of stealthy hyperuniform processes, and establish the veracity of this conjecture. In particular, we prove that

Theorem 1.1. *Let Ξ be a stealthy hyperuniform point process. Let $B(x;r)$ be the ball with centre x and radius r. Then there exists a positive number r_0 such that $\mathbb{P}[|\Xi \cap B(x;r_0)| = 0] = 0$.*

Furthermore, we establish the explicit dependence of r_0 on the size of the gap b in the spectrum, and show that it satisfies an inverse power law.

Theorem 1.2. *The quantity r_0, as in the statement of Theorem* 1.1, *can be chosen to be Cb^{-1}, where b is the radius of the maximal ball (centered at the origin) that is contained in the gap of the structure function S, and C is a universal constant.*

The proofs of Theorems 1.1 and 1.2 exploit an anti-concentration property for particle numbers of stealthy hyperuniform processes:

Theorem 1.3. *Let Ξ be a stealthy hyperuniform point process on $\mathbb{R}^d$ or $\mathbb{Z}^d$ with one point intensity ρ and b the radius of the largest ball around the origin (in the wave space) on which the structure function of Ξ vanishes. There exists numbers $C, c > 0$ (independent of all parameters of Ξ) such that, the number of points of Ξ in any given d-dimensional cube of side-length Cb^{-1} is a.s. bounded above by $c\rho b^{-d}$.*

The fact that holes in SH processes cannot be bigger than a deterministic size suggests a high degree of crystalline behaviour in these processes. In this paper, we also establish a remarkable maximal rigidity property of these ensembles. We can, in fact, do this in the setting of GS processes. Before stating our main result, we need to discuss the concepts of rigidity and maximal rigidity. This we will do in the context of $\mathbb{R}^d$, noting that these concepts in the setting of $\mathbb{Z}^d$ are completely analogous.

For a point process (more generally, a random field or a random measure) Ξ on $\mathbb{R}^d$ and a bounded domain $D \subset \mathbb{R}^d$, statistic Ψ defined on Ξ restricted to D is said to be rigid if Ψ is completely determined by (that is, a deterministic function of) the process Ξ restricted to $D^{\complement}$. To put things in perspective, a point process having rigidity is in notable contrast to the Poisson process, where the process inside and outside of D are statistically independent. Rigidity phenomena for particle systems have been investigated quite intensively in the last few years, and it has been shown to appear in many natural models which are, nonetheless, far removed from being crystalline. Key examples include the Ginibre ensemble [GhP, OsS], Gaussian zeros [GhP], the Dyson log gas [Gh-1, GhL], Coulomb systems [GhL, AiMa] and various determinantal processes related to random matrix theory [Bu]. In [Gh-2], rigidity phenomena were shown to be intimately connected to mutual singularity properties of various Palm measures of a point process, which is a topic of independent interest in particle systems theory. In [GhK], a one parameter family of Gaussian random series were introduced, whose zeros exhibit rigidity of increasing number of moments, as the parameter is allowed to vary.

In this paper, we show that GS random measures on $\mathbb{R}^d$ or $\mathbb{Z}^d$ exhibit maximal rigidity: namely, for any domain $D \subset \mathbb{R}^d$, the random measure $[\Xi]_{|D^{\complement}}$ determines completely the measure $[\Xi]_{|D}$ (that is, the latter is a deterministic measurable function of the former). Stated in formal terms, we prove:

Theorem 1.4. *Let Ξ be a generalized stealthy random measure on $\mathbb{R}^d$ or $\mathbb{Z}^d$. Then for any bounded domain D, the random measure $[\Xi]_{|D}$ is almost surely determined by (i.e., is a measurable function of) the random measure $[\Xi]_{|D^{\complement}}$.*

This leads to the following conclusion:

Corollary 1.5. *Let Ξ be a generalized stealthy random measure on $\mathbb{R}^d$ or $\mathbb{Z}^d$. Then Ξ is completely determined by (that is, measurable with respect to) its tail sigma field.*

We further show that, to have maximal rigidity in the sense discussed above, it suffices that the structure function vanishes faster than any polynomial at the origin in the wave space:

Theorem 1.6. *Let Ξ be a random measure on $\mathbb{R}^d$ or $\mathbb{Z}^d$ such that the structure function of Ξ vanishes faster than any polynomial at the origin in the wave space. Then for any bounded domain D, almost surely the random measure $[\Xi]_{|D}$ is determined by (i.e., is a measurable function of) the random measure $[\Xi]_{|D^{\complement}}$. Consequently, Ξ is completely determined by (that is, measurable with respect to) its tail sigma field.*

In the special case of the 1D lattice where the random variable takes only discrete values, that is $\mathbb{Z}$-valued processes on $\mathbb{Z}$, our results can also be seen as a consequence of a remarkable recent theorem of Borichev et al. [BoSW]. They show that, in the 1D setting above, the Fourier transform of the covariance function vanishing on a subinterval of $\mathbb{T}$ (which is the wave space in this case) implies that the process is essentially periodic. However, the methods of [BoSW], which rely strongly on complex analysis of one variable, seem to be rather specially suited for the 1D discrete case, and in [BoSW] the authors raise the question as to whether one can expect similar results in the discrete setting in higher dimensions (that is, $\mathbb{Z}$-valued processes on $\mathbb{Z}^d$ for $d > 1$). This leaves out of its purview point processes on $\mathbb{R}^d$ (i.e., in the continuum) for $d \geq 1$. For continuum random fields or random measures on $\mathbb{Z}^d$ or $\mathbb{R}^d$, periodicity is not true for stealthy processes as evidenced by the key example of Gaussian processes, already noted in [BoSW] and discussed in Sect. 5. This is an open question for particle systems - just how random can SH processes be, say in the sense of entropy or decay of correlations; see [ToZS] for physical arguments and numerical evidence regarding the existence of SH disordered particle systems.

The question of inference about a stochastic process from its diffraction spectrum has a long history in diffraction theory, and we believe the results in the present article would be of interest to that body of literature. We refer the reader who is interested in further exploration of this direction to [BaBM, BaKM, BaG]. Note that the central Bragg peak $\rho^2\delta_{k=0}$ is not included in $S(k)$. The quantity $S(k)$ is also referred to as the Bartlett spectrum [DaV].

2. Setup and Basic Notions

In this section, we formally define the setup in which we are working.

At the very outset, let us introduce the notation that Ω denotes the wave space. This would imply that $\Omega = \mathbb{T}^d$ when the physical space is $\mathbb{Z}^d$, and $\Omega = \mathbb{R}^d$ when the physical space is $\mathbb{R}^d$. What Ω is, will usually be clear from the context. On a related note, we will denote functions that live on the wave space with a cap, and we will denote functions that live on the physical space without a cap.

Let $\mathcal{S}$ denote the space of complex Borel measures on $\mathbb{R}^d$ (resp., $\mathbb{Z}^d$). Then a random measure on $\mathbb{R}^d$ is a random variable that takes values in the space $\mathcal{S}$. Since a locally finite point configuration can be identified with its counting measure, a point process is a special case of a random measure (for a detailed treatment of point processes, we refer the interested reader to [DaV]). It also covers locally L_1 random fields on $\mathbb{R}^d$, by identifying a random field $\{X_u\}_{u\in\mathbb{R}^d}$ with the random measure $X_u\mathrm{d}u$ (where $\mathrm{d}u$ is the Lebesgue measure on $\mathbb{R}^d$). If Ξ is a random measure, then by $\mathrm{d}[\Xi]$ we will denote a specific realization of the random variable Ξ.

A random measure is characterized by the distribution of its linear statistics, namely the random variables $\int \varphi(x)\mathrm{d}[\Xi](x)$ for C_c^∞ test functions φ on $\mathbb{R}^d$. A random measure Ξ is called translation invariant if, for any $u \in \mathbb{R}^d$, the distribution of the measure $\mathrm{d}[\Xi](\cdot - u)$ is the same as that of $\mathrm{d}[\Xi](\cdot)$.

Translation invariance of Ξ implies that $\mathbb{E}\left[\mathrm{d}[\Xi]\right] = \rho\mathrm{d}u$, where ρ is the intensity of the random measure and $\mathrm{d}u$ denotes the Lebesgue measure.

The covariance functional β of a random field Ξ is defined by the equation, for any two test functions φ and ψ:

$$\mathrm{Cov}\left[\int \varphi\mathrm{d}[\Xi], \int \psi\mathrm{d}[\Xi]\right] = \int\int \varphi(s)\psi(t)\beta(s-t)\mathrm{d}s\mathrm{d}t.$$

In general, the covariance functional $d\beta(s,t)$ will be a measure (that is a constant times the Lebesgue measure on the sets $s-t=$ const. because of translation invariance). We assume here for simplicity that this measure, in fact, has a density with respect to the Lebesgue measure $dsdt$ on $\mathbb{R}^d \times \mathbb{R}^d$.

As a special case of this, for a translation invariant point process Ξ, we denote by G the fully truncated pair correlation function, given by

$$G(x-y) = \rho_2(x-y) - \rho^2 + \rho\delta(x-y),$$

where ρ_2 is the two point correlation function of Ξ and ρ is its (one point) intensity [GhL, MaY]. It can be verified via a simple computation that G is the same as β as discussed above.

The Fourier transform $\hat{\beta}$ of the covariance functional β is the structure function $S(k)$ (or, more generally, the structure measure $dS(k)$).

We now formally define hyperuniform and stealthy hyperuniform random measures.

Definition 1. A point process is called *hyperuniform* if the structure function S vanishes at the origin.

Equivalently, it can be shown that a point process is hyperuniform if and only if the variance of the particle number in a domain D grows slower than the volume $|D|$ (and not like the volume, as would be the case for Poisson and other extensive point processes), see [GhL, L].

We will focus on *Stealthy* hyperuniform processes, which are defined as follows:

Definition 2. A point process is called *stealthy hyperuniform* (SH) if the structure function S vanishes in a neighbourhood of the origin.

Finally, we have *Generalized Stealthy* processes:

Definition 3. A point process (or random measure) on $\mathbb{R}^d$ (resp., $\mathbb{Z}^d$) is called a generalized stealthy (GS) process if its structure function S vanishes in an open set $U \subset \mathbb{R}^d$ (resp. $\mathbb{T}^d$).

The crucial difference between stealthy and generalized stealthy processes is that, for a generalized stealthy process, the gap U in the diffraction spectrum need not contain the origin.

3. Universal Bounds on Hole Sizes

Our goal in this section is to prove Theorem 1.1. Let Ξ be a stealthy process on $\mathbb{R}^d$ or $\mathbb{Z}^d$, whose structure function vanishes in a neighbourhood U of the origin in the wave space. Let b denote the radius of the largest ball, with centre at the origin, that is contained in U.

3.1. Linear statistics. We consider an L_2 function φ, and the linear statistic $I(\varphi) = \int \varphi d[\Xi]$ corresponding to φ. We consider the variance of the linear statistic which, for translation invariant point processes, can be written as

$$\mathrm{Var}[I(\varphi)] = \int_{\Omega} |\hat{\varphi}(\xi)|^2 S(\xi) d\xi. \tag{1}$$

Now, if $S(\xi)$ vanishes in a neighbourhood $U \subset \Omega$, and if φ is such that $\hat{\varphi}$ is supported on U, we have $\mathrm{Var}[I(\varphi)] = 0$. This implies that a.s. we have

$$\int \varphi \mathrm{d}[\Xi] = \mathbb{E}[I(\varphi)] = \rho\hat{\varphi}(0). \tag{2}$$

This is a crucial observation in the context of stealthy processes with important implications, as we shall see in the ensuing sections.

3.2. Anti-concentration of particles in stealthy processes. In this section, we show a rather surprising "anti-concentration" property (namely, Theorem 1.3) of the particles in a stealthy hyperuniform point process. Though it arises contextually in the proof of boundedness of holes in such processes, we believe that this property is interesting in its own right.

Proof of Theorem 1.3. We observe that, since the process Ξ is translation invariant, it suffices to consider the cube (in which we want to prove the anti-concentration) to be centered at the origin.

We first choose an auxiliary test function $\hat{\psi}$ on Ω such that $\hat{\psi}$ is a C_c^∞ non-negative radial function supported on the ball of radius 1/3 centered at the origin. This has several consequences, which we enumerate below:

- Since $\hat{\psi}(\xi)$ is invariant under the map $\xi \mapsto -\xi$, we deduce that ψ is real valued.
- Since $\hat{\psi}$ is a radial function, so is ψ.
- Since $\hat{\psi}$ is non-negative, we have $\psi(0) = \int \hat{\psi}(\xi)\mathrm{d}\xi > 0$.
- Since $\hat{\psi}$ is C_c^∞, we deduce that ψ is a Schwarz function, and hence in fact $\psi > 0$ on a neighbourhood of the origin.
- Dividing $\hat{\psi}$ by a positive constant if necessary, we may assume that $\psi \geq 1$ on a cube of side $a > 0$ centered at the origin.

We now select φ to be the function such that $\hat{\varphi}(\xi) = (\hat{\psi} * \hat{\psi})(\xi/b)$. Clearly, $\varphi(x) = b^d[\psi(bx)]^2$.

We observe that:

- $\hat{\psi} * \hat{\psi}$ is supported inside the ball of radius 2/3 (centered at the origin) in the wave space. Consequently, $\hat{\varphi}$ is supported in the ball of radius $\frac{2}{3}b$ centered at the origin, and hence its support $\subset U$ (which contains a ball having center at the origin and radius b).
- φ is a non-negative radial Schwarz function.
- $\hat{\varphi}(0) = \int \varphi(x)\mathrm{d}x > 0$.
- $\varphi \geq b^d$ on a cube B centred at the origin and having side length ab^{-1}, where a is as defined while enumerating the properties of ψ.

Due to the fact that the support of $\hat{\varphi}$ is contained inside U, the function φ satisfies (2) a.s. This, together with the fact that $\varphi \geq 0$, implies that

$$\sum_{x \in \Xi \cap B} \varphi(x) \leq \rho\hat{\varphi}(0) = \rho \ \hat{\psi} * \hat{\psi}(0).$$

This, in turn, implies that

$$|\Xi \cap B| \cdot b^d \leq \rho\hat{\psi} * \hat{\psi}(0).$$

Consequently, we have

$$|\Xi \cap B| \leq \hat{\psi} * \hat{\psi}(0) \cdot \rho b^{-d},$$

as desired. □

3.3. Proof of Theorem 1.1. In this section, we use the anti-concentration results of the previous section in order to establish boundedness of holes for stealthy hyperuniform point processes.

For the rest of this section, we will work on the event (which by Theorem 1.3 occurs with probability 1) where anti-concentration of particle number holds simultaneously for all cubes of side length ab^{-1} having centers with co-ordinates that are rational multiples of $\sqrt{d}$.

In what follows, we will denote by $A(r,s)$ the L_∞ annulus $\{x \in \mathbb{R}^d : r \leq \|x\|_\infty < s\}$. By θ we will denote Cb^{-1}, where C and b are as in the statement of Theorem 1.3. Then for a parameter $R > 0$, to be thought of as large, we will consider the decomposition

$$\mathbb{R}^d = A(0, R\theta) \bigcup \cup_{j=1}^{\infty} A(R\theta + j\theta, R\theta + (j+1)\theta).$$

In the remainder of this section, we will denote $A(0, R\theta)$ by A_0 and $A(R\theta + j\theta, R\theta + (j+1)\theta)$ by A_j for $j \geq 1$.

We consider the Eq. (2) for φ (chosen as in Sect. 3.2), and observe that we can rewrite the same as

$$\left(\sum_{x \in \Xi \cap A_0} \varphi(x)\right) + \sum_{j=1}^{\infty} \left(\sum_{x \in \Xi \cap A_j} \varphi(x)\right) = \rho \hat{\varphi}(0). \tag{3}$$

Let R be such that the event $F_R := \{|\Xi \cap A_0| = 0\}$ has positive probability. Then, on the event F_R, (3) reduces to

$$\sum_{j=1}^{\infty} \left(\sum_{x \in \Xi \cap A_j} \varphi(x)\right) = \rho \hat{\varphi}(0). \tag{4}$$

Since φ is a Schwarz function, we have $|\varphi(x)| \leq C_\varphi |x|^{-d-1}$. We now proceed to understand the quantity C_φ for our choice of φ. We begin with the observation that $C_\varphi = \||x|^{d+1}\varphi\|_\infty$. Going to the Fourier space, this can be bounded above as

$$C_\varphi = \||x|^{d+1}\varphi\|_\infty \leq \|D^{d+1}(\hat{\varphi})\|_1,$$

where D is the radial derivative. But recall that

$$\hat{\varphi}(\cdot) = (\hat{\psi} * \hat{\psi})(\cdot/b),$$

so $\|\hat{\varphi}\|_1 = b^d \|\hat{\psi} * \hat{\psi}\|_1$, and $\|D^{d+1}\hat{\varphi}\|_1 = b^{-1}\|D^{d+1}(\hat{\psi} * \hat{\psi})\|_1$. Here we recall the fact that ψ is a universal choice (that is, not dependent on any parameters of the process Ξ), and hence all norms of $\hat{\psi} * \hat{\psi}$ are universal constants in the context of our problem.

Putting together the above observations, we have $|\varphi(x)| \le Cb^{-1}|x|^{-d-1}$, where C is a universal constant. Consequently,

$$\left|\sum_{x\in\Xi\cap A_j}\varphi(x)\right| \le Cb^{-1}|\Xi\cap A_j|[(R+j)\theta]^{-d-1},$$

which implies that on the event F_R we have

$$0 < \rho\hat{\varphi}(0) \le Cb^{-1}\sum_{j=1}^{\infty}|\Xi\cap A_j|[(R+j)\theta]^{-d-1} \tag{5}$$

Let $\Gamma(x,\theta)$ denote the cube with centre $x\in\mathbb{R}^d$ and side-length θ. We can then write $A_j = A((R+j)\theta, (R+(j+1))\theta)$ as a disjoint union

$$A_j = \bigcup_{i=1}^{N}\Gamma(x_i,\theta),$$

such that the centres x_i have co-ordinates that are rational multiples of $\sqrt{d}$ and $N = C[(R+j)\theta]^{d-1}$ for some positive number C.

But the anti-concentration property of particle numbers (Theorem 1.3) implies that $|\Xi\cap\Gamma(x_i,\theta)| \le c\rho b^{-d}$, which in turn implies that $|\Xi\cap A_j| \le c\rho b^{-d}N$. This implies that

$$0 < \rho\hat{\varphi}(0) \le Cb^{-1}\sum_{j=1}^{\infty}c\rho b^{-d}N[(R+j)\theta]^{-d-1} \le c_1 b^{-d-1}\rho R^{-1}\theta^{-d-1} = c_2\rho R^{-1}, \tag{6}$$

where, in the last step, we have used the fact that $\theta = Cb^{-1}$. Consequently, $R \le c_3$, for some universal constant c_3. Hence the size of any hole is bounded above by $R\theta \le \kappa b^{-1}$ for some universal constant κ, as desired.

4. Rigidity

In this section, we prove Theorem 1.4.

4.1. Point processes. It suffices to prove the theorem when the domain in question is a ball. This is because, any bounded set is a subset of a large enough ball, and it is easy to see that if Theorem 1.4 holds for a domain D, then it holds for any subset of D. By way of notation, let $B(x;r)$ be the ball of radius r and centre x. We want to prove the theorem for the domain $B(0;a)$. We recall the notation that $\Gamma(x,r)$ is the cube with center $x\in\Omega$ (where Ω denotes the phase space) and side-length $r>0$.

We consider $\mu\in\Omega$ and $\beta>0$ small enough so that $\Gamma(2\mu,2\beta)\subsetneq U$ (where the structure function S vanishes on the open set U). Moreover, for $\theta\in\mathbb{R}^d$ such that $\|\theta\|_\infty$ is small enough, we have $\Gamma(2\mu,2\beta)+2\theta\subset U$. Let $\mathbf{1}_A$ be the normalized indicator of a set A. Namely, $\mathbf{1}_A = \frac{1}{|A|}\chi_A$, where χ_A is the usual indicator function of the set A, and $|A|$ is the Euclidean volume of A. Now let $\varphi(x,\theta,\beta)$ be a such that $\hat{\varphi}(\xi,\theta,\beta) := \mathbf{1}_{\Gamma(\mu,\beta)+\theta}\star\mathbf{1}_{\Gamma(\mu,\beta)+\theta}(\xi)$ (where the Fourier transform is with respect to the variable x).

Then $\hat{\varphi}(\cdot, \theta, \beta)$ is supported inside U for all θ, β small enough. Hence (2) is satisfied a.s. We will work on the event (of probability 1) where all $\varphi(\cdot, \theta, \beta)$ (with (θ, β) as above and having all co-ordinates rational ; a set that we will denote by Q) satisfy (2) simultaneously.

We notice that φ as above has the explicit expression

$$\varphi(x, \theta, \beta) = e^{2\pi i \langle 2\mu + 2\theta, x\rangle} \prod_{i=1}^{d} \left(\frac{\sin \pi \beta x(i)}{\pi \beta x(i)} \right)^2,$$

where $x = (x(1), \ldots, x(d)) \in \mathbb{R}^d$. (2) can be rewritten for $\varphi(\cdot, \theta, \beta)$ as

$$\sum_{x \in \Xi \cap B(0;a)} \varphi(x, \theta, \beta) = \rho \hat{\varphi}(0, \theta, \beta) - \sum_{x \in \Xi \cap B(0;a)^{\complement}} \varphi(x, \theta, \beta). \tag{7}$$

Let $X_1, \ldots, X_N$ be the points inside $B(0; a)$. We deduce that the point configuration Υ outside $B(0; a)$ determines the quantities (for each $(\theta, b) \in Q$)

$$\sum_{j=1}^{N} e^{2\pi i \langle 2\mu + 2\theta, X_j\rangle} \prod_{i=1}^{d} \left(\frac{\sin \pi \beta X_j(i)}{\pi \beta X_j(i)} \right)^2 = F(\Upsilon, \theta, \beta) \tag{8}$$

where F is a measurable function.

Equation (8) is true for all (θ, β) with rational co-ordinates that lie in a small neighbourhood of the origin. Since the left hand side of (8) is continuous in (θ, β) for a fixed realization $(X_1, \ldots, X_N)$, (8) is true for $(\theta, 0)$ for all θ such that $(\theta, \beta) \in Q$. for some $\beta > 0$. This leads us to

$$\sum_{j=1}^{N} e^{2\pi i \langle 2\mu + 2\theta, X_j\rangle} = F(\Upsilon, \theta) \tag{9}$$

for a measurable function F.

Since the left hand side of (9) is continuous in θ, we can take limits to deduce that (9) is true for all θ in a neighbourhood of 2μ. But this determines the set $(X_1, \ldots, X_N)$ completely. A simple way to see this is that the left hand side of (9) is the characteristic function of the (compactly supported) counting measure corresponding to the set $\{X_1, \ldots, X_N\}$.

4.2. Random fields and random measures. The techniques of Sect. 4.1, although introduced for point processes, can be extended to the setting where Ξ is a translation invariant random field, or even a translation invariant random measure on $\mathbb{R}^d$ or $\mathbb{Z}^d$. In what follows, we discuss the natural extensions of the key concepts used in Sect. 4.1 to the setting of random fields and random measures.

Following the argument for point processes, we consider $\beta > 0$ small enough so that $\Gamma(2\mu, 2\beta) \subsetneq U$, and let $\varphi(x, \theta, \beta)$ be such that $\hat{\varphi}(\xi, \theta, \beta) := \mathbf{1}_{\Gamma(\mu,\beta)+\theta} \star \mathbf{1}_{\Gamma(\mu,\beta)+\theta}(\xi)$, where the Fourier transform is in the x variable. Then $\hat{\varphi}(\cdot, \theta, \beta)$ is supported inside U for all θ, β small enough. Clearly, φ as above has the explicit expression

$$\varphi(x, \theta, \beta) = e^{2\pi i \langle 2\mu + 2\theta, x\rangle} \prod_{i=1}^{d} \left(\frac{\sin \pi \beta x(i)}{\pi \beta x(i)} \right)^2,$$

where $x = (x(1), \ldots, x(d)) \in \mathbb{R}^d$.

We note that for each (θ, β) the function $\varphi(\cdot, \theta, \beta) \in L_1(\mathrm{d}x)$, hence the linear statistic $\int \varphi \mathrm{d}[\Xi]$ is well-defined and finite a.s. Exploiting the support properties of φ, we can write down the following analogue of (2):

$$I(\varphi(\cdot, \theta, \beta)) = \int \varphi(x, \theta, \beta)\mathrm{d}[\Xi] = \mathbb{E}[I(\varphi)] = \rho\hat{\varphi}(0, \theta, \beta). \tag{10}$$

We immediately obtain an analogue of (7) as

$$\int_{B(0;a)} \varphi(x, \theta, \beta)\mathrm{d}[\Xi] = \rho\hat{\varphi}(0, \theta, \beta) - \int_{B(0;a)^{\complement}} \varphi(x, \theta, \beta)\mathrm{d}[\Xi]. \tag{11}$$

Let Ξ^{in} and Ξ^{out} respectively denote the restrictions of the random measure Ξ to $B(0; a)$ and $B(0; a)^{\complement}$. Then we want to infer about Ξ^{in} given Ξ^{out}. Proceeding as in the case of point processes, we eventually deduce that

$$\int e^{2\pi i \langle 2\mu + 2\theta, x\rangle}\mathrm{d}[\Xi^{\text{in}}] = F(\Xi^{\text{out}}, \theta) \tag{12}$$

for all θ in a neighbourhood of 0, F being a measurable function. But a.s. Ξ^{in} is a finite Borel measure supported on the compact set $B(0; a)$. Consequently, the characteristic function $\int e^{2\pi i \langle 2\mu + 2\theta, x\rangle}\mathrm{d}[\Xi^{\text{in}}]$ admits an extension to a holomorphic function in $\theta \in \mathbb{C}^d$. But for a holomorphic function, its values in any (real) neighbourhood determines the function completely. Hence (12) completely determines the characteristic function of Ξ^{in} (for any given realization), and hence completely determines the measure Ξ^{in}.

4.3. Rigidity under fast decay of the structure function. In this section, we show that if the structure function of an invariant point process (or more generally, an invariant random field or random measure) on $\mathbb{Z}^d$ or $\mathbb{R}^d$ vanishes at 0 faster than any polynomial, then the process in question exhibits maximal rigidity.

Our main tool will be the following proposition:

Proposition 4.1. *Let Ξ be an invariant point process (or more generally, an invariant random field or random measure) on $\mathbb{Z}^d$ or $\mathbb{R}^d$ whose structure function $S(k)$ vanishes at $k = 0$ faster than any polynomial. Then, for any Schwarz function φ, positive integer m and $L > 1$, the random integral $\int \varphi(z/L)\mathrm{d}[\Xi](z)$ satisfies*

$$\mathrm{Var}[\int \varphi(z/L)\mathrm{d}[\Xi](z)] \le C(\varphi, m)L^{-m}, \tag{13}$$

for some positive number $C(\varphi, m)$.

Proof. Let $\epsilon_m < 1$ be such that $|S(\xi)| \le C_m|\xi|^m$ for $|\xi| \le \epsilon_m$. We recall the notation that $\varphi_L(z) = \varphi(z/L)$, so that

$$\mathrm{Var}[\int \varphi(z/L)\mathrm{d}[\Xi](z)]] = \int_{\mathbb{R}^d} |\hat{\varphi_L}(\xi)|^2 S(\xi)\mathrm{d}\xi. \tag{14}$$

Let $m_0 > 0$ be such that $\int_{\mathbb{R}^d} |S((\xi)|^2(1+|\xi|)^{-m_0}\mathrm{d}\xi < \infty$ (in other words, the covariance functional β belongs to the Sobolev class $-m_0$).

Now, $\hat{\varphi_L}(\xi) = L^d \hat{\varphi}(L\xi)$. Consequently, because φ is a Schwarz function, we have

$$|\hat{\varphi}(\xi)|^2 \leq \frac{C_j(\varphi)}{(1+|\xi|)^j},$$

and thus we can choose j such that

$$|\hat{\varphi_L}(\xi)|^2 \leq C(\varphi, m)(1+L|\xi|)^{-\frac{1}{2}m_0 - \frac{1}{2}d - m}. \tag{15}$$

This implies, in particular, that

$$|\hat{\varphi_L}(\xi)|^2 \leq C(\varphi, m)(1+|\xi|)^{-m_0/2} L^{-m} |\xi|^{-\frac{1}{2}d - m}. \tag{16}$$

and

$$|\hat{\varphi_L}(\xi)|^2 \leq C(\varphi, m) L^{-m} |\xi|^{-m}. \tag{17}$$

We estimate the integral in (14) from above by breaking it into two regimes: $|\xi| < \epsilon_m$ and $|\xi| \geq \epsilon_m$.

For $|\xi| \geq \epsilon_m$, using (17) we upper bound the integral as

$$\begin{aligned}
&\int_{|\xi|\geq\epsilon_m} |\hat{\varphi_L}(\xi)|^2 |S(\xi)| \mathrm{d}\xi \\
&\leq C(\varphi, m) L^{-m} \left(\int_{|\xi|\geq\epsilon_m} |S(\xi)|^2 (1+|\xi|)^{-m_0} \mathrm{d}\xi \right)^{1/2} \left(\int_{|\xi|\geq\epsilon_m} |\xi|^{-d-2m} \mathrm{d}\xi \right)^{1/2} \\
&\leq C(\varphi, m) L^{-m} \left(\int_{\mathbb{R}^d} |S(\xi)|^2 (1+|\xi|)^{-m_0} \mathrm{d}\xi \right)^{1/2} \left(\int_{|\xi|\geq\epsilon_m} |\xi|^{-d-2m} \mathrm{d}\xi \right)^{1/2} \\
&\leq C'(\varphi, m) L^{-m}.
\end{aligned}$$

For $|\xi| < \epsilon_m$, using the fact that $|S(\xi)| \leq C_m |\xi|^m$ and (17), we upper bound the integral by

$$\int_{|\xi|<\epsilon_m} |\hat{\varphi_L}(\xi)|^2 |S(\xi)| \mathrm{d}\xi \leq C(\varphi, m) C_m L^{-m} \int_{|\xi|<\epsilon_m} |\xi|^m \cdot |\xi|^{-m} \mathrm{d}\xi \leq C''(\varphi, m) L^{-m}.$$

Combining the two bounds, we deduce the statement of the proposition. □

Now we are ready to prove Theorem 1.6.

Proof of Theorem 1.6. Let D be a bounded domain in $\mathbb{R}^d$. We want to show that the point process (or the random field) outside D determines the field inside D with probability 1.

To this end, we consider a bump function φ that is $\equiv 1$ on D. Clearly, the scalings $\varphi_L(z) := \varphi(z/L)$ all satisfy these properties. For $[k] \in \mathbb{Z}_+^d$ and $x \in \mathbb{R}^d$, we denote by $x^{[k]}$ the monomial $x_1^{k_1} \cdots x_d^{k_d}$, where x_i and k_i are respectively the co-ordinates of x and k. We also denote by $|k|$ the quantity $\sum_{i=1}^d k_i$. We will apply Proposition 4.1 to the Schwarz functions $\varphi^{[k]}(x) = x^{[k]}\varphi(x)$.

For large enough L, depending only on φ, $[k]$ and Ξ, we have

$$\mathrm{Var}[\int \varphi_L^{[k]} \mathrm{d}[\Xi]] \leq C(\varphi, k) L^{-2k-1}.$$

This implies that for large enough L, depending only on φ, $[k]$ and Ξ, we have

$$\mathrm{Var}[\int L^{|k|}\varphi_L^{[k]}\mathrm{d}[\Xi]] \le C(\varphi,k)L^{-1}.$$

Therefore, if L is allowed to vary over the sequence $2^n, n \in \mathbb{Z}$, we can invoke the Borel Cantelli lemma and deduce that $\left(\int L^{|k|}\varphi_L^{[k]}\mathrm{d}[\Xi] - \mathbb{E}[L^{|k|}\varphi_L^{[k]}\mathrm{d}[\Xi]]\right) \to 0$ almost surely. Observe that

$$\int_{\mathbb{R}^d} L^{|k|}\varphi_L^{[k]}\mathrm{d}[\Xi] = \int_D L^{|k|}\varphi_L^{[k]}\mathrm{d}[\Xi] + \int_{D^{\complement}} L^{|k|}\varphi_L^{[k]}\mathrm{d}[\Xi].$$

But on the set D, the function $L^{|k|}\varphi_L^{[k]}(x)$ is simply $x^{[k]}$. This enables us to deduce that

$$\left(\int_D x^{[k]}\mathrm{d}[\Xi] + \int_{D^{\complement}} L^{|k|}\varphi_L^{[k]}\mathrm{d}[\Xi] - \mathbb{E}\left[L^{|k|}\varphi_L^{[k]}\mathrm{d}[\Xi]\right]\right) \to 0$$

almost surely. But both the second and the third terms in the display above can be computed exactly, given the field outside D. This completes the proof of maximal rigidity. $\square$

5. Origins of Maximal Rigidity: Gaussian Processes on $\mathbb{Z}^d$

To explore the origin of the maximal rigidity imposed by the gap, we consider generalized stealthy (GS) Gaussian processes on $\mathbb{Z}^d$. We will consider such a process as a limit of finite "stealthy" Gaussian processes. By a finite Gaussian process, we mean a Gaussian process on $\mathbb{Z}_n^d$; its wave space being identified with the discrete torus $\mathbb{T}_n^d$. In this context, "GS" would imply the structure function vanishing on a subset $U_n \subset \mathbb{T}_n^d$. The idea is that, for large n, $\mathbb{T}_n^d$ approximates $\mathbb{T}^d$ and U_n approximates a neighbourhood in $\mathbb{T}^d$.

Let $(\xi_n(x))_{x\in\mathbb{Z}_n^d}$ be a Gaussian random variable with covariance

$$C(x-x') = \sum_{k=0}^{n-1} S\left(\frac{k}{n}\right)\exp\left(\frac{2\pi i k}{n}(x-x')\right).$$

It follows that this Gaussian field (that lives on $\mathbb{Z}_n^d$) has the structure function $S(\cdot)$ (that lives on $\mathbb{T}_n^d$). It also follows that this Gaussian random variable has a density (with respect to the Lebesgue measure on $\mathbb{R}^n$) that is given by

$$\mu(\xi) = \frac{1}{(2\pi)^{d/2}|\mathrm{Det}(C)|^{1/2}}\exp\left(-\frac{1}{2}\sum_{x,x'\in\mathbb{T}_n^d}\xi(x)C^{-1}(x-x')\xi(x')\right)$$

Let U be a fixed open set in $\mathbb{T}^d$. We choose S to be a non-negative function on $\mathbb{T}^d$ that vanishes on U; note that by the Bochner correspondence, such a function gives rise to a (unique up to translations) Gaussian process on $\mathbb{Z}^d$. To emulate a (generalized) stealthy Gaussian process in the finite setting, we simply consider the restriction of S to $\mathbb{T}_n^d$ (identified canonically as a subset of $\mathbb{T}^d$).

We now define the Gaussian vector $\widetilde{\xi}$ (on $\mathbb{T}_n^d$) by

$$\widetilde{\xi}(k) = \frac{1}{n^{d/2}} \sum_{x \in \mathbb{Z}_n^d} \xi(x) \exp(ikx),$$

where $k = \frac{2\pi j}{n}$, $j \in \mathbb{Z}_n^d$. Observe that for each k, this is in particular, a linear combination of $\xi(x)_{x \in \mathbb{Z}_n^d}$. Using the Fourier transform, a simple computation shows that

$$\mathbb{E}[|\widetilde{\xi}(k)|^2] = S(k),$$

and hence $\mathbb{E}[|\widetilde{\xi}(k)|^2] = 0$ for $k \in U$, which implies $\widetilde{\xi}(k) = 0$ a.s. for such k. Since the number of such indices k is $|U| \cdot n^d$, we conclude that the condition of being GS imposes $|U| \cdot n^d$ linear constraints on the Gaussian process ξ living on the physical space $\mathbb{Z}_n^d$. Because of these linear constraints, the values of the Gaussian process on any domain (in $\mathbb{Z}_n^d$) of size $|U| \cdot n^d$ is determined exactly by the rest of the values. This gives a version of maximal rigidity for finite n.

Thus, for finite n, a GS Gaussian process exhibits degeneracy (i.e., mutual algebraic dependence of its values), and the degree of this degeneracy is, heuristically speaking, proportional to the size of the gap in the spectrum. The maximal rigidity can be viewed as a limit of this degeneracy in the finite situation.

Acknowledgements The work of J.L.L. was supported in part by the AFOSR Grant FA9550-16-1-0037. The work of S.G. was supported in part by the ARO Grant W911NF-14-1-0094, the NSF Grant DMS-1148711 and the NUS Grant R-146-000-250-133. J.L.L. thanks the Systems Biology group of the Institute for Advanced Studies for their hospitality during this work. The authors thank Salvatore Torquato for valuable discussions, and the anonymous referees for their insightful comments and suggestions.

References

[AiMa] Aizenman, M., Martin, P.: Structure of Gibbs states of one-dimensional Coulomb systems. Commun. Math. Phys. **78**(1), 99–116 (1980).

[BaG] Baake, M., Grimm, U.: Aperiodic Order, Vol. 1. Cambridge University Press, Cambridge (2013)

[BaBM] Baake, M., Birkner, M., Moody, R.: Diffraction of stochastic point sets: explicitly computable examples. Commun. Math. Phys. **293**(3), 611–660 (2010)

[BaKM] Baake, M., Koesters, H., Moody, R.: Diffraction theory of point processes: systems with clumping and repulsion. J. Stat. Phys. **159**(4), 915–936 (2015)

[BoSW] Borichev, A., Sodin, M., Weiss, B.: Spectra of stationary processes on Z. In: Baranov, A., Kisliakov, S., Nikolski, N. (eds.) 50 Years with Hardy Spaces. Operator Theory: Advances and Applications, vol. 261. Birkhäuser, Cham (2017)

[Bu] Bufetov, A.I. : Rigidity of determinantal point processes with the Airy, the Bessel and the Gamma kernel. Bull. Math. Sci. **6**(1), 163–172 (2016)

[BuFN] Burcaw, L., Fieremans, E., Novikov, D.: Mesoscopic structure of neuronal tracts from time-dependent diffusion. NeuroImage **114**, 18 (2015)

[ChDZCT] Chertkov, E., DiStasio, R.A. Jr.., Zhang, G., Car, R., Torquato, S.: Inverse design of disordered stealthy hyperuniform spin chains. Phys. Rev. B **93**, 064201 (2016)

[DaV] Daley, D., Vere-Jones, D.: An Introduction to the Theory of Point Processes, Vols. I (2nd edn., 2003) and II. Springer, Berlin (2007)

[DeSMRPRJTB] DegIInnocenti, R., Shah, Y.D., Masini, L., Ronzani, A., Pitanti, A., Ren, Y., Jessop, D.S., Tredicucci, A., Beere, H.E., Ritchie, D.A.: THz quantum cascade lasers based on a hyperuniform design. Proc. SPIE **9370**, 93700A (2015)

[FlTS] Florescu, M., Torquato, S., Steinhardt, P.: Designer disordered materials with large complete photonic band gaps. Proc. Natl. Acad. Sci. USA **106**, 20658 (2009)

[Gh-1] Ghosh, S.: Determinantal processes and completeness of random exponentials: the critical case. Probab. Theory Relat. Fields **163**(3-4), 643–665 (2015)

[Gh-2] Ghosh, S.: Palm measures and rigidity phenomena in point processes. Electron. Commun. Probab. **21** (2016)

[GhL] Ghosh, S., Lebowitz, J.: Number rigidity in superhomogeneous random point fields. J. Stat. Phys. **166**(3-4), 1016–1027 (2017)

[GhP] Ghosh, S., Peres, Y.: Rigidity and tolerance in point processes: Gaussian zeros and Ginibre eigenvalues. Duke Math. J. **166**(10), 1789–1858 (2017)

[GhK] Ghosh, S., Krishnapur, M.: Rigidity hierarchy in random point fields: random polynomials and determinantal processes. arXiv:1510.08814

[GoLSp] Goldstein, S., Lebowitz, J., Speer, E.: Large deviations for a point process of bounded variability. Markov Process. Relat. Fields **12**, 235–256 (2006)

[Gi] Ginibre, J.: Statistical ensembles of complex, quaternion, and real matrices. J. Math. Phys. **6**(3), 440–449 (1965)

[HaMS] Haberko, J., Muller, N., Scheffold, F.: Direct laser writing of three dimensional network structures as templates for disordered photonic materials. Phys. Rev. A **88**, 043822 (2013)

[HeL] Hexner, D., Levine, D.: Hyperuniformity of critical absorbing states. Phys. Rev. Lett. **114**, 110602 (2015)

[HeCL] Hexner, D., Chaikin, P., Levine, D.: Enhanced hyperuniformity from random reorganization. Proc. Natl. Acad. Sci. **114**(17), 4294–4299 (2017)

[JaLMa] Jancovici, B., Lebowitz, J., Manificat, G.: Large charge fluctuations in classical Coulomb systems. J. Stat. Phys. **72**(3), 773–787 (1993)

[JiT] Jiao, Y., Torquato, S.: Maximally random jammed packings of Platonic solids: hyperuniform long-range correlations and isostaticity. Phys. Rev. E **84**, 041309 (2011)

[JiLHMCT] Jiao, Y., Lau, T., Hatzikirou, H., Meyer-Hermann, M., Corbo, J.C., Torquato, S.: Avian photoreceptor patterns represent a disordered hyperuniform solution to a multiscale packing problem. Phys. Rev. E **89**, 022721 (2014)

[L] Lebowitz, J.: Charge fluctuations in Coulomb systems. Phys. Rev. A **27**, 1491–1494 (1983)

[MaST] Marcotte, E., Stillinger, F., Torquato, S.: Nonequilibrium static growing length scales in supercooled liquids on approaching the glass transition. J. Chem. Phys. **138**, 12A508 (2013)

[MaY] Martin, Ph., Yalcin, T.: The charge fluctuations in classical Coulomb systems. J. Stat. Phys. **22**(4), 435–463 (1980)

[OsS] Osada, H., Shirai, T.: Absolute continuity and singularity of Palm measures of the Ginibre point process. Probab. Theory Relat. Fields **165**(3-4), 725–770 (2016)

[To-1] Torquato, S.: Random Heterogeneous Materials: Microstructure and Macroscopic Properties. Springer, New York (2002)

[To-2] Torquato, S.: Hyperuniformity and its generalizations. Phys. Rev. E **94**(2), 022122 (2016)

[ToSt] Torquato, S., Stillinger, F.: Local density fluctuations, hyperuniformity, and order metrics. Phys. Rev. E **68**, 041113 (2003)

[ToZS] Torquato, S., Zhang, G., Stillinger, F.H.: Ensemble theory for stealthy hyperuniform disordered ground states. Phys. Rev. X **5**, 021020 (2015)

[ZhST-2] Zhang, G., Stillinger, F.H., Torquato, S.: Ground states of stealthy hyperuniform potentials: I. Entropically favored configurations. Phys. Rev. E **92**, 022119 (2015)

[ZhST-1] Zhang, G., Stillinger, F.H., Torquato, S.: Ground states of stealthy hyperuniform potentials: II. Stacked-slider phases. Phys. Rev. E **92**, 022120 (2015)

[ZhST-3] Zhang, G., Stillinger, F.H., Torquato, S.: Transport, geometrical, and topological properties of stealthy disordered hyperuniform two-phase systems. J. Chem. Phys. **145**, 244109 (2016)

[ZhST-4] Zhang, G., Stillinger, F.H., Torquato, S.: Can exotic disordered "stealthy" particle configurations tolerate arbitrarily large holes?. Soft Matter **13**, 6197–6207 (2017). arXiv:1705.04415

Communicated by H. Spohn

www.ingramcontent.com/pod-product-compliance
Lightning Source LLC
Chambersburg PA
CBHW081220060126
37764CB00006B/330